Ecology

Ecology

Evolution, Application, Integration

David T. Krohne

Wabash College

New York   Oxford

OXFORD UNIVERSITY PRESS

Oxford University Press is a department of the University of Oxford.
It furthers the University's objective of excellence in research,
scholarship, and education by publishing worldwide.

Oxford   New York
Auckland   Cape Town   Dar es Salaam   Hong Kong   Karachi
Kuala Lumpur   Madrid   Melbourne   Mexico City   Nairobi
New Delhi   Shanghai   Taipei   Toronto

With offices in
Argentina   Austria   Brazil   Chile   Czech Republic   France   Greece
Guatemala   Hungary   Italy   Japan   Poland   Portugal   Singapore
South Korea   Switzerland   Thailand   Turkey   Ukraine   Vietnam

For titles covered by Section 112 of the US Higher Education
Opportunity Act, please visit www.oup.com/us/he for the
latest information about pricing and alternate formats.

Published by Oxford University Press
Oxford University Press
198 Madison Avenue, New York, New York 10016
http://www.oup.com

Library of Congress Cataloging-in-Publication Data
Krohne, David T.
Ecology: Evolution, Application, Integration / David T. Krohne, Wabash College.
    pages cm
  Includes bibliographical references and index.
  ISBN 978-0-19-975745-9 -- ISBN 978-0-19-975747-3 (instructor's edition)
1. Ecology. 2. Nature--Effect of human beings on. I. Title.
  QH541 K747 2015
  577--dc23
                                                                    2014047651

Printing number: 9 8 7 6 5 4 3 2

Printed in the United States of America
on acid-free paper

To Dave, Eric, and John

About the Author

David Krohne received his BA from Knox College in 1974 and his PhD from the University of California, Berkeley, in 1979. Dr. Krohne was the Norman E. Treves Professor of Biology at Wabash College, where his ecology and evolution courses included field experiences for students in the Everglades, the Florida Keys, the Galápagos, the rainforest, and Yellowstone. In 2003 Dr. Krohne received the McLain-McTurnan-Arnold Excellence in Teaching Award. Dr. Krohne's research interests include the ecology and genetics of populations of small mammals, the population dynamics and conservation biology of rare prairie plants, the structure and organization of tallgrass prairie plant communities, and a 25-year rephotography study of the Yellowstone fires of 1988. He retired from Wabash College in 2010.

Contents in Brief

Contents

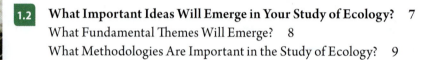

Chapter 4   Terrestrial Communities   63

Chapter 5   Freshwater and Marine Communities   97

Chapter 6   Behavioral Ecology   117

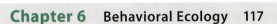

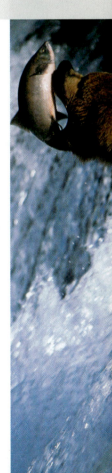

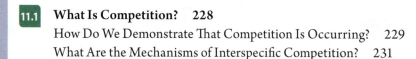

PART 3   COMMUNITIES: INTERACTIONS AMONG SPECIES   225

Preface

The natural world—the Earth and its myriad plant and animal communities—is inherently interesting to many students. Ecology, the scientific analysis of these systems, is a fascinating but dauntingly complex collection of interactions. The sheer volume of information, terms, and principles can be overwhelming for a student confronting it all for the first time. This is especially true given that the vast majority of the students taking ecology courses come from a wide range of life science majors. Many students rightfully wonder, "Why is learning ecology important to me?" Therefore, my goal has been to provide an approach that transforms ecology from a collection of loosely associated pieces of information into an integrated, concept-driven, and comprehensible system of understanding. I do this while also shedding light on the impact humans have on ecosystems. To accomplish this, the text presents ecological theory and applications within an evolutionary context. Importantly, the text also develops scientific reasoning skills by teaching students not just what we know about the field, but how we know what we know.

More Evolution

This text *integrates modern evolutionary theory throughout*. Why is evolution pedagogically valuable? Evolution has great explanatory power—most ecological interactions are rooted in adaptive evolution. Moreover, as students learn to place ecological problems in an evolutionary context, they find they can memorize less because they are able to deduce key ideas. Once they have seen the process of adaptive evolution in one context, they can anticipate its role in the next, replacing rote learning with insight and logic. In every chapter, both through integrated textual discussions and also the "The Evolution Connection" boxes, I direct students' attention to the reciprocal relationship between ecology and evolution, highlighting that evolution is driven by ecology and ecology is driven by evolution and the organism's adaptations. Chapter 13, for example, The Structure of Communities, not only discusses "community heritability and structure" in a box called The Evolution Connection, but explores other aspects of evolution and adaptation throughout the chapter, such as the effect of coevolution on community structure, and how analysis of the genetic relationships among related species influences their ability to coexist.

More Application

The audience for general ecology courses is made up of a wide array of life science majors, ranging from future ecologists to pre-medical and health majors. Students learn more effectively when the topic of study has relevance to their experience. This text addresses the needs of this diverse set of majors by introducing engaging, real-world examples, especially those that examine the relationship between ecology and its application in the form of environmental science. In addition to chapter-opening questions and stories, and full chapters on subjects such as conservation biology (Chapter 18) and human global ecology (Chapter 20), all chapters include "The Human Impact" boxes, which highlight important environmental issues. Importantly, the text continually asks the student to think critically and quantitatively within these applied contexts; therefore, quantitative topics are presented within the context of real-world applications and examples. Examples of such discussions are woven throughout the text, touching on such topics as human-induced evolution as a result of insect pesticides (Chapter 2); changes to coral reefs (The Human Impact, Chapter 3); competition from invasive species transported by people (The Human Impact, Chapter 13), and the way in which forest roads can prevent the normal movement patterns of forest species such as monkeys (Chapter 18).

More Process of Science

In order to train students not only in what we know about ecology, but also how we know what we know about the subject, the text employs a question-based pedagogy. Each chapter begins with a fundamental ecological question. The sections of the chapter are designed around a logical sequence of smaller questions, the answers to which eventually enable the student to answer the chapter's main question. This approach models the process of science—researchers address fundamental ecological questions in the same way. As students gain experience with this approach, they can apply it to new problems and questions. Also, every chapter features "On the Frontline" boxes highlighting recent studies that illustrate the application of the scientific method to the subject of the chapter. In these boxes, the hypotheses and the predictions that follow are clearly presented and the data that support or reject those hypotheses are explained. Finally, because developing quantitative insight and problem-solving skills is a core objective for the course, the text presents quantitative topics clearly and within applied contexts. More advanced mathematical elements are isolated and explained in boxes called "Do the Math." Examples include the discussion of population change (Chapter 8) and ways to measure the relative contributions of the environment and genetics in behavior (Chapter 6).

Teaching and Learning Features

- **"The Evolution Connection" boxes:** Some relevant evolutionary principles require supplementary discussion. These boxes illuminate additional evolutionary concepts that apply to the ecological content of the chapter.
- **"The Human Impact" boxes:** This box applies important principles from the chapter to significant environmental issues.
- **"On the Frontline" boxes:** This box highlights a recent study that illustrates the application of the scientific method to the subject of the chapter. The hypotheses and the predictions that follow are clearly presented, and the data that support or reject these hypotheses are explained.
- **"Do the Math" boxes:** Ecology is a quantitative science. This box supplements the chapter by explaining additional quantitative topics and equations, giving the student an opportunity to focus on the mathematics at the heart of ecological study.
- **Active learning features:** Students are encouraged to engage in reading through several question-based features. "Thinking About Ecology" questions, found throughout the chapters, ask students to apply their understanding of key concepts. "The Evolution Connection" and "The Human Impact" boxes conclude with probing queries. Importantly, most figures also include an interpretive question to help the student apply visual data.

Visual Guide to the Book

Ecology: Evolution, Application, Integration is distinguished by the following approaches:
- Integrates modern evolutionary theory throughout
- Highlights applications and connections
- Emphasizes inquiry and the process of science
- Presents quantitative topics clearly and in real-world applied contexts

These approaches thoroughly inform the scholarship, narrative, and features of the book.

Chapter Introductions

- Each chapter opens with a summary of the **Concepts** being discussed and the **Features** used to bring these topics home.

CONCEPTS

4.1 What Determines Species Distributions? p. 63

4.2 What Are the Fundamental Types of Terrestrial Communities? p. 72

FEATURES

Do the Math
The Concept of Statistical Significance p. 65

The Evolution Connection
Historical Effects on Species Distributions p. 69

On the Frontline
Fire and Grazing in the Prairie Community p. 71

The Human Impact
Global Warming and Ecological Communities p. 76

- A vivid **Case Study** introduces the topic of each chapter.

Chapter 4

Terrestrial Communities

In 1802 the German explorer and naturalist Alexander von Humboldt ascended the volcano Chimborazo in Ecuador, then thought to be the highest mountain in the world. As he climbed, Humboldt made detailed measurements of temperature, moisture, and air pressure. Along the way he made a simple but profound observation: the **vegetation**, the form of the plant life, changes dramatically with elevation. Almost one hundred years later, the biologist C. H. Merriam made a dramatic journey—from the bottom of the Grand Canyon through the Painted Desert to the summit of San Francisco Peak near Flagstaff, Arizona (Figure 4.1). Like Humboldt, Merriam recognized that the vegetation occurs in distinct zones associated with specific elevations. He referred to these bands of vegetation as **life zones**. He reasoned that each is determined by the combination of temperature and moisture, both of which change with elevation. Moreover, he noticed that the bands of vegetation on the slopes of the mountain are similar to the bands of vegetation that occur from the southwestern United States northward into Canada and Alaska. The alpine meadows of San Francisco Peak resemble the tundra of the high Arctic. His names for the mountain life zones reflect their latitudinal counterparts.

Humboldt and Merriam identified two universal principles of plant distribution: plants occur in *patterns*, groups of species that live together, and these patterns are associated with a specific climate. Together these principles lead us to the fundamental question of this chapter: *How are species organized into communities?*

4.1 What Determines Species Distributions?
In Chapter 3 we explored the concept of tolerance limits—the range of physical conditions within which the species can survive. When conditions exceed the species' tolerance limits, the organism must adapt

CONCEPTS

4.1 What Determines Species Distributions? p. 63

4.2 What Are the Fundamental Types of Terrestrial Communities? p. 72

FEATURES

Do the Math
The Concept of Statistical Significance p. 65

The Evolution Connection
Historical Effects on Species Distributions p. 69

On the Frontline
Fire and Grazing in the Prairie Community p. 71

The Human Impact
Global Warming and Ecological Communities p. 76

Inside the Chapter

4.1 What Determines Species Distributions?
In Chapter 3 we explored the concept of tolerance limits—the range of physical conditions within which the species can survive. When conditions exceed the species' tolerance limits, the organism must adapt

- The **Core Questions** posed at the beginning of the chapter are reflected in the titles of each major section in the chapter.

KEY CONCEPTS 4.1

- Climate is the result of a combination of geographic features: latitude, altitude, aspect, and proximity to large bodies of water.
- The substrate and soil also determine species' presence or absence, especially in plants and sessile animals. It is a product of weathered parent material, external inputs, and organic matter.
- A species' potential geographic range is determined by the combination of climate, substrate, and local biological interactions. Dispersal and time determine whether the species reaches a favorable region.

QUESTION:

Why are biotic interactions near the base of the flow chart in Figure 4.10?

- The end of each section is capped with a summary of **Key Concepts** and a **Question** to encourage the application of learning material.

Thematic Boxes

- Four types of thematic boxes are used throughout to highlight key themes and principles of the book.

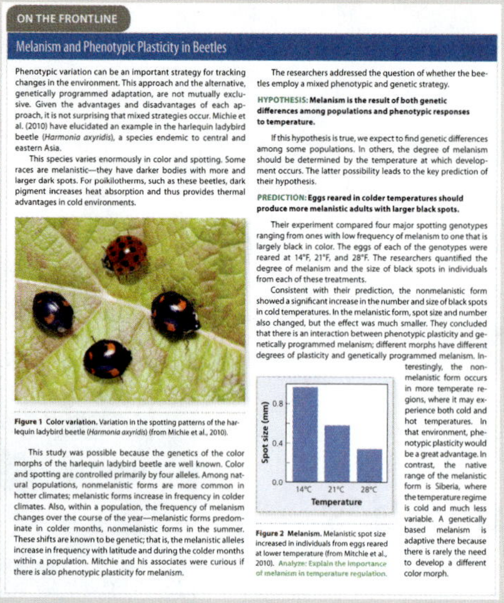

Melanism and Phenotypic Plasticity in Beetles

Phenotypic variation can be an important strategy for tracking changes in the environment. This approach and the alternative, genetically programmed adaptation, are not mutually exclusive. Given the advantages and disadvantages of each approach, it is not surprising that mixed strategies occur. Michie et al. (2010) have elucidated an example in the harlequin ladybird beetle (*Harmonia axyridis*), a species endemic to central and eastern Asia.

This species varies enormously in color and spotting. Some races are melanistic—they have darker bodies with more and larger dark spots. For poikilotherms, such as these beetles, dark pigment increases heat absorption and thus provides thermal advantages in cold environments.

Figure 1 Color variation. Variation in the spotting patterns of the harlequin ladybird beetle (*Harmonia axyridis*) (from Michie et al., 2010).

This study was possible because the genetics of the color morphs of the harlequin ladybird beetle are well known. Color and spotting are controlled primarily by four alleles. Among natural populations, nonmelanistic forms are more common in hotter climates; melanistic forms increase in frequency in colder climates. Also, within a population, the frequency of melanism changes over the course of the year—melanistic forms predominate in colder months, nonmelanistic forms in the summer. These shifts are known to be genetic; that is, the melanistic alleles increase in frequency with latitude and during the colder months within a population. Michie and his associates were curious if there is also phenotypic plasticity to melanism.

The researchers addressed the question of whether the beetles employ a mixed phenotypic and genetic strategy.

HYPOTHESIS: Melanism is the result of both genetic differences among populations and phenotypic responses to temperature.

If this hypothesis is true, we expect to find genetic differences among some populations. In others, the degree of melanism should be determined by the temperature at which development occurs. The latter possibility leads to the key prediction of their hypothesis.

PREDICTION: Eggs reared in colder temperatures should produce more melanistic adults with larger black spots.

Their experiment compared four major spotting genotypes ranging from ones with low frequency of melanism to one that is largely black in color. The eggs of each of the genotypes were reared at 14°C, 21°C, and 28°C. The researchers quantified the degree of melanism and the size of black spots in individuals from each of these treatments.

Consistent with their prediction, the nonmelanistic form showed a significant increase in the number and size of black spots in cold temperatures. In the melanistic form, spot size and number also changed, but the effect was much smaller. They concluded that there is an interaction between phenotypic plasticity and genetically programmed melanism; different morphs have different degrees of plasticity and genetically programmed melanism. Interestingly, the nonmelanistic form occurs in more temperate regions, where it may experience both cold and hot temperatures. In that environment, phenotypic plasticity would be a great advantage. In contrast, the native range of the melanistic form is Siberia, where the temperature regime is cold and much less variable. A genetically based melanism is adaptive there because there is rarely the need to develop a different color morph.

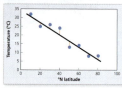

Figure 2 Melanism. Melanistic spot size increased in individuals from eggs reared at lower temperature (from Michie et al., 2010). **Analyze:** Explain the importance of melanism in temperature regulation.

- **On the Frontline:** Uses a "hypotheses and predictions" structure to illustrate the format of scientific inquiry, engaging the student in an interpretive view of recent cases of groundbreaking ecological research.

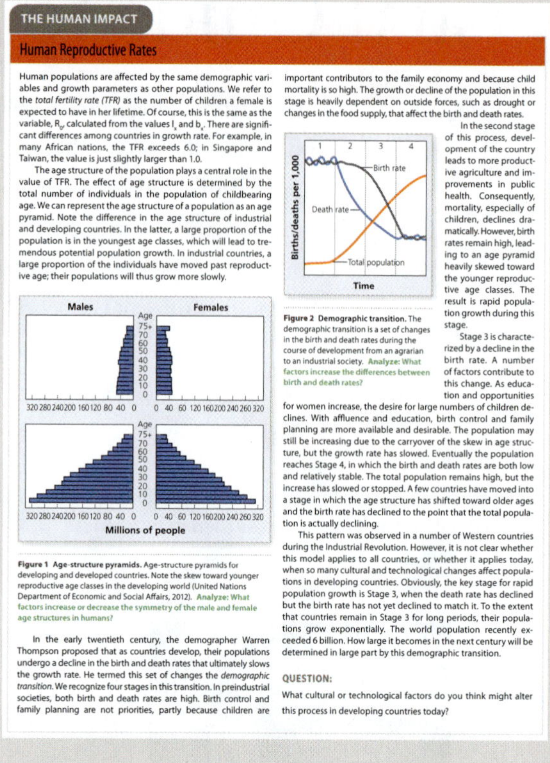

Human Reproductive Rates

Human populations are affected by the same demographic variables and growth parameters as other populations. We refer to the *total fertility rate (TFR)* as the number of children a female is expected to have in her lifetime. Of course, this is the same as the variable, R_0, calculated from the values l_x and b_x. There are significant differences among countries in growth rate. For example, in many African nations, the TFR exceeds 6.0; in Singapore and Taiwan, the value is just slightly larger than 1.0.

The age structure of the population plays a central role in the value of TFR. The effect of age structure is determined by the total number of individuals in the population of childbearing age. We can represent the age structure of a population as an age pyramid. Note the difference in the age structure of industrial and developing countries. In the latter, a large proportion of the population is in the youngest age classes, which will lead to tremendous potential population growth. In industrial countries, a large proportion of the individuals have moved past reproductive age; their populations will thus grow more slowly.

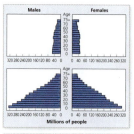

Figure 1 Age-structure pyramids. Age-structure pyramids for developing and developed countries. Note the skew toward younger reproductive age classes in the developing world (United Nations Department of Economic and Social Affairs, 2012). **Analyze:** What factors increase or decrease the symmetry of the male and female age structures in humans?

In the early twentieth century, the demographer Warren Thompson proposed that as countries develop, their populations undergo a decline in the birth and death rates that ultimately slows the growth rate. He termed this set of changes the *demographic transition*. We recognize four stages in this transition. In preindustrial societies, both birth and death rates are high. Birth control and family planning are not priorities, partly because children are

important contributors to the family economy and because child mortality is so high. The growth or decline of the population in this stage is heavily dependent on outside forces, such as drought or changes in the food supply, that affect the birth and death rates.

In the second stage of this process, development of the country leads to more productive agriculture and improvements in public health. Consequently, mortality, especially of children, declines dramatically. However, birth rates remain high, leading to an age pyramid heavily skewed toward the younger reproductive age classes. The result is rapid population growth during this stage.

Stage 3 is characterized by a decline in the birth rate. A number of factors contribute to this change. As education and opportunities for women increase, the desire for large numbers of children declines. With affluence and education, birth control and family planning are more available and desirable. The population may still be increasing due to the carryover of the skew in age structure, but the growth rate has slowed. Eventually the population reaches Stage 4, in which the birth and death rates are both low and relatively stable. The total population remains high, but the increase has slowed or stopped. A few countries have moved into a stage in which the age structure has shifted toward older ages and the birth rate has declined to the point that the total population is actually declining.

This pattern was observed in a number of Western countries during the Industrial Revolution. However, it is not clear whether this model applies to all countries, or whether it applies today, when so many cultural and technological changes affect populations in developing countries. Obviously, the key stage for rapid population growth is Stage 3, when the death rate has declined but the birth rate has not yet declined to match it. To the extent that countries remain in Stage 3 for long periods, their populations grow exponentially. The world population recently exceeded 6 billion. How large it becomes in the next century will be determined in large part by this demographic transition.

Figure 2 Demographic transition. The demographic transition is a set of changes in the birth and death rates during the course of development from an agrarian to an industrial society. **Analyze:** What factors increase the differences between birth and death rates?

QUESTION:

What cultural or technological factors do you think might alter this process in developing countries today?

- **The Human Impact:** Applies important principles from the chapter to a significant environmental issue.

Linear Regression

Many physical and biological features of marine systems change with depth or latitude. For example, ocean temperature declines with latitude and light decreases with depth. Salinity increases with distance from the mouth of a river where it enters the ocean. In Chapter 2 we used measures of central tendency such as the mean or median to describe a data set. Here, our interest is not the average temperature, light, or salinity. Rather, it is the change in those variables with others such as depth or latitude. How can we quantify the relationship between two variables?

Imagine that you measure the ocean surface temperature at several points from south to north along a single longitudinal line. You obtain the data shown in Figure 1. The relationship between temperature and latitude appears to be linear. How can we quantify this relationship? One method used in ecology is *linear regression*. Linear regression determines the parameters of the line that best fit the data. These parameters, the slope and y-intercept, quantify the nature of the relationship between latitude and temperature and can be used to compare data sets.

But how do we determine the line that best fits these data? The standard process is called *least squares regression*, the procedure used by most calculators and computer regression programs. The computations for least squares regression are tedious but the concept underlying them is straightforward. We define the "best line" through the data points as that line which lies closest to each of the data points. The computational procedure generates a line for which the sum of the squared distances between each point and the line is smallest—hence *least squares* regression. The procedure uses the squared distance between each point and the line because some points lie below the line; others lie above. If we simply used the absolute summed distances, the negative distances would cancel the positive ones and we could not be sure that the total distance from all points to the line is minimal.

We might also want to ask, "How strong is the association between temperature and latitude?" We measure this with the *correlation coefficient*. The most common correlation measure is the Pearson product-moment correlation, measured with the variable r. The value of the correlation coefficient is calculated as

$$r = \frac{\sum_{i=1}^{n}(X_i - \bar{X})(Y_i - \bar{Y})}{\sqrt{\sum_{i=1}^{n}(X_i - \bar{X})^2}\sqrt{\sum_{i=1}^{n}(Y_i - \bar{Y})^2}}$$

where x_i is the *i*th value of one of the variables, y_i is the *i*th value of the other, and $\bar{X}$ and $\bar{Y}$ are the means of the two groups. The value of r ranges from −1.0 to +1.0. Negative values indicate that an increase in one variable is associated with a decrease in the other. Positive values mean that as one variable increases, the other does as well. A correlation coefficient of 0 means that the two variables are not associated; each changes randomly with respect to the other.

We can evaluate the strength of the correlation with the value r^2. The square of the correlation coefficient is a measure of the proportion of the variation in x that is explained by variation in y. The correlation coefficient between temperature and latitude in Figure 1 is −0.96. (Its square is 0.92.) Thus, 92 percent of the variation in temperature is explained by latitude. If the value of r^2 is low, other factors are probably important.

Figure 1 Latitude and temperature. A hypothetical relationship between latitude and temperature. **Analyze:** Why do most points not fall exactly on the line?

Figure 2 Correlation. Three possible correlations among temperature and latitude: (a) positive, (b) zero, and (c) negative. **Analyze:** If temperature and latitude are negatively correlated, does this necessarily mean that latitude causes the temperature change?

- **Do the Math** focuses on the quantitative aspects of ecology, where the student can focus specifically on the mathematics relevant to the ecological questions at hand.

Oxygen and the Origin of Complex Life Forms

We know that the chemical composition of the Earth has changed over geologic and evolutionary time. Moreover, the dynamic nature of the Earth's abiotic systems is a direct result of the activity of living organisms. In fact, the biotic and abiotic components of the Earth's ecosystems have been engaged in an intimate feedback system from the time life first appeared.

The first organisms were simple, composed of one or a few cells. Initially, all were heterotrophs dependent on organic material in the seas or on other organisms. When photosynthesis evolved, it provided a major new source of organic energy and released molecular oxygen into the atmosphere. Ferric iron deposits—that is, oxidized iron rocks—are first found at this time, confirming the presence of oxygen in the atmosphere. It was at this point that the carbon cycle came to resemble more closely the modern pattern, in which carbon is fixed by photosynthesis and released by respiration.

An oxygen-rich atmosphere radically changed the biochemistry and energy processes in living things. The origin of eukaryotes occurred in this new, oxygen-rich world. One of the most significant biochemical differences between the eukaryotes and prokaryotes is that the former use oxygen as a final electron acceptor in respiration, a difference that generates far more energy. In fact, many scientists believe that one of the most profound expansions of animal life was powered by oxygen and the energy it provided (Payne et al., 2009; Dahl et al., 2010).

Approximately 530 million years ago, in an event known as the Cambrian explosion, the diversity of organisms suddenly increased by an order of magnitude. Within a few million years, most of the modern animal phyla had emerged. These new taxa were larger and more complex; their internal structure now composed of tissues and organs. The energetic demands of these new forms for development, activity, and maintenance were served by the energy from photosynthesis and the efficient respiration possible because of the availability of oxygen. In addition, oxygen may have contributed to larger, more complex organisms because it plays an important role in the production of complex polymers, such as collagen, that provide structural support in larger animals.

Figure 1 Devonian predators. Large predatory fish evolved in the Devonian period, when the atmosphere and aquatic systems had higher oxygen concentrations.

A second major increase in the size of organisms occurred approximately 400 million years ago in the Devonian Period. This, too, was associated with a jump in atmospheric oxygen. The Devonian oxygen increase is correlated with the diversification of vascular plants, which likely contributed importantly to this change. Estimates place the oxygen concentration as high as 36 percent—much higher than the present 21 percent. The Devonian is known for the origin of large predatory fish—a body type and ecology that requires a great deal of energy to sustain (Figure 1). Fish are relatively intolerant of hypoxia. Thus, it is not surprising that the origin of large predatory fish coincides with evidence of higher marine oxygen levels (Dahl et al., 2010). Of course, these predators would also require a significant prey base. The numbers and sizes of organisms lower in the food chain also increased at this time.

QUESTION:

How is oxygen production via vascular plants on land connected to the oxygen concentration of marine systems?

- **The Evolution Connection** reminds the student of relevant evolutionary theory and applies it to the ecological content of the chapter. This feature supplements the evolutionary theory that is already woven throughout each chapter.

Figure 7.8 Subspecies. The subspecies of *Ensatina escholtzii* are found in the Coast Range and Sierra Nevada. Each race is morphologically distinct and found in a particular portion of the species' range, yet all can interbreed. **Analyze:** What evolutionary forces do you think cause the color differences among subspecies?

THINKING ABOUT ECOLOGY:

Diagram a system of blood flow in which the direction of flow in the two vessels is not opposite but parallel. How does the flow of heat change compared to a countercurrent system?

- **Thinking About Ecology** questions are found throughout each chapter to enable students to apply what they've learned to situations with which they might be familiar.

- **Analyze**. Most figure captions conclude with a probing question that ensures that the student fully understands a figure or can relate it to other concepts in the chapter. Answers to all of the questions are provided in an answer key at the back of the book.

THINKING ABOUT ECOLOGY:

Consider a population with the following genotype frequencies:

genotype	AA	Aa	aa	Sum
number	14	45	40	= 99

a. Is this population in Hardy-Weinberg equilibrium?
b. What are the values of p and q?
c. If this population mates at random and no other factors affect it, what will be the genotype frequencies in the next generation?
d. Will the values of p and q change?

Long-Track Tornadoes (1950–2013)
25-mile, or longer, tracks, by intensity

— Weak (F/EF 0–1)
— Strong (F/EF 2–3)
— Violent (F/EF 4–5)

- **Visual program**. A rich ensemble of relevant photographs, maps, and figures underscore the key points of each chapter.
- Photos
- Maps
- Figures

Universe ↑ Galaxies ↑ Solar systems ↑ Planets ↑ Earth ↑ Ecosphere ↑ Ecosystems ↑ Communities ↑ Populations ↑ Organisms ↑ Organ systems ↑ Organs ↑ Tissues ↑ Cells ↑ Protoplasm ↑ Molecules ↑ Atoms ↑ Subatomic particles

Realm of Ecology

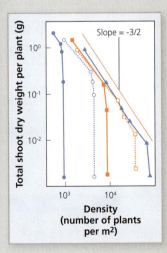

Slope = -3/2

Total shoot dry weight per plant (g)

10^0
10^{-1}
10^{-2}

10^3 10^4

Density (number of plants per m²)

BIOME 3
DESERT

Figure 3.1 Desert

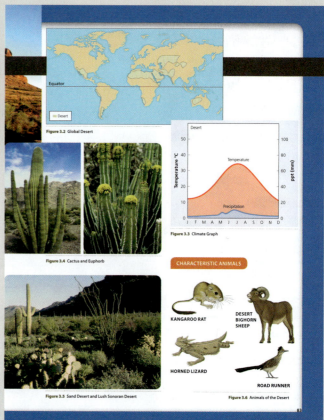

Figure 3.2 Global Desert

Figure 3.3 Climate Graph

Figure 3.4 Cactus and Euphorb

CHARACTERISTIC ANIMALS

KANGAROO RAT

DESERT BIGHORN SHEEP

HORNED LIZARD

ROAD RUNNER

Figure 3.5 Sand Desert and Lush Sonoran Desert

Figure 3.6 Animals of the Desert

KEY FEATURES

Deserts are defined climatically as regions where evaporation exceeds precipitation. The world's deserts vary widely in severity and in the pattern of rainfall. Parts of the Atacama Desert in Peru and Northern Chile have never recorded rain. In contrast, parts of the Sonoran Desert in the United States receive more than 300 millimeters of precipitation each year. Some, like the Sonoran, have a pronounced rainy season or monsoon. In all deserts, the variation in precipitation is large relative to the mean.

The substrate of deserts consists of bare rock, sand, alkali flats, or gravel. They also range from the white gypsum sands of White Sands, New Mexico to black lava in Death Valley, California. The surface of many desert soils consists of a matrix of algae, fungi, and lichens that bind soil particles into a fragile surface mat (Biome Figure 3.4).

Desert vegetation is highly variable and dependent on both the substrate and the pattern of precipitation. The flora consists of species adapted to high temperature and low moisture and includes cacti, euphorbs, grasses, and shrubs.

Deserts are among the most diverse of the biomes, because they are defined climatically rather than by the structure of the vegetation. The dominant forms of the plant life and the amount of plant production are tightly correlated with precipitation. Thus, the wide range of rainfall patterns and amounts lead to a wide range of vegetation types and biomass. In addition, the variety of substrate types and chemical composition selects for divergent types of vegetation and specifically adapted flora. Also, some high-elevation deserts such as in Mongolia and the Great Basin of the United States experience long periods of sub-freezing temperatures.

Many desert plants demonstrate the principle of evolutionary convergence—different evolutionary lines with similar adaptations. The cacti of the New World deserts closely resemble plants in a different family, the euphorbiaceae, of Africa. Both have spines and photosynthetic stems but no leaves. Many desert grasses adopt the annual habit and produce seeds highly resistant to drought. The seeds remain in the substrate, sometimes for decades, until it rains. They then germinate quickly, flower, set seed, and die (Biome Figure 3.5).

SIGNIFICANCE TO HUMANS

Humans have typically thought of deserts as "wastelands," useful only for mineral extraction and other exploitative uses. Globally, deserts are expanding as a result of human activities, including overgrazing, poor agricultural practices, and climate change

WHAT EVOLUTIONARY ADAPTATIONS ARE FOUND IN DESERT PLANTS?	
PLANT CHARACTERISTIC	ADAPTIVE SIGNIFICANCE
Allelopathy.	Reduces nearby competition for water.
Reduced leaves.	Reduce surface area for water loss.
Hairy leaves.	Increase shade and reduce wind speed at leaf surface and thus reduce water loss.
Annual habit.	Can exploit brief and unpredictable periods of moisture.

82

83

BIOME 8
TUNDRA

Figure 8.1 Arctic Tundra

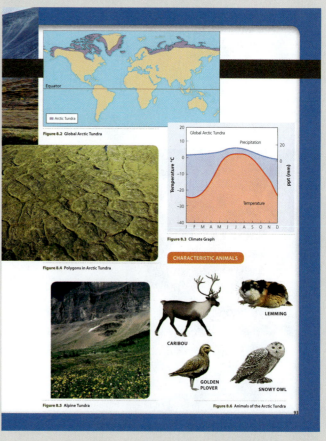

Figure 8.2 Global Arctic Tundra

Figure 8.3 Climate Graph

Figure 8.4 Polygons in Arctic Tundra

CHARACTERISTIC ANIMALS

LEMMING

CARIBOU

GOLDEN PLOVER

SNOWY OWL

Figure 8.5 Alpine Tundra

Figure 8.6 Animals of the Arctic Tundra

KEY FEATURES

The tundra biome is characterized by low, perennial grasses, sedges, shrubs, mosses, and lichens. A few tree species occur in the tundra, but their growth form is radically different—they grow prostrate with their branches horizontal along the ground.

The tundra is characterized by extreme seasonal variation in climate. Summer temperatures reach as high as 35°C. During the twenty-four hours of daylight. In winter the sun shines only briefly or not at all and the temperature often drops below −35°C. The tundra is climatically desert-like—much of this biome receives less than twenty-five centimeters of precipitation per year. However, because much of the region is underlain by permafrost, water does not drain away from the surface. Still, the desiccating cold, wind, and low precipitation have led to many adaptations similar to desert plants.

The short growing season of the tundra results in a rich, spectacular period of abundance and reproduction for both plants and animals. Migratory birds, caribou, musk ox, and lemmings occur locally in high numbers in the summer. In the same brief window, most plants flower and set seed resulting in remarkable floral displays.

The arctic tundra takes many forms depending upon latitude, elevation, and topography. Grassland vegetation is found in drier sites. Low, wet areas contain cotton grass, sedges, and sphagnum moss. Fell fields are exposed rocky habitats with sparse vegetation. In some, lichens are the only plant cover. Climate and permafrost combine to cause an array of local topographic changes. For example, the freeze-thaw cycle causes soil to rise or fall depending on the accumulation of water and the presence or absence of underlying permafrost. This cycle also sorts gravel and small rocks by size. One result is the development of polygons—small domes of fine particles surrounded by strips of heavier particles. Because so many plant species are near their range limits, small changes in elevation and drainage have profound effects on their distribution. The center of a polygon may be only a meter higher than the edge but supports a distinctly different vegetation and flora (Biome Figure 8.4).

Alpine tundra occurs at high elevation in the major mountain ranges. The flora and vegetation of the alpine tundra is similar to that of the arctic tundra. Indeed, many species are closely related. However, there are also important ecological differences. Most alpine tundra is not underlaid with permafrost. It experiences neither twenty-four hours of summer daylight nor twenty-four hours of winter darkness. Snow accumulation is significant. As this snow gradually melts over the summer, it provides a constant supply of water to the vegetation (Biome Figure 8.5).

SIGNIFICANCE TO HUMANS

This biome represents some of the last large tracts of remaining wilderness. Some parts of the tundra contain significant oil and gas deposits, leading to conflicts over the development of these resources.

WHAT EVOLUTIONARY ADAPTATIONS ARE FOUND IN TUNDRA PLANTS?	
PLANT CHARACTERISTIC	ADAPTIVE SIGNIFICANCE
Hairy leaves and stems.	Reduce water loss.
Dark or red color.	Absorbs solar radiation and warms plant.
Rapid flowering and seed set.	Permits reproduction during brief summer.
Heliotropic flowers.	Warm flower and perhaps attracts pollinators.

92

93

- Visual Guide to Earth's Biomes. A special feature of Chapter 4 is a highly visual guide to Earth's major terrestrial biomes.

Support Package

Oxford University Press offers a comprehensive ancillary package for instructors who adopt *Ecology: Evolution, Application, Integration.* The Ancillary Resource Center (ARC), located at www.oup-arc.com/krohne, contains the following teaching tools:

- **Digital Image Library:** Includes electronic files in PowerPoint format of every illustration, graph, photo, figure caption, and table from the text in both labeled and unlabeled versions
- **Lecture Notes:** Editable lecture notes in PowerPoint format for each chapter help make preparing lectures faster and easier than ever. Each chapter's presentation includes a succinct outline of key concepts and featured research studies, and incorporates the graphics from the chapter.
- **Computerized Test Bank:** Test item file, written by the author, includes 400 multiple-choice, fill-in-the blank, and short-answer questions

Contact your local OUP sales representative or visit www.oup-arc.com/krohne to learn more and gain access to these instructor resources.

Acknowledgments

A project like this does not come to fruition without the help of many people. I have had the benefit of teaching with colleagues who not only have a deep understanding of science but have devoted their lives to communicating their understanding to students. For this I'm grateful to the Wabash College Biology Department, especially David Polley, John Munford, and Eric Wetzel. Our conversations about teaching shaped not only my own classroom work but the approach I've adopted in this text. I also benefited from my interactions with a generation of students at Wabash College who helped me test the approach and structure of this book. It was from them that I learned the pedagogical value of asking good questions. I thank Greg Hoch for the ongoing conversations that have challenged and deepened my understanding of ecology and evolution. I first learned to appreciate the beauty of ecology and its relationship to evolution from Peter Schramm of Knox College, a gift I cannot repay.

I am grateful to the team at Oxford University Press for helping me turn an idea into a book. It was Jason Noe, senior editor, who guided me from a concept to an effective, coherent structure and organization. Thom Holmes, the developmental editor on this project, keenly edited my writing and brought his wisdom and creativity to bear on the pedagogical features in this book. Assistant editor Andrew Heaton somehow managed to keep track of a complex art manuscript and ensure that it meshed with the text, and he and editorial assistants Katie Naughton and Caitlin Kleinschmidt worked to review the text manuscript of my book as it progressed through several drafts. Development interns Allison Pratt and Ayesha Khan contributed mightily to the final preparation of the manuscript. In production, art director Michele Laseau, production manager Lisa Grzan, production editor Jane Lee, and the team at Precision Graphics worked together on a tight deadline to turn my manuscript into a visually appealing textbook. I would also like to thank Patrick Lynch, editorial director; John Challice, vice president and publisher; Frank Mortimer, director of marketing; David Jurman, marketing manager; Elizabeth Geist and Christine Naulty, marketing assistants; Jolene Howard, director of market development; Meghan Daris, market development associate; and Bill Marting, national sales manager.

I'm grateful to the friends and family who contributed to this project in indirect but crucial ways. I thank my wife, Sheryl, for moral support and encouragement but especially for her reminders to focus on the true audience for this text—the students. Andrew Hoth, Jim Walker, Betsy Shirah, and Susan Hargreaves contributed with their interest, curiosity, and enthusiasm. Luna kept me on task.

Manuscript Reviewers

We have greatly benefited from the perceptive comments and suggestions of more than seventy talented scholars and instructors who reviewed the manuscript, illustrations, and supplements of *Ecology: Evolution, Application, Integration*. Their insight and suggestions contributed immensely to the published work. We are especially grateful to the following reviewers for their insight and support on the project:

David Baumgardner,
  Texas A&M University
Eric Blackwell,
  Delta State University
Robert Bode,
  Illinois Wesleyan University
Victoria Borowicz,
  Illinois State University
Kenneth M. Brown,
  Louisiana State University
Lauren B. Buckley,
  University of Washington
Julia A. Cherry
George R. Cline,
  Jacksonville State University
Gretchen C. Coffman,
  University of San Francisco
Scott Connelly,
  University of Georgia
Lorelei Crerar,
  George Mason University
Mark S. Demarest,
  University of North Texas
Paul Dijkstra,
  Northern Arizona University
John J. Dilustro,
  Chowan University
Danielle Dixson,
  Georgia Institute of Technology
Thomas Ford,
  Concord University
Brett Goodwin,
  University of North Dakota
Christiane Healey,
  University of Massachusetts
  Amherst
Malcolm Hill,
  University of Richmond
John C. Jahoda,
  Bridgewater State University
Thomas W. Jurik,
  Iowa State University
Ragupathy Kannan,
  University of Arkansas–
  Fort Smith
Tigga Kingston,
  Texas Tech University
Catherine Kleier,
  Regis University

Jamie Kneitel,
  California State University,
  Sacramento
Suzanne Koptur,
  Florida International University
Kate Lajtha,
  Oregon State University
Lynn A. Mahaffy,
  University of Delaware
D. Nicholas McLetchie,
  University of Kentucky
Sean B. Menke,
  Lake Forest College
Daniel Moon,
  University of North Florida
Scott Newbold,
  Colorado State University
Onesimus Otieno,
  Oakwood University
Kristen Page,
  Wheaton College
Wiline M. Pangle,
  Ohio State University
Mitchell Pavao-Zuckerman,
  University of Arizona
Mark Pyron,
  Ball State University
Laurel Roberts,
  University of Pittsburgh
Irene Rossell,
  University of North Carolina at
  Asheville
Nathan Sanders,
  University of Copenhagen
Susan Schwinning,
  Texas State University
Inna Sokolova,
  University of North Carolina at
  Charlotte
Lara Souza,
  The University of Oklahoma
Linda Brooke Stabler,
  University of Central Oklahoma
Kathleen Sullivan Sealey,
  University of Miami
Stephen Sumithran,
  Eastern Kentucky University
Keith Summerville,
  Drake University

Carol Thornber,
  University of Rhode Island
Heather Throop,
  New Mexico State University
David Vandermast,
  Elon Unversity
Mitch Wagener,
  Western Connecticut State
  University
Deborah A. Waller,
  Old Dominion University
Jane M. Wattrus,
  College of St. Scholastica

J. Wilson White,
  University of North Carolina at
  Wilmington
Robert S. Whyte,
  California University of
  Pennsylvania
Benjamin A. Zamora,
  Washington State University
Jennifer Zettler,
  Armstrong Atlantic
  State University

Part 1

Organisms, the Environment, and Evolution

Horseshoe crabs are one of the most ancient and persistent species we know. They evolved some 450 years ago, yet they persist today in vast numbers in the shallow seas to which they are adapted. In Part 1 we explore the process of adaptive evolution—the responses of organisms to their physical and biological environments. We will examine the mechanisms of evolution, the basic environments that organisms inhabit, and the ways these environments shape their adaptations. Chapter 6 explores adaptations to members of their own species. In Chapter 7 you will see that intraspecific variation is both a requirement for adaptive evolution and a result of it.

Chapter 1

Introduction to Ecology

It is early spring in the Swan Valley in northwest Montana. Much snow remains, but scattered patches of bare ground have emerged on south-facing slopes exposed to the sun. A snowshoe hare (*Lepus americanus*) nibbles the buds of a young alder along a creek bottom. The hare, in its pure-white winter pelage, stands out against the brown snow-free background (Figure 1.1). Another hare feeds nearby, this one mottled brown and white in transition from its winter to its summer coat. One hare is conspicuous; the other is camouflaged.

L. Scott Mills of the University of Montana studies the ecology of snowshoe hares in this valley. By following the fates of animals wearing radio collars, he has found that hare mortality is highest in the spring and fall. In this habitat many predators take hares, including lynx, pine martens, coyotes, wolves, and raptors. Mismatches—brown hares in a white habitat or white hares in a snow-free landscape—may be responsible for the high spring and fall mortality. Mills's research asks fundamental *ecological* questions about this system: How and why does coat color change in snowshoe hares?

1.1 What Is Ecology?

What makes the change in coat color in snowshoe hares an *ecological* question? **Ecology** is the study of the interactions between an organism and its biological and physical environment. The fate of hares is determined by their predators, by their coat color, and by the amount and pattern of snow cover. But it is also determined by the distribution of the alders, willows, and the other species hares eat. These plants not only provide food for hares, they also provide cover and protection from predators. Moreover, they affect the amount of snow that accumulates and the rate at which it melts. The key element of this system is *interaction*, the web of relationships among hares, their predators, their food supply, and the physical environment, including climate and soil (Figure 1.2).

Our definition of ecology cannot possibly convey the richness of the field. That is a task that requires the next 19 chapters. However, we

Figure 1.1 Snowshoe hares. Snowshoe hares molt from brown to white in winter. Mismatched hares, white in summer or brown in winter, are probably more conspicuous to predators.

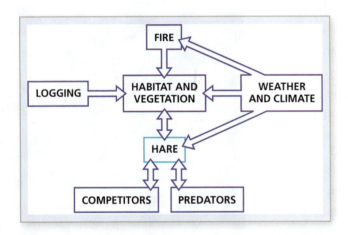

Figure 1.2 Ecological interactions. Snowshoe hares are part of a web of ecological interactions that includes humans, other species of plants and animals, and physical factors. **Analyze: What other interactions among components of this system might occur?**

ecology The study of the interactions between an organism and its biological and physical environment.

proximate factor A direct or immediate cause of a biological process or phenomenon.

ultimate factor The deeper cause of a phenomenon that explains why it occurs.

can begin the process by exploring more fully the question of coat color in the snowshoe hare. To do so, we must answer a number of questions (Figure 1.3). The answer to each reveals new interactions. The questions also probe the subject at different levels. For example, the direct effects of external stimuli on the hare's physiology are **proximate factors**, direct or immediate causes. Photoperiod and temperature are proximate factors in this system. Changes in photoperiod, increases in the spring and decreases in the fall, initiate the molt. Abnormally high spring temperatures accelerate the molt; lower temperatures slow it. Another level of inquiry addresses this question: *Why* do hares have genetic and physiological mechanisms for color change? The answers are higher-order causes known as **ultimate factors**. Ultimate factors often have an evolutionary basis. The evidence we have thus far is consistent with the idea that coat color confers some protection from predation. If predation is an important source of mortality, natural selection should favor hares whose coat color camouflages them from predators.

As we work through these questions, we see just how complex the hare's environment really is. The interaction of coat color and predation is just the starting point. We cannot fully understand that relationship in isolation. For example, we know that hares are naturally infected with nematode parasites. The hare's parasite load affects its vulnerability to predation. Also, the parasite load is higher in hares whose nutritional status is compromised. Hares eat the young shoots of aspen, willow, and alder. Fire increases the numbers of these species and their nutrient content. Other herbivores, including animals of vastly different size such as moose and voles, exploit the same plants and thus compete with hares for essential food resources (Figure 1.4). And so the web of interaction grows—to include predation, browsing, parasitism, and competition. Humans, too, are part of the web. Hare mortality is higher in clearcuts than in intact forest because logging removes plant cover that protects hares from predators. Humans affect the system indirectly as well. There is evidence that human industrial activity is causing global warming. As a result, snow arrives later and melts sooner in some places. If this results in more frequent mismatches between hares and their background and thus higher predation, we too play a role in the ecology of coat color in hares.

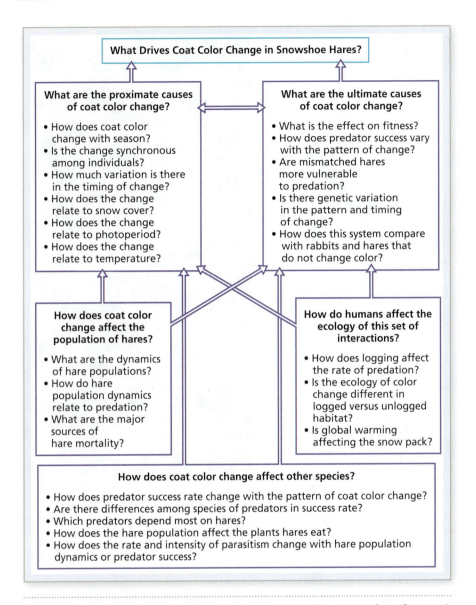

What Drives Coat Color Change in Snowshoe Hares?

What are the proximate causes of coat color change?

- How does coat color change with season?
- Is the change synchronous among individuals?
- How much variation is there in the timing of change?
- How does the change relate to snow cover?
- How does the change relate to photoperiod?
- How does the change relate to temperature?

What are the ultimate causes of coat color change?

- What is the effect on fitness?
- How does predator success vary with the pattern of change?
- Are mismatched hares more vulnerable to predation?
- Is there genetic variation in the pattern and timing of change?
- How does this system compare with rabbits and hares that do not change color?

How does coat color change affect the population of hares?

- What are the dynamics of hare populations?
- How do hare population dynamics relate to predation?
- What are the major sources of hare mortality?

How do humans affect the ecology of this set of interactions?

- How does logging affect the rate of predation?
- Is the ecology of color change different in logged versus unlogged habitat?
- Is global warming affecting the snow pack?

How does coat color change affect other species?

- How does predator success rate change with the pattern of coat color change?
- Are there differences among species of predators in success rate?
- Which predators depend most on hares?
- How does the hare population affect the plants hares eat?
- How does the rate and intensity of parasitism change with hare population dynamics or predator success?

Figure 1.3 Research questions. Ecological research proceeds by asking a series of specific research questions. In this example, the answer to each of these questions ultimately answers the broader question: What causes coat color change in snowshoe hares?

Analyze: Can you devise an experiment to address one of these questions?

Figure 1.4 Competition. Other herbivores such as moose (left) and voles (right) compete with hares for food.

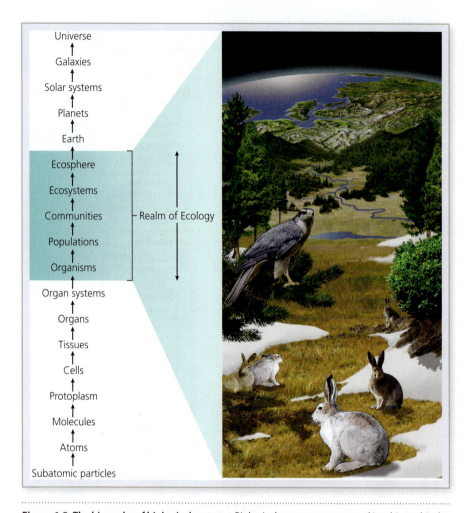

Figure 1.5 The hierarchy of biological systems. Biological systems are arranged in a hierarchical organization in which each level is more inclusive than the one below. Often, new properties emerge in the transition from a lower to a higher level. **Analyze: Can you give an example of a property that emerges in this way?**

organismal level The level of the biological hierarchy in which the focus is on the individual.

population ecology The study of the interactions between a group of individuals of a given species and the environment.

ecological community A group of coexisting species.

ecosystem All the biotic and abiotic components of a community.

biosphere (ecosphere) All the ecosystems of the Earth.

The ecology of this system encompasses several levels of biological organization. Ecology falls at the upper levels of the traditional hierarchical organization of biological systems (Figure 1.5). Ecology itself is hierarchically subdivided as well. Some ecologists work at the **organismal level**. They study interactions between individuals and their environment. Physiological ecology and behavioral ecology traditionally address organism-level interactions. **Population ecology** focuses on the dynamics of a group of individuals of a single species. The analysis of the causes of fluctuations in the numbers of hares is an example of a population-level inquiry. Groups of coexisting species constitute an **ecological community**. The hares, their food plants, their predators, parasites, and competitors are part of a community. The next most inclusive level of organization in ecology is the **ecosystem**. The ecosystem is composed of all the interacting living (biotic) and nonliving (abiotic) components of a community. Abiotic components include the geology, soils, and climate in the ecosystem. The summed ecosystems of the Earth—that is, all life and its interactions with the physical environment—constitute the **biosphere** or ecosphere.

The example of the hares and the relationships depicted in Figure 1.3 illustrate a fundamental aspect of the conceptual structure of ecology. Ecologists focus on the organism, the physical environment, and the systems of interactions that arise among them. Indeed, these three elements are present in each chapter of the text. However, the relative emphasis on each shifts depending on the element under consideration. In Part 1, "Organisms, the Environment, and Evolution," we emphasize the direct interaction between the organism and the physical environment. In Parts 2 and 3, "Populations" and "Communities: Interactions Among Species," the emphasis shifts to the webs of interactions among populations and species. Part 4 addresses ecosystem processes, the most inclusive set of interactions, where we see the intimate interaction between the physical environment and systems of populations and communities.

Ecology is connected to a number of other fields. **Natural history** is the observational study of plants and animals in their natural environment. These observations often lead to ecological questions. For example, natural historians described the change in hare coat color long ago. This observation led naturally to the ecological questions—how and why does the change occur? The primary difference between natural history and ecology is that ecology is usually driven by a specific hypothesis. The observation that hares change color led ecologists to speculate on its ultimate causes and to generate hypotheses about its advantages. Careful natural history work, especially meticulous observation, is essential to the study of ecology. The best ecologists are also fine natural historians. Not only do their observations raise new questions, they inform the hypotheses and experiments they devise.

There is a close connection between ecology and the related field of **environmental science**, the study of the impact humans have on the environment. Environmental science is an applied science. Its goal is to identify the causes of human environmental impacts, analyze their effects, and devise solutions to them. Of course, ecology is crucial to this analysis. In each chapter we will examine some of the important applied aspects of the ecological principles we address. By the time you reach Chapter 20 ("Human Global Ecology") you will have a firm grasp of the principles and concepts of ecology. With that information in hand we then further explore the relationship between ecology and environmental science.

The resource management sciences are applied fields that are also closely connected to ecology (Figure 1.6). **Wildlife management** is the science of ensuring the persistence of populations of game and nongame species for hunters and other wildlife enthusiasts. **Forestry**, **range management**, and **fisheries biology** are sciences devoted to managing and husbanding resources important for human use and consumption. These fields use ecological principles and concepts to ensure that human activities such as hunting, grazing, logging, and fishing are sustainable.

natural history Qualitative or observational study of organisms in their natural environment.

environmental science The study of the human impact on the environment.

wildlife management The study of the methods and principles by which we can maintain viable populations of wildlife species.

forestry The study of management practices that ensure the sustainable harvest and ecological health of forests.

range management The study of management practices that ensure the health and viability of grassland habitats used by domestic or wild animals.

fisheries biology The study of management practices that ensure the sustainable harvest and ecological health of fish populations.

Figure 1.6 Resource management sciences. The resource management sciences develop procedures to manage resources like game, fish, and timber. Game birds like this pheasant are managed to provide sustainable hunting opportunities.

KEY CONCEPTS 1.1

- Ecology is the study of the interactions between an organism and its biological and physical environment.
- Ecologists study interactions at multiple levels of organization: the organism, the population, the community, and the ecosystem.
- Natural history observations raise ecological questions and frame hypotheses.
- Other applied fields, such as environmental science and the management sciences, apply ecological principles to resources important to humans.

QUESTION:

How are the ecology and evolution of snowshoe hares connected?

1.2 What Important Ideas Will Emerge in Your Study of Ecology?

As you progress through the rest of the text, you will find some recurring ideas—concepts that have wide application in ecology. Ecology, like all sciences, is developing at a rapid pace. New data and new analyses constantly modify our understanding of ecological interactions. However, some principles underlie both classic and the most recent studies. They provide a conceptual framework that allows us to make sense of the overwhelming volume of new information ecologists are generating. Two basic types of concepts will emerge: First, you will discover a set of fundamental ecological principles. These represent the overarching themes that unite and explain ecological systems. Second, that set of principles has been elucidated by the specific application of the scientific method to the unique challenges posed by ecological questions.

What Fundamental Themes Will Emerge?

Perhaps the single most important theme in ecology is its intimate connection with the theory of evolution. We devote an entire chapter to evolutionary biology, but you will find the principles of evolution woven into every chapter. Implicit in our discussion of hare coat color change is the idea that the change is an evolutionary **adaptation**. An adaptation is a trait or combination of characteristics of an individual that increases its **evolutionary fitness**—that is, its survival and reproduction. The living and nonliving components of the environment determine which individuals have the highest fitness. In the case of hares, the pattern of snow cover and predators are environmental determinants of a hare's fitness. These two factors (along with many others) constitute **selection pressure** on hares. That is, they determine which hares survive and which die and how many offspring they produce. Those hares with genes that cause them to change color with the season live longer and ultimately produce more offspring.

The hares are also selective agents on the organisms with which they interact. For example, in some parts of the hare's range, especially where hares occur in great numbers, birch trees incorporate phenols in their tissue. These compounds make the trees less palatable to hares, thus protecting them from browsing. Each new adaptation in one species creates new selection pressures on others. The evolution-ecology connection is pervasive and continuous. It encompasses the three key elements of ecology—the organism, the physical environment, and systems of interactions.

A second important theme is that ecological systems do not necessarily achieve equilibrium. In this way ecology differs from other areas of biology. The difference arises because ecology lies in the upper tier of the biological hierarchy depicted in Figure 1.5. An important change occurs above the level of the individual. Cells, tissues, organs, and individuals exhibit the property of **homeostasis**, regulatory mechanisms that maintain their physiological parameters within narrow limits. An individual whose temperature rises too high or low or whose internal salt balance or pH is too extreme will die. Thus, natural selection favors homeostatic mechanisms that maintain a dynamic equilibrium within fairly narrow bounds. This selective force operates at the level of the individual but not at higher levels of organization. Fitness is a property of the *individual*. And natural selection determines which *individuals* live or die. Populations and communities do not have fitness, and natural selection does not favor some populations or communities over others. Thus, there is no imperative for homeostasis at the higher levels of the population, community, or ecosystem.

The ecological phenomenon comparable to homeostasis in organisms is known as the "balance of nature." The balance of nature was an important principle early in the history of ecology. Ecologists now appreciate that there is no logical imperative for the balance of nature. In addition, empirical studies document that many ecological systems either do not achieve an equilibrium state or it is ephemeral. We now understand that although there are mechanisms that regulate ecological interactions and move the system toward equilibrium, they do not operate in all systems and they can be overwhelmed by random forces. Modern ecologists recognize the importance of both equilibrium and nonequilibrium ecological processes.

The human impact on the environment and on ecological processes is another important theme that will emerge. As noted earlier, ecology and environmental science are distinct fields. Still, they are intimately connected. The human impact

adaptation A trait that increases an individual's fitness in a specific environment.

evolutionary fitness The survival and reproduction of a particular individual as determined by its characteristics.

selection pressure The environmental factors (biotic and abiotic) that determine fitness.

homeostasis Regulatory mechanisms that maintain an organism's physiological parameters within specific limits.

on the environment is so pervasive that virtually no ecological system on Earth is unaffected by human activity. Humans modify two of the main elements of ecology—the organisms and the physical environment. We are responsible for the elimination of some species and the spread of others; we modify the physical environment by altering the climate as well as the chemical composition of the land, water, and atmosphere. As a result, we radically alter many ecological interactions. Today we cannot assume that any ecological process is pristine; all are affected in some way by human activity. Of course, the human connection is also important because we have an inherent interest in our own health, welfare, and quality of life, all of which are affected by the environment.

What Methodologies Are Important in the Study of Ecology?

The concept of the testable hypothesis is central to the process of science in general and ecology in particular. Again, the ecology of coat color in hares illustrates this idea. Professor Mills made two observations: (1) hare mortality is highest in spring and fall, and (2) predation is responsible for many of these deaths. You might conclude that these observations are ultimately driven by predation on mismatched hares. However, this logic generates not a conclusion but a *hypothesis*. The data are consistent with the predation hypothesis but they do not prove it.

What data would be needed to test the hypothesis? Good hypotheses lead to testable **predictions**. A prediction is an observation or result we expect if our hypothesis is true. A good hypothesis makes specific, logical predictions that can be tested. Moreover, a good hypothesis is **falsifiable**: it must be possible that some data or observation could unequivocally show that the hypothesis is false. In the case of the hares, the predation hypothesis predicts that the excess spring and fall mortality is due to predation on mismatched hares (Figure 1.7). This prediction allows us to potentially falsify the hypothesis. We can test it by measuring the relative numbers of matched and mismatched hares in the population and the rates of predation each experiences. If it turns out that mismatched hares are no more likely to be captured and eaten than camouflaged hares, we reject the predation hypothesis—it has been falsified. On the other hand, if mismatched hares are preyed upon more frequently

prediction The result or observation we expect if a hypothesis is true.

falsifiable Describes a hypothesis that can be proven incorrect by data or observation.

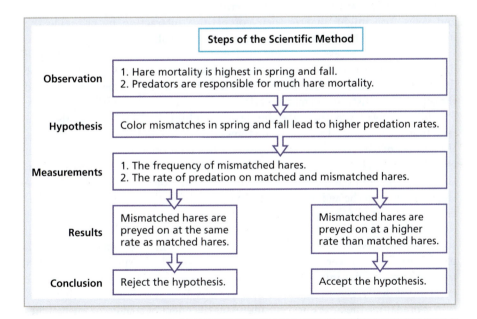

Figure 1.7 Hypotheses. The development of a hypothesis begins with an observation. The hypothesis is tested by collecting data that could potentially show the hypothesis to be false. The results allow the researcher to either accept or reject the hypothesis. **Analyze: If we accept the hypothesis, is it necessarily true?**

Figure 1.8 Model species. Stickleback fish are an example of a model species—one that has characteristics that lend themselves to ecological research.

than matched hares, we accept the hypothesis. Note that although we can reject a hypothesis as false, we cannot definitively prove it to be true. It is always possible that some other unknown factor correctly explains the pattern of hare mortality. We accept as correct those hypotheses that we cannot prove false.

You will discover that ecologists study a wide range of organisms. The choice of the study system and the organism is not random—and in fact the choice is crucial to the success of the study. What criteria are important in the choice of the study system? Some species are chosen because we know their basic biology and natural history so well. The researcher can address sophisticated questions and hypotheses without investing time and energy collecting basic natural history information. Sticklebacks, anoles, monkey flowers, columbines, Darwin's finches, and voles are examples of important model organisms used to address important questions (Figure 1.8). Sometimes the researcher chooses a species and system because it lends itself to the analysis of a specific question or hypothesis. Thus, Professor Mills chose snowshoe hares to test hypotheses about climate and predation because that system—hares in the Swan Valley—is ideal for his investigation. Some systems are chosen because they are amenable to experimental manipulation. For example, there are many sessile organisms that live on boulders in rocky intertidal marine ecosystems. Researchers can move rocks (and thus organisms) within the system or remove certain species relatively easily. Consequently, the rocky intertidal has been the system of choice in many classic ecological experiments.

The text will also illustrate the importance of mathematics and quantitative analysis in ecology. Mathematics and statistics are especially valuable in larger, more inclusive systems such as populations or communities where the results of experiments are often quantitative. For example, snowshoe hare mortality is greater in the spring and fall. But this pattern is not absolute; mortality does occur at other times. It is the *frequency* of death, a quantitative parameter, which changes over the course of the year. Quantitative results such as these reflect the fact that many causal factors contribute to the ecological variables we measure. Ecology, by its very definition, entails interaction and complex webs of causation. Mathematical analysis allows us to tease apart the relative importance of multiple interacting factors.

Imagine you want to do a set of experiments on competition between hares and voles. You might analyze their competitive interaction by changing the number of voles in the habitat and measuring the population response of the hares. However, in the process you might also alter other interactions in which both hares and voles participate. The effect you measure on the hares is unlikely to be due solely to the direct effect of the experiment; indirect effects also contribute to the outcome. Statistical and mathematical analyses help overcome these obstacles. Figure 1.9 shows a hypothetical example of a graphical analysis that reveals an interaction effect (Figure 1.9). The graph shows that as the number of voles increases, the number of hares declines. The slope of the line quantifies the competitive effect of voles on hares. But notice that the relationship changes with the age of the forest. In both young and old forests, hares decrease when voles are abundant. But the slopes differ with the age of the forest. In other words, there is an interaction between forest age and the effect of voles on the population of hares.

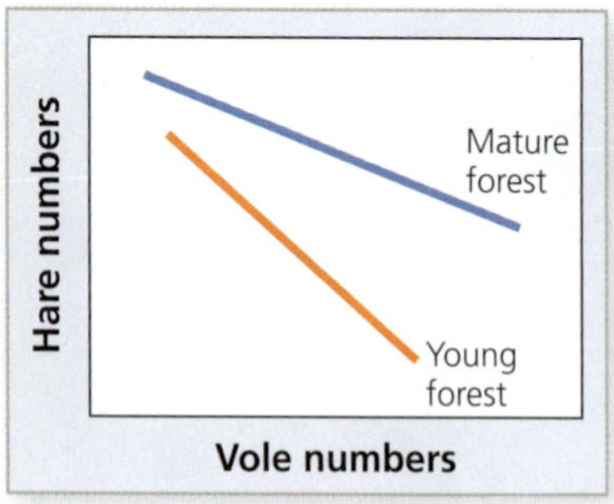

Finally, questions we ask about the natural world are crucial to the process of science. The initial study of hare mortality asked, "In which season is hare mortality highest?" The answer suggested another question: "Is the spring-fall mortality increase due to color mismatches?" To answer that question, we must ask if the mortality is due largely to predators or to other factors. If the answer to that question is predators, the next question addresses the effect of coat color mismatches on the predation rate. Our understanding of the natural world progresses as a network of questions, answers, and new questions. The best scientists are skilled at identifying the next logical question. The structure of this text reflects the importance of this skill. Each chapter is organized around a fundamental question

Figure 1.9 Interactions. A hypothetical interaction between the numbers of hares and voles and the age of the forest habitat. When vole numbers increase, hare numbers decline. However, the rate of decline depends on the age of the forest. **Analyze: What would this graph look like if there were no interaction between hares, voles, and the age of the forest?**

such as "How are populations regulated?" or "What is the effect of competition?" Those questions are much too broad to address directly. But if we ask a series of smaller questions, we can begin to develop an answer to the larger question. In this way, your learning reflects the scientific process.

KEY CONCEPTS 1.2

- Three important themes will emerge in your study of ecology:
  a. Ecology and evolution are intimately connected.
  b. Ecological systems do not necessarily achieve an equilibrium state.
  c. Humans have a significant impact on the Earth's ecological systems.
- Ecology employs the scientific method to answer questions about ecological systems.
- Ecology is a quantitative science that relies on mathematical and statistical analyses.
- The questions we ask about the natural world determine the hypotheses we generate and the analyses we perform.

QUESTION:

How does the subject matter of ecology, which is high in the biological hierarchy, affect the way we study ecological systems?

Summary

1.1 What Is Ecology?

- Ecology is the study of the interaction between an organism and its biotic and abiotic environment.
- Ecology is subdivided into hierarchical categories: the organism, the population, the community, and the ecosystem.
- Ecology is closely connected to natural history, environmental science, and the management sciences.

1.2 What Important Ideas Will Emerge in Your Study of Ecology?

- Ecology and evolution are intimately related.
- Ecological systems do not necessarily exhibit the property of homeostasis.
- Quantitative analysis is an important tool in modern ecology.
- In ecology, as in other fields of science, the questions we ask about nature shape our analyses and experiments.

Key Terms

adaptation p. 8
biosphere p. 6
ecology p. 3
ecological community p. 6
ecosystem p. 6
environmental science p. 7
evolutionary fitness p. 8

falsifiable p. 9
fisheries biology p. 7
forestry p. 7
homeostasis p. 8
natural history p. 7
organismal level p. 6
population ecology p. 6

prediction p. 9
proximate factor p. 4
range management p. 7
selection pressure p. 8
ultimate factor p. 4
wildlife management p. 7

Review Questions

1. Why is ecology so simple to define yet such a deep and rich field?

2. What components of the scientific process are illustrated by the snowshoe hare coat color example?

3. What is the significance of ecology's place in the biological hierarchy?

4. Why are questions so important in science?

5. What do we mean by the term "ecological interaction"?

6. Explain the relationship between ecology and evolution in the snowshoe hare coat color example.

7. What is the relationship between proximate and ultimate causation in ecology and evolution?

Chapter 2

Adaptation and Evolution

In October 1835 the young naturalist Charles Darwin stood on a rocky shore in the Galápagos and threw a marine iguana (*Amblyrhynchus cristatus*) as far as he could into the sea. It immediately swam back to shore and Darwin repeated the toss. Over and over he threw the iguana into the surf and watched it return to shore: "It invariably returned in a direct line to the spot where I stood. It swam near the bottom, with a graceful and rapid movement, and occasionally aided itself over the uneven ground with its feet" (Darwin, 1839).

Although Darwin thought the iguanas ugly ("It is a hideous-looking creature, of a dirty black colour, stupid, and sluggish in its movements"), he was also fascinated by them. He described them in great detail, noting that their tails are flattened sideways and their feet are webbed. In the water their sluggish movements change to graceful undulations that propel them rapidly. He opened the stomachs of several and found them packed with the marine alga *Ulva*. Their feeding behavior, too, interested the young Darwin. He noted that, on returning from a feeding excursion, they lie on the black lava rock, their bodies oriented for maximum exposure to the sun. Darwin observed that iguanas enter the ocean only reluctantly except to eat. He speculated that this reluctance "may be accounted for by the circumstance that this reptile has no enemy whatsoever on shore, whereas at sea it must often fall a prey to the numerous sharks. Hence, probably, urged by a fixed and hereditary instinct that the shore is its place of safety . . . it there takes refuge" (Darwin, 1839).

Darwin was practicing what today we would call *evolutionary ecology*. He was observing the characteristics of the iguana in relation to its environment. Moreover, this work was not based on preserved specimens in a museum; it was conducted in the field, in the iguanas' natural environment. Note, too, that in his speculation about their reluctance to enter the water he was drawing inferences about the *origin* of their behavior. This hypothesis exemplifies the concept of ultimate causes (Chapter 1).

The set of unique characteristics that Darwin described facilitate the marine iguana's success in the Galápagos environment. As Darwin noted, its flattened tail and webbed feet—unknown in other iguanas—aid in swimming. We now understand, too, that its dark color and habit of resting on black lava oriented toward the sun is important in two ways. First, iguanas are poikilotherms; that is, they do not physiologically maintain a constant body temperature, and so their body temperature drops while they are foraging in the cold waters of the Galápagos. Their dark color absorbs rather than reflects sunlight, restoring a higher internal temperature. The iguana's body temperature is also important for digestion. Like many herbivores, iguanas depend on symbiotic bacteria to digest plant matter. Basking maintains a temperature beneficial to their gut bacteria and digestion (Fields et al., 2008).

fitness The ability of an individual to survive and reproduce relative to other individuals in the population.

adaptation A trait that increases an individual's fitness in a specific environment.

The characteristics of an iguana determine its **fitness**, its relative ability to survive and reproduce in the Galápagos environment. Each of these characteristics is an **adaptation**, a trait that improves the fitness of the organism in a specific environment. Darwin's great insight, developed from countless observations, was that adaptations arise over time by an evolutionary process. Moreover, he developed the theory of natural selection as the central mechanism of this evolutionary process. In the example of the marine iguana, we see the intimate connection between evolution and ecology. The adaptations of this reptile evolved in a specific ecological situation: an herbivorous creature inhabiting a set of volcanic islands, nearly devoid of land predators, surrounded by cold marine waters. The evolution of the marine iguana makes little sense except in this ecological context, and full understanding of the ecology of this species is impossible without understanding the evolution of its adaptive responses to the environment. Thus, we begin this chapter with an evolutionary question crucial to all of ecology: *How do organisms adapt to their environment?*

2.1 How Did Darwin Develop the Theory of Evolution by Natural Selection?

Charles Darwin was invited to join the voyage of the HMS *Beagle* as ship's naturalist and gentleman companion to the captain, Robert Fitzroy (Figure 2.1). The *Beagle* was sent on a voyage around the world to improve the maps of many places, especially the coast of South America.

How Did the Voyage of the *Beagle* Change Darwin's Thinking?

The *Beagle* was away from England for five years (Figure 2.2). A number of factors combined to influence Darwin's thinking. First, Darwin was fascinated by geology.

Early in the trip he made careful observations of geological features, especially the volcanic Cape Verde Islands in the Atlantic. He began to understand the length of time associated with the origin of volcanic islands in the deep ocean. He carried on board with him an important contemporary treatise on geology, Charles Lyell's *Principles of Geology*. This book was significant in that it advocated a new view of geology, namely the theory of *uniformitarianism*. According to this view, modern landforms arose slowly by gradual processes rather than by rapid, cataclysmic events. Moreover, observing processes at work today is the key to understanding what occurred in the past. Darwin recognized the relevance of this concept to the growing science of biology.

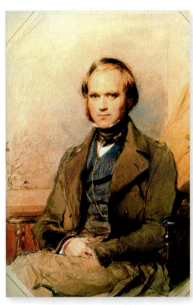

Figure 2.1 Charles Darwin. Darwin developed the theory of evolution by natural selection.

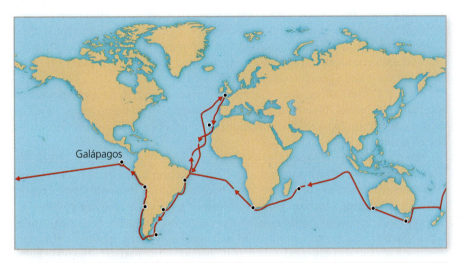

Figure 2.2 The voyage of the HMS *Beagle* 1831 to 1836. On this voyage, Darwin observed the patterns of geographic variation in plants and animals. He was especially intrigued by the unusual fauna of the Galápagos.

Second, Darwin had the opportunity to collect and observe plants and animals across large stretches of the South American continent. As the *Beagle* coasted north along Chile, Peru, and Ecuador, Darwin noticed changes in the organisms—his specimens from different sites along the coast were clearly related, but they varied with the geography. Among Darwin's most famous collections were those from the Galápagos Islands, where he made some crucial observations. He noted that many of the plants and animals of the Galápagos were similar to species he had seen on the mainland, yet they were subtly different. In addition, he noted that some groups had diversified into a variety of forms on the islands. The most famous examples are the finches we now know as Darwin's finches (Figure 2.3) and the Galápagos tortoise. Darwin conceived a mechanism to explain the variation in finches and tortoises. He reasoned that an ancestral colonist from the mainland had, over time, diverged into the variants that now occupy each island.

Darwin did not develop the theory of natural selection during the voyage of the *Beagle*. After his return to England he spent years studying his collections or the analyses of them done by other experts. He also embarked on a series of extraordinarily detailed studies of organisms including barnacles, earthworms, and orchids. He opened his first notebook on the subject of evolution in 1837. From this point on, his notes show clearly that he was moving toward the great idea that formed the basis of his classic work, *On the Origin of Species by Means of Natural Selection*.

Figure 2.3 Galápagos species. Many of the islands of the Galápagos have finches and tortoises that are unique to that island but are clearly related to those on other islands.

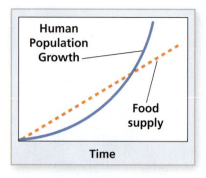

Figure 2.4 Food competition. The economist Robert Malthus argued that the human population increases exponentially (solid line) whereas the food supply (dashed line) can only grow arithmetically. Consequently, there will inevitably be competition for food (Malthus, 1798). **Analyze: Where in this graph does competition begin?**

natural selection The increase in the frequency of individuals with inherited traits that increase their fitness relative to other individuals.

What Was Darwin's Logical Argument?

The observations Darwin made on the voyage of the *Beagle* suggested that evolutionary change had occurred. However, change was not sufficient to advance a theory of evolution. He also required a mechanism. *On the Origin of Species* is an enormously complex and comprehensive treatment of evolutionary change. Nevertheless, Darwin made a simple, cogent argument for his mechanism. It is based on three empirical observations and a logical deduction that follows from them.

First, Darwin documented the wide array of variation within species. His own collections and observations showed that within any species and even within a local population, there are variants—individuals that differ in some way. This was no new insight. Amateur and professional collectors had appreciated it for years. In fact, collectors prized unusual specimens. However, Darwin studied and documented variation more thoroughly and systematically than anyone had before. He also studied the phenomenon in domestic animals. Darwin understood that much of this variation is inherited. In today's terminology we say it has a genetic basis.

Second, Darwin recognized that most plants and animals have prodigious reproductive potential. He noted, however, that virtually no organism achieves its maximum output. External factors reduce the reproductive rate or the survival of the offspring.

Next, Darwin understood the intensity of competition in nature. His thinking was influenced by an essay by the economist Robert Malthus. Malthus argued that there would always be a lower class because human reproduction increases the population exponentially. However, the food supply increases only arithmetically because increases occur only by the addition of new lands to cultivation. Consequently, the human population will inevitably outstrip the food supply (Figure 2.4). Darwin saw evidence of this process in the natural world. He knew that competition among individuals limits reproduction and survival.

These observations led Darwin to a crucial conclusion: any individual whose inherited characteristics make it more competitive will persist; others will die out. Over time, individuals possessing these advantageous characteristics increase in frequency in a given population relative to others. This is the central tenet of his concept of **natural selection**: individuals with inherited traits that increase their reproductive output or survival will increase in frequency in the population relative to other competing individuals.

How Do We View Darwin's Theory Today?

Darwin's insight and his careful work documenting the phenomenon of natural selection place him among the intellectual giants in the history of science. By the middle of the nineteenth century, many scientists already accepted the idea of evolution. Knowledge of the fossil record grew rapidly and geologists suggested older and older ages for the Earth. Darwin's key contribution was a mechanism for evolutionary change that operated on an equally ancient time scale.

One gaping hole in Darwin's idea was the mechanism of inheritance. Darwin devised a theory of blended inheritance in which offspring acquire a mix of characteristics of both parents. It wasn't until the early twentieth century that a valid mechanism of genetics was applied to Darwin's work. Subsequently, biologists united the growing understanding of inheritance with Darwin's mechanism of change. The result, known as the modern synthesis, brought evolutionary thinking to the center of the science of biology.

Modern molecular biology has contributed importantly to the development of evolutionary theory. As we will see in this chapter, there are other mechanisms of evolutionary change besides natural selection, but because they depend on an understanding of genetics and molecular biology, Darwin could not have conceived of them. Thus, it is fair to assert that Darwin was correct—but incomplete. The process of natural selection he proposed has been well documented in case after case. But as we shall see, the process of evolutionary change is richer and more complex than Darwin could have imagined.

In the time since Darwin, advances in geology and biology have substantially increased our confidence in the concepts of evolutionary change and the force of natural selection. We have a much more accurate and sophisticated understanding of the age of the Earth and the time span over which evolution has occurred. We also have a vastly more comprehensive history of life from the fossil record that clearly documents the details of change over time. Natural selection itself has been observed in operation in living species and can be inferred from careful analyses of fossils. In sum, the theory of evolution by natural selection has enormous empirical support. Evolution is central to modern biology as both an organizing principle and a basis for experimental analysis.

KEY CONCEPTS 2.1

- On the voyage of the *Beagle* Darwin observed and documented geographic patterns of variation within species.
- Three empirical observations (heritable variation within species; high but unrealized reproductive potential; competition for scarce resources) led Darwin to a crucial logical conclusion: those variants that leave more offspring or survive at a higher rate will increase in frequency over time.
- Darwin's theory was correct but incomplete. He did not understand the mechanism of inheritance. And he did not envision other mechanisms of evolutionary change that we now understand contribute to adaptive evolution.

QUESTION:

Why is the connection between inheritance and variation crucial to Darwin's mechanism of natural selection?

2.2 What Is Evolution?

Our description of evolution has thus far been based on the concept as understood in Darwin's time—changes in species over time and the emergence of new species. This concept, while correct, is somewhat vague and certainly nonquantitative. What exactly is meant by *change over time*? To answer this question we must understand how genes and alleles are organized in a population.

How Are Genes Organized in Populations?

The **phenotype** is the characteristic morphology, physiology, and behavior of the organism. The phenotype is the product of the organism's **genotype**, the sum total of its **genes** (as well as interaction of the genotype with the environment—see Section 2.3). A gene is a sequence of DNA that codes for the amino acid sequence that constitutes a specific protein. Some of these proteins are structural elements of the organism. Others are enzymes that produce other structural elements or that catalyze the biochemical reactions that constitute the organism's physiology. The variation among individuals so important to Darwin's theory arises by **mutation**, a random change in the DNA sequence of a gene. Mutations change the amino acid structure of the protein product of the gene and hence its structure and function. Most mutations are deleterious but a few result in a more successful phenotype. It is these new, beneficial variants that provide the basis for natural selection.

phenotype The characteristics including morphology, physiology, and behavior of an individual.

genotype The genetic makeup of the individual that, in concert with the environment, determines the phenotype.

gene A sequence of DNA that encodes the amino acid sequence of a specific protein.

mutation A random change in the DNA sequence of a gene.

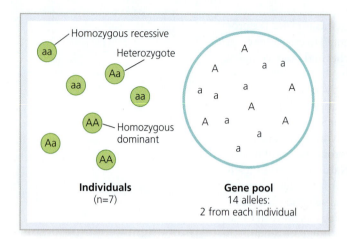

Figure 2.5 The gene pool. The gene pool consists of all the alleles in all the individuals in a population. In this population of seven individuals, the genotypes are distributed as shown in the left side of the figure. The gene pool consists of all of the alleles of those seven individuals. **Analyze:** What is the difference between a genotype and the allele frequency?

population genetics A field of genetics that analyzes the dynamics of genes in an entire population.

gene pool The sum total of all alleles in a population.

allele frequencies The proportion each allele represents in the gene pool.

Mendelian inheritance and the molecular biology of the gene focus on the characteristics of the individual. However, evolution by natural selection affects a *population* of individuals. The study of these changes is the province of **population genetics**. In population genetics the focus is on the entire population and its **gene pool**, the sum total of all alleles in the population (Figure 2.5). We characterize the gene pool for any given trait by the **allele frequencies**, the proportion each allele represents in the population. For example, in Figure 2.5 there are 7 individuals with the genotypes shown in the figure. This means that there are 14 alleles for this trait in the population (2 alleles per individual $\times$ 7 individuals), as shown in the right side of Figure 2.5. In this gene pool 6/14 alleles are A (0.43), whereas 8/14 (0.57) are a. We traditionally assign the variable p to the frequency of the dominant allele, A in this case. The frequency of the recessive allele (a) is denoted by q. If there are just two alleles at a locus, $p + q = 1.0$.

How Do We Model Changes in the Gene Pool?

Population genetics allows us to formalize the concept of evolution: evolution is a change in allele frequencies, the values of p and q, over time. We mathematically model evolutionary change using a system known as the Hardy-Weinberg equilibrium. Consider a population of 100 individuals with the following genotype frequencies:

AA	Aa	aa		Sum
36	48	16	=	100.

To calculate p, the frequency of the A allele, we need to compute the total number of A alleles in the population divided by the total number of alleles. Each homozygous dominant carries two A alleles; each heterozygote contains one. If there are 100 individuals in the population, there are 200 total alleles at this locus. Thus, the value of p is calculated

$$[(36 \times 2) + (48)]/200 = 120/200 = 0.6.$$

We calculate the value of q in similar fashion:

$$[(16 \times 2) + 48]/200 = 80/200 = 0.4.$$

If there are only two alleles, $p + q = 1.0$. Thus, we can also calculate q

$$0.6 + q = 1.0,$$
$$q = 0.4.$$

Now imagine that the individuals in this population mate at random and exactly replace themselves in the next generation. We can calculate the expected distribution of genotypes that results. Homozygous dominant individuals result from two A alleles coming together. An easy way to think of the probabilities is to imagine a bag of alleles (the gene pool) made up of 60 percent A alleles and 40 percent a alleles. The probability of reaching into the bag and extracting two A alleles is

$$0.6 \times 0.6 = 0.36.$$

This calculation is based on the product rule of probability: the probability of two independent events occurring simultaneously is the product of their independent probabilities. Thus, the probability of two A alleles coming together is

$$p \times p = p^2 = (0.6)^2 = 0.36.$$

Similarly, for the homozygous recessives, the probability is

$$q \times q = q^2 = (0.4)^2 = 0.16.$$

The heterozygote calculation is based on similar reasoning. Imagine that you reach into the gene pool and pull out an allele at random for the gene contributed by the mother, the maternal contribution. The probability that it is A is 0.6. Now the probability that the second allele (the paternal contribution) is a is 0.4. So it is possible to generate a heterozygote in this way with probability

$$0.6 \times 0.4 = 0.24$$

or

$$p \times q.$$

However, there is a second way to generate a heterozygote. The first allele (maternal) could be a (with probability 0.4). If so, the probability that the paternal allele is A (giving a heterozygote) is 0.6. This order occurs with the probability

$$0.4 \times 0.6 = 0.24$$

or

$$q \times p.$$

So, overall the probability of forming a heterozygote is

$$(0.6 \times 0.4) + (0.4 \times 0.6) = 0.24 + 0.24 = 0.48$$

or

$$(p \times q) + (q \times p) = 2pq.$$

Thus the expected genotype frequencies for the population are

$$
\begin{array}{ccc}
AA & Aa & aa \\
p^2 & 2pq & q^2.
\end{array}
$$

And because there are only these three possible genotypes,

$$p^2 \quad 2pq \quad q^2 = 1.0.$$

So, in a population of 100 individuals, the genotype frequencies are

Genotypes	AA	Aa	aa	Sum
Proportion	0.36	0.48	0.16	= 1.0
Number	36	48	16	= 100.

A population with genotype frequencies that satisfy the equation $p^2 + 2pq + q^2 = 1$ is said to be in **Hardy-Weinberg equilibrium**. The importance of this concept is that in the random mating example we just conducted, there was no change in the genotype or allele frequencies from one generation to the next. If evolution

Hardy-Weinberg equilibrium A mathematical representation of the genotype frequencies of a population in which the allele and genotype frequencies are not changing.

THINKING ABOUT ECOLOGY:

Consider a population with the following genotype frequencies:

genotype	AA	Aa	aa	Sum
number	14	45	40	= 99

a. Is this population in Hardy-Weinberg equilibrium?
b. What are the values of p and q?
c. If this population mates at random and no other factors affect it, what will be the genotype frequencies in the next generation?
d. Will the values of p and q change?

is defined as a change in allele frequencies, the Hardy-Weinberg equilibrium represents the case in which no evolution has occurred. In other words, a population in Hardy-Weinberg equilibrium is not evolving. This leads to the obvious question: What mechanisms change the Hardy-Weinberg equilibrium such that allele frequencies do change?

KEY CONCEPTS 2.2

■ Evolution proceeds by changes in the genetic composition of a population.
■ The gene pool consists of the sum total of all alleles in the population. Allele frequencies are key aspects of the gene pool.
■ The Hardy-Weinberg equilibrium describes the genotype frequencies of a population whose allele frequencies are not changing.

QUESTION:

What is the relationship between the allele frequencies (p and q) and the genotype frequencies in the Hardy-Weinberg equilibrium?

2.3 What Are the Mechanisms of Evolution?

Recall that evolution is defined as a change in allele frequency in a population. In the previous section, we showed that a population that is in H-W equilibrium will remain so. Thus, a population in H-W equilibrium is not evolving. However, H-W equilibrium only occurs if certain conditions hold. Under other circumstances, the equilibrium breaks down and allele frequencies change. By definition, such a population is evolving. What are these crucial conditions?

In the example of a population in H-W equilibrium above, we made a set of implicit assumptions.

First, we assumed that none of the genotypes has an advantage in terms of survival or reproduction compared to the others. If, for example, only half of the homozygous recessives survived to reproduce, their frequency would no longer be described by the value q^2. Second, we assumed that the population behaves like a much larger, in fact infinite, population. This assumption is based on the operation of the laws of probability. In very small populations, chance events may result in deviations from the expectations of probability theory. Third, we assumed that there was no net immigration or emigration of particular genotypes. For example, if all the homozygous dominant individuals emigrate, the genotype and allele frequencies will change. Finally, we assumed that there are no new mutations producing novel alleles or changing the dominant allele to recessive or vice versa.

If these four key conditions hold, a population in Hardy-Weinberg equilibrium will remain there with no change in allele frequencies and hence no evolution. Now we see the importance of the Hardy-Weinberg model: its assumptions provide us with potential mechanisms of evolutionary change. If any of these conditions does not hold, evolution occurs.

Violation of the first assumption leads to evolution by natural selection: differential survival or reproduction of different genotypes. Imagine that a large proportion of the homozygous recessives die before reproducing. The allele frequencies will shift to a higher proportion of the dominant allele (Figure 2.6). The second assumption requires a large population size so that the laws of probability operate as expected. If this assumption does not hold, random changes in allele frequency occur. Random changes in allele frequencies are known as **genetic drift** (Figure 2.7). Violation of the third assumption leads to the net gain or loss of certain alleles by movement of individuals. We refer to this as **gene flow** (Figure 2.8).

genetic drift Random changes in allele frequencies.

gene flow The net movement of alleles to or from a population.

mutation pressure Changes in allele frequency due to the origin of new alleles in the population.

selection coefficient (s) The proportion of a genotype that is not represented in the next generation due to death or reproductive failure.

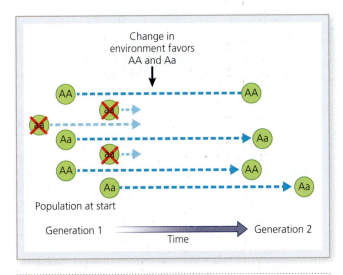

Change in environment favors AA and Aa

Population at start

Generation 1 → Time → Generation 2

Figure 2.6 Natural selection and heredity. Natural selection occurs over more than one generation. If the homozygous recessives in this population are at a selective disadvantage, their frequency will decrease from one generation to the next. When this occurs the frequency of the recessive allele (*q*) decreases and the frequency of the dominant allele (*p*) increases. **Analyze:** Is it possible for the dominant allele to decrease by natural selection?

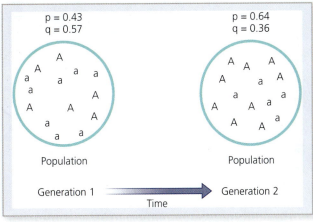

Population   Population

Generation 1 → Time → Generation 2

Figure 2.7 Effect of genetic drift on allele frequencies. We expect alleles to be distributed in gametes and in the next generation in the same frequency as generation 1. But in small populations, chance events result in some alleles being under- or overrepresented. Allele frequencies have changed only by chance. **Analyze: Does the value of *p* always increase by drift?**

ASSUMPTION	MECHANISM IF ASSUMPTION IS VIOLATED
No differential success of genotypes	Natural selection
Population is infinitely large (random mating)	Genetic drift
No net immigration or emigration by genotype	Gene flow
No new mutations	Mutation pressure

TABLE 2.1 The assumptions of the Hardy-Weinberg equilibrium and the mechanism of evolution that results if they are violated.

Finally, new mutations change allele frequencies. The evolutionary change resulting from new mutants is known as **mutation pressure**. Because mutations are so rare, they have a very small impact on overall allele frequencies and we can largely ignore their impact; the effects of selection, drift, and gene flow are generally much larger. Of course, mutations are important in that they generate novel alleles on which the other three mechanisms act.

An important point arises from this analysis: there are in fact *four* distinct mechanisms of evolution. Darwin is known for his explication of natural selection. Here is an example of the way in which Darwin was correct but incomplete. He of course knew nothing about genes, alleles, or the models developed by Hardy and Weinberg. Let us examine each of these mechanisms in more detail.

How Does Selection Change Allele Frequencies?

Our current understanding of natural selection is based on the Darwinian logic discussed above. Selection occurs when a particular genotype does not survive or reproduce in the same proportion as others. We quantify this effect with the **selection coefficient**, *s*: the proportion of a particular genotype that is not represented in the next generation. A selection coefficient of 1.0 means that the allele is lethal. Figure 2.9 shows the impact of different values of *s* on the change in allele frequency during selection.

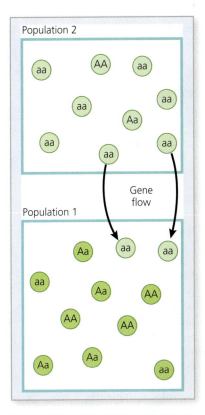

Figure 2.8 Gene flow. Gene flow occurs when there is net immigration (or emigration) of certain genotypes into (or out of) a population. The movement of alleles changes the allele frequency in the gene pool. **Analyze: What is the significance of "net" movement?**

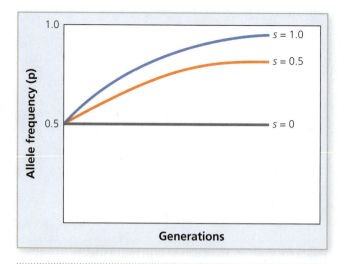

Figure 2.9 Allele frequency. The rate of change in allele frequency by natural selection is determined by the value of s, the selection coefficient. If $s = 1.0$, the allele is lethal and all homozygous recessives die. If $s = 0$, there is no selective advantage or disadvantage to the allele. **Analyze: How would these curves change if the dominant allele were lethal?**

directional selection A form of selection in which one tail of the phenotypic bell curve is favored.

stabilizing selection A form of selection in which the central portion of the phenotypic bell curve is favored.

THINKING ABOUT ECOLOGY:

Figure 2.9 shows the change in allele frequency under different selection coefficients. In the case of the lethal recessive ($s = 1.0$), the frequency of the dominant allele (p) approaches but never reaches 1.0. Why doesn't selection cause the frequency of the fixation of the dominant allele to reach 1.0?

evolutionary trade-offs The idea that many traits that confer a fitness advantage with respect to one aspect of the environment can also have a fitness cost relative to another.

disruptive selection A form of selection in which the two tails of the phenotypic bell curve are favored.

This process has now been described in detail for a number of organisms and ecological situations. One of the most thoroughly documented examples is the pattern of selection on the bills of Darwin's finches on the Galápagos (Grant and Grant, 1993, 2009). On the island of Daphne, the medium ground finch (*Geospiza fortis*) eats seeds of many sizes but prefers smaller seeds that are easier to crack open. When a major drought occurred in the Galápagos, seed production was severely reduced. As a result, the small seeds preferred by the majority of finches were rapidly depleted. The remaining seeds, mostly the large, hard seeds of *Tribulus cistoides*, are difficult to crack open. Nearly 85 percent of the finches on Daphne died because their bills were not large and strong enough to crack open these seeds. There was strong selection for deeper, more massive bills, and the proportion of individuals with such bills increased rapidly (Figure 2.10).

Consider this process in terms of the variation in bill sizes before and after the selection event. Bill size is a quantitative trait; that is, bills differ among finches in measurable characteristics (depth, mass, etc.). In many natural populations variable quantitative traits are distributed in a bell curve (Figure 2.11). When the environment abruptly changed, those individuals with bill sizes at the right edge of the bell curve were favored. Individuals with smaller bills made up most of the 85 percent that died. As a result, the entire distribution shifted to the right to larger average bill sizes. This form of selection, in which one tail of the bell curve is favored, is known as **directional selection**.

As shown in Figure 2.10, selection can also favor other portions of the bell curve. **Stabilizing selection** occurs when individuals in both tails of the curve are at a selective disadvantage and selection favors individuals with intermediate characteristics. The net result is a narrower distribution. In many species body size is subject to stabilizing selection. Given these factors, it is unlikely that

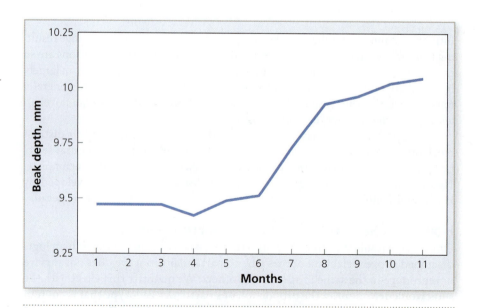

Figure 2.10 The rapid shift in beak depth of finches on Daphne Island following a drought. When the supply of small seeds was depleted, only birds with bills large enough to crack open the large seeds survived (Boag and Grant, 1988). **Analyze: Why is beak depth an important feature of the finch bill and its ability to open seeds?**

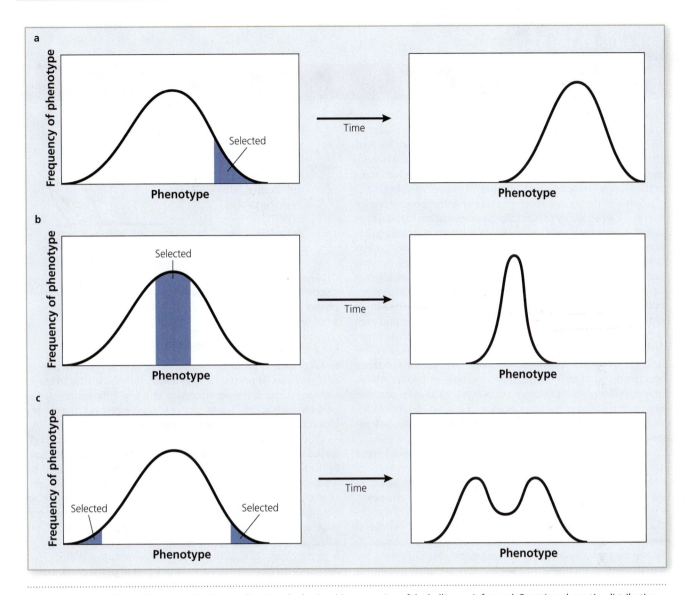

Figure 2.11 Three forms of natural selection. In directional selection (a) one portion of the bell curve is favored. Over time the entire distribution shifts in that direction. Stabilizing selection (b) occurs when the intermediate phenotypes are more fit than the extremes. In disruptive selection (c) the extreme phenotypes are favored compared to the intermediate phenotypes. **Analyze: Is it important whether these phenotypes are genetically determined?**

directional selection will continue to push a species to larger and larger *or* to smaller and smaller sizes. Intermediate sizes will contribute proportionally more to the next generation than extreme sizes. There are advantages to large size and to small size, but each also has a cost. This type of selection also illustrates the concept of **evolutionary trade-offs**; that is, a feature that confers an advantage in one respect may have a cost in some other important respect.

Disruptive selection occurs when the tails of the distribution are favored over the intermediate phenotypes. The monkey flower, *Mimulus luteus*, produces red nectar guides that lead its pollinators, insects and hummingbirds, to the nectar reward and the stamens and anthers, where they effect pollination. Note the variation in the size and shape of these nectar guides (Figure 2.12). Research on pollinator preferences in this species showed that insects prefer large guides that point directly toward the center of the flower. In contrast, hummingbirds prefer smaller, heart-shaped nectar guides. Because both

Figure 2.12 The color patterns in *Mimulus* flowers guide their pollinators to the nectar. There is variation in the morphology of these guides. Insects prefer large nectar guides; hummingbirds prefer smaller, heart-shaped guides. The result is disruptive selection on the morphology of the guides (Carezza et al., 2003).

The Evolution of Resistance in Pathogens and Pests

Given the size of the human population and the magnitude of our impact on the natural world, it is no surprise that humans constitute an important selective force on other species. In essence we have changed the environment for many species, both natural and domesticated. As a result, human activity has provided us with some of the most well-documented examples of natural selection.

Some of the most important cases of anthropogenic evolution occur when we try to control organisms that are harmful to us or to the plants and animals we raise. Antibiotics directed against human pathogens and pesticides designed to control crop pests demonstrate our impact on the evolutionary process. Bacteria gradually develop resistance to each new antibiotic we develop. Because they reproduce so rapidly and occur in such large numbers, the probability of a mutation conferring resistance to an antibiotic is significant. Moreover, many bacterial strains and even different species can transfer genes horizontally—that is, to other, unrelated individuals rather than simply to their descendants. The result is the rapid spread of mutations that confer resistance and the origin of strains resistant to multiple antibiotics. MRSA, or methicillin-resistant *Staphylococcus aureus*, is an example of an organism that has acquired resistance to a variety of antibiotics, including the penicillin group (which includes methicillin, dicloxacillin, and nafcillin) as well as the cephalosporin group.

Weed and insect pest resistance to pesticides used on crops are among the most well-studied examples of human-induced evolution. New pesticides are embraced by farmers because their immediate impact on pests and weeds is so great. However, the longer these compounds are in the environment, the more likely their targets are to develop resistance. Notice the rise of insecticide-resistant species in recent decades. Cotton is attacked by many insects, and consequently cotton farmers are among the highest users of pesticides. Approximately 40 percent of all the insecticides used in the United States are applied to cotton. As insects acquire resistance, larger doses of these compounds are required. For example, in just five years the dose of endrin required to control the cotton bollworm increased from 0.01 mg/gram of insect larva to 0.13 mg/gm insect larva. In the same period the required dose of the combination of toxaphene and DDT increased by a factor of 10.

Resistance to antibiotics or pesticides can be conferred by a variety of mechanisms. Consequently, mutations in a number of biochemical, physiological, and morphological traits may confer resistance. For example, DDT resistance in insects can occur by:

- An increase in lipid content, which allows DDT to be sequestered in fat tissue
- Enzymes that metabolize DDT, yielding less toxic products

- Changes in the nervous system that reduce DDT toxicity
- Reduction of the permeability of the cuticle to DDT
- Behaviors that reduce contact with DDT

The specific genes responsible for these phenotypes are scattered among all of the chromosomes in *Drosophila*. None of these traits completely protects the insect from DDT. But each incrementally increases its resistance. One reason for the increase in resistance over time is the accumulation of many different resistance genes in individuals. These super-resistant individuals have very high fitness and thus high reproductive success.

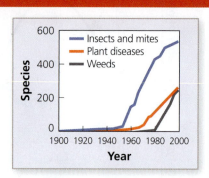

Figure 1 Insecticide and selection. The application of insecticides selects for resistant phenotypes. These resistant forms have increased rapidly after the initial exposure (Weber, 1994).

The development of herbicide-resistant crops has also accelerated the evolution of pest resistance. Strains of corn and soybeans resistant to the herbicide glyphosate are widely used by American farmers. Some 90 percent of the soybeans and 70 percent of the corn planted in the United States are resistant to glyphosate. These resistant crop strains are favored because large amounts of glyphosate can be applied to control weeds without harming the crops. However, prolonged exposure to high concentrations of glyphosate accelerates the evolution of resistance by the weeds. At least one weed, pigweed (*Ameranthus palmeri*), is now fully resistant to glyphosate. Pigweed can grow up to 80 mm per day, attains a height of more than two meters, and can damage harvest equipment.

The relevant principles of evolution are clear: natural selection acting on rapidly reproducing pests and pathogens inevitably results in the evolution of resistance to virtually any chemical treatment we develop. At present our strategy has been the development of new compounds. However, this must be coupled with the judicious use of antibiotics and pesticides that reduces the exposure that accelerates the evolution of resistance.

QUESTION:

How would you describe the evolution of resistance in terms of the selection coefficient, *s*?

---

non-Darwinian evolution Genetic drift.

effective population size (N$_e$) The subset of the total population that mates at random.

types of pollinators exert selective pressure on the flowers, their summed impact represents disruptive selection: flowers with intermediate characteristics were the least successful—they were not visited by either birds or insects. The result is that monkey flowers tend to have either large guides or heart-shaped guides (Medel et al., 2003).

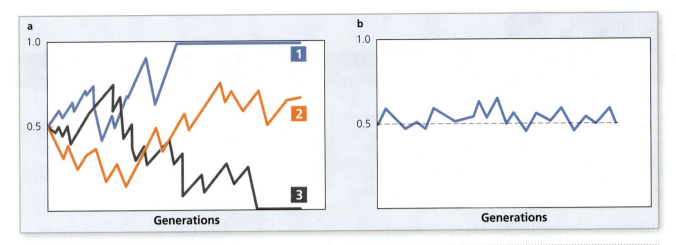

Figure 2.13 Genetic drift. The effects of genetic drift are more pronounced in small populations (a) than in large populations (b). In (a) each line represents a single, small population. In population 1, drift led to fixation of the allele. Population 3 lost the allele. In population 2, the allele persisted until the end of the experiment. **Analyze: Can loss or fixation occur in a large population?**

How Does Genetic Drift Change Allele Frequencies?

Genetic drift is the random change in allele frequency. If, for example, we expect homozygous dominants to occur with the frequency p^2, it is entirely possible that the actual frequency will be lower or higher than this value simply by chance. The phenomenon is more pronounced in small populations. Consider the analogy of flipping a coin. We expect 50 percent heads and 50 percent tails. But if you flip a coin only four times, it's entirely possible to get three heads and a tail or even four heads and no tails just by chance. It is far less likely to obtain all heads if you flip the coin 100 times. In a population in which allele frequencies change by drift, evolution is occurring but not by means of natural selection. Thus, genetic drift is also known as **non-Darwinian evolution**.

A number of factors contribute to chance deviation from expectations. However, all are the result of some factor that causes the population to depart from purely random mating in which the laws of probability play out as expected. We define the **effective population size (N_e)** as the subset of the total population that mates randomly. Any characteristic of the population that reduces random mating reduces the effective population size. As N_e becomes smaller than the actual population size, drift becomes more pronounced. For example, if the sex ratio, the proportion of the population that is male and female, is heavily skewed toward one sex, the population will not behave like a large, randomly mating group. Consider a population of 100 individuals. If 90 are female and only 10 are male, those few males must mate with all the females to produce the next generation. In effect the population is not 100 randomly mating individuals; all the offspring are the product of just 10 males, and because of their small numbers, chance events will skew their genetic contribution, leading to shifts in allele frequency.

If we follow a series of populations in which drift occurs over time, some reach a p value of 1.0 whereas others fall to 0 (Figure 2.13). When $p = 1.0$ the recessive allele has been lost. In this case we say that the dominant allele has become fixed. Similarly, when $p = 0$ the dominant allele has been lost and the recessive has been fixed. Loss and fixation of alleles are the inevitable consequence of genetic drift. Because drift is a random process, it is unpredictable whether any particular allele will be lost or fixed. But given enough time, one or the other will occur (Figure 2.14). The effect is evolutionarily important because it means that over time genetic drift, and thus loss and fixation, will reduce the genetic variation in the population. And of course it is the variation inherent in the population on which selection operates.

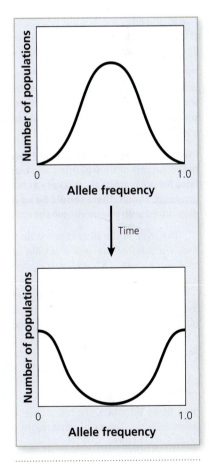

Figure 2.14 Genetic drift leads inevitably to the loss or fixation of alleles. Over time, a group of populations will shift from a normal distribution of allele frequencies to populations in which the allele has either been fixed or lost. **Analyze: Why is the number of populations with fixation eventually the same as the number with loss?**

Genetic Drift in a Desert Plant

We know that evolution can proceed by genetic drift as well as by natural selection. However, teasing apart the relative importance of these processes is not easy, especially in a natural population. The desert annual flower *Linanthus parryae* illustrates the problem. This species contains two morphs, a white and a purple morph that are spatially separated across a ravine. Schemske and Bierzychudek (2007) addressed the origin of this pattern and two alternative explanations for it.

HYPOTHESIS 1: The spatial distribution of color morphs is the result of selection favoring the white morph in one microhabitat and the purple morph in the other.

HYPOTHESIS 2: The spatial distribution of color morphs is the result of genetic drift acting on the local frequency of the color morph.

The first hypothesis suggests that each color morph is adaptive in one environment but not the other. Under the second hypothesis the color morphs are independent of the environment. Thus, these hypotheses lead to specific predictions:

PREDICTION OF HYPOTHESIS 1: Purple morphs transplanted into the microhabitat that contains the white morph should be less successful there than the white morph. In the reciprocal transplant, the white morph should be less successful than the purple morph.

PREDICTION OF HYPOTHESIS 2: There should be no difference in the success of transplanted and resident morphs. Also, alleles for other traits should vary at random across the habitat gradient. That is, there should be no consistent sets of alleles associated with one environment or the other.

The logic of the second prediction is that if drift is operating, the two sides of the ravine act like small populations in which random changes in allele frequencies occur. If so, the alleles, including those for color, should show no consistent spatial pattern.

Schemske and Bierzychudek used both transplant experiments and molecular techniques to test their hypotheses. Transplant experiments showed that each color morph is more successful in its own habitat. This was especially true in a dry year in which the blue- flowered plants produced 68 percent more seed in their home habitat than transplanted white-flowered plants. Similarly, white-flowered-plants produced 17 percent more seed in their home habitat than transplanted blue-flowered plants. The researchers were able to detect soil differences across the ravine that might account for the differential selection pressure.

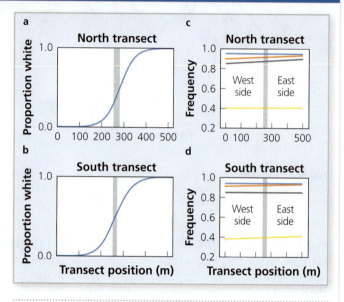

Figure 1 Flower color. There is a rapid shift in the flower color morph across the ravine (a and b). The ravine is shown by the gray line. However, across this same boundary, no shifts in allele frequencies for four other genes were noted (c and d). Each line represents the allele frequency of a different gene (Schemske and Bierzychudek, 2007). **Analyze:** What would the graphs c and d look like if allele frequencies shifted along with flower color?

To test the drift hypothesis, tissue samples were analyzed for the alleles present at a number of allozyme markers. Allozymes are alternative forms of a gene (alleles) that can be detected by the differences in the chemical properties of their amino acids. Amino acid differences among proteins result in differences in the allozymes' size and charge. These can be detected by placing samples in an electrophoretic gel. An electric current is passed through the gel and the allozymes migrate according to their size and charge. In this way, a large sample of genes can be analyzed for the patterns of their alleles. Schemske and Bierzychudek used this technique to test the prediction of the drift hypothesis.

The molecular results clearly support the predictions of the selection hypothesis. Whereas the shift in color morph is sharp at the ravine, no other allele shows any consistent pattern of difference across the ravine. However, a key question remains: Is flower color itself the target of selection? Or are the genes for blue and white flowers linked in some way to other genes important to success on different sides of the ravine?

What Are the Effects of Gene Flow?

Gene flow is the net movement of a particular allele to or from the population. If the movement of individuals (and their alleles) in and out of the population is random, there is no effect—that is, no gene flow. But if certain genes disproportionately enter or leave, allele frequencies change. Gene flow may reinforce or oppose the changes that occur by natural selection and drift. If the same genes that are entering the population are those that are increasing randomly within the population by genetic drift, the effect of drift will be more pronounced. Perhaps the most important effect of gene flow occurs when it opposes the changes driven by natural selection. Consider the case in Figure 2.15. Selection favors the dominant allele. However, there is net movement of the *a* allele into the population in the form of homozygous recessives. Clearly, the frequency of *A* will not increase in the way it would under selection alone; gene flow decreases the effect of natural selection. Thus, gene flow has an important impact on evolution by selection: the more isolated the population, the less gene flow and the more effective natural selection can be.

KEY CONCEPTS 2.3

- Evolution is defined as change in allele frequencies in a population.
- A population in Hardy-Weinberg equilibrium experiences no change in allele frequency; it is not evolving.
- A population in H-W equilibrium will remain there if (1) there is no differential success of genotypes, (2) the population is large, (3) there is no net movement of alleles in or out of the population, and (4) there are no new mutations.
- The violations of these four respective conditions represent potential mechanisms of evolution: (1) natural selection, (2) genetic drift, (3) gene flow, and (4) mutation pressure.

QUESTION:

What is the difference between the concepts of allele frequency and genotype frequency?

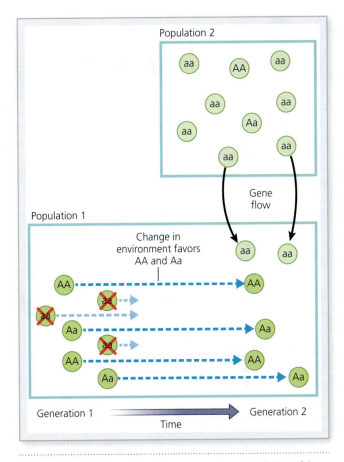

Figure 2.15 Gene flow. Gene flow can oppose natural selection if the same alleles that are selected against move into the population from elsewhere. In this example, selection favors the AA and Aa genotypes. However, gene flow by the immigration of aa individuals maintains their frequency despite their selective disadvantage. **Analyze: Can gene flow accelerate natural selection?**

2.4 How Do Adaptations Arise?

With this background in evolution, we can return to the origin of adaptations, traits that improve the success of the individual in a particular environment. We can now appreciate Darwin's observations of marine iguanas more fully. Darwin's insight was the intimate relationship between the organism and its environment. His observations of iguanas documented the features that improve their success in their unique environment and way of life.

How Does Selection Lead to Adaptive Change?

The success of a genotype is measured by its fitness. Fitness determines the relative ability of a genotype to obtain genetic representation in the next generation. The frequency of a genotype is reduced by selection against it—that is, by the value of the selection coefficient (s). Those genotypes that are less fit decrease in frequency by the amount s each generation. Formally, the Hardy-Weinberg analysis includes the selective coefficient and fitness as follows:

Genotype	AA	Aa	aa
Frequency before selection	p^2	$2pq$	q^2
Fitness	1	1	$1 - s$.

DO THE MATH

Modeling the Rate of Change by Selection

The Hardy-Weinberg equations allow us to model the process of evolution by natural selection. As presented in Section 2.2, the results are qualitative; they show that a particular allele will decline in frequency if selected against. However, we can also use the Hardy-Weinberg model to ask quantitative questions about selection. For example, what determines the rate of change of an allele frequency under selection?

Consider a population in which the homozygous recessives are selected against. Imagine that the selection coefficient represents the proportion that dies each generation. We model this as before but we algebraically include the effect of natural selection.

	GENOTYPES			POPULATION SIZE
	AA	Aa	aa	
Initial frequency				1
Fitness	p^2	$2pq$	q^2	
Proportions after selection	1	1	$1 - s$	$1 - sq^2$
(Determined by multiplying the initial genotype frequency by its fitness)	p^2	$2pq$	$q^2(1 - s)$	Population size is reduced by the loss of homozygous recessives.

In order to convert the proportions to genotype frequencies that add up to 1, we divide each by the total population size.

$$(p^2/1 - sq^2) + (2pq/1 - sq^2) + (q^2(1 - s)/1 - sq^2) = 1$$

Now let us consider what happens to the A allele under selection. We expect it to increase, but how fast? To answer this question, we need to calculate the change in p—that is, Δp. To do this we need to find the difference between the initial and new frequency. $\Delta p = p^1 - p$ where p^1 is the new frequency of A and p is its initial frequency. The value of p is based on the total number of A alleles divided by the total number of alleles in the population. Remember, each homozygous dominant contains two A alleles (that is, $2p^2$) and each heterozygote contains one (that is, $2pq$). The total number of alleles in the population is two times the population size ($2(1 - sq^2)$). So we calculate p^1

$$p^1 = 2p^2 + 2pq/2(1 - sq^2).$$

Because $p + q = 1$, we can algebraically calculate

$$p^1 = 2p^2 + 2pq/2(1 - sq^2) = p(p + q)/1-sq^2 = p/1 - sq^2.$$

Then, to calculate the change in p,

$$\Delta p = p^1 - p,$$

$$\Delta p = p - p(1 - sq^2)/1 - sq^2 = spq^2/1 - sq^2.$$

Now, examine this equation to see what we learn from it. First, because p and q are less than 1, Δp must be positive. That makes sense: if the recessive is selected against, p should increase. It also makes sense that according to this model the rate of change due to selection increases as the selection coefficient increases. But you might not have predicted that Δp also depends on the initial values of p and q. In other words, the rate of selection varies with the initial allele frequencies. Specifically, when the value of p is small (and thus q is large), the rate of evolution (Δp) is larger. The rate of increase of a beneficial mutant is higher when it is new—that is, at low frequency. This is one value of such mathematical models: they can reveal relationships that may not be intuitively obvious to us. And, importantly, we have an equation that allows us to predict quantitatively the effect of selection.

In this example, there is complete dominance so heterozygotes have the same phenotype and fitness as the homozygous dominants. Note that fitness is a relative term; that is, we measure it in relation to the fitness of other genotypes.

If no other factors are operating, natural selection sorts the genotypes in the population according to their fitness. Because the environment imposes the selective force, the ecology of the organism is central to the adaptive process. The ancestors of modern marine iguanas that had flattened tails and webbed feet were more successful at foraging in the ocean and left more offspring. The result, over many generations, is the set of specialized morphological adaptations we see today.

What Is the Relationship Between Variation and Natural Selection?

Variation among individuals in a population was central to Darwin's logical argument for natural selection. If every individual in the population is identical, there is nothing on which selection can act and adaptive change cannot occur. The population geneticist R. A. Fisher showed mathematically that there is a quantitative relationship between variation and natural selection. According to Fisher's Fundamental Theorem, the increase in fitness of the population is directly proportional to the amount of genetic variation in the population (Fisher, 1930).

According to Fisher, most populations are in the process of evolution; they have not come to equilibrium. New favorable mutations arise but have not yet been fixed by natural selection. Consequently, there are fitness differences among individuals on which selection acts.

The variation Darwin emphasized in his theory must be inherited; that is, it must have a genetic basis. However, there is another important type of variation in populations. **Phenotypic plasticity** is developmental or physiological variation among phenotypes induced directly by the environment. For example, humans born and raised at high elevation differ physiologically from those who live at sea level. They have more red blood cells and hemoglobin, larger lung capacity, and other physiological changes that allow them to function at altitude. If those same individuals developed at sea level, they would not exhibit these changes.

The variation among individuals in a population thus has two possible bases: genetic differences among individuals and phenotypic plasticity in response to differences in the environment. The **heritability** of a trait is the proportion of the phenotypic variation that is due to genetic differences among individuals. Formally, it is the proportion of the phenotypic variation in a population that results from the additive effects of genes. The higher the heritability, the tighter the genetic control of development. The rate of evolution depends on the combination of the intensity of selection and the heritability of the trait. Specifically, the evolutionary response to selection, R, is calculated

$$R = h^2 s.$$

R is typically measured as the proportional change in a character over time.

Selection and variation are causally related in other important ways. Selection leads to geographic variation within species. If a species has a geographic range that encompasses an array of different environments, local selection can lead to significant variation among those locales and environments. Genetically distinct populations that are locally adapted to a particular environment are known as **ecotypes**. The beach mouse, *Peromyscus polionotus*, inhabits a range of environments from coastal Florida to inland sites in Alabama. Soil color varies across this range from the white beach sands in Florida to dark, loamy soil in Alabama. Coat color in *Peromyscus* varies accordingly. Each local population represents an ecotype—a genetically distinct population whose coat color is locally adaptive (Mullen and Hoekstra, 2008).

Two aspects of the environment determine the probability that local ecotypes will evolve. First, the spatial shift in selection pressure must be relatively abrupt. If selection changes gradually across a large geographic area, it is less likely that a discrete ecotype will arise—the variation in selection pressure is too small. In addition, when the selection pressure is strong, its local effect is more pronounced and there is a greater probability that it will result in a local evolutionary response. Plants, whose individuals are rooted in the ground, exemplify these conditions. Abrupt changes in soil type can lead to ecotypic differentiation over very short distances. For example, mine tailings, the waste piles of rock from which ore has been extracted, often contain high concentrations of heavy metals such as lead or arsenic. Not only do heavy metals constitute a strong selective challenge to the plant, there is often a sharp boundary between the tailings and adjacent nontoxic soil. Local ecotypes have also been demonstrated in populations inhabiting serpentine soil. Serpentine soil is a difficult environment for plants due to its poor water holding capacity and low concentrations of nitrogen, phosphorus, and potassium. It is also high in magnesium and some heavy metals. Serpentine occurs as outcrops surrounded by other soil types. In the sunflower, *Helianthus exilis*, serpentine ecotypes grow within just a few meters of populations on more normal soil. Sambatti and Rice (2006) used molecular techniques to document the sharp genetic differences between serpentine ecotypes and adjacent populations.

phenotypic plasticity The ability of an organism to produce different phenotypes in different environments.

heritability A measure of the proportion of the phenotypic variation for a trait that is determined by additive effects of its genes.

ecotype A genetically distinct population that is adapted to local environmental conditions.

THINKING ABOUT ECOLOGY:

A population of sunflowers inhabits a small outcrop on serpentine soil, surrounded by a much larger population on fertile soil. A mutation arises that greatly improves survival on serpentine soil. Selection strongly favors this allele. How would the increase in the frequency of this new allele change if the sunflowers are (a) wind-pollinated or (b) pollinated by bees that move only from one flower to the next adjacent plant?

THE EVOLUTION CONNECTION

Phenotypic Plasticity

Both natural selection and phenotypic plasticity are processes that shape the phenotype. However, they differ in important ways. Natural selection modifies the phenotype by shifting the frequencies of alleles. It operates through the success or failure of individuals. We measure its effect at the population level—as a change in allele frequencies. Also, selection operates over a longer time scale than phenotypic plasticity. Whereas selection requires more than one generation, phenotypic plasticity occurs during the development of the individual.

Phenotypic plasticity is most common in species in which individuals experience more than one environment. For example, the morphology of the olive (*Olea europaea*) varies with the wind environment to which the tree is exposed. Individuals that develop high in the canopy where they are exposed to stronger winds develop smaller leaves and thicker twigs than individuals protected from wind (Garcia-Verdugo et al., 2009). The Trinidad guppy (*Poecilia reticulata*) inhabits some streams that contain predatory fish and others that do not. The probability that it will live long enough to reproduce is lower in streams with predators. The presence of predators triggers a developmental change in guppies: in streams where predators occur, guppies become sexually mature at smaller sizes (Dowdall et al., 2012). In both these cases phenotypic plasticity allows the *individual* to respond appropriately to its environment.

Despite the fundamental differences between natural selection and phenotypic plasticity, both are related to adaptive evolution in important ways. First, phenotypic plasticity is itself an adaptation that arises by natural selection. For example, many plants respond to shading by competing plants with longer stems that increase the exposure of their leaves to light. The light-dependent physiological mechanisms that elongate the internodes, and thus stem length, are the result of natural selection (van Leunen and Fischer, 2005). The environment directly determines the phenotype—but in a way that adapts the plant to its environment.

Second, phenotypic plasticity can either facilitate or retard adaptive evolution by natural selection. For example, phenotypic plasticity can shield the organism from natural selection. If the plastic response produces a moderately successful phenotype, natural selection will have less impact on the gene pool and new adaptations may arise more slowly (de Jong, 2005; Crispo, 2008). Alternatively, plasticity may facilitate colonization of novel and challenging environments. By producing phenotypes tolerant of the new conditions, the organism may persist long enough for natural selection to shape additional adaptations. Salt marsh populations of sticklebacks (*Gasterosteus aculeatus*) putatively exemplify this process. In the estuaries where they live, populations of sticklebacks experience radically different levels of salinity depending on whether they occur near shore where freshwater enters or near the open ocean where salinity is higher. In the estuary of the St. Lawrence River, the freshwater population evolved relatively recently from an ancestral salt-tolerant population. The ancestral population retains considerable phenotypic plasticity for salt tolerance. In contrast, the freshwater population is genetically adapted to the freshwater environment. Apparently the ancestral plastic species was able to colonize and persist in freshwater, where a genetically distinct freshwater ecotype eventually arose (McCairns and Bernatchez, 2009).

QUESTION:

Explain how phenotypic plasticity affects the selection coefficient (s) and thus the effect of natural selection.

How Do We Explain Imperfection?

One of the most fascinating things about the study of biology in general and ecology in particular is the remarkable adaptations we observe in nature. The match between organism and environment is so elegant and complex, its study is compelling. Certainly this was Darwin's interest, as so much of his career was devoted to understanding adaptive evolution. He was even able to make predictions about adaptations not yet known to science. The Madagascar star orchid (*Angraecum sesquipedale*) is perhaps the most famous example. Darwin observed that the nectar in this flower is located at the bases of extremely long tubes (Figure 2.16). He predicted that an insect pollinator with an extraordinarily long proboscis exists on the island. Nearly 100 years later a hawk moth with precisely the requisite proboscis, *Xanthopan morganii praedicta*, was finally discovered there.

Although we tend to focus on these examples of the remarkable fit between organisms and their environment, it is important to recognize that not all organisms are perfectly adapted. Imperfection is common in the natural world—there are many species whose features are clearly not optimally suited for their environment. For example, the vertebrate eye is not optimally efficient because the

Figure 2.16 Mutually beneficial adaptations. In the Madagascar star orchid the nectar is found at the base of a very long tube. Darwin predicted the existence of a hawk moth with a proboscis long enough to reach the nectar of the star orchid. It wasn't until 100 years after Darwin's description that the hawk moth was discovered.

retina lies beneath the photoreceptor cells, and the optic nerve must pass through the retina, leading to a blind spot near the center of the retina. An engineer would design this system differently. And in fact, many invertebrate eyes have a more efficient arrangement.

This kind of imperfection is expected under the theory of evolution. Several factors contribute to non-optimal design features. First, adaptations are derived from the modification of existing structures. In the giant panda (*Ailerupoda melanoleuca*) a sixth digit or "thumb" is used to strip the leaves from bamboo during feeding. This "thumb" is derived from the radial sesamoid, a bone that in most mammals is part of the wrist (Figure 2.17). The panda's "thumb" is a contraption, a feature derived from the elaboration of a preexisting part. It works well but is not as efficient as the primate thumb derived from the first digit (Gould, 1980).

Another reason for imperfections in nature is that evolution is an ongoing process. We should not imagine that evolution has reached its final state, an equilibrium of perfect adaptation. Not only is the environment continuously changing, organisms are still in the process of reacting to those changes. This concept has led to the **Red Queen hypothesis**, which states that for some organisms the environment changes faster than adaptations can arise by natural selection. The term is derived from Lewis Carroll's *Through the Looking Glass*, in which the Red Queen had to run faster and faster just to keep in place. This hypothesis is often applied to host-parasite systems in which a rapidly reproducing parasite infests a more slowly reproducing host. Because the parasite goes through many generations so much faster, it can evolve more rapidly than its host. Consequently, the host is often a step behind the evolution of its parasite.

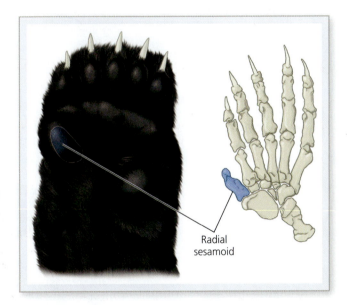

Figure 2.17 Panda thumbs. In the giant panda, the radial sesamoid bone develops into a "thumb." This "thumb" is actually a sixth digit, with a different developmental origin than the primate thumb.

Red Queen hypothesis   The idea that the environment changes faster than adaptations can arise by natural selection.

Figure 2.18 Too big to be real. Land creatures as large as the fictitious Godzilla cannot evolve because the weight that must be supported by the legs is too great. The weight of the individual relative to the cross section of the supporting legs results in an upper limit to the size of land animals.

adaptive landscape A graphical representation of the fitnesses associated with different genotypes in a population.

Finally, there are formal constraints on the adaptive response to the environment. No amount of selection can violate the basic laws of physics and chemistry. For example, there is a limit to how large a vertebrate land animal can be. If an animal is scaled up to larger size, there is a point at which its legs can no longer support the mass of its body. This occurs because the mass of the animal increases as the cube of its linear dimensions (that is, with its volume), whereas the strength of the legs increases only as the square of the linear dimension (the diameter of the legs). This places a maximum size on land vertebrates (Figure 2.18). Similarly, there is a lower limit to the size of a bird or mammal. As an animal gets smaller, the ratio of its surface area, from which it loses heat, increases rapidly compared to its volume, which generates heat. We do not find birds smaller than hummingbirds or mammals smaller than shrews because such an animal would lose heat so rapidly that it could not consume enough food to maintain a constant body temperature. Thus, the laws of physics constrain adaptive evolution.

What Is the Adaptive Significance of Genetic Drift?

Genetic drift is called non-Darwinian evolution because the random changes that occur by drift do not necessarily increase the fitness of the organism. However, like gene flow, genetic drift may oppose the effects of natural selection or it may reinforce selection.

Sewall Wright (1931) proposed another, more direct effect of genetic drift on adaptive evolution. His idea is based on the concept of an **adaptive landscape** (Figure 2.19). Imagine a plane determined by allele frequencies at two different

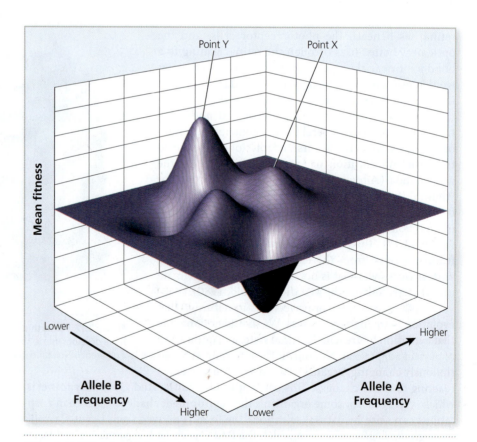

Figure 2.19 Wright's adaptive landscape. Wright's adaptive landscape is based on a three-dimensional map of the fitness associated with particular genotypes. A population at point X cannot reach the higher fitness at point Y by selection. It requires the random shift in allele frequency of drift to get to the saddle between the peaks, where selection can move it toward Y (Wright, 1931). **Analyze: Why can't selection move the allele frequencies to the saddle between point X and point Y?**

genetic loci. Each point in that plane represents a unique combination of the two alleles. Now let the third dimension represent the fitness of each combination. The result is something like a topographic map of the genetic landscape in which the heights of the peaks represent fitness. Now consider a population at point X on the landscape. It is at a minor peak in fitness. But there is a combination of alleles with higher fitness. How can selection move the population to that higher peak? Any incremental shift in allele frequencies *lowers* the population fitness. Selection will never favor such a change; it can only move the population uphill, to higher fitness. Selection alone leaves the population stuck on the minor fitness peak. Now imagine that this is a small population in which drift is important. Drift can increase or decrease fitness—its effect is random. Thus, drift could potentially take the population "downhill" to the saddle between points X and Y, where selection could quickly move it up to the higher peak. Wright suggested that adaptive evolution will occur most rapidly in small populations due to this phenomenon. Well-documented examples of this process are not known, but as you can imagine, this is a very difficult scenario to study in the field.

THINKING ABOUT ECOLOGY:

Drosophila resistance to DDT is affected by multiple genes controlling a number of physiological and biochemical processes (see "The Human Impact"). How does that phenomenon relate to the concept of Wright's adaptive landscape as shown in Figure 2.19? What does this mean for the process of evolution of DDT resistance in insects?

KEY CONCEPTS 2.4

- Fitness is the relative ability of a genotype to obtain genetic representation in the next generation by virtue of its survival and reproduction.
- Natural selection requires genetic variation on which to act.
- The existence of imperfection and examples of poorly adapted phenotypes are consistent with the theory of evolution.

QUESTION:

What are the major factors that constrain the process of natural selection?

Putting It All Together

It has been more than 150 years since the publication of *The Origin of Species*, in which Darwin laid out his theory of evolution by natural selection. It has been even longer since he spent an afternoon tossing a marine iguana into the sea to observe its swimming ability. We can now describe the process that led to the features of the iguana that Darwin observed. Characteristics such as a flattened tail and webbed feet are adaptations to the marine environment. Behavioral tendencies, such as the iguana's habit of basking on dark rocks exposed to the sun, are adaptations that increase the iguana's core temperature after a bout of foraging in the cold ocean.

We also understand now that these traits have a genetic basis. The ancestors of modern iguanas were characterized by variation in the shape of their tails and the morphology of their feet. Over time, selection favored those variants better able to swim and forage in the sea. The adaptive process may have been accelerated or retarded by random genetic change, particularly in relatively small island populations. The geographic isolation of the Galápagos reduced gene flow from other ancestral iguana populations that might have retarded the process of adaptation.

This process, played out across myriad environments, led to the stunning biological diversity of the Earth. Moreover, the adaptations that arise by the process are intimately connected to ecology. It was the ecology of the marine environment that shaped the evolution of the marine iguana. At the same time, it is the adaptations of the iguana that determine its ecological role in the Galápagos, that allow it to forage on algae in the cold waters surrounding the islands. Thus, ecology and evolution are intimately connected. The evolution of each species is a response to its ecology; ecological processes are the direct result of the adaptive evolution of coexisting species.

There are also many ecologically relevant evolutionary processes that we have not addressed. For example, the mechanisms by which new species arise are important to an understanding of the ecology of species diversity. The evolution of altruistic behavior plays a key role in behavioral ecology. And the history of extinction on Earth is relevant to the ecology of endangered species. These and many other aspects of evolution will come up when they are important to the ecological context. Our treatment of the process of evolution begins with this introduction but continues throughout our study of ecology.

Summary

2.1 How Did Darwin Develop the Theory of Evolution by Natural Selection?
- Darwin studied the characteristics of plants and animals in their natural contexts.
- The voyage of the *Beagle* exposed Darwin to the variation within and among species.
- Many scientists in Darwin's time were convinced that evolution occurs, but until Darwin developed the theory of natural selection, no mechanism explained change over time.
- Darwin's concept of natural selection was based on a logical argument:
  - There is heritable variation among individuals.
  - The organism's potential reproductive output is not realized in nature.
  - There is intense competition for scarce resources.
  - Those variants whose traits increase their fitness, their relative ability to survive and reproduce, leave more offspring.
- Darwin's view of evolution by natural selection was correct but incomplete.

2.2 What Is Evolution?
- Evolution is defined as genetic change in a population, specifically a change in allele frequencies.
- Population genetics is based on the concept of a gene pool, the sum of all alleles in a population.

- The Hardy-Weinberg equilibrium explains the conditions under which evolution (Δp) does not occur.

2.3 What Are the Mechanisms of Evolution?
- Violations of the conditions for Hardy-Weinberg equilibrium represent potential mechanisms of evolution:
  - Natural selection = differential reproduction and survival of genotypes
  - Genetic drift = random changes in allele frequencies
  - Gene flow = the net movement of alleles into or out of a population
  - Mutation pressure = novel mutations in the population
- There are three basic modes of selection: directional, stabilizing, and disruptive.
- Genetic drift is more pronounced when the effective population (N_e) is small.

2.4 How Do Adaptations Arise?
- Fitness is the relative ability to obtain genetic representation in the next generation.
- Genetic variation is the raw material on which the mechanisms of evolution act. The amount of variation limits the rate of evolutionary change.
- Not all features of organisms are perfectly adapted. There are many imperfections in biology. These are expected in evolutionary theory.

Key Terms

adaptation p. 14
adaptive landscape p. 32
allele frequencies p. 18
directional selection p. 22
disruptive selection p. 23
ecotype p. 29
effective population size (N_e) p. 25
evolutionary trade-offs p. 23
fitness p. 14

gene p. 17
gene flow p. 20
gene pool p. 18
genetic drift p. 20
genotype p. 17
Hardy-Weinberg equilibrium p. 19
heritability p. 29
mutation p. 17
mutation pressure p. 21

natural selection p. 16
non-Darwinian evolution p. 25
phenotype p. 17
phenotypic plasticity p. 29
population genetics p. 18
Red Queen hypothesis p. 31
selection coefficient p. 21
stabilizing selection p. 22

Review Questions

1. How were Darwin's ideas influenced by people in other fields?

2. Why were the Galápagos finches and tortoises so important to Darwin's thinking?

3. Why is "genetic change" a useful criterion for evolution?

4. How do natural selection, genetic drift, and gene flow interact?

5. How can gene flow enhance or retard evolution by natural selection?

6. Explain the relationship between the genotype, phenotype, and heritability.

7. What is the relationship between "effective population size" and random mating?

8. Can evolution occur without a change in fitness?

9. What is the relationship between the selective coefficient (s) and fitness?

Further Reading

Grant, P.R., and B.R. Grant. 2002. Unpredictable evolution in a 30-year study of Darwin's finches. *Science* 296:707.

This paper addresses the relationship of chance and predictability in evolution. Some aspects of evolutionary change are predicable; others are not. The interaction of the predictable and unpredictable is complex and cannot be understood except in long-term studies of natural populations.

Hoeck, E.A., et al. 2010. Differentiation with drift: a spatio-temporal genetic analysis of Galápagos mockingbirds populations (*Mimus spp.*). *Philosophical Transactions of the Royal Society B* 365(1543):1127–1138.

This modern study of the mockingbirds of the Galápagos uses molecular techniques to analyze the relative importance of population size, drift, and gene flow on the patterns of genetic variation. It is an excellent example of the interaction of these forces in shaping the genetics of a species group.

Weaver, T.D., et al. 2007. Were Neandertal and modern human cranial differences produced by natural selection or genetic drift? *Journal of Human Evolution* 53:135–145.

This study examines the origin of the cranial differences between humans and Neandertals. One hypothesis is that the robust skulls of Neandertal were adaptations for using the teeth as tools, whereas the more delicate craniums of modern humans were an adaptation for speech. The alternate hypothesis is that the two populations were small and isolated and the differences arose by genetic drift.

Weiner, J. 1994. *The Beak of the Finch*. New York: Vintage Books.

This book chronicles the long-term studies of the Galápagos finches by Peter and Rosemary Grant. Not only does the book explain the wide array of evolutionary questions addressed by the Grants' work, it is an excellent introduction to the nature of fieldwork in evolutionary ecology.

Chapter 3

Adaptations to the Physical Environment

Bands of bright color radiate outward from the edge of a hot spring in Yellowstone National Park. At first glance you might imagine that these bands are mineral deposits, places where compounds have precipitated from the boiling water as it cools. However, they are biofilms, aggregates of microorganisms that adhere to one another and the rock surface. Each color represents a different species of bacteria or archaea. Remarkably, the water covering these films would scald your skin. In fact, some of these **thermophilic**, or heat-loving, species are growing quite happily near the boiling point of water. So far the record holder for high temperature is *Pyrobacterium brockii*, whose optimum growth occurs at 102–105°C and which can survive at 115°C!

Snow falls on the ocean just offshore from Antarctica. There is no wind and the snow slowly accumulates on dead-calm water. It doesn't melt because the saltwater, at –2°C, is colder than the snow falling on top of it. Swimming actively in these waters is the fish *Macropterus maculates*. Fish are **poikilotherms**, animals whose internal temperature varies. Most poikilotherms are **ectotherms**; their body temperature is determined primarily by the external environment. Somehow the tissues of *Macropterus* do not freeze even when the surrounding water is well below the freezing point. The whales and seals that inhabit these waters are **homeotherms**, whose body temperature is regulated within narrow limits. Most homeotherms are **endotherms**—their internal temperature is maintained by metabolic activity.

Figure 3.1 The Couch's spadefoot toad lies dormant in the sand until rain stimulates its emergence and breeding.

■ **thermophile** An organism that tolerates high temperature.

■ **poikilotherm** An organism whose internal temperature varies, often in response to the external temperature.

■ **ectotherm** An organism whose body temperature is determined by the external environment.

■ **homeotherm** An organism whose body temperature is regulated within narrow limits.

■ **endotherm** An organism whose internal temperature is maintained by metabolic activity.

In the Sonoran Desert of southern Arizona, rain has not fallen for months. Near a shallow depression, adult Couch's spadefoot toads (*Scaphiophus couchii*) lie buried in the sand (Figure 3.1). They are torpid, with their metabolism slowed significantly. Nitrogenous waste in the form of ammonia and urea accumulates in with their tissues. This high solute concentration draws water osmotically from the sand around them, keeping them hydrated. When a thunderstorm approaches, the vibration of thunder activates the adults, which quickly emerge, mate, and lay eggs as the rain begins to fill the small temporary pond. The eggs hatch in just 15 hours; the larvae metamorphose in as few as nine days. As the pond dries, the adults burrow into the sand, where they may remain inactive for months.

We know that the physical environment represents an important selective force that shapes physiology, morphology, and even behavior. Each of these organisms faces a set of challenging environmental conditions—high temperature, subfreezing conditions, and drought. How is each able to tolerate these physiological challenges? Although these particular species are remarkable for the extremes they can tolerate, the question applies to all species. If you consider the wide range of habitats where life occurs—from the poles to the tropics, aquatic and terrestrial, saltwater and freshwater—the ability of life to persist in the face of physical extremes is remarkable. Across the wide array of environments there are species that tolerate extremes of temperature, salinity, O_2, CO_2, water, sunlight, and even radiation and pressure. We are still learning about the extremes to which living things can adapt. This leads to the fundamental question we will address in this chapter: *How do species adapt to the physical environment?*

■ **physical (abiotic) factors** The physical conditions that affect an organism's growth and survival.

■ **physical resources** The energy and inorganic materials an organism requires.

■ **tolerance, law of** The concept that there are upper and lower bounds to the physical factors within which an organism can survive.

3.1 How Do Environmental Factors Limit Growth and Survival?

All organisms interact with the physical environment in two important ways. First, they obtain **physical resources**, the inorganic materials or energy they require for existence. For example, plants require CO_2, light, and water for photosynthesis. **Physical (or abiotic) factors** are physical conditions such as salinity or temperature that affect growth and survival. In reality, the line between physical resources and physical factors is indistinct. For example, water may be a crucial resource, but if too abundant it can become a limiting factor.

The effect of the physical environment on growth and survival was among the first ecological principles elucidated. The **law of tolerance** states that there are upper and lower limits to the physical factors an organism can tolerate

(Shelford, 1913). Species differ in the factors that limit their growth and in their ranges of tolerance. Furthermore, the deleterious effects that occur outside the range of tolerance depend on the specific physical factor. For example, most enzymes have a range of conditions such as pH, osmotic potential, and temperature at which they can function. Extreme conditions may significantly reduce the rate at which they catalyze reactions or may destroy them entirely. At low temperature, enzymatic function may slow to the point that crucial cellular functions are limiting or lethal. If the temperature is high enough, function may cease irreversibly because protein enzymes denature.

When the physical conditions exceed the organism's tolerance limits, its fitness is compromised. Thus, an important component of fitness is homeostasis, the ability to maintain physiological systems within certain limits across a range of external conditions (Figure 3.2). Consequently, there is strong selection for mechanisms that ensure that the tolerance limits match the abiotic conditions the organism typically faces.

We know from Chapter 2 that many phenotypic traits vary among individuals. If these differences have a genetic basis, they provide the raw material on which natural selection acts. Now consider a population like that shown in Figure 3.3. The tolerance curve represents the tolerance limit of the entire population. The individuals that comprise this curve differ due to variation in their physiology, morphology, or other phenotypic traits. If the environment changes, some individuals may be better able to tolerate the new range of conditions. Directional selection then shifts the population tolerance limit.

The adaptive process is complicated by the fact that many of the physical challenges faced by species are not constant. Some change in predictable fashion. In temperate and Arctic/Antarctic regions, temperature, moisture, and light change with the season. At high latitudes these changes can be extreme. For example, the summer mean high temperature in parts of northern Alaska is more than 50°C higher than the winter mean. Alaskan summers are also characterized by 24 hours of daylight, whereas the sun does not rise for many weeks in the winter. In other regions, such as parts of Asia and the Sonoran Desert of the southwestern United States, seasonal rains known as monsoons result in temporal variation in the moisture available to plants and animals. Many tropical forests experience a relatively constant temperature regime but pronounced wet and dry seasons.

Temporal variation occurs on many time scales. In coastal marine systems, tides rise and fall twice each day, often with profound effects. Tide pools, small depressions in a rocky shoreline, exemplify the environmental extremes tides can produce (Figure 3.4). At low tide, seawater remains in tide pools, where it provides a refuge for animals that cannot tolerate desiccation. However, the physical conditions in those refuges change over time. At low tide and full sunlight the water temperature gradually rises, and as water evaporates the salinity increases. Animals such as starfish, barnacles, and limpets that inhabit tide pools experience a challenging range of physical conditions that change by the hour.

Temporal variation in the physical environment poses a significant challenge to the organism's fitness. In general it is much simpler to adapt to a harsh but constant environment than to one that is constantly changing. Moreover, unpredictable changes pose a greater challenge than predictable ones (Figure 3.5). The organism must not only assess the nature of its environment in some way, it must

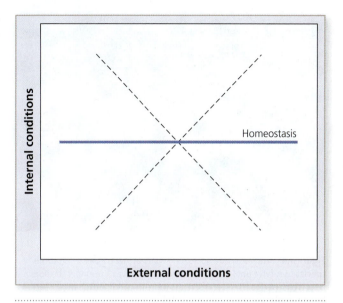

Figure 3.2 Homeostasis. Homeostasis is the ability to maintain constant internal conditions despite changes in the external environment. The solid line represents homeostasis. The dashed lines represent species whose internal conditions change with the external conditions. **Analyze: What would the graphs of temperature look like for a killer whale and a salmon in the North Pacific Ocean?**

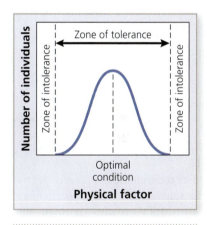

Figure 3.3 Tolerance. Organisms show a bell-shaped curve of tolerance of physical conditions. Outside the boundaries of tolerance, the organism cannot survive. **Analyze: Why is this curve bell-shaped?**

THINKING ABOUT ECOLOGY:

The individual tolerance curve is not the same as the tolerance curve for the whole population. What is the role of each in the processes of natural selection and adaptation?

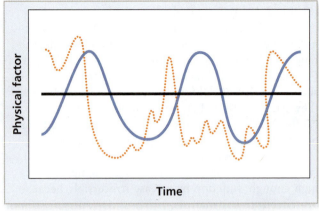

Figure 3.4 Tide pool. Tide pools hold water at low tide, allowing some species to survive until the tide rises again. However, the longer the tide pool is exposed to the air and sun, the more its conditions change. The water temperature may increase, and evaporation may increase the salinity.

Figure 3.5 Environmental patterns. Some environments are constant (solid line). Others change predictably (dashed line) and some change unpredictably (blue line). **Analyze:** How do these patterns affect the organism's adaptations to the environment?

then counteract those changes, in some cases very quickly. The spadefoot toad faces an environment in which sufficient moisture for reproduction is not only rare, it is unpredictable. Its adaptive solution must be different from the common leopard frog, for example, that thrives in an abundantly moist environment and whose reproductive cycle is predictably seasonal.

The adaptive responses to the physical environment fall into two broad categories: (1) avoidance of harsh physical conditions or (2) adaptations that match the organism's tolerance limits to the physical conditions it faces. We examine each in the sections that follow. As you proceed through this material, note that adaptations to the physical environment illustrate two important evolutionary principles introduced in Chapter 2. First, they demonstrate the concept of evolutionary trade-offs. For any potential adaptive solution to a physical challenge there are both costs and benefits. If, for example, a warbler migrates to the tropics to avoid the northern winter, it incurs a significant energetic cost and risk of mortality. For such a strategy to evolve, the benefits must exceed these costs. The **principle of allocation** states that adaptations to one challenge may preclude or reduce adaptations to others (Levins, 1968). Thus, for example, if an organism adapts to high temperature, its ability to tolerate cold may be compromised. Second, physiological adaptations illustrate the principle of formal constraints on adaptive evolution. For example, a set of fundamental laws of physics govern the transfer of heat from one object to another. These laws constrain the morphological and physiological adaptations of animals and plants to extreme temperatures.

allocation, principle of   The concept that an adaptation to one selective factor may preclude or reduce adaptations to others.

KEY CONCEPTS 3.1

- Each species is limited by one or more physical factors and resources.
- The distinction between physical factors and physical resources is sometimes blurred.
- Each species has an upper and lower tolerance limit for physical factors.
- Variable and unpredictable physical conditions pose especially difficult problems.

QUESTION:

What special adaptive challenges are posed by variable environments?

DO THE MATH

Working with Bell Curves

The *bell curve* or *normal curve* is central to many aspects of ecology. Its basic form is a plot of the frequency distribution of a series of observations, measurements, or other variables. Recall from Chapter 2 the importance of the bell curve in the process of natural selection. This distribution represents potential fitness differences among individuals. Natural selection shifts the distribution based on these fitness values. We also see bell curves in the organism's responses to the physical environment. Shelford's law of tolerance is based on the observation that not only are there abiotic limits to survival, there is also variation in survival within those limits.

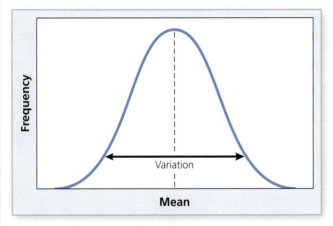

Figure 1 Bell curve. A bell curve is characterized by a mean value and variation around that mean. **Analyze: How does the variance measure variation around the mean?**

Many of the physical factors we consider in this chapter exhibit a bell curve. For example, if we measure the maximum temperature on a particular day over many years, the distribution will have the form of a bell curve. Although bell curves have a standard form, they differ in some important characteristics. How do we quantify these differences?

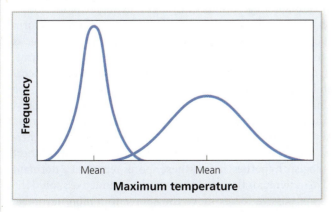

Figure 2 Comparing bell curves. These curves have similar shape but differ in the amount of variation around the mean. **Analyze: Which has high variance? Which has low variance?**

We characterize data sets such as these with two basic statistical measures: a measure of *central tendency* and a measure of variation. Measures of central tendency include the *mean* and *median*. The mean ($\bar{x}$) is simply the average value in the data set

$$\bar{x} = \frac{\sum_1^n x_i}{n}$$

where *n* is the total number of observations, or *sample size*, and x_i is the *ith* value in the data set. The median is the value in the data set for which the number of values larger and smaller is equal. If we array all the observations from smallest to largest, the median is the midpoint value. Both the mean and median are useful ways of characterizing a data set. If we report that the mean temperature at the North Pole is –12°C, this single value is one way to describe the nature of the temperature at the pole. The mean and median differ in one important way: the mean value is affected much more by extreme values than the median (Table DTM-1).

TABLE DTM-1 The mean and median of two sets of temperature measurements. Note that the mean is affected by the extreme value in Set 2; the median is not.

	SET 1	SET 2
	21	21
	21	21
	22	22
	23	23
	25	25
	26	26
	27	27
	29	29
	29	29
	29	29
	30	68
Mean	25.6	29.1
Median	26	26

It is also useful to quantify the variation in a data set. Clearly the distributions shown in Figure 3.1B differ. To quantify that variation, we employ one of the measures of variation, all of which are related to the *variance*. We calculate the variance with the following equation:

$$S^2 = \frac{\sum_1^n (\bar{x} - x_i)^2}{n-1} .$$

continued

DO THE MATH *continued*

Working with Bell Curves

Consider how this equation measures variation. We subtract each value in the data set from the mean, square it, and sum these values. Then we divide that sum by $n - 1$. In essence this equation calculates a value analogous to the average difference between each value and the mean. It is not precisely an average because we square the differences and divide the sum by $n - 1$ rather than simply n. Still, high values of S^2 occur when there is a wide range of values; we obtain small values of S^2 when most of the values are very near the mean. There are two other important measures of variation, each of which is related to S^2. The *standard deviation* (S) is the square root of the variance. The *standard error* is calculated as

$$SE_{\bar{x}} = \frac{S}{\sqrt{n}}.$$

There are two important uses of these measures of variation. First, if our data set is based on a set of sample measurements, they measure how confident we are in the mean value we derive from the samples. If the variation is large, it might be that by chance we have acquired some extreme values that skew the mean. If so, the mean is not a very good representation of the population. The standard error is particularly useful to analyze this possibility. Specifically, we can use the standard error to calculate a 95 percent confidence interval for the mean. In other words, in our sample there is a 95 percent probability that the actual mean of the population is equal to

$$\bar{X} \pm 1.96SE.$$

The second importance of the measures of variation is that in nature the variation is often as important, or perhaps more important, than the mean. This , is particularly apt in this chapter. Imagine, for example, that we find that a species of cactus is absent from a particular desert. We learn that this species cannot tolerate temperatures below freezing. When we measure the temperature regime in that habitat, we find that the mean temperature is 29°C, far above freezing. However, further analysis shows that the variance associated with that mean is large. In fact, temperatures well below freezing are common. In this case, the mean temperature has little importance to the cactus; it is the variation around that mean that limits its survival and thus its presence.

3.2 What Adaptations Avoid Harsh Conditions?

Avoidance entails two possible strategies: the organism can enter an inactive, resistant state until conditions improve or it can move somewhere where conditions are more favorable.

How Does Inactivity Protect the Species?

The essence of this strategy is to avoid difficult conditions by reducing the level of activity or exposure to the environment. Seeds exemplify the concept. They may remain in the ground through long periods of drought, extreme cold, or other conditions that would kill seedlings or adults. Many bacteria and protists form protective spores that can remain inactive for years. Some insects overwinter as pupae or larvae deep in the soil or other protected places. These inactive forms, especially seeds and spores, remain viable despite extreme temperature, desiccation, toxins, and even high doses of radiation. When conditions improve, the seeds or spores germinate and the rest of the life cycle is completed. We are just beginning to understand the magnitude of this phenomenon. For example, Arctic marine sediments receive as many as 10^8 spores per square meter per year of thermophilic bacteria that cannot possibly grow in that cold environment (Hubert et al., 2009). Microbiologists hypothesize that these spores represent a dormant microbial "seed bank" of individuals that could one day become active should the environment change or if passive forces move the spores to a more favorable environment.

Some plant seeds have mechanisms to ensure that germination does not occur prematurely or at the wrong time. Surprising as it may seem, there are some types

THE HUMAN IMPACT

Profound Changes to Coral Reefs

According to the law of tolerance, there are upper and lower bounds within which each species can survive. A tolerance curve exists for each important physical factor a species encounters, but some factors are more critically limiting than others. All contribute to the species' growth and survival and ultimately its presence or absence in a particular environment. The ability of a species to adapt and survive depends largely on how much time it has to do so. Evolution is a slow process, one that often cannot respond effectively if environmental changes come at a furious pace.

Humans have so profoundly altered the physical environment that we have created physical conditions beyond the tolerance limits for many species. Natural selection requires multiple generations to effect adaptive change. Consequently, rapid changes in the environment may preclude adaptive evolution. Moreover, anthropogenic environmental changes have compounding effects because it is rare that we affect a single factor in any environment (see also "On the Frontline" for a discussion of interactions among factors). This interactive effect is illustrated by profound changes that are occurring in coral reefs, one of the most productive and diverse types of ecosystem on Earth.

Figure 1 Coral reefs. Coral reefs are among the most diverse communities on Earth.

These reefs are composed of the accumulated growth of corals. The surface layer of the structure contains the living coral; the bulk of the reef is composed of the skeletons of older corals. Living corals are inhabited by a group of symbiotic algae known as zooxanthellae. The coral provides protection and some nutrients; the zooxanthellae provide the corals with energy from photosynthesis. Corals and their zooxanthellae thrive under very precise conditions. Coral reefs occur in regions where the ocean temperature lies between 20°C and 30°C and the water is shallow enough that sufficient sunlight penetrates to drive photosynthesis.

Human activity has shifted the physical environment beyond the corals' boundaries of tolerance. This occurs due to both global and local effects. As the Earth's temperature has increased from greenhouse warming, shallow marine systems have experienced a temperature increase of nearly 2°C. When the water temperature exceeds 30°C, the corals eject their zooxanthellae. Without their primary energy source, the corals die. Since it is the algae that give the corals their brilliant colors, the result is *coral bleaching*—transformation to a reef composed of dead, white coral. In addition, much of the excess CO_2 that enters the atmosphere from burning fossil fuels dissolves in the ocean, where it forms carbonic acid. As the pH of the ocean drops, the formation of the calcium carbonate central to coral structure is inhibited. As humans alter the temperature and pH regimes near reef systems, they shift the conditions beyond the coral's tolerance.

Local impacts are also important. Human density is high and increasing in coastal communities worldwide. One consequence is the release of excess nutrients from agricultural runoff or from sewage. These nutrients stimulate algal growth in the water, limiting light penetration to the zooxanthellae and compounding the effect of increased temperature.

Presumably corals and their associated zooxanthellae could adapt to these changes in physical factors. However, these challenges have been imposed on the species in just a few generations—changes so rapid that adaptive evolution probably cannot respond quickly enough.

QUESTION:

Humans are changing the thermal and chemical environment faced by corals. Do our activities affect the physical resources or physical factors, or both?

of seeds that must be exposed to a minimum period of cold conditions before they will germinate. While this strategy may seem counterintuitive, it actually ensures that seeds produced during one growing season do not germinate until favorable conditions return in the next. We have stressed that unpredictable conditions pose a significant challenge. One adaptive response is seeds or spores that are resistant to abiotic challenges but remain viable for long periods of time. They can persist in a dormant state and then germinate when favorable conditions return.

Figure 3.6 Temperature affects activity level. Birds such as poorwills (top) and hummingbirds (bottom) can become torpid at low temperature.

Figure 3.7 Hibernation. Ground squirrels are obligate hibernators. They must hibernate for some portion of the winter. Bears are facultative hibernators that move into and out of hibernation as the temperature changes.

■ **torpor** A state of decreased physiological function during periods of harsh conditions.

■ **hibernation** An extended form of torpor.

■ **aestivation** A period of torpor or hibernation to avoid heat and water stress.

A tactic employed in vertebrates is **torpor**, a state of slowed body functions. The various forms of torpor address two problems common in harsh conditions—the condition itself, such as cold or desiccation, and the fact that under such conditions energy is often limiting. In torpor the heart rate slows, the body temperature drops, and the metabolic rate decreases, all of which conserve energy. Birds such as hummingbirds, poorwills, and swifts enter torpor on cold nights (Figure 3.6).

Hibernation is an extended form of torpor in which the physiological changes that occur in torpor continue over a long period, sometimes the entire winter. **Aestivation** refers to periods of torpor or hibernation during the summer to avoid heat and water stress. There is variation among hibernators in the degree to which bodily functions are depressed. Some species, such as ground squirrels, are classified as **obligate hibernators**—they must enter hibernation each year (Figure 3.7). Ground squirrels accumulate significant fat reserves during the summer months. Cues from ambient temperature and photoperiod induce obligate hibernators to enter hibernation. Their low metabolic rate is maintained by stored body fat. During hibernation the basal metabolic rate drops to 2–4 percent of normal and body temperature is maintained within a few degrees above ambient temperature. When the arctic ground squirrel (*Spermophilusparryii*) hibernates, temperatures in parts of the body drop below freezing. Most physiological functions are dramatically reduced: heart rate may drop as low as three beats per minute and respiration to as few as five breaths per minute. Kidney function often ceases altogether.

Facultative hibernators enter a torpid state, but they can be quickly aroused by external stimuli. Bears and bats exemplify this strategy. Most facultative hibernators do not reach as deep a state of physiological depression as true hibernators. For example, bears' body temperature drops only about 6°C and they wake much more frequently. Some, including polar bears (*Ursus arctos*), complete gestation and give birth in the winter den.

Although the proximate advantage of hibernation may be overcoming harsh physical conditions, other benefits accrue to hibernators. Some ground squirrels remain underground for many months. While in hibernation their survival is higher than at other times of the year because they are protected from threats such as predators. This may explain why even some tropical mammals hibernate. For example, the Madagascan fat-tailed dwarf lemur (*Cheirogaleus medius*) hibernates for seven months of the year even though winter temperatures exceed 30°C (Dausman et al., 2004). Figure 3.8, a comparison of hibernators with physically and ecologically similar nonhibernators, shows the survival benefits enjoyed by hibernating species (Figure 3.8).

Torpor and hibernation are found in many avian and mammalian taxa. Hibernating mammal species are found interspersed among taxonomic groups that do not hibernate. The physiological changes during hibernation result from the differential expression of a set of genes common to all mammals. This, in conjunction with the presence of torpor in birds, suggests that hibernation is an ancestral trait that has been continued in some groups and abandoned in others (Carey et al., 2003).

What Movement Patterns Are Protective?

The second fundamental strategy for avoiding harsh physical conditions is **migration**, the seasonal movement from one region to another and back. Like hibernation, migration may be obligate or facultative. This strategy is advantageous when the abiotic challenges are seasonal and thus predictable or there are appropriate cues that foretell a change in conditions.

Migration requires considerable energy and entails a high risk of mortality. Thus, migration is adaptive only if the cost of not migrating is higher than the cost of movement. Migratory behavior has clearly evolved multiple times among many taxonomic groups. For example, more than 30 species of bats are known to migrate, but these species are scattered across a number of lineages that include nonmigratory species (Bisson et al., 2009). The migratory behavior of birds is a combination of genetically programmed tendencies modified by specific environmental factors. Northern Hemisphere migratory birds respond to decreasing photoperiod, the length of sunlight per day, by accumulating energy reserves to fuel migration and with an increasing restlessness and orientation toward the south. Photoperiod cues interact with weather to trigger the actual move. Conditions such as strong north winds, dropping temperatures, and snow are the proximate triggers for many species.

Most birds fuel their migration with fat. If they stop to replenish their energy reserves, the food they take in is quickly converted to fat and stored in adipose tissue. Some migratory birds deposit fat loads as high as 50 percent of their body mass (McWilliams et al., 2004). Although some replenishment of energy reserves may be possible along the way, some species depend entirely on stored energy reserves. For example, some hummingbirds migrate across the Gulf of Mexico. Obviously, they must accomplish this using only stored energy.

Mammals undertake long-distance migrations as well. The wildebeest (*Connochaetes spp.*) of the Serengeti migrate hundreds of kilometers to find surface water during the dry season, then return to areas of abundant forage when the rains return. In northern Alaska the Porcupine caribou herd migrates some 400 km or more from the interior mountains to the coast during the calving season. This movement ensures adequate forage and takes advantage of coastal winds that provide relief from insects. The pronghorn antelope of the southern Yellowstone area and Grand Teton National Park migrate several hundred miles south to the valley of the Green River, where they find less snow and more forage than on their northern high-elevation summer range (Figure 3.9).

obligate hibernator An organism that must seasonally hibernate.

facultative hibernator An organism that does not have to become torpid but does so during harsh conditions and can arouse quickly if conditions change.

migration A seasonal movement from one region to another and back.

THINKING ABOUT ECOLOGY:

Facultative hibernation is adaptive in some species; obligate hibernation is adaptive in others. What do you imagine are the relative costs and benefits of facultative versus obligate hibernation?

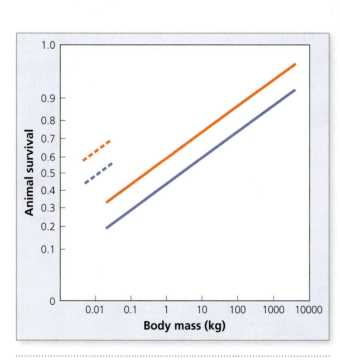

Figure 3.8 Hibernators vs. nonhibernators. Survival of hibernating and non-hibernating mammals. Orange lines: hibernators; blue lines: non-hibernators; dashed lines: bats; solid lines: other mammals (from Turbill et al., 2011). **Analyze: What do you conclude about the effect of hibernation on survival?**

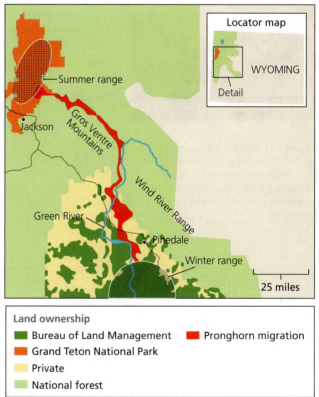

Figure 3.9 Pronghorn antelope migration. Pronghorn antelope undertake one of the longest migrations of any mammal in North America. They travel from their summer range in Yellowstone National Park some 600 km to their winter range in lower-elevation sagebrush habitats near the Wind River Range in Wyoming.

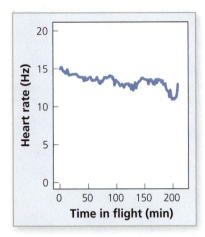

Figure 3.10 **Tracking by transmitter.** Technical advances such as the small radio transmitter allow researchers to monitor birds' heart rates during migration (from Robinson et al., 2010). **Analyze: What factors would you expect to affect heart rate during flight?**

■ **behavioral thermoregulation** Behaviors that allow the organism to seek and use external factors such as sunlight or warm rocks to change their internal temperature.

Figure 3.11 **Marine iguanas.** Marine iguanas, like many poikilothermic reptiles, bask in the sun to increase their internal temperature. They absorb heat from the environment rather than generate it internally.

Figure 3.12 **Basking.** This graph shows passive warming of the dunnart by basking. The heat acquired from the environment speeds arousal from torpor (from Warnecke et al., 2008). **Analyze: What would be the advantage and disadvantage of relying on passive warming?**

Migrating animals assume a number of mortality risks. Young calves in the Porcupine caribou herd are preyed upon by wolves that follow the caribou migration. Migrating birds are also vulnerable to predation in flight or when they stop to rest and feed. Storms may push migrants far off course, resulting in energy costs that reduce survival during flight or the ability to reach an appropriate wintering area. The energetics and ecology of migration is an area of active research. Technical advances in telemetry equipment, geolocators attached to birds, radar, and satellite imagery have significantly accelerated research on the ecology and physiology of bird migration (Robinson et al., 2010) (Figure 3.10).

Humans sometimes increase migration mortality as well. Wind turbines are typically situated in areas with strong, predictable winds. Often these same regions are traditional migration routes because the birds make use of the same prevailing wind patterns. Consequently, there is considerable research interest in the extent to which wind turbines contribute to mortality during migration. The antelope Wyoming migration passes through several narrow gaps in the mountains. Natural gas development in these choke points is of particular concern to conservation biologists because it might block migration.

Migration and hibernation are strategies that avoid harsh conditions that occur over large spatial or temporal scales. However, sometimes the harsh conditions are local or only last a short while. In this case, avoidance may be as simple as moving about to a more favorable spot in the local environment. For example, some reptiles employ **behavioral thermoregulation**. As ectothermic poikilotherms, their activity may slow so much during cold weather that they cannot forage or escape their predators. They raise their internal temperature behaviorally by the process of basking, positioning their bodies in places where the external temperature is higher or where they receive direct sunlight that warms their tissues (Figure 3.11). They may orient their bodies in specific ways to maximize heat absorption, or they may choose warmer sites and backgrounds with a higher temperature. Behavioral thermoregulation may affect other aspects of the life history as well. For example, female eastern fence lizards (*Scleroporus undulates*) actively seek nest sites with higher ambient temperature where their offspring will develop more rapidly (Angilletta et al., 2009).

Behavioral thermoregulation can also reduce the energetic cost of raising the body temperature at the end of a bout of torpor. For example, a small Australian

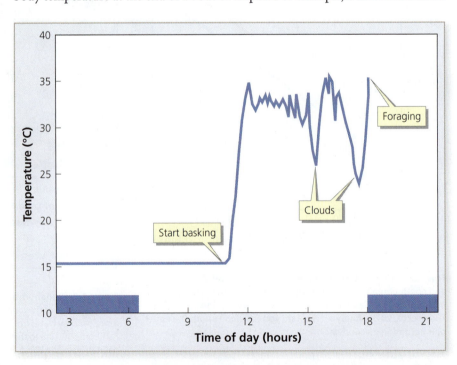

marsupial, the dunnart (*Sminthopsis crassicaudata*), uses basking behavior to rewarm following torpor. Passive warming of this type significantly increases body temperature at low energy costs (Figure 3.12).

Other behaviors connected to local movement patterns contribute to homeostasis. For example, nocturnal behavior protects many desert rodents from thermal and water stress. During the heat of the day they retire to a cool burrow, where the relative humidity is significantly higher. Active movement is reserved for the night, when temperatures are less extreme.

These strategies for avoiding difficult physical conditions are not necessarily mutually exclusive. For example, hummingbirds migrate to avoid winter cold and the absence of their main food sources. However, when they arrive back on their breeding grounds in the spring, they may encounter occasional periods of cold weather. When this occurs, hummingbirds become torpid, thereby reducing their energy needs.

KEY CONCEPTS 3.2

- Organisms adapt to physical stress with one of two basic strategies: avoidance (migration, hibernation, torpor, resistant seeds or spores) and adjustment of the tolerance limit to the prevalent physical condition.
- The temporal and spatial patterns of variation in the environment determine the type of avoidance strategy a species adopts.
- Behavior is an important component of some avoidance strategies.

QUESTION:

Under what spatial or temporal environmental patterns would you expect avoidance to be most common?

3.3 How Do Physiological Adaptations Alter the Organism's Tolerance Limits?

If hibernation or migration strategies are not available to an organism, it may permanently leave an environment, die, or develop physiological adaptations to help it survive. Each species' tolerance limit is determined by its unique biochemical, physiological, and morphological characteristics.

The adaptive physiological challenges species face are as diverse as the array of habitats found on Earth. Thus, there is an enormous diversity of limiting physical factors. Some, like temperature, water, and salinity, are limiting for many species. Others are rare, challenges peculiar to a specific environment or ecology. For example, marine animals that live or dive deep in the ocean experience tremendous changes in external pressure during deep dives. Other environments, such as Mono Lake in California, are known for their unusual chemistry. Mono Lake is highly alkaline (Figure 3.13). The bacteria and brine shrimp that inhabit its waters must tolerate a pH greater than 10. Some environments even have extreme levels of ionizing radiation. Where this occurs, radio-resistant species such as the bacterium *Deinococcus radiodurans* and some bdelloid rotifers can survive doses of radiation 100 times greater than most other organisms.

Despite the many abiotic challenges organisms face, the potential impacts of these challenges are similar: changes in external conditions such as temperature or chemistry may cause the organism's internal conditions to change as well. If the new internal conditions exceed the organism's tolerance limits, the impact is detrimental. The physiological adaptations to difficult abiotic conditions fall into two broad categories: (1) some species' physiology tolerates a wide range of

Figure 3.13 Challenging environments.
Mono Lake is a highly alkaline environment in which only a few species, such as brine shrimp (inset), can survive.

internal conditions, and (2) some species employ mechanisms that counteract the external challenge so that internal conditions remain within narrow limits. The latter approach often has a high energetic cost. Among the many physical challenges species face, two are nearly universal in their importance—temperature and water affect many species in some way. Because of their general importance we will use them to illustrate the nature of the two main categories of physiological adaptation to the challenges of the abiotic environment.

How Does the Tolerance Limit Vary Within and Between Individuals?

The first category—increased tolerance of internal variation—implies that tolerance limits can be shaped by natural selection. For example, some organisms, such as the wood frog (*Rana sylvatica*), are especially tolerant of desiccation. Whereas most animals cannot survive if they lose 30 percent of their body's water, the wood frog can survive even if it loses 60 percent of its total body water (Churchill and Storey, 1993). Desert plants such as cacti are also known for their desiccation resistance.

In general, poikilotherms tolerate greater variation in their internal temperature than homeotherms. In fact, poikilothermy and homeothermy constitute a continuum: some homeotherms allow their internal temperature to vary more than others. For example, the antelope ground squirrel (*Ammospermophilus leucurus*) inhabits the Mojave and Sonoran Deserts of the southwestern United States, where it experiences daytime temperatures as high as 70°C. The squirrel forages in the hot sun, gradually accumulating heat until its internal temperature sometimes rises as high as 43°C. It then retires to its burrow and sprawls with its belly on the cool ground until its temperature drops.

For some species the ability to tolerate difficult physical conditions also depends on the history of their exposure. **Acclimation** is the process in which an individual physiologically adjusts to challenging abiotic conditions. Most species can tolerate more difficult conditions if they are exposed to them gradually. This suggests that some tolerance limits are not rigidly fixed but are modified as a result of exposure to conditions near the upper and lower bounds of tolerance. An extreme example is the community of lichen species that inhabit polar regions. They tolerate rapid exposure to temperatures as low as −78°C, but if they are cooled slowly they can survive temperatures as low as liquid nitrogen (−196°C). In most species, populations that encounter greater variation in their abiotic environment acclimate more readily than other populations of the same species

acclimation An individual's physiological adjustment to challenging abiotic conditions.

whose environment is constant. For example, the acclimation ability of populations of the rufous-collared sparrow (*Zonotrichia capensis*) varies with latitude. High-latitude populations that typically encounter greater environmental temperature variation are better able to acclimate than low-latitude populations that experience less variation (Figure 3.14).

How Do Organisms Adapt to the Thermal Environment?

If we are to understand the physiological adaptations to the thermal environment, we must first understand the effect of temperature on chemical reactions, including the biochemical reactions important to organisms. The rate of a chemical reaction, whether in a test tube or inside the cell of a microorganism, plant, or animal, is determined by the following equation:

$$R_2 = R_1 (Q_{10})^{\frac{T_2 - T_1}{10}}$$

where R_2 is the reaction rate at temperature 2 (T_2) and R_1 is the reaction rate at temperature 1 (T_1). The magnitude of the temperature effect is measured by the parameter Q_{10}, the increase in the reaction rate for each 10°C increase in temperature. Thus, if a 10° increase in temperature doubles the reaction rate, $Q_{10} = 2.0$.

These changes have profound effects on the organism. As biochemical reactions slow at low temperature, important cellular processes may be compromised. If the internal cellular temperature reaches 0°C, ice crystals may form inside the cell, causing physical damage to organelles and the cell membrane. There is also a maximum temperature, beyond which cellular and organismal death occurs. Proteins and DNA denature at temperatures above 40°C in most organisms. Cell membrane integrity is also compromised at high temperature. In addition, oxygen consumption usually increases as temperature increases. If the demand for oxygen exceeds the ability of the organism to acquire and distribute it, the effect may be lethal.

For poikilotherms and microorganisms whose internal temperature varies with their environment, biochemical adaptations to temperature stress are crucial. Many of the enzymes in poikilotherms function across a relatively wide range of temperatures. Other specialized biochemical mechanisms may also be protective. For example, Antarctic fish that inhabit subfreezing water protect their tissues from ice formation by producing antifreeze glycoprotein molecules (AFGPs). Glycoproteins are amino acid chains with short chains of sugars attached. The sugar component of the glycoproteins binds to the planes of ice crystals and inhibits their growth (Evans et al., 2010).

Many plants and animals produce a class of proteins known as **heat shock proteins (HSPs)** when the organism experiences a sudden increase in temperature. They stabilize the three-dimensional structure of important proteins, protecting them until the temperature decreases and the protein can function normally. Analogous **cold-acclimation proteins**, or **CAPs**, consist of some 20 proteins synthesized by bacteria grown at low temperature (Feller and Gerday, 2003). Their functions are poorly understood but seem to mitigate the effect of low temperature on DNA and RNA structure.

In microorganisms, biochemical adaptations to high temperature must counteract the lethal effect of protein and DNA denaturation at high temperature. For example, the DNA of bacteria and archaea that inhabit hot springs is composed of an unusually high proportion of G-C base pairs because the triple bonds of this pairing add stability to the DNA molecule relative to the double bonds of A-T pairs. The species-specific Q_{10} value of enzymes also reflects the species' thermal environment.

Adaptive changes such as these can be produced in the laboratory. Most strains of *E. coli* grow optimally at 37°C but poorly at 20°C. When researchers exposed a series of cultures to 20°C for 2,000 generations, strains emerged whose

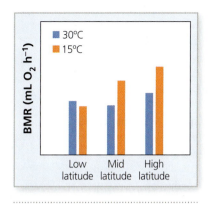

Figure 3.14 High-elevation metabolism. The metabolic rate of rufous-collared sparrows changes with elevation and the temperatures they have been exposed to. Birds from high-elevation populations acclimate to low temperature better than those from low elevation (Cavieres and Sabat, 2009). **Analyze: Why would you expect high-elevation populations to acclimate better to temperature changes?**

heat shock proteins (HSPs) Proteins that protect the organism from sudden increases in temperature.

cold-acclimation proteins (CAPs) Proteins that protect the organism from sudden decreases in temperature.

THINKING ABOUT ECOLOGY:

Some of the low-temperature strains in Bennett and Lenski's experiment grew better at 20°C than others. Some strains adapted to 20°C could still grow slowly at 37°C. What do these results suggest about the number of genes affected by this selection regime?

heat Thermal energy that can be transferred from one body to another.

calorie The amount of heat required to raise one gram of water 1°C.

specific heat capacity The amount of heat a substance requires to raise one gram 1°C.

conduction The transfer of heat between two bodies in physical contact.

radiation The transfer of heat between two objects not in physical contact.

convection The transfer of heat from an object to a moving fluid or gas.

heat of vaporization The decrease in an object's heat content by the evaporation of water.

emissivity The tendency of an object to emit radiation.

absorptivity The tendency of an object to absorb radiation.

critical temperature The maximum and minimum temperature an organism can experience without expending energy to maintain a constant internal temperature.

thermal neutral zone The range of temperature between the upper and lower critical temperatures.

nonshivering thermogenesis The use of brown fat to increase heat production.

countercurrent heat exchanger An arrangement of vessels such that the flow is in opposite directions, maximizing the temperature differential between the fluids and thus maximizing heat exchange between them.

THINKING ABOUT ECOLOGY:

Diagram a system of blood flow in which the direction of flow in the two vessels is not opposite but parallel. How does the flow of heat change compared to a countercurrent system?

optimal growth occurs at the lower temperature. Moreover, the shift to the lower optimal temperature was generally accompanied by a decrease in their ability to grow at 37°C (Bennett and Lenski, 2007). It is not yet clear precisely what biochemical changes occurred, but the trade-off between growth at high versus low temperature represents an example of Levin's principle of allocation.

The physical mechanisms of heat transfer determine the potential adaptations to extreme temperature. **Heat** is thermal energy that can be transferred from one body to another. The **calorie** is the amount of heat required to raise one gram of water 1°C. Other substances require different amounts of heat, their **specific heat capacity**, to achieve this 1°C increase. Substances with high heat capacity, such as water, tend to gain and lose heat slowly.

Heat is gained or lost by an object by four processes: **conduction**, **radiation**, **convection**, and the evaporation of water. Conduction is the transfer of heat between two bodies in physical contact. Convection is the transfer of heat from an object to a moving fluid (or gas). Radiation is the transfer of heat between two objects not in physical contact. Any object above absolute zero (0°K) emits electromagnetic radiation, which can transfer heat across air or through a vacuum to another object. The **emissivity** of an object is its tendency to emit radiation; **absorptivity** is its tendency to absorb the radiation it receives. Conduction, convection, and radiation transfer heat only from high-temperature to low-temperature objects. Heat is required for water to evaporate. Thus, evaporation from a surface decreases the object's heat content by an amount known as the **heat of vaporization**. At 22°C, 584 calories are required to change one gram of water to vapor. This is an important means of cooling for many animals, but of course it comes with a high cost in water. Moreover, evaporation is the only means of lowering the temperature of a body if it is warmer than its environment.

The heat content of a plant or animal H_{total} is determined by the equation

$$H_{total} = H_m \pm H_c \pm H_{cv} \pm H_r - H_e$$

where H_m is metabolic heat production, H_c is conductive heat gain or loss, H_{cv} is convective gain or loss, H_r is radiative heat gain or loss, and H_e is evaporative heat loss. Animals in cold climates maximize metabolic heat production and minimize the avenues of heat loss. In hot environments, heat accumulates from the animal's own metabolism and from external inputs. Their adaptations minimize these external gains. Insofar as possible they also maximize the transfer of heat to the environment, although the physics of heat transfer make this difficult when the ambient temperature is high.

In Chapter 2 we defined formal constraints on adaptive evolution: physical laws that constrain the potential adaptive response to the environment. The heat content equation represents an example of this important phenomenon. The equation states that there are five physical processes for heat gain or loss. The morphological, physiological, or behavioral adaptations to thermal stress are limited to these five mechanisms. As we examine the thermal adaptations of animals and plants, you will see that although there are important differences between their adaptations, there are fundamental similarities among them because all are based on this same equation. This is especially true of mobile poikilotherms and homeotherms. Although poikilotherms do not maintain a constant body temperature, they too must regulate their internal temperature within some limits. They employ the same kinds of morphological and physiological adaptations that operate in homeotherms.

Thermal Adaptations of Animals

At low and high temperatures homeotherms must expend energy to regulate their internal temperature. The body temperatures at which the metabolic rate increases to maintain a constant internal temperature are known as lower and upper **critical temperatures**. The **thermal neutral zone** is the range of temperatures between these upper and lower bounds where no additional metabolic activity is required to maintain a constant internal temperature. Let us consider thermal adaptations of homeotherms in relation to the mechanisms of heat gain and loss.

In most homeotherms the basal metabolic rate increases when the ambient temperature is low and decreases when it is high (Figure 3.15). Some animals, such as bats and some rodents, have a metabolic adaptation called **nonshivering thermogenesis**. They use the metabolic potential of brown fat to increase heat production (Hill and Wyse, 1989). Brown fat is high in cytochrome c, which facilitates its metabolism and thus the heat produced from it. Shivering and physical movement also raise the internal temperature.

Morphological adaptations such as hair, feathers, and fat affect the rate of heat gain and loss because they reduce the rates of conduction and convection. Feathers are evolutionarily derived from reptilian scales. They are believed to have first evolved as insulation in dinosaurs and later were modified to aid in flight. The insulation provided by feathers allows small birds such as kinglets to inhabit cold climates without the energetic cost of migration. A kinglet, one of the smallest birds of the northern forests, can maintain an internal body temperature 60°C higher than its environment. Because aquatic animals are particularly vulnerable to conductive heat loss, homeotherms such as whales and seals that inhabit cold waters have thick layers of blubber or subcutaneous fat that insulate them against conductive heat loss. Penguins make use of the insulative quality of feathers. At densities of as many as 70 feathers per square inch, these birds can swim in subfreezing water and withstand air temperatures as low as −40°C. Both hair and feathers also reduce air flow near the skin, thereby reducing convection.

Radiative heat transfer is affected by morphology and behavior. Size and shape are particularly important determinants of the thermal stress an organism experiences. Small organisms have a high ratio of surface area relative to their volume. In addition, the relative shapes of limbs, ears, and other appendages play a role. Any aspect of shape that exposes a larger proportion of the body to the external environment increases the potential transfer of heat by radiation (and by convection and conductance as well). Basking behavior increases radiative heat gain by exposing a large surface area to the sun or warm rocks. Color also affects radiative heat transfer. Dark skin, feathers, or hair maximize heat gain; lighter colors minimize it. Consequently, many desert organisms are lighter in color than related species or subspecies in cooler regions.

Countercurrent heat exchangers, common in many animals, are specialized morphological adaptations of the blood vessels that raise or lower the internal temperature relative to the environment. The flippers of dolphins have a large surface area from which heat is lost to the environment. The arrangement of blood vessels mitigates these losses by passing warm blood from the interior along vessels containing cooler blood from the periphery (Figure 3.16). The flow of blood in the two vessels is in

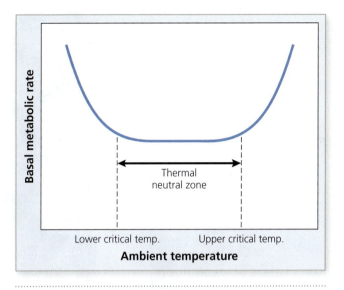

Figure 3.15 The basal metabolic rate as a function of ambient temperature. The physiological mechanisms to regulate temperature require energy and thus affect the BMR. The temperature range in which the animal's metabolic rate does not increase is the thermal neutral zone. **Analyze: What behaviors might help keep an animal in its thermal neutral zone?**

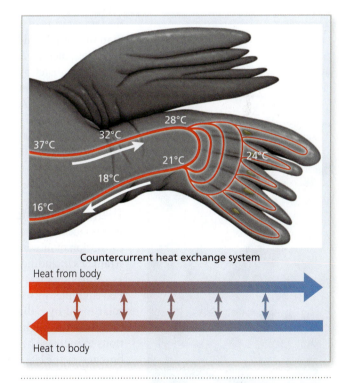

Figure 3.16 Countercurrent heat exchange systems. In countercurrent heat exchange systems, warm blood (red) flows past cool blood (blue) flowing in the opposite direction. This arrangement ensures that at every place along the parallel vessels, the temperature differential of the two vessels is maximized. This helps the body retain heat. **Analyze: How would this change if the vessels ran in the same direction (a concurrent exchange)?**

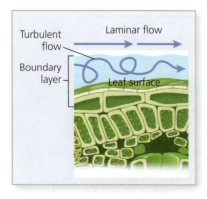

Figure 3.18 Heat loss in leaves. Turbulent air flow in the boundary layer near the surface of a leaf minimizes convective heat loss because less heat is carried away. **Analyze: Why would laminar flow carry away more heat?**

■ **thermogenic** Describes plants that can raise their internal temperature above ambient.

■ **gular fluttering** Rapid throat movement in birds to lose heat by evaporative cooling.

Figure 3.19 Turbulent flow. The arctic lousewort is covered with small hairs that minimize heat loss by creating turbulent flow at the surface of the plant.

Figure 3.17 Dry-climate plants. Plants that inhabit hot, dry climates have adaptations such as vertical leaves, shiny leaves, and hairs on the stem and leaves. In others, the leaves are greatly reduced and most of the photosynthesis occurs in the stem. All these adaptations reduce water loss.

opposite directions. This means that the temperature of blood from the interior is always warmer than the blood returning from the periphery. Note that even though the temperature of the interior blood gradually drops, at each point it is still warmer than the peripheral blood. This means that the heat transfer continues even as the peripheral blood warms.

Heat transfer by evaporative water loss is exploited by many animals that inhabit hot environments because this is the only mechanism that can transfer heat from the organism to a warmer environment. For mammals, sweating and panting are the most common mechanisms for evaporative cooling. Birds increase the cooling effect of panting by extremely rapid contractions of the throat known as **gular fluttering**. There is a limit to the use of evaporative cooling in desert environments where water is scarce because panting or sweating may shift the limiting factor from heat to water.

Thermal Adaptations of Plants

Plants face the same thermal challenges as poikilotherms. Adult plants do not have the option of moving to more favorable temperatures. They make use of avoidance strategies including seed design, dormancy, and dispersal mechanisms.

Morphological adaptations are especially important in plants (Figure 3.17). For example, in some desert plants, vertically oriented leaves minimize direct solar input by radiation. In other environments, some plants have shiny leaves that reflect sunlight, thereby reducing leaf temperature. Plants also have adaptations to increase heat loss by convection. Convective heat loss is a function of air flow across the leaf surface. Laminar flow—air that moves in a straight line across the leaf surface—carries heat away rapidly. If the moving air encounters surface irregularities, eddies create turbulence near the leaf surface in a region known as the boundary layer (Figure 3.18). The more turbulent the flow, the less heat is lost by convection. In hot environments where convective heat loss is advantageous, leaves tend to be small and smooth; in cold environments leaves tend to be larger, with more surface irregularities. In Arctic and alpine regions, the leaves and stems of some plants are covered with hairlike filaments to decrease convective heat loss (Figure 3.19).

The vast majority of plants are ectothermic. However, a few **thermogenic** species are able to raise their temperature above the ambient air temperature. For example, skunk cabbage (*Symplocarpus fenifolius*) can raise its temperature up to 5°C above ambient. Flowers of this early-blooming plant are often seen protruding through the snow (Figure 3.20). Thermogenic plants produce heat by increased mitochondrial energy production. Both the density of mitochondria in cells and the respiratory enzyme activity are higher in thermogenic plants (Ito-Imaba et al., 2009).

How Do Organisms Adapt to Water Stress?

Water is, of course, crucial for life. Many biochemical pathways require the input of water to function. In addition, the water content of cells or tissues determines the concentration of solutes. Cells and tissues function within narrow ranges of internal solute concentration. Consequently, the organism must manage the flow of water in or out of the tissues.

The movement of water is governed by the **water potential**, an energy gradient between two systems caused by their relative water and solute concentrations. Water flows from high energy (high water potential) to low energy (low water potential). Three processes determine the total water potential. First, water moves from regions of low to high solute concentration by osmosis. A system with a high solute concentration has a low water potential—water will flow toward that lower potential. Second, **pressure potential** is the water energy associated with pressure exerted when water is forced from one place to another by muscles or other mechanisms. Because their cell walls are rigid, plant cells develop pressure potential known as **turgor pressure** when water enters the cell. Finally, attraction and adhesion of water molecules on surfaces causes **matric pressure**. For biological systems the total water potential is determined by the equation

$$\varphi_{total} = \varphi_{osmosis} + \varphi_{pressure} + \varphi_{matric}.$$

The water balance of any organism is determined by the value of the water potential between the organism and its environment. If the water potential of the organism is higher than its surroundings, it will tend to lose water by osmosis. This is the challenge faced by species that live in saltwater (Figure 3.21). In freshwater environments, where the external water potential is high, water will tend to enter the organism. This may require physiological mechanisms to prevent the tissue water content from rising too high. Just as the adaptations to counteract heat stress are based on the physical processes of heat transfer, adaptations to water stress center around the components of water potential.

Species differ in the degree of water homeostasis they maintain. For example, some species, such as starfish and oysters, are **osmoconformers**—they allow their internal water balance to vary with the external conditions. Other species maintain a more narrow range of solute concentration. **Osmoregulators**, such as crabs and some annelid worms, employ mechanisms to counteract the osmotic gain or loss of water. The costs and benefits of these two strategies are similar to those of poikilothermy and homeothermy: osmoconformers expend less energy maintaining a narrow range of solute concentration, but they must tolerate a greater range of internal conditions. Osmoregulators must expend energy moving water against the osmotic gradient.

The primary environmental dichotomy driving water potential in aquatic species is the osmotic difference between freshwater and saltwater. Most marine

Figure 3.20 Thermogenic plants. Some plant species, such as this skunk cabbage, are thermogenic. They produce heat that can even melt snow as the plant emerges in spring.

water potential An energy gradient between two systems caused by their relative water and solute concentration. Water flows from high energy (high water potential) to low energy.

pressure potential The water energy due to pressure exerted as water is forced from one place to another.

turgor pressure Pressure potential caused by the influx of water to a cell with a rigid cell wall.

matric pressure The attraction and adhesion of water on surfaces.

osmoconformers Organisms that allow their internal water balance and solute concentration to vary with the external conditions.

osmoregulators Organisms that maintain their internal water balance and solute concentration within narrow limits.

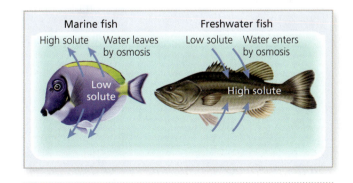

Figure 3.21 Osmosis in fish. Freshwater fish tend to gain water by osmosis because their tissues are hypertonic relative to their environment. Marine fish tend to lose water by osmosis because their tissues are hypotonic relative to saltwater. **Analyze:** What would be the effect of placing a freshwater fish in saltwater?

fish are osmoregulators. They generally employ one of two strategies to counteract the loss of water to the environment. Some, such as the hagfish (*Myxine spp.*), actively concentrate salts to reduce their hypotonicity relative to saltwater. Most teleost fish drink copious amounts of seawater and expel the salt through the gills. The water they take in counteracts the inevitable loss of water to the environment. The energetic cost of this strategy is high because they must actively transport salt from the gills to a region of high salt concentration, the seawater.

Freshwater organisms face the opposite problem: water enters their tissues by osmosis. Fish and amphibians have similar adaptations to this environment. They produce copious amounts of very dilute urine, thus reducing their internal water content. Because the urine carries some important ions with it, they must actively transport ions from the water into their tissues.

The fundamental physiological challenge of terrestrial organisms is water loss. Desiccation was a crucial challenge for the first plants and animals to colonize the land some 400 million years ago. Evaporation is the most important avenue by which water is lost to the air. An array of adaptations minimize the effect of these losses. Some minimize evaporation; others maximize the efficiency with which available water is used.

For terrestrial animals, barriers to surface water loss, such as the keratinized scales of reptiles, are important adaptations. Behaviors such as nocturnality and selection of cooler, moister microhabitats, such as burrows, further prevent water loss. The means by which nitrogenous wastes are excreted are intimately connected to water balance. Nitrogen can be excreted in a variety of chemical forms (Table 3.1) that differ in toxicity and solubility.

In general, the more toxic forms such as ammonia are more soluble and are typically excreted in highly dilute form. Species such as fish and amphibians that excrete ammonia must invest more water in the elimination of nitrogen than birds that excrete nitrogen as uric acid. However, there are energetic costs as well. Urea and uric acid are produced from ammonia at considerable energetic cost.

metabolic water Water produced by oxidative metabolism.

Oxidative metabolism produces **metabolic water** that can be used by the organism. For each molecule of sugar that is oxidized, six water molecules are produced:

$$C_6H_{12}O_6 + 6O_2 \rightarrow 6CO_2 + 6H_2O.$$

The metabolism of glycogen and fats also produces metabolic water; the exact amount depends on the hydrogen content of the molecule. Some specialized desert animals, such as kangaroo rats, can survive solely on metabolic water along with avoidance mechanisms such as nocturnal behavior and the use of relatively humid burrows.

TABLE 3.1 Chemical forms by which nitrogenous waste is excreted

NITROGENOUS WASTE	ORGANISMS	SOLUBILITY	TOXICITY
Ammonia	Crocodiles Amphibians Fish Some invertebrates	High	High
Uric acid	Birds Snakes and lizards Gastropods Insects Turtles Crocodiles	Low	Low
Urea	Mammals Some teleost fishes	Moderate	Low

THE EVOLUTION CONNECTION

Plant Adaptations to Life on Land

The colonization of land by plants and animals constitutes a seminal event in the history of life on Earth. We know that life first arose in an aquatic environment and persisted there for hundreds of millions of years. Plants first colonized the land during the Ordovician Period, some 434 million years ago. Aquatic algae are the putative ancestors of modern land plants.

Modern plants occupy a wide array of harsh habitat conditions. But for aquatic organisms colonizing land, the terrestrial environment constituted an environment that was truly extreme. Obviously desiccation was a significant problem in this new environment. But other fundamental aspects of the organism were affected. Water supported the mass of the plant; on land no such external support was available. Temperature fluctuations also tend to be greater in air than water.

Given these extreme challenges, what advantages could be exploited on land? Water absorbs light energy much more effectively than air. Consequently, aquatic plants were limited to shallow waters where sufficient light penetrated to power photosynthesis. On land, solar energy is far more abundant—both in intensity and the wavelengths available. The partial pressures of oxygen and carbon dioxide are also much higher on land than in the water. Plant photosynthesis was CO_2-limited in water. On land, light and CO_2 increased the energy that the plant could channel into other aspects of the life history, including growth, resistance to grazers, reproduction, and competition with other plants. Population pressure, competition, and grazing by animals in the crowded aquatic environment have also been suggested as important selective factors that favored the transition to land (McFarland et al., 1985).

Plant adaptations to land fall into four main categories. First, there are adaptations such as waxy cuticles that prevent tissue desiccation. Stomata solve the problem of gas exchange across leaves covered with impervious wax. Second, mechanisms of water transport within the plant enabled plants to transport water from the roots where it is absorbed to the aboveground parts of the plant. Third, aquatic reproduction was completely dependent on the movement of motile gametes in water. The first land plants were restricted to moist environments where the gametes could still move through water. Eventually, mechanisms of gamete fusion independent of water arose. These are highly developed in the angiosperms, in which the male gametes are found in pollen and the egg is protected in an ovary. Mechanisms of pollen movement that employed wind and animals made angiosperm reproduction completely independent of water. Finally, mechanisms of support arose very early in terrestrial plants. Lignin, a polymer that provides rigidity to cells, can be found in 400-million-year-old fossils.

The first plants that adapted to this new extreme environment found themselves in a world rich in resources and low in competition from other plants. Over time, these early colonists diverged and adapted to the myriad new habitats and ecological possibilities on land. The first angiosperms arose approximately 125 million years ago. They now represent the most diverse group

Figure 1 Land plants. A wide array of land plants, including the ferns (a), angiosperms (b), and gymnosperms (c), arose from the first plants to colonize land.

Figure 2 Angiosperms. Angiosperms now inhabit a wide range of ecological conditions and are found on every continent.

of land plants and are found in virtually every terrestrial habitat. The history of land plants is an example of an *adaptive radiation*: a few initial colonists in a new habitat that adapt and gradually diversify into many species. This has been an important process in the history of life on Earth. And it is the basis for the rich ecological diversity of many systems. Some radiations, such as the terrestrial plants, occurred across millions of years and over many continents. Others, like Darwin's finches on the Galápagos, are restricted to a few species in a small area over a relatively short period of time. But all exemplify the close connection between ecology and evolution.

QUESTION:

Which of the physical challenges to plant colonization of land would you expect to have affected the first animals that colonized land?

Plants lose water primarily by evaporation from the leaves and stems. Thus, the higher the ambient temperature, the higher the rate of evaporative water loss. Of course, high temperature and low soil moisture frequently occur together in desert environments and thus compound plants' water stress. The fate of plants exposed to dry conditions depends on the length and intensity of water stress. Mortality due to water stress occurs by three main mechanisms (McDowell et al., 2009). If the plant must close the stomata to limit water loss for long periods, it may experience carbon starvation. High-intensity water stress leads to tissue desiccation and eventually failure of the hydraulic mechanisms that transport water and nutrients in the plant. Finally, resistance to biotic agents such as pathogens and parasites may break down in a plant under water stress. Plant responses to drought are species specific, even among plants in the same habitat. During a severe drought in New Mexico, the mortality rate of piñon pine (*Pinus edulis*) approached 97 percent, whereas less than 1 percent of junipers (*Juniperus monosperus*) died (McDowell et al., 2009).

Plants that inhabit such environments are known as **xerophytes**. The adaptations of xerophytes include two important tactics: extracting as much moisture as possible from the soil and reducing evaporation. The former typically involves the expenditure of large amounts of energy to actively transport water into the roots. Morphological adaptations also reduce evaporative water loss. Some involve physical barriers to evaporation, such as waxy cuticles, whereas others are based on minimizing the surface area from which water evaporates. Other adaptations include small hairs that shade the leaf or stem surface and reduce air movement, which in turn reduces evaporation.

Plants obtain water primarily from the soil via the roots. Water integrates into the soil matrix in the small spaces between soil particles. Water leaves the soil by drainage, evaporation, and plant uptake. The water-holding capacity of soil, specifically its matric potential, depends on its structure and chemistry. Sandy soils have larger pores and relatively low adhesion of water to the particles. Thus, water passes through sandy soils, resulting in low moisture availability for plants. Clays have much smaller pores and surface ions that bind the polar water molecules relatively tightly. The amount of water held in the soil is its **field capacity**. The same properties that determine how much water soil holds also determine how easy it is for plant roots to absorb it. Clays hold water more tightly than sandy soils. The **permanent wilting point** occurs when the water potential of the soil falls so low that the remaining molecules, tightly bound to soil particles, cannot be extracted by the roots.

The salt content of the soil determines how much energy the plant must expend to take in water. In saline soils, the plant must use energy to actively transport water against the osmotic gradient. Saline soils occur in some deserts and in coastal habitats that experience tides. The plants that inhabit these environments face additional water stress, particularly in the desert, where evaporative water loss is also high. Not surprisingly, relatively few species inhabit these environments.

Even if the plant surface area is small and protected by barriers to evaporation, the plant remains vulnerable to evaporative water loss from the stomata. The stomata must be open for the uptake of CO_2 for photosynthesis (Figure 3.22). However, when they are open, the internal tissues are exposed, significantly increasing the potential for evaporative water loss. This poses a physiological dilemma for the plant: close the stomata to prevent water loss but limit photosynthesis. This dilemma excludes many plants from hot, dry climates.

xerophyte A plant that tolerates hot, dry environments.

field capacity The amount of water held in the soil.

permanent wilting point When the water potential of the soil is so low that water cannot be extracted by the roots.

C_4 pathway An alternate photosynthetic pathway in which carbon is fixed as a 4-carbon molecule by an enzyme with high affinity for CO_2.

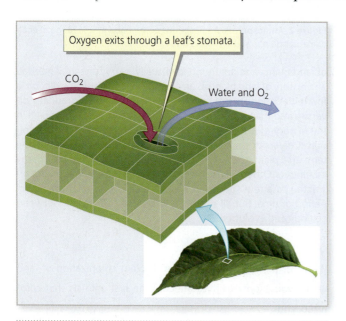

Oxygen exits through a leaf's stomata.

CO_2

Water and O_2

Figure 3.22 Leaf exchange. Leaf stomata are pores in the leaf surface that allow the plant to take in CO_2. However, when the stomata are open, they account for most of the water a plant loses.

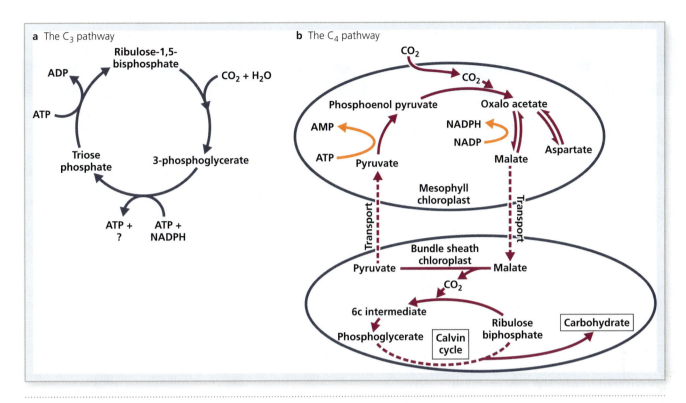

Figure 3.23 Photosynthesis. The C_3 and C_4 photosynthetic pathways differ in the molecules that react with CO_2. In the C_3 pathway CO_2 reacts with ribulose 1,5 bisphosphate to form a 3-carbon molecule, 3-phosphoglycerate. In C_4 photosynthesis CO_2 reacts with phosphoenolpyruvate to form the 4-carbon molecule oxaloacetate. The enzyme that fixes carbon in the C_4 pathway has a higher affinity for CO_2. (a) The C_3 pathway; (b) The C_4 pathway. **Analyze: What is the significance of the higher enzyme affinity for CO_2 in the C_4 pathway?**

The **C_4 pathway** is a biochemical adaptation found in some species that allows the plant to take in CO_2 while limiting evaporative losses through the stomata. It accomplishes this by opening the stomata at night when the temperature, and thus evaporative water loss, is lower. In order to understand this adaptation, we must explore more fully the physiological dilemma faced by plants that don't have the C_4 pathway.

The first photosynthetic pathway elucidated, known as the Calvin cycle, uses CO_2 to build carbohydrate molecules (using light energy). In the Calvin cycle the enzyme RuBP carboxylase catalyzes the reaction of CO_2 and ribulose bisphosphate to form phosphoglycerate, a 3-carbon molecule (Figure 3.23). Thus, the Calvin cycle is also known as the **C_3 pathway**. RuBP carboxylase is highly sensitive to the CO_2 concentration; if the concentration falls, the enzyme instead picks up O_2, in a process known as **photorespiration**. Photorespiration produces no energy and consumes much of the carbon fixed in the Calvin cycle—altogether it constitutes a huge energy drain on the plant. Thus, the plant must maintain high leaf concentrations of CO_2 to prevent photorespiration. It can do this only by keeping the stomata open, which causes the loss of significant amounts of water.

C_4 photosynthesis differs importantly in that the enzyme that fixes CO_2, phosphoenolpyruvate carboxylase (PEP carboxylase), has a much higher affinity for CO_2 than RuBP carboxylase. This means it can fix CO_2 at much lower concentrations. Moreover, fixation of CO_2 is physically separate from the Calvin cycle. Fixation occurs in the mesophyll cells; the Calvin cycle occurs in the bundle-sheath cells.

C_3 pathway A photosynthetic pathway in which carbon is fixed as a 3-carbon molecule by RuBP carboxylase.

photorespiration A pathway that occurs in C_3 plants when the CO_2 levels fall so low that RuBP carboxylase picks up O_2 instead of CO_2.

Figure 3.24 Corn roots. Corn produces adventitious roots near the soil surface. They allow the plant to take in oxygen and nutrients in saturated soil.

The Effect of Salinity on Photosynthesis

Physical factors limit the growth and distribution of all species. Although it is easier to explain individual challenges and the adaptive responses to them, rarely does an organism face a single limiting physical factor. Thus, an important question arises: How do the interactions among physical factors affect fitness?

Mangroves, a collection of trees found in low-latitude coastal environments, are ideal organisms to address this question. Mangroves occur in discrete bands from the shore inland, each species occupying a specific region of soil moisture and salinity. López-Hoffman et al. (2007) used the strong environmental gradient experienced by mangroves to examine the interaction of salinity and light. These factors constitute a logical pairing because each potentially affects the flow of CO_2 into the plant through the stomata. Saline conditions limit the plant's water uptake, which in turn leads the plant to close the stomata to reduce water loss. However, this also reduces the uptake of CO_2 for photosynthesis and hence the plant's response to variation in light intensity.

HYPOTHESIS: The negative effect of salinity on net photosynthesis is greater under high light conditions.

The researchers reasoned that at low light, photosynthesis is limited primarily by light, but when light is plentiful, leaf CO_2 becomes the most limiting factor.

PREDICTION 1: When subjected to a range of salinities and light regimes in the laboratory, mangroves should achieve the highest rate of net photosynthesis at low salinity and the lowest rate of net photosynthesis when salinity is high.

PREDICTION 2: Similar results should occur in the field when seedlings are transplanted into an array of light-salinity levels.

The researchers chose the black mangrove (*Avicennia germinans*) for their study because in their system it is the most salt-tolerant species. In the greenhouse they subjected seedlings to three salinity levels (20 percent, 70 percent, 120 percent full

seawater) and four light levels (6 percent, 12 percent, 25 percent, 50 percent of photosynthetically active radiation) for a total of 12 possible combinations. The net rate of photosynthetic assimilation was measured on sample plants after four months of treatment. Plants were harvested after 197 or 276 days and a series of morphological measurements taken that allowed the researchers to calculate the relative growth rate, net assimilation rate, and leaf area. In the field they established plots of 12 light levels under each of two salinity conditions. Seedlings were transplanted into each plot; nine months later the plants were harvested to measure growth and survival.

The predictions of their hypothesis were met in both the laboratory and the field. Net photosynthesis increased much more rapidly with light at low salinity than at high salinity. They measured the rate of stomatal conductance of CO_2, a measure of the rate of transpiration. The results confirmed that high salinity limits leaf transpiration. Similar results were obtained in the field. Both survival and growth were affected by the combination of light and salinity. The response to increased light depended on the salinity: at low salinity, both growth and survival increased rapidly as light intensity increased; at high salinity, the positive effect of light intensity was significantly reduced. Together the greenhouse and field studies supported the hypothesis that CO_2 limits photosynthesis when light is abundant but salinity reduces stomatal conductance. These results confirm the logical prediction that physical factors do not operate independently of one another. Instead, the impact of each is affected by others. Of course, light and salinity are not the only important physical factors in this system. There is also variation in nutrient availability as a function of tidal influx, storm damage, drought, and soil moisture. The structure and dynamics of mangrove ecosystems are driven by the interactions among these factors.

Figure 1   The black mangrove. The black mangrove lives in saturated soil with high salinity. It thus faces significant physiological challenges in water balance.

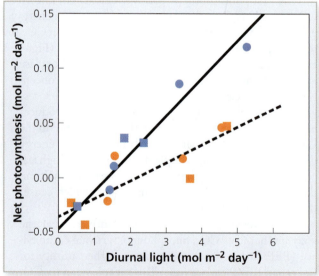

Figure 2   Net photosynthesis. Net photosynthesis increased more rapidly with light at low salinity (solid line) (López-Hoffman et al., 2007).   **Analyze: Does this result support the hypothesis?**

In C_4 photosynthesis, the stomata open at night and carbon is fixed as a 4-carbon molecule by PEP carboxylase. Even when the stomata close during the day, the enzyme can continue to fix carbon as the internal leaf CO_2 concentration falls. The 4-carbon molecules produced enter the bundle-sheath cells, where they release carbon in the form of CO_2. This generates locally high concentrations of CO_2, which allow RuBP carboxylase to fix this carbon and enter the Calvin cycle.

Too much water also poses physiological problems for plants. The most important physiological consequence of flooding occurs not as a result of excess water per se but the effect of water on the plant's oxygen status. Plants absorb oxygen gas primarily through the roots from the tiny air pockets in the interstices among soil particles. As water fills the pores, less gaseous oxygen is available and death from anoxia may ensue (Kazlowski, 1984). Plant adaptations to flooding include two basic tactics. First, biochemical adaptations in the root allow them to survive periods of anoxia by decreasing metabolic activity and increasing the use of stored energy. Second, waterlogged soil may stimulate the development of **adventitious roots** near the soil surface, where there is more available oxygen (Figure 3.24). Species such as mangroves, which inhabit permanently waterlogged soil, produce aerial roots that absorb oxygen directly from the air.

adventitious roots Roots produced at or above the soil surface.

....

KEY CONCEPTS 3.3

- Temperature and water are fundamentally important components of the physical environment.
- Adaptive responses to temperature make use of and are constrained by the physics of heat transfer (conductance, radiation, convection, and evaporation).
- Water stress affects other key physiological processes such as temperature regulation, osmotic homeostasis, and oxygen absorption.

QUESTION:

How is the interaction of water and temperature stress similar and different for plants and animals?

....

Putting It All Together

The many physical environments on Earth pose significant adaptive challenges to organisms. Nevertheless, we find species in virtually every environment on the planet, from boiling hot springs to the brutal temperatures of Antarctica and from the driest deserts to the wettest rainforests. In one sense the adaptive responses to the physical environment fall into a small number of basic strategies: resistant or inactive life cycle stages, movement to more favorable conditions, or mechanisms that increase tolerance. In another sense, there are millions of unique adaptations—not only does each species employ one or more of the basic strategies but their specific morphological, physiological, biochemical, and behavioral responses are unique.

We can begin to understand this complexity when we analyze the nature of the variation in the physical environment. Consider a series of questions regarding the organism and the challenges imposed by the physical environment:

Which physical factor is most limiting?
Is that factor constant or variable?
Is the variation in the physical environment predictable or unpredictable?
Does the limiting factor interact with others?
Is the organism mobile or sessile?

The answers to these questions shape and constrain the adaptive responses to an environmental challenge. The thermophilic bacterium *Pyrobacterium brockii* and the Antarctic fish *Macropterus maculates* are adapted to radically different environments. But their adaptations are more similar to each other than to the spadefoot toad. *Pyrobacterium* and *Macropterus* each experience an extreme environment with relatively little variation. Each has adaptations to protect its cellular machinery from temperature damage. In contrast, the spadefoot toad must adjust its physiology and life cycle to the unpredictable appearance of water and the impossibility of crossing desert in search of moisture. The toad employs two fundamental physiological strategies: it tolerates extreme desiccation and urea concentrations and it persists in inactive form, ready to exploit unpredictable and ephemeral rains.

▪ Summary

3.1 How Do Environmental Factors Limit Growth and Survival?
- Physical resources are inorganic nutrients or energy resources; physical factors are physical conditions that affect growth and survival.
- According to the law of tolerance, there are upper and lower bounds to the physical factors an organism can tolerate.
- Homeostasis is the set of mechanisms that maintain the organism's internal physiology within set limits.
- Abiotic factors vary across time and space. Both the mean value and the pattern of variation are important to the adaptive process.
- The adaptive responses to the abiotic environment fall into two broad categories:
  - Mechanisms by which the organism avoids harsh conditions
  - Adaptations that alter the organism's physiology to counteract the physical conditions

3.2 What Adaptations Avoid Harsh Conditions?
- The principle of allocation states that adaptations to one challenge may preclude or reduce adaptations to others. Each adaptation has a cost; each is a trade-off.
- Organisms avoid harsh conditions by reducing their activity during difficult times. Resistant structures such as spores and seeds allow the organism to persist until conditions are favorable.
- Many physical challenges require considerable energy expenditure to counteract. Periods of inactivity such as hibernation or torpor avoid these energy costs.
- Hibernation represents a spectrum of levels of inactivity.
- Migration—large-scale seasonal movements—avoids difficult physical conditions, energy shortage, or even insect pests.
- Migration entails significant mortality costs due to its energetic demands and exposure to external sources of mortality such as predation.

- Small-scale movements, including behavioral thermoregulation, allow the animal to exploit local favorable conditions.

3.3 How Do Physiological Adaptations Alter the Organism's Tolerance Limits?
- If the physical challenge cannot be avoided by inactivity or migration, the only other alternatives are increased mortality and physiological adaptation to the conditions.
- Each habitat on Earth poses unique physical challenges. A large array of abiotic factors limits organisms and their distributions.
- Water and temperature illustrate the concepts central to adaptations to these many challenges.
- Some species, such as poikilotherms and osmoconformers, allow their internal conditions to vary widely.
- The rate of biochemical reactions is affected by temperature. Important biomolecules such as DNA and protein denature at high temperatures. Biochemical adaptations protect the function and integrity of reactions and cellular structures.
- Heat is gained or lost according to the equation

$$H_{total} = H_m \pm H_c \pm H_{cv} \pm H_r - H_e.$$

- These mechanisms of heat transfer shape and constrain the physiological and morphological adaptations to thermal stress. Each of the factors on the right side of the equation plays a role in these adaptations.
- Water flows from high to low water potential. The components of water potential are shown in this equation:

$$\varphi_{total} = \varphi_{osmosis} + \varphi_{pressure} + \varphi_{matric}.$$

- As is the case with temperature, some organisms tolerate a wider range of osmotic and water conditions (osmoconformers) than others (osmoregulators).
- Freshwater and saltwater represent a fundamental dichotomy in the physiological challenges organisms face with respect to water and osmosis.

Key Terms

absorptivity p. 50
acclimation p. 48
adventitious roots p. 59
aestivation p. 44
allocation, principle of p. 40
behavioral thermoregulation p. 46
C_3 pathway p. 57
C_4 pathway p. 56
calorie p. 50
cold-acclimation proteins (CAPs) p. 49
conduction p. 50
convection p. 50
countercurrent heat exchange p. 50
critical temperature p. 50
ectotherm p. 38
emissivity p. 50

endotherm p. 38
facultative hibernator p. 45
field capacity p. 56
gular fluttering p. 52
heat p. 50
heat of vaporization p. 50
heat shock proteins (HSPs) p. 49
hibernation p. 44
homeotherm p. 37
matric pressure p. 53
metabolic water p. 54
migration p. 45
nonshivering thermogenesis p. 50
obligate hibernator p. 44
osmoconformers p. 53
osmoregulators p. 53

permanent wilting point p. 56
photorespiration p. 57
physical (abiotic) factors p. 38
physical resources p. 38
poikilotherm p. 38
pressure potential p. 53
radiation p. 50
specific heat capacity p. 50
thermal neutral zone p. 50
thermogenic p. 52
thermophile p. 38
torpor p. 44
tolerance, law of p. 38
turgor pressure p. 53
water potential p. 53
xerophyte p. 56

Review Questions

1. What is the relationship between physical factors and physical resources?

2. What are the adaptive advantages of avoidance strategies? What are the costs?

3. Explain, using specific examples, the interactions among physical factors.

4. What are the differences and similarities among hibernation, torpor, and aestivation?

5. What role does temporal variation and predictability play in the adaptive response to the physical environment?

6. Explain, using specific examples, a biochemical, a morphological, and a physiological adaptation to temperature stress.

7. What is the dilemma of water and temperature stress faced by plants? How does C_4 photosynthesis address that problem?

8. How do animals make use of the physical processes of heat transfer to adapt to heat stress?

9. What is the relationship between water and nitrogen excretion?

10. Why is temperature so important to biochemical processes?

Further Reading

Amils, R., et al. (eds.). 2007. *Life in extreme environments.* New York: Springer.

This book reviews the strategies to cope with extreme environments in bacteria, plants, animals, and humans. It also considers the topic from the point of view of astrobiology—that is, the possibility of life on other planets.

Karasov, W.H., and C. Marinez del Rio. 2007. *Physiological ecology: how animals process energy, nutrients, and toxins.* Princeton, NJ: Princeton University Press.

This book focuses on energy ecology—the acquisition, digestion, and energetics of food. Diet is a fundamental component of the ecology of the organism, and each type of diet has significant physiological implications.

Larcher, W. 2001. *Physiological plant ecology: ecophysiology and stress physiology of functional groups.* New York: Springer.

This comprehensive book explores the details of plant physiological ecology. It covers a broad range of physical resources and factors, including light and minerals. It also discusses the pattern of carbon allocation to roots, shoots, and reproductive structures, a crucial component of the plant's physiology.

Smith, S.D., et al. 1997. *Physiological ecology of North American desert plants.* New York: Springer.

This classic book begins with an exploration of the stresses placed on desert plants by temperature, light, and moisture. It then considers the adaptive responses of different plant forms (shrubs, cacti, grasses, etc.). In the process the book compares the diverse adaptive solutions by different plant groups to similar challenges.

Chapter 4

Terrestrial Communities

In 1802 the German explorer and naturalist Alexander von Humboldt ascended the volcano Chimborazo in Ecuador, then thought to be the highest mountain in the world. As he climbed, Humboldt made detailed measurements of temperature, moisture, and air pressure. Along the way he made a simple but profound observation: the **vegetation**, the form of the plant life, changes dramatically with elevation. Almost one hundred years later, the biologist C. H. Merriam made a dramatic journey—from the bottom of the Grand Canyon through the Painted Desert to the summit of San Francisco Peak near Flagstaff, Arizona (Figure 4.1). Like Humboldt, Merriam recognized that the vegetation occurs in distinct zones associated with specific elevations. He referred to these bands of vegetation as **life zones**. He reasoned that each is determined by the combination of temperature and moisture, both of which change with elevation. Moreover, he noticed that the bands of vegetation on the slopes of the mountain are similar to the bands of vegetation that occur from the southwestern United States northward into Canada and Alaska. The alpine meadows of San Francisco Peak resemble the tundra of the high Arctic. His names for the mountain life zones reflect their latitudinal counterparts.

Humboldt and Merriam identified two universal principles of plant distribution: plants occur in *patterns*, groups of species that live together, and these patterns are associated with a specific climate. Together these principles lead us to the fundamental question of this chapter: *How are species organized into communities?*

4.1

What Determines Species Distributions?

In Chapter 3 we explored the concept of tolerance limits—the range of physical conditions within which the species can survive. When conditions exceed the species' tolerance limits, the organism must adapt

■ **vegetation** The form of the plant life in a region.

■ **life zone** A band of vegetation associated with a specific altitude on a mountain.

■ **geographic range** The region in which a particular species is found.

■ **adiabatic cooling** The decrease in air temperature that occurs at higher elevation or altitude.

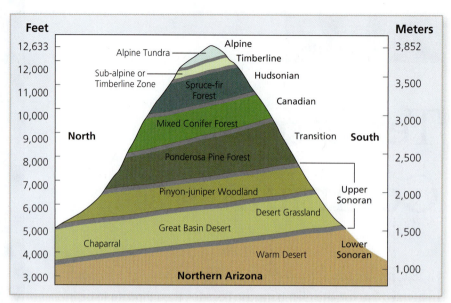

Figure 4.1 Vegetation zones. C. H. Merriam identified a series of vegetation zones, each associated with a specific elevation. He recognized that these "life zones" are determined by the physical conditions characteristic of each elevation. **Analyze: Why do the zones slant toward lower elevation on the north side of the peak?**

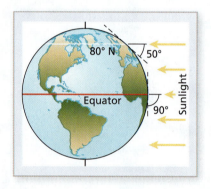

Figure 4.2 Angles of sunlight. Sunlight strikes the Earth at different angles at different latitudes. **Analyze: Explain why the angle of impact is important.**

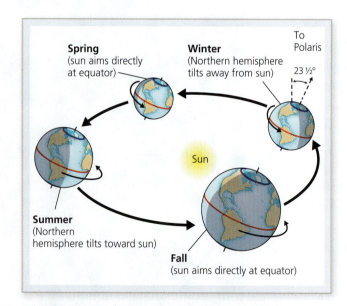

Figure 4.3 The Earth's axis. The axis of the Earth changes relative to the sun over the course of the year. In the northern hemisphere summer the northern hemisphere tilts toward the sun; in the winter it tilts away. **Analyze: How do these changing angles affect temperature and day length?**

to the new conditions, move to a more favorable site, or enter a resistant state. If these strategies fail, the result is local extinction. Superimposed on these physical limits are the effects of other species—interactions that either facilitate or inhibit the species' presence. Together the physical and biological environments determine the species' **geographic range**, the area where it is normally found. If we are to understand the physical limits to a species' geographic range, we must first understand the determinants of the abiotic conditions.

What Determines the Abiotic Conditions in Terrestrial Systems?

Climate is a key component of the physical environment. One of the most important determinants of climate is the change in temperature with latitude and altitude. The movement of cold, dense air at high latitude and warm, less dense air at the equator set global air masses in motion. The equatorial regions have a higher mean temperature than those farther north or south due to the angle at which sunlight strikes the Earth. The angle is 90° at the equator and becomes shallower toward the poles (Figure 4.2). There is also greater seasonal temperature variation at high latitude due to the tilt of the Earth's axis and its orbit around the sun (Figure 4.3). Day length varies little at the equator but greatly at higher latitudes. Beyond the Arctic and Antarctic circles, the sun does not rise for much of the winter and does not set during the corresponding portion of the summer.

The altitudinal temperature gradient is due to the phenomenon of **adiabatic cooling**. Adiabatic cooling is the process by which temperature decreases as air pressure decreases. It occurs because as air pressure decreases, the gases expand. When this occurs, the

The Concept of Statistical Significance

In the last chapter we investigated some of the statistics used to characterize a data set—the measures of central tendency and measures of variation. We have already compared groups using these measures. For example, in Chapter 2 we discussed the difference in mean bill size among Galápagos finches. Consider two sets of summer temperature measurements taken at different latitudes. Note that the mean temperatures and their distributions are not identical. They might differ because of the difference in latitude. Or they might differ simply because of sampling effects—perhaps one set of samples happened to contain a number of low values and the other one had an excess of high values. How do we statistically evaluate the significance of these observed differences?

We begin by explicitly phrasing the question in the form of a hypothesis. One possibility is that the two means do not really differ; that is, there is no actual effect of latitude on temperature. The means simply *appear* to differ because of chance variation in the samples that were taken. The hypothesis that these two means differ only as a result of chance sampling error is known as the *null hypothesis (H_o)*. The alternative hypothesis (H_a) is that latitude has an effect on the mean temperature. How do we distinguish between these two hypotheses?

To address this question, we calculate a *test statistic*, a variable that measures the size of the difference between the two groups. *The Student's t test* is a useful test statistic for the comparison of means. We calculate the value of *t*

$$t = \frac{\bar{x}_1 - \bar{x}_2}{S_x \sqrt{\frac{1}{n_1} + \frac{1}{n_2}}}$$

where $\bar{x}_1$ and $\bar{x}_2$ are the two means, n_1 and n_2 are the sample sizes of the two populations, and S_x is the pooled variance of the two populations S_1 and S_2. The latter is calculated

$$S_x = \sqrt{\frac{(n_1 - 1)S_{x_1}^2 + (n_2 - 1)S_{x_2}^2}{n_1 + n_2 - 2}}$$

Now consider how the value of *t* changes under different outcomes of an experiment. The value of *t* is large when the difference between the means is large, when the variation in the two data sets (S_x) is small, and when the sample sizes (n_1 and n_2) are large. In other words, very different means derived from large samples of populations with little variance results in a large value of *t*. How do we evaluate the size of *t*?

Imagine that the null hypothesis is true: the means are actually the same but the two sample means differ only because of chance effects. Now imagine under that assumption that you repeat the sampling and the calculation of the means and the *t*-statistic many times. If we plot the frequency distribution of these values of *t*, we obtain a bell curve with a mean value of 0. By

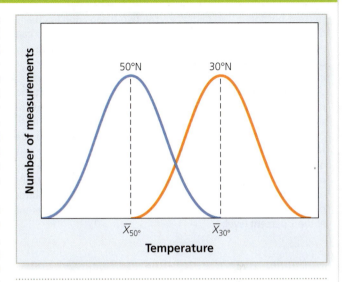

Figure 1 Temperature measurements. The frequency distribution of temperature measurements at two different latitudes. The mean temperatures differ.   **Analyze: What is the significance of the overlap of these bell curves?**

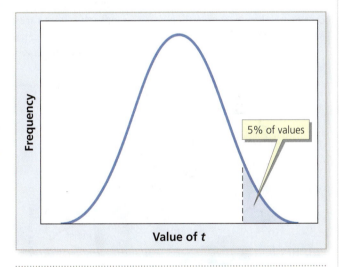

Figure 2 Values of t. The frequency distribution of values of *t* under the null hypothesis. Five percent of all values lie in the tail to the right of the dotted line.   **Analyze: Explain the significance of large values of t.**

chance a few samples will generate large values of *t*. This distribution allows us to evaluate the meaning of the single value of *t* we calculated. Note that when H_o is true, only 5 percent of the *t* values fall in the tail of the curve shown in the figure. This provides a mechanism to evaluate a single value of *t* from one experiment. Imagine that the value of *t* in our samples falls in that 5

continued

DO THE MATH *continued*

The Concept of Statistical Significance

percent portion of the tail. When the two means don't actually differ, this should occur only 5 percent of the time. Thus, we can accept the alternative hypothesis that the two means do actually differ—and there is only a 5 percent chance this conclusion is incorrect.

When the value of *t* lies in that 5 percent tail, we reject the null hypothesis and accept the alternative hypothesis. By convention, biologists adopt the 5 percent value as the basis for *statistical significance*. We say in this case that *the two means are significantly different at the 5 percent level*. In rejecting H$_o$ there is only a 5 percent chance that we have committed a *type I error:* rejecting the null hypothesis when it is true. If the value of *t* falls even farther out on the tail, say where only 1 percent of *t* values occur under H$_o$, we say that the means are significantly different at the 1 percent level.

The Student's *t* test can be used if three assumptions are true: (1) the two populations are independent of one another—that is, measurements from one group do not also appear in the other; (2) the variances of the two populations are equal; and (3) the two populations have a normal (bell-shaped) distribution.

Throughout the text we will encounter experiments and comparisons where we note that differences are statistically significant. The Student's *t* test is just one of many statistics employed by ecologists. Other types of data, such as frequencies, regression lines, and correlations, require a specific type of statistic. Some are used when the assumptions of another test do not hold. Each type of test is evaluated in conceptually the same way—the comparison of the test statistic to the results expected under the null hypothesis. Statistical significance of each is based on the chance that the actual result could occur if the null hypothesis were true.

---

maritime climate The climate near a large body of water, usually characterized by a narrow range of temperature.

continental climate The climate in regions far from large bodies of water, usually characterized by large temperature fluctuations.

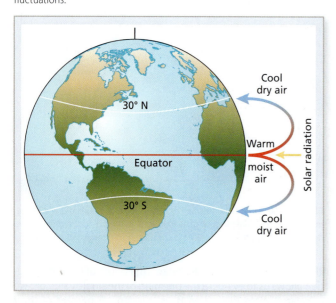

Figure 4.4 Mid-latitude desertification. Mid-latitude desertification occurs when sunlight at the equator causes warm, moist air to rise. As it cools, precipitation occurs near the equator. The cool, dry air moves away from the equator and returns to the surface, where it warms. The result is dry climate at 30° north and 30° south. **Analyze:** Why is precipitation so much lower in the air that returns to the Earth at 30° N and 30° S?

molecules, their kinetic energy, and the heat of the system occupy a larger volume and thus the temperature decreases. In mountainous regions the temperature drops approximately 3.5°C (6.5°F) per 1,000 m increase in elevation.

Air temperature is determined in part by the temperature of the water or land over which it passes. The temperature of air that passes over a large body of water onto the shore varies less than air that moves across land. This occurs because water has a high heat capacity—it requires a large amount of heat to increase its temperature, and conversely it cools slowly compared to land. Consequently, the temperature of large bodies of water, and the air passing over them, changes more slowly and varies less than air over large tracts of land. We distinguish between a **maritime climate**, the climate near a large body of water, in which the seasonal and daily temperature change is relatively small, and a **continental climate**, in which the temperature variation is much larger.

Dry climates occur as a consequence of both global and local geographic effects. The general pattern of global air circulation leads to a phenomenon known as **mid-latitude desertification** (Figure 4.4). When solar radiation strikes the tropical regions, it heats the surface air, causing it to rise. Because cool air can hold less moisture than warm air, precipitation occurs. This is one of the causes of high rainfall in the tropics. This cool, dry air moves north and south away from the equator. Cool air is also heavier than warm air, so these air masses sink back to the Earth's surface at approximately 30° north and south latitude. Precipitation is now greatly reduced because not only has this air lost its moisture over the tropics, the warm air holds its moisture. Many of the world's driest regions are found at 30° north and south latitude.

Mountain ranges also affect the pattern of precipitation. In North America prevailing west winds pick up moisture over the Pacific Ocean. When that moist air reaches the West Coast, it encounters the Cascade or

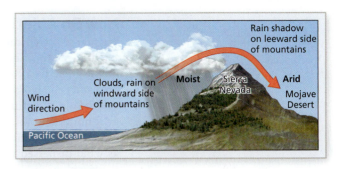

Figure 4.6 Rain shadows. Rain shadows occur on the lee side of major mountain ranges. As moist air rises on the windward side of the range, it cools and its moisture precipitates as rain or snow. As the dry air passes over the mountains, it descends and warms. **Analyze:** How is this phenomenon similar to mid-latitude desertification?

Figure 4.5 Mt Baker. Mt. Baker receives heavy snowfall each winter. Moisture-laden air from the Pacific Ocean hits the mountain, rises, and cools. This causes the moisture to precipitate, mostly in the form of snow.

Sierra Nevada mountain ranges. As the air rises, it cools and precipitation ensues. Mt. Baker is a Cascade volcano that stands immediately east of the waters of the Pacific and Puget Sound (Figure 4.5). It is the first high peak to intercept the moist air from the Pacific. The west slope of the mountain receives 180–360 cm of precipitation each year. In 1999 it set the all-time record for snowfall in a single season—2,895 cm!

As the air passes over the mountains, it descends again and warms. This air mass, like that at 30° north or south latitude, is both dry and warm. The result is that the lee side of the mountains receives far less precipitation than the windward side. This phenomenon is known as a **rain shadow**. Many deserts are located in a rain shadow (Figure 4.6).

Mountains affect the local climate in other important ways. The angle and duration of sunlight varies with the direction, or **aspect**, of the slope. In the Northern Hemisphere, the daily arc of the sun is across the southern sky. The result is that north-facing slopes receive much less direct sun than south-facing slopes. East-facing slopes receive early-morning sun, which has less impact in the cooler morning hours than direct sun on west-facing slopes in the afternoon. Thus, in the Northern Hemisphere, north- and east-facing slopes tend to be cooler and moister; south- and west-facing slopes tend to be hotter and drier. Of course, this phenomenon is reversed in the Southern Hemisphere.

The other key component of the abiotic conditions in terrestrial ecosystems is the nature of the substrate, especially the soil. The fundamental physical properties of soil are the result of the **parent material**, the rock and mineral substrate from which soil develops. Physical forces such as freezing and thawing and erosion expose and break down the parent material, which determines the two key components of soil—its nutrient content and its texture. As discussed in Chapter 3, soil texture determines the water-holding capacity of soil. Weathered or deposited material determines the chemical nature of soil. For example, in the southwestern United States, many low-lying basins contained lakes during the Pleistocene. When the climate became warmer and drier, these lakes evaporated, concentrating salts in the developing soil (Figure 4.7). Living organisms and dead plant and animal material are also key constituents of soil. The living organisms decompose the dead biomass, releasing nutrients into the soil. These complex organic molecules interact with the inorganic compounds derived from the parent material. The rate at which this occurs is the result of climate; thus, soils differ regionally in the amount and nature of organic content.

mid-latitude desertification A phenomenon in which warm, dry air is found at approximately 30° N and 30° S, resulting in a desert climate.

rain shadow The tendency of the lee side of mountain ranges to be drier than the windward side.

aspect The direction a mountain slope faces.

parent material The rock and mineral substrate underlying a region.

Figure 4.7 Desert valley. The remnants of lakes in the desert are high in salt and alkali. As the climate in this region became warmer and drier, ancient lakes evaporated, concentrating the salt and alkali they contained.

How Do Abiotic Conditions Affect Species' Distributions?

Natural selection adjusts the tolerance limits to the conditions typically encountered. If the tolerance limits do not match the abiotic conditions, the organism might move to a more favorable environment. This requires mobility and the proximity of favorable places. But if dispersal to a more favorable region is not possible, local extinction ensues. Thus, the combination of dispersal and tolerance limits determines the boundaries of the potential geographic range. Species with a large tolerance limit and high mobility (either of adults or propagules such as wind-dispersed seeds) typically have large ranges; less mobile species with narrow physiological tolerance have smaller ranges. We should not assume that present geographic ranges are fixed. Some are still expanding, especially those of invasive species that humans have spread into new regions.

Any physical factor or combination of factors can determine the range limit. For terrestrial organisms, especially plants, temperature and moisture are the most common limiting factors. The saguaro cactus (*Carnegiea gigantea*) cannot tolerate freezing temperatures for more than 36 hours. Its northern range limit occurs where freezes longer than 36 hours occur (Figure 4.8).

A number of tree species reach altitudinal or latitudinal limits known as **tree line**, where the climate is just too harsh for their survival. Near the tree line, the growth form, the physical form of the tree, changes in response to the physical challenges imposed by cold, snow, and wind. Low temperatures and blowing snow both desiccate and freeze the buds on branch tips, affecting the tree's shape. Frozen soil moisture is unavailable, causing further desiccation. Near tree line, many conifers develop what is known as a **Krumholz growth form**, characterized by a length of bare trunk above a mat of horizontal branches near the ground (Figure 4.9). Wind, cold, and blowing snow account for the bare trunk just at the average snow level. Branches below the snow are relatively protected and grow horizontally. Farther north or higher on a mountain, shoots no longer survive and the tree develops a prostrate growth form. The tree line occurs where even this growth form cannot survive.

For many plants, both the northern and southern range limit is set by climate. A number of temperate tree species are limited by temperature at both their northern and southern limits (Morin et al., 2007). For example, sugar

■ **tree line** The altitude or latitude at which trees can no longer survive.

■ **Krumholz growth form** A growth form of trees at high elevation characterized by a mat of branches near the ground and a single, often bare, shoot at snow level.

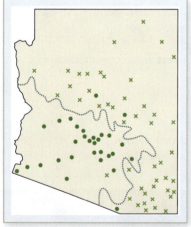

Figure 4.8 Saguaro cactus range limit. The range limit of saguaro cactus is shown in relation to sites where periods of freezing more than 36 hours long occur. The dots represent sites with no records of periods of freezing longer than 36 hours; X = sites where freezes last more than 36 hours. The dotted line shows the northern range limit of saguaro cactus (Hastings and Turner, 1965). **Analyze: Why doesn't the mean temperature at each site explain the saguaro distribution?**

Figure 4.9 Krumholz trees. Krumholz trees are stunted by the harsh conditions near tree line.

THE EVOLUTION CONNECTION

Historical Effects on Species Distributions

In this chapter we focus on the fundamental types of terrestrial communities. If community composition is determined by the species that inhabit the same region, species range limits are central to understanding the distribution and nature of communities. Both are the product of the ecological interaction between the species and the physical and biological environments.

However, each species also has an evolutionary history that plays a role in its current distribution.

Historical shifts in climate have affected the range limits of living species. For example, magnolias are currently found in just two regions—the southeastern United States and Asia. We know from the fossil record that this group was once much more widespread. But as the global climate changed, most species went extinct; the extant species persist only where the climate remained appropriate. Similarly, the fossil record shows that redwoods were once widespread in North America. As the continent became drier after the retreat of the Pleistocene glaciers, redwoods retreated to their current distribution in the moist forests of the northern Pacific coast.

Geological history is an important component of biogeographic history. The North American flora and fauna changed markedly when the Bering Land Bridge connected Alaska and Siberia during the Pleistocene glaciations, permitting the migration of new species to North America. Marsupial mammals evolved early in the evolutionary history of the mammals, when the continents were still united in one supercontinent, Pangaea. When Pangaea broke up, eventually forming the separate continents that we know today, the marsupials already present in Australia continued to evolve into many new species. Later, when placental mammals arose in Asia, Australia was no longer connected to other land masses, and thus there are no native terrestrial placental mammals there.

Figure 1 Pangaea. The supercontinent Pangaea gradually broke up, forming the continents we know today. **Analyze: How might evolution likely differ on one large continent compared to many smaller ones?**

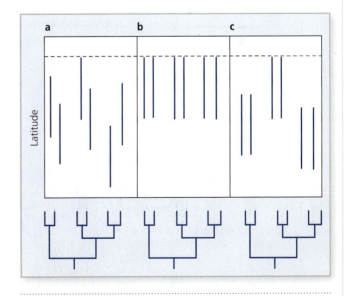

Figure 2 Clades. The three possible patterns of ranges of members of a clade. Each line in the upper box shows the latitudinal range of a species. The phylogenetic trees below show the branching pattern that led to the species' ranges (Kaustuv et al., 2009). **Analyze: Explain how these phylogenies would lead to different species ranges.**

continued

Historical Effects on Species Distributions

There are other, more direct evolutionary effects on a species' distribution. Dispersal ability interacts with phylogenetic history to determine its geographic presence or absence. Each *clade*, a species and its descendants, also has a geographical distribution. A clade begins with a range limit defined by a single ancestral species. As the clade diversifies over time, its range limit changes. Dispersal carries species to new regions. As the constituent species shift their geographic distributions, they adapt to some regions and go extinct in others.

The distribution of species within a clade follows one of three patterns. First, the range limits of closely related species sometimes diverge during or after speciation. Second, in other clades the range limits of closely related species remain similar. And third, in some clades the range limits of closely related species remain similar to one another but different from other groups of species. The second and third possibilities constitute *phylogenetic conservatism*, in which the range limit of a species is constrained by its evolutionary history. This implies that the range limit is not simply a product of current ecological interactions between the organism and its environment. It is also a product of the species' evolutionary history.

QUESTION:

How might the range limits of newly evolved species differ from older ones?

---

maple (*Acer saccharum*) and white ash (*Fraxinus americana*) reach their northern limit where the growing season is not long enough for their fruits to mature. The southern range limit occurs where high temperatures disrupt the normal pattern of dormancy such that flowering occurs so late that the fruits are at risk of frost in the fall.

Although the combination of dispersal and tolerance determines the *potential* geographic range, the actual range occupied is modified by other factors. For example, disturbance factors such as fire may reduce the actual range relative to the potential range. There is sufficient moisture in the eastern Great Plains to support trees. However, because of recurring disturbance by fire, much of this region supported grasslands with only scattered trees. Biological interactions may also play a role. Some interactions, such as an insect that pollinates a specific plant species, are positive—one species facilitates the presence and success of another. Others are negative. Predators and competitors that negatively impact other species may restrict their ranges. Ecological interactions among species are the subject of much of the rest of this book. Thus, although we will not explore them here, they are fundamentally important components of community composition and range limits.

In sum, a species' geographic range is the result of physical and biological factors (Figure 4.10). Mobility and the process of dispersal determine the geographic range available to the organism. Each locale the species can potentially reach falls either within or beyond its physiological tolerance limits. Thus, the physical environment filters species, ultimately determining their *potential* range limits. Positive and negative biological factors then determine the *actual* distribution, the species' presence or absence. Of course, the logical tree in Figure 4.10 represents these processes at one moment in time. The other relevant dimension to this process is evolutionary time, because the physical and biological environments are constantly changing. Thus, geographic ranges are dynamic—they change with the climate, adaptive responses to climate change, and dispersal.

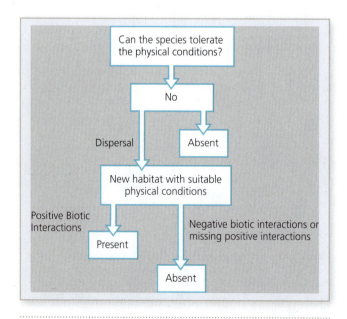

Figure 4.10 Geographic range. The factors that determine a species' presence or absence in a particular region (Hastings and Turner, 1965). **Analyze: Why is the species' abiotic tolerance the first question?**

Fire and Grazing in the Prairie Community

Fire and grazing are environmental disturbances that affect many plant communities. Some ecologists refer to fire as a "global herbivore" because it converts organic material into inorganic products, changes the structure of a community, and constitutes a selective force on the vegetation. Both processes have been important components in the tallgrass prairies of North America. Bison, once numbering in the tens of millions, were the principle grazers in this community. Historical records verify that fires were frequent and extensive when large tracts of undisturbed prairie still existed. Are these two disturbance factors similar? Can we distinguish between their effects on the prairie plant community?

Spasojevic et al. (2010) addressed these questions at the Konza Prairie Biological Station, a 3,500 ha native prairie in the Flint Hills of Kansas. Konza is one of the National Science Foundation Long-Term Ecological Research study sites. Watersheds on the property have been exposed to different grazing and fire treatments for more than 20 years. This history allowed the researchers to devise a hypothesis that answers their research questions:

Figure 1 Konza Prairie. Konza Prairie is a long-term ecological research site. In this photo, the prairie on the left was burned prior to the growing season. The prairie on the right and foreground was not burned.

HYPOTHESIS: Fire and grazing have similar ecological impacts on the structure of the prairie plant community.

This hypothesis led to the following testable prediction:

PREDICTION: The plant species composition and community structure of plots exposed to fire and to grazing should converge.

In order to test this prediction, Spasojevic and his colleagues compared the species composition and the sets of plant functional traits in communities exposed to different fire and grazing regimes over a 22-year period. Functional traits of the prairie community include the types of plants and their growth forms. They include characteristics such as the frequencies of C_4 grasses, clonal species, perennial plants, annual plants, legumes, shrubs, and various combinations of these traits (see table).

Trait Group	Description
1	perennial C_4 grasses
2	perennial C_3 grasses that are clonal
3	perennial non-leguminous forbs that are short and clonal
4	perennial non-leguminous forbs that are short and not clonal
5	perennial non-leguminous forbs that are medium to tall and clonal
6	perennial non-leguminous forbs that are medium to tall and not clonal
7	perennial legumes that are short to medium and not clonal
8	deciduous shrubs that are not clonal
9	deciduous shrubs that are clonal
10	annual C_4 grasses
11	annual and perennial C_3 grasses that are not clonal
12	annual non-leguminous forbs that are short and not clonal
13	annual/biennial non-leguminous forbs that are medium/tall and not clonal
14	perennial legumes that are clonal
15	evergreen shrubs

They found that grazing and fire have complex effects on species composition. The number and diversity of species was higher in grazed than in burned treatments. In contrast, trait composition converged over the 22-year study. However, the degree of convergence depended on fire frequency: sites that burned every four years were most similar to grazed sites. Annually burned sites were the most different from the other treatments. These results supported the prediction of their hypothesis: that community trait structure should be similar under fire and grazing disturbance. It is interesting that the degree of convergence was highest when the fire interval was four years. Historical records suggest that the mean interval between fires in the prairie was about four years. Thus, grazing and a natural fire regime have similar effects on the trait structure of a community but not on the specific species present.

We know that climate and soils determine the broad geographic patterns of plant communities. This study demonstrates

continued

ON THE FRONTLINE   *continued*

Fire and Grazing in the Prairie Community

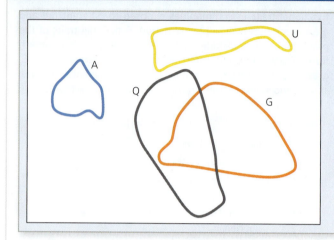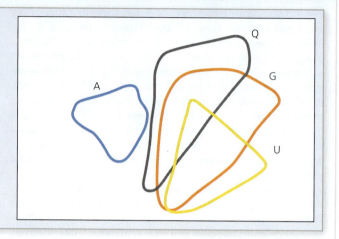

Figure 2 Community similarity. Two-dimensional plots of indices of community similarity. Lines surround groups with similar experimental treatment. Communities that are similar plot in the same portion of the two-dimensional graph. Plot (a) is for community composition; Plot (b) is for trait groups. Symbols: A = annual burn; Q = four-year burn; U = ungrazed; G = grazed (Spasojevic et al., 2010) **Analyze: What parts of this figure indicate that sites that burned every four years were most similar to grazed sites?**

that within a community, biotic and abiotic factors determine the patterns of local species' presence and absence and the frequencies of different functional groups of plants. It also raises a new set tof questions—how and why do grazing and fire determine the functional structure of the community? Or, more generally, what mechanisms structure local assemblages of species? We address this important question in Chapter 12.

THINKING ABOUT ECOLOGY:

Would you expect range limits to be more permanent in the Arctic or the tropics?

KEY CONCEPTS 4.1

- Climate is the result of a combination of geographic features: latitude, altitude, aspect, and proximity to large bodies of water.
- The substrate and soil also determine species' presence or absence, especially in plants and sessile animals. It is a product of weathered parent material, external inputs, and organic matter.
- A species' potential geographic range is determined by the combination of climate, substrate, and local biological interactions. Dispersal and time determine whether the species reaches a favorable region.

QUESTION:

Why are biotic interactions near the base of the flow chart in Figure 4.10?

4.2   What Are the Fundamental Types of Terrestrial Communities?

Early biogeographers recognized regions of ecologically and evolutionarily related animals. Wallace, Darwin's contemporary and co-discoverer of the concept of natural selection, was particularly interested in places where there are sharp transitions in the fauna from one region to another. He noted that one such transition occurs along a line between Borneo and Sulawesi and continuing to the southwest between the islands of Bali and Lombok (Figure 4.11). Bali and Lombok are separated by only a few miles, but the fauna of Bali is related to the fauna of Asia; that of Lombok is related to the fauna of Australia and New Guinea. A deep water trench separates the two islands. Thus, the islands have been separated for a long time—even during periods when the ocean level was low enough that the island of Bali was connected to the Asian mainland and Lombok was connected to Australia.

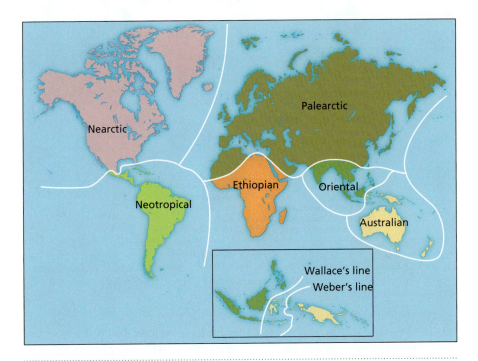

Figure 4.11 Biogeographical realms. The biogeographical realms are regions with similar flora and fauna. Wallace's line (inset) delineates the shift from the Oriental to the Australian realm.
Analyze: What hypothesis could explain the separation of the Oriental and Palearctic realms?

These observations led Wallace to synthesize the knowledge of animal taxonomy and biogeography in a set of **biogeographical realms**. Biogeographical realms share many related species and a similar geographic history. Each is roughly associated with a particular continent or land mass, and each is presently or historically more or less isolated from the others. For example, the South American realm is currently connected to the Nearctic only by the narrow Isthmus of Panama. At times of higher sea level, this connection did not exist. This explains why the fauna of South America is so distinct from that of the Nearctic. Australia has perhaps the most unique fauna due to its isolation. It is characterized by a diverse marsupial fauna, the absence of placental mammals, and relatively few species of fish, amphibians, and reptiles.

The large-scale patterns of biogeography separate both plants and animals according to their evolutionary history, the geological connections among regions, and regional climate. On a finer scale, we recognize sets of distinct plant communities that occupy specific regions. These systems are known as **biomes** (Figure 4.12). A biome is characterized by specific types of vegetation, the *form* of the plant life, such as grassland or deciduous forest. The **flora**, the specific plant species, determines the nature of the vegetation. Thus, the same biome in Europe and North America is characterized by the same vegetation type even though the floras are quite distinct.

Some early ecologists, especially F. C. Clements (1916), asserted that the community is a type of natural unit. He maintained that these communities have characteristics of superorganisms. He believed that they are highly integrated

biogeographical realms Regions that share species and geographic history.

biome The basic plant community types, characterized by a specific vegetation, that occur in a particular region.

flora The plant species of a region.

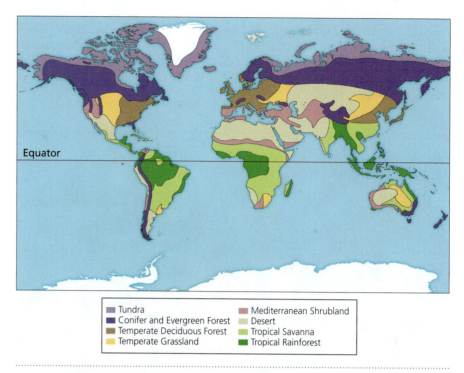

Figure 4.12 Biomes. The biomes of the world. Each biome is characterized by its vegetation and climate. **Analyze: What features of this map suggest that climate is the crucial factor in the distribution of the vegetation?**

sets of species that have evolved together for long periods. In Clements's view these communities grow and develop much like an individual organism. Disturbance by storm or fire initiates the development of the community appropriate to the climate. Ultimately, a final stage known as a climax occurs. In the Clementsian view the biomes represent these final climax communities.

At about the same time another view of the community emerged. Gleason (1926) rejected the notion that communities are natural units. He emphasized an individualistic view of plant communities. Instead of the cohesive, coevolved units of Clements, Gleason believed that plant communities are simply random assemblages of species that just happen to be able to tolerate the same physical conditions. Each species in a community is an independent ecological and evolutionary unit. Yes, some are intimately connected to others. But each is responding to its own selective pressures and ecological interactions. In Gleason's model, disturbance is not an anomaly that temporarily disrupts the natural unit. Rather, it is a frequent and common ecological process that contributes to the random nature of community composition.

The concept of the biome is somewhat antiquated—today we understand that the world's vegetation does not fit neatly into such a small set of types. The distribution of species associated with a particular biome does not always align precisely with the boundaries of the biome. For example, Figure 4.13 shows the range limits of several species characteristic of the North American prairie (grassland) as well as the boundaries of that biome. Note that all these species are found in regions beyond the prairie biome. Moreover, we have abandoned the Clementsian idea that biomes are cohesive, integrated natural units of vegetation with organism-like characteristics. Nevertheless, the biome concept is useful in a general, descriptive sense to identify some basic habitat types.

The distribution of plants is determined in large part by physical factors, especially temperature and moisture. A **climate diagram** helps visualize the temperature and moisture conditions (Figure 4.14). In a climate diagram the monthly temperature and precipitation are depicted together. The vertical axis

THINKING ABOUT ECOLOGY:

How does Figure 4.13 reject the Clementsian view of biomes?

climate diagram A graphical representation of climate showing the seasonal temperature and precipitation changes of a region.

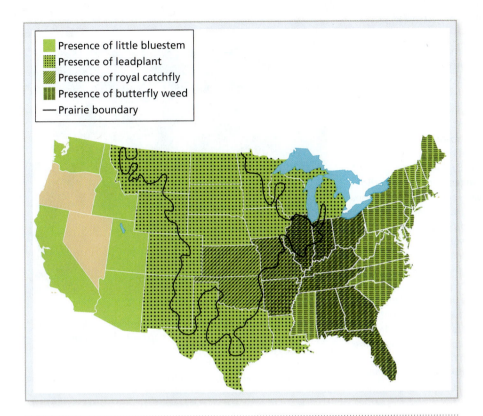

Figure 4.13 Prairie plant distribution. The geographic ranges of several prairie plants and the boundary of the prairie biome. Note that none of these species' range falls entirely within the prairie biome. **Analyze: How would you explain the fact that these prairie plants have different geographic ranges?**

is scaled such that 10°C is the same increment as 20 mm of precipitation. This convention allows us to represent visually the availability of water: when the temperature line falls above the precipitation line, evaporation exceeds precipitation. When the reverse is true, moisture is sufficient for most plant life. We can see the effects of these patterns as determinants of the major global vegetation types (Figure 4.15).

THINKING ABOUT ECOLOGY:

In Canada the biomes occur in east-west bands. In the United States the biomes are oriented as north-south bands. What hypothesis can you devise to explain this pattern?

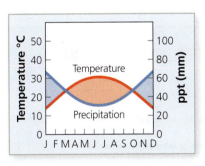

Figure 4.14 Climate diagram. An example of a climate diagram showing the mean monthly temperature and mean monthly precipitation. The axes are scaled to show when there is sufficient (shaded blue) and insufficient moisture for plants (shaded red). **Analyze: What adaptations would allow a plant to survive in this climate?**

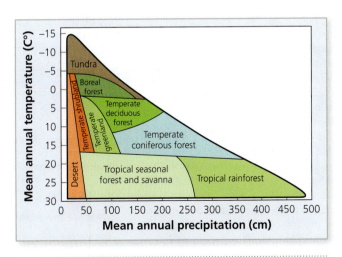

Figure 4.15 Vegetation types. The Earth's major vegetation types are determined by the interaction of moisture and temperature. **Analyze: Why do you suppose the shapes of these blocks of vegetation types differ?**

Global Warming and Ecological Communities

The phenomena of global warming and climate change currently receive a great deal of attention. The evidence is now overwhelming that the emission of carbon dioxide and other gases is causing the mean temperature of the planet to rise. It is important to understand the physical processes underlying this phenomenon.

Solar radiation comes to Earth as short-wavelength energy. Some of the energy from that radiation is absorbed by the Earth, heating it; some is reflected back to space by particles in the atmosphere or objects on the surface. The Earth radiates some of the absorbed energy as long-wavelength radiation. Carbon dioxide is transparent to incoming short-wavelength solar radiation. However, it absorbs and reflects the long-wavelength radiation emitted by the rocks, soil, buildings, and other objects on the Earth. As a result, heat is trapped by atmospheric CO_2.

Climate scientists distinguish between *global warming* and *climate change*. Global warming underlies the broader phenomenon of climate change. The distinction is that as the temperature of the Earth increases, the regional effects on the climate vary in complex ways. Some regions will experience more precipitation; others will experience less. In some cases it is the variation in temperature or precipitation that will change most drastically. Some places will even receive more snowfall than previously.

One consequence of these changes is the emergence of what are called *no-analog climates* (Williams and Jackson, 2007). No-analog climates are combinations of precipitation and temperature patterns unlike any current climate. Obviously, the development of no-analog climates poses a problem for established plant and animal communities. We have generally assumed that as the climate changes, communities will simply shift their locations to new regions where the new climate suits them. However, recent studies of past plant communities suggest that the future is less certain. For example, during the Pleistocene glaciations 17,000 to 12,000 years ago, parts of the midwestern United States supported a forest of spruce, ash, oak, and hornbeam. No modern community contains those species; it is called a *no-analog community*.

Climate models predict that as the global temperature increases, as much as 39 percent of the terrestrial planet will experience no-analog climates. At the same time perhaps half of the world's existing climates will disappear. We cannot assume that today's plant communities will simply shift their geographic ranges, because future no-analog climates may not support them. For example, the southeastern United States is predicted to become hotter and drier during the summer months. If so, the frequency and intensity of fire will likely increase as well, which in turn will impact some forest associations there that are not well adapted to fire. We can expect significant changes in the patterns, composition, and function of plant communities over the next decades.

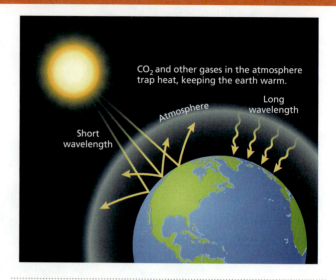

Figure 1 **Greenhouse effect.** The greenhouse effect occurs because CO_2 and other gases in the atmosphere trap heat. Short-wavelength light can pass through the CO_2 in the atmosphere. When this radiation strikes the Earth it is radiated back as longer-wavelength energy, which is retained by CO_2 and other greenhouse gases. **Analyze: What are the natural and manmade sources of CO_2 in the atmosphere?**

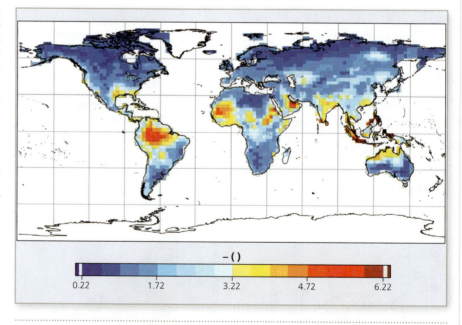

Figure 2 **No-analog climates.** This map shows the set of no-analog climates under greenhouse warming. The amount of change from the current climate is shown in each region by the color scale (Williams et al., 2007). **Analyze: Why are these changes greater at lower latitude?**

QUESTION:

What are the important ecological and evolutionary differences between human-caused climate change and the other changes that have occurred over the course of geological history?

This section concludes with a reference guide to the eight key terrestrial biomes of the Earth (see following page).

KEY CONCEPTS 4.2

- Each terrestrial biome occurs in a region with specific climatic conditions, particularly temperature and moisture.
- Each biome is characterized by a specific form of vegetation.

QUESTION:

Which of the terrestrial biomes appear to be the most variable in terms of climate and vegetation?

TROPICAL RAINFOREST

Figure 1.1 Tropical Rainforest

KEY FEATURES

Tropical rainforest is one of the most extensive biomes on the planet. Its high temperature and rainfall make its climate especially conducive to life. As a result, tropical rainforest is characterized by lush, diverse forest vegetation. The diversity of both plants and animals is astonishing. In some tropical rainforests more than 250 species of trees have been recorded in a single hectare.

These forests are also characterized by many epiphytic plants—species that grow on other plants. Some are parasitic; others simply use another plant for support or to position themselves where they can absorb nutrients and water (Biome Figure 1.4).

Rainforests have a distinct physical structure. There is a canopy of tall trees whose branches intermix and cast deep shade. The canopy is broken by a few super canopy trees that stand above the canopy where they can obtain sunlight. Below the canopy is a complex layer of epiphytes, vines, and small trees.

The soils of tropical rainforests are very old and highly weathered. Many of the nutrients derived from bedrock have been released and the high precipitation leeches nutrients from the soil. Most of the nutrients in rainforest are found in living organisms rather than in soil.

Given the geographic extent of tropical rainforest, it is not surprising that there is variation in both the climate and the ecology of these forests. Rainforests differ in the degree of seasonality, particularly in the amount of rainfall. A distinct type of rainforest known as a **cloud forest** occurs at elevations between 1,000 m and 2,500 m. At this altitude, the air is cool enough that the moist tropical air forms dense fog and clouds (Biome Figure 1.5). Cloud forests are not as lush as lowland rainforest, but they often contain very high animal species diversity, especially birds.

SIGNIFICANCE TO HUMANS

Many prescription drugs are derived from rainforest plants. These lush forests store large quantities of carbon, which is released into the atmosphere if rainforest is cleared and burned, adding to global warming.

WHAT EVOLUTIONARY ADAPTATIONS ARE FOUND IN TROPICAL RAINFOREST?	
PLANT CHARACTERISTIC	ADAPTIVE SIGNIFICANCE
Leaf morphology with "drip tips" that channel water off of leaves.	Reduces nutrient loss by leeching; reduces growth of bacteria and fungi on leaf surface.
Trunk buttresses.	Provides support for tall canopy trees in shallow soil.
Shallow root systems.	Absorb scarce nutrients before the nutrients are carried away by water or captured by another plant.
Epiphytic habit.	Uses other plants for support and competitive access to light and nutrients.

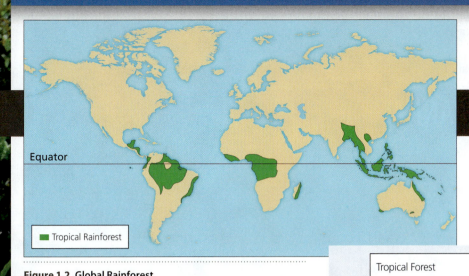

Equator

■ Tropical Rainforest

Figure 1.2 Global Rainforest

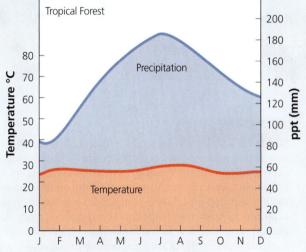

Tropical Forest

Precipitation

Temperature

Figure 1.3 Climate Graph

Figure 1.4 Rainforest Epiphytes

CHARACTERISTIC ANIMALS

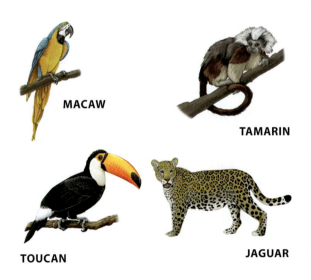

MACAW

TAMARIN

TOUCAN

JAGUAR

Figure 1.5 Cloud Forest

Figure 1.6 Animals of the Tropical Rainforest

BIOME 2
TROPICAL SAVANNA

Figure 2.1 Tropical Savanna

KEY FEATURES

Tropical dry forests are seasonal forests with pronounced wet and dry seasons. One of the most extensive forms of dry forest, known as savanna, occurs in latitudinal bands 20° north and south of the equator where the dry season is especially pronounced. Savannas are characterized by open grasslands containing isolated trees. Although savannas occur in South America and extensively in Australia, those in Africa are most familiar because they are famous for large herds of herbivores such as giraffes, antelope, wildebeests, and zebras, as well as predators such as lions, cheetahs, and wild dogs. The pronounced wet and dry seasons lead to major migrations of animals in search of water and green forage.

Fire plays an important ecological role in savanna. During the dry season, lightning-caused fires are common. Fire kills young trees but favors the grasses, which can quickly resprout from the base or from germinating seed. Thus, fire maintains the vegetation structure of this biome.

Unlike forest biomes, there is little vertical structure to the vegetation in a savanna. Instead, the important structure is horizontal. Most of the grasses of savanna are bunch grasses. Consequently, the open grassland is characterized by distinct patchiness at the ground surface—clumps of dense grass separated by relatively open soil. Savanna trees create local isolated microclimates that are important to both plants and animals. In their shade, temperature and evaporation levels are lower. This favors colonization by shrub species which cannot survive in open sun. Animals also seek out this shade, and their feces and urine further enrich the micro-environment for shrubs (Biome Figure 2.4).

Savannas occur as transitional habitat between forest and grassland in many parts of the world. In North America, the boundary between the grasslands and the eastern deciduous forest contained many small patches of savanna.

SIGNIFICANCE TO HUMANS

The fossil evidence indicates that the first bipedal primate ancestors of modern humans lived in the African savanna. This biome has been important to humans as pasture and rangeland for domestic animals for thousands of years.

WHAT EVOLUTIONARY ADAPTATIONS ARE FOUND IN TROPICAL DRY FORESTS?	
PLANT CHARACTERISTIC	**ADAPTIVE SIGNIFICANCE**
Deciduous habit.	Reduces water loss during dry season.
Grasses re-sprout from roots.	Plants re-grow quickly following fire.
Thorns and spines.	Protect plant from grazers.

Equator

■ Tropical Savanna

Figure 2.2 Global Savanna

Figure 2.4 Savanna Tree

Savanna   Precipitation

Temperature

Figure 2.3 Climate Graph

CHARACTERISTIC ANIMALS

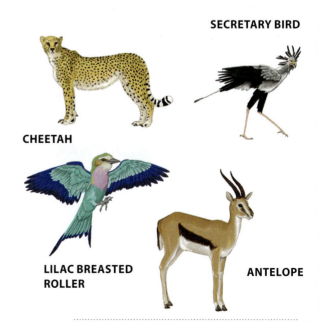

SECRETARY BIRD

CHEETAH

LILAC BREASTED
ROLLER

ANTELOPE

Figure 2.5 Herd of Zebra

Figure 2.6 Animals of the Tropical Savanna

DESERT

Figure 3.1 Desert

KEY FEATURES

Deserts are defined climatically as regions where evaporation exceeds precipitation. The world's deserts vary widely in severity and in the pattern of rainfall. Parts of the Atacama Desert in Peru and Northern Chile have never recorded rain. In contrast, parts of the Sonoran Desert in the United States receive more than 300 millimeters of precipitation each year. Some, like the Sonoran, have a pronounced rainy season or monsoon. In all deserts, the variation in precipitation is large relative to the mean.

The substrate of deserts consists of bare rock, sand, alkali flats, or gravel. They also range from the white gypsum sands of White Sands, New Mexico to black lava in Death Valley, California. The surface of many desert soils consists of a matrix of algae, fungi, and lichens that bind soil particles into a fragile surface mat (Biome Figure 3.4).

Desert vegetation is highly variable and dependent on both the substrate and the pattern of precipitation. The flora consists of species adapted to high temperature and low moisture and includes cacti, euphorbs, grasses, and shrubs.

Deserts are among the most diverse of the biomes, because they are defined climatically rather than by the structure of the vegetation. The dominant forms of the plant life and the amount of plant production are tightly correlated with precipitation. Thus, the wide range of rainfall patterns and amounts lead to a wide range of vegetation types and biomass. In addition, the variety of substrate types and chemical composition selects for divergent types of vegetation and specifically adapted flora. Also, some high-elevation deserts such as in Mongolia and the Great Basin of the United States experience long periods of sub-freezing temperatures.

Many desert plants demonstrate the principle of evolutionary convergence—different evolutionary lines with similar adaptations. The cacti of the New World deserts closely resemble plants in a different family, the euphorbiaceae, of Africa. Both have spines and photosynthetic stems but no leaves. Many desert grasses adopt the annual habit and produce seeds highly resistant to drought. The seeds remain in the substrate, sometimes for decades, until it rains. They then germinate quickly, flower, set seed, and die (Biome Figure 3.5).

SIGNIFICANCE TO HUMANS

Humans have typically thought of deserts as "wastelands," useful only for mineral extraction and other exploitative uses. Globally, deserts are expanding as a result of human activities, including overgrazing, poor agricultural practices, and climate change

WHAT EVOLUTIONARY ADAPTATIONS ARE FOUND IN DESERT PLANTS?	
PLANT CHARACTERISTIC	ADAPTIVE SIGNIFICANCE
Allelopathy.	Reduces nearby competition for water.
Reduced leaves.	Reduce surface area for water loss.
Hairy leaves.	Increase shade and reduce wind speed at leaf surface and thus reduce water loss.
Annual habit.	Can exploit brief and unpredictable periods of moisture.

Equator

Desert

Figure 3.2 Global Desert

Figure 3.4 Cactus and Euphorb

Desert

Figure 3.3 Climate Graph

CHARACTERISTIC ANIMALS

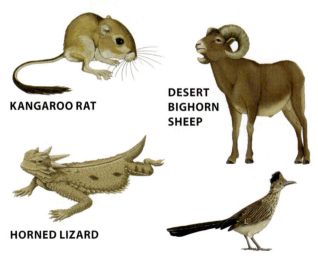

KANGAROO RAT

DESERT
BIGHORN
SHEEP

HORNED LIZARD

ROAD RUNNER

Figure 3.6 Animals of the Desert

Figure 3.5 Sand Desert and Lush Sonoran Desert

MEDITERRANEAN SHRUBLAND

Figure 4.1 Mediterranean Shrubland

KEY FEATURES

This biome is characterized by both climate and vegetation structure. The climate, a "Mediterranean climate," is defined by dry summers and winter rain. The vegetation consists of a floristically diverse group of shrubs and small trees. This biome is also known as **chaparral.** It is found on every continent except for Antarctica.

Fire is an important ecological feature of this biome. Many species are not only adapted to fire but require fire for germination and vigorous growth. Some species contain volatile chemicals that are highly flammable. Thus, chaparral fires are especially hot and intense. Fire opens up space in the dense vegetation, which is quickly usurped by root sprouts of adults and new seedlings.

Most of the species in this biome are related to tropical species. They probably evolved when dry summer climates arose in the subtropical margins of the ancestral forests. The structure of the vegetation is similar among the examples of this biome found in Africa, Europe, North America, South America, and Australia. The flora comprising each of these five systems is quite different; their morphological similarities are due to evolutionary convergence. In fact, most Mediterranean shrubland species are endemic, that is, are found in one place but nowhere else. For example, 40 percent of the species in California chaparral are endemic.

Many Mediterranean shrublands are found in regions with current and historically high human population density. Consequently, much of this biome has been destroyed or significantly altered. In Southern Spain and Portugal a long history of grazing and human selection for the native species of oaks that produce cork has modified the structure and floral composition of the shrubland. Chaparral is the dominant vegetation type in many of the hills and mountains of Southern California. The high human density and valuable real estate have led to the development of extensive housing tracts in chaparral habitat. This places human habitation at significant risk from the recurring fires characteristic of this biome.

SIGNIFICANCE TO HUMANS

The soils of Mediterranean shrublands tend to be rocky with low fertility. Thus, this biome is not highly prized for agriculture although grazing occurs in some systems.

WHAT EVOLUTIONARY ADAPTATIONS ARE FOUND IN MEDITERRANEAN SHRUB PLANTS?	
PLANT CHARACTERISTIC	ADAPTIVE SIGNIFICANCE
Re-sprout from roots.	Plant grows quickly following fire.
Flammable oils (a few species such as chamise).	Promotes fire that reduces local competition.
Hard, needle-like (sclerophyllous) leaves with few stomata.	Reduces water loss.

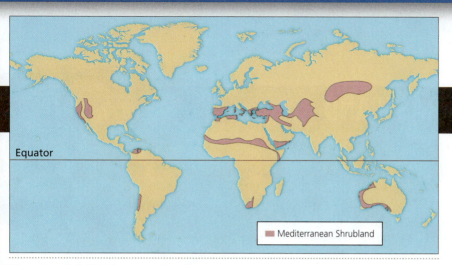

Equator

■ Mediterranean Shrubland

Figure 4.2 Global Mediterranean Shrubland

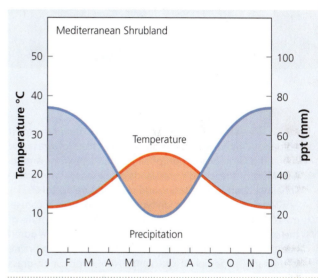

Mediterranean Shrubland

Temperature

Precipitation

Figure 4.3 Climate Graph

CHARACTERISTIC ANIMALS

GRAY FOX

CALIFORNIA CONDOR

CALIFORNIA QUAIL

BRUSH RABBIT

Figure 4.4 Animals of the Mediterranean Shrubland

BIOME 5

TEMPERATE GRASSLAND

Figure 5.1 Temperate Grassland

KEY FEATURES
Grassland was once one of the most ubiquitous biomes in the world. Large tracts of grassland covered a large proportion of all continents except Antarctica. Most of the world's grasslands occur where strong seasonality in both precipitation and temperature occurs.

The dominant vegetation type in this biome is, of course, grass. However, grasslands differ markedly in the nature of the grasses. For example, most African grasslands are dominated by annual grasses, whereas the North American prairie is primarily a perennial grassland. Non-woody, non-grass species known as **forbs** contribute much to the floral diversity of grasslands.

The grassland biomes supported large populations of ungulates such as bison in North America and horses and antelope in Asia. Thus, grazing has been an important selective force in these communities. Fire is an important component of most of the world's grasslands. Many of these systems occur in regions where precipitation is sufficient to support trees or shrubs but recurring fire has reduced or eliminated trees.

The North American grasslands vary significantly floristically and ecologically. There is a strong moisture gradient—low precipitation in the rain shadow of the Rocky Mountains and high precipitation in the Midwest—that affects the structure and composition of the grassland. In the West, the short grass prairie is dominated by bunch grasses such as grama and little bluestem. The eastern or tallgrass prairie is dominated by sod-forming grasses such as big bluestem and Indian grass, species that sometimes stand more than three meters tall at maturity. Other unique grassland systems are scattered across North America where the climate and fire favor grasses. For example, the Palouse prairie of the northern intermountain region (Eastern Washington, Northeast Oregon, and Western Idaho) consists of bunch grasses grading into sage brush steppe habitat. Unique coastal prairies with many endemic species occur on the California and Texas coasts.

SIGNIFICANCE TO HUMANS
Grasslands are sometimes known as the "bread basket of the world." The lush, productive grass is important for grazing. The North American prairie soil is among the most fertile soils in the world.

WHAT EVOLUTIONARY ADAPTATIONS ARE FOUND IN GRASSLAND PLANTS?	
PLANT CHARACTERISTIC	**ADAPTIVE SIGNIFICANCE**
Meristem located underground.	Protects growing point from fire and grazing.
Deep roots.	Provide access to nutrients and deep ground water.
Narrow leaves.	Reduce water loss.
High silica content.	Deters grazing.

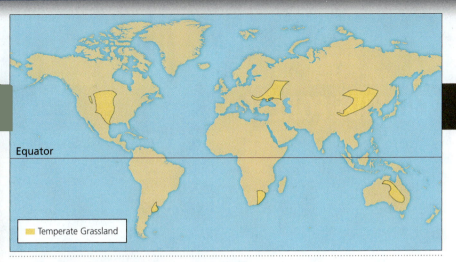

Equator

Temperate Grassland

Figure 5.2 Global Temperate Grassland

Figure 5.4 Prairie Fire

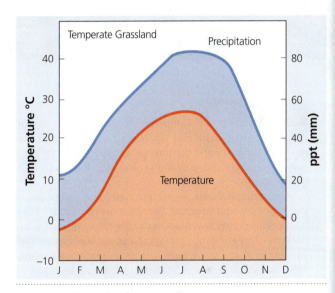

Figure 5.3 Climate Graph

CHARACTERISTIC ANIMALS

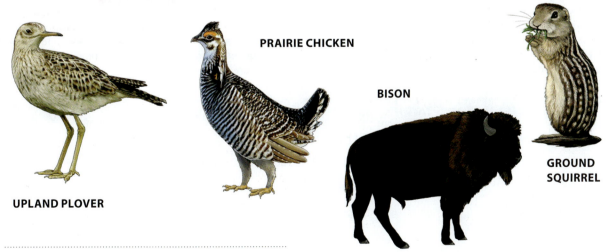

PRAIRIE CHICKEN

BISON

GROUND
SQUIRREL

UPLAND PLOVER

Figure 5.5 Animals of the Temperate Grassland

TEMPERATE DECIDUOUS FOREST

Figure 6.1 Temperate Deciduous Forest

KEY FEATURES

Temperate deciduous forests occur primarily between 30° and 50° N on the east sides of the northern hemisphere continents. Deciduous trees predominate where the growing season is at minimum four months long.

Like tropical forest, temperate deciduous forest has a consistent physical structure. There is a closed canopy of the interlocking leaves and branches of the tallest trees. An understory layer contains saplings of canopy trees as well as some shrubs and small trees that never reach the canopy. There is a shrub layer and finally an herbaceous layer just at the ground level. In old growth forest the canopy is closed and little light reaches the lower layers. Thus, the species that occur there must be tolerant of low light conditions.

There is growing evidence that fire played a role in the ecology of deciduous forest. The higher precipitation and humidity probably limited most of these fires to relatively small ground fires.

The temperate forests of eastern North America comprise a complex set of **tree associations,** combinations of tree species with similar moisture and soil requirements that commonly co-occur. In general, maples and beeches are **mesic species** that require more moisture; oaks and hickories are dry adapted or **xeric species**.

Two physical processes are ecologically important in this biome. First, the dense canopy of old growth forest significantly limits light penetration below this uppermost layer. The pattern of species diversity, individual growth, and replacement of canopy trees is the result of competition for light. Old growth forests with an intact canopy often have an open understory. Second, the accumulation and eventual decay of organic matter is a central ecological process in these forests. The annual leaf fall adds a significant carbon and nutrient input to soil. Dead wood—branches or entire trunks—supplements this material. The release and rapid uptake of nutrients supports the large woody biomass of this system (Biome Figure 6.4 about here].

SIGNIFICANCE TO HUMANS

The deciduous forests have been important sources of timber, especially hardwoods. In addition, most occur on relatively fertile soils. Consequently, large portions of this biome were cleared for agriculture.

WHAT EVOLUTIONARY ADAPTATIONS ARE FOUND IN DECIDUOUS FOREST PLANTS?	
PLANT CHARACTERISTIC	ADAPTIVE SIGNIFICANCE
Deciduous habit.	Reduces water loss during winter dormancy.
Leaves with large surface area.	Increase competitive access to light.
Herbs flower early.	Completes the life cycle before trees leaf out and cast shade.

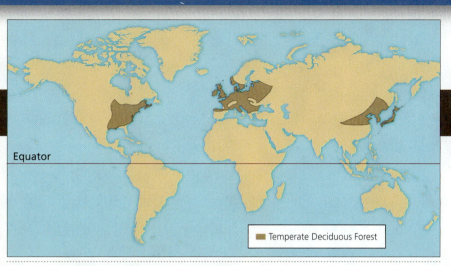

Figure 6.2 Global Temperate Deciduous Forest

Figure 6.4 Old Growth Forest and Open Understory

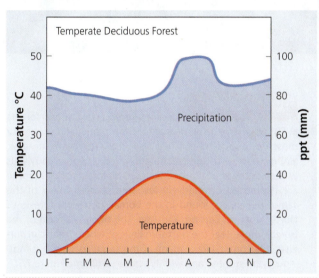

Figure 6.3 Climate Graph

CHARACTERISTIC ANIMALS

BLACK BEAR

WHITE TAILED DEER

SCARLETT TANAGER

PILEATED WOODPECKER

Figure 6.5 Animals of the Temperate Deciduous Forest

BIOME 7

CONIFER FOREST

Figure 7.1 Boreal Forest

KEY FEATURES

Conifers replace deciduous forests at 50° N, where the winters are too long, cold, and snowy for most deciduous trees. They also occur in the mountains, along coasts with high precipitation, and on poor, sandy soils in the Southeastern United States. The forests of this biome are composed of relatively few tree species. Many of the same species are found in the boreal (northern) forests of North America, Europe, and Asia. The **taiga** is a form of the boreal forest near the northern limit of where trees can survive. It is composed of many of the same species, but individuals are stunted and widely scattered (Biome Figure 7.5).

Fire is common in the boreal and montane conifer forests. Some species, such as lodgepole pines in the mountains and jack pines in the boreal forest, are adapted to fire. Their cones open and release the seeds when exposed to high temperatures.

The dominance of conifers in the boreal and montane forests is also due to the nature of the soils in these regions. Conifers can survive in the thin and relatively infertile soils that developed since the retreat of the Pleistocene glaciers in the last 10,000 years. The slow decomposition of organic matter in this cold, dry climate also reduces the nutrient of these soils.

Permafrost, ground still frozen from the Pleistocene glacial advances, persists on north-facing slopes at high latitude and prevents deep penetration of roots.

SIGNIFICANCE TO HUMANS

The conifers of this biome are economically important for lumber and paper pulp. Relatively little old growth forest remains, especially at lower latitudes and in the montane conifer forests of the West.

WHAT EVOLUTIONARY ADAPTATIONS ARE FOUND IN CONIFER FOREST PLANTS?	
PLANT CHARACTERISTIC	**ADAPTIVE SIGNIFICANCE**
Evergreen habit.	Immediate photosynthesis when sunlight and temperature increase.
Needle-like leaves.	Reduce surface area for water loss.
Tree shape narrows at top; drooping branches.	Shed snow.

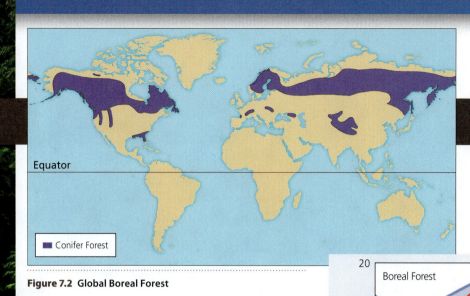

Figure 7.2 Global Boreal Forest

Equator

■ Conifer Forest

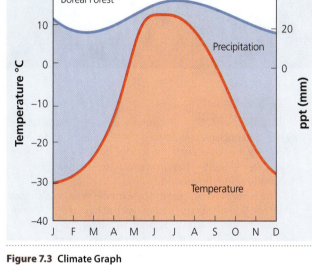

Figure 7.3 Climate Graph

Boreal Forest

Precipitation

Temperature

J F M A M J J A S O N D

Figure 7.4 Montane and Coastal Conifer Forest

CHARACTERISTIC ANIMALS

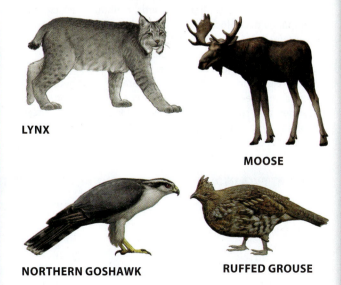

LYNX

MOOSE

NORTHERN GOSHAWK

RUFFED GROUSE

Figure 7.5 Taiga

Figure 7.6 Animals of the Boreal Forest

BIOME 8

TUNDRA

Figure 8.1 Arctic Tundra

KEY FEATURES

The tundra biome is characterized by low, perennial grasses, sedges, shrubs, mosses, and lichens. A few tree species occur in the tundra, but their growth form is radically different—they grow prostrate with their branches horizontal along the ground.

The tundra is characterized by extreme seasonal variation in climate. Summer temperatures reach as high as 35°C. During the twenty-four hours of daylight. In winter the sun shines only briefly or not at all and the temperature often drops below −35°C. The tundra is climatically desert-like—much of this biome receives less than twenty-five centimeters of precipitation per year. However, because much of the region is underlain by permafrost, water does not drain away from the surface. Still, the desiccating cold, wind, and low precipitation have led to many adaptations similar to desert plants.

The short growing season of the tundra results in a rich, spectacular period of abundance and reproduction for both plants and animals. Migratory birds, caribou, musk ox, and lemmings occur locally in high numbers in the summer. In the same brief window, most plants flower and set seed resulting in remarkable floral displays.

The arctic tundra takes many forms depending upon latitude, elevation, and topography. Grassland vegetation is found in drier sites. Low, wet areas contain cotton grass, sedges, and sphagnum moss. Fell fields are exposed rocky habitats with sparse vegetation. In some, lichens are the only plant cover. Climate and permafrost combine to cause an array of local topographic changes. For example, the freeze-thaw cycle causes soil to rise or fall depending on the accumulation of water and the presence or absence of underlying permafrost. This cycle also sorts gravel and small rocks by size. One result is the development of polygons—small domes of fine particles surrounded by strips of heavier particles. Because so many plant species are near their range limits, small changes in elevation and drainage have profound effects on their distribution. The center of a polygon may be only a meter higher than the edge but supports a distinctly different vegetation and flora (Biome Figure 8.4).

Alpine tundra occurs at high elevation in the major mountain ranges. The flora and vegetation of the alpine tundra is similar to that of the arctic tundra. Indeed, many species are closely related. However, there are also important ecological differences. Most alpine tundra is not underlaid with permafrost. It experiences neither twenty-four hours of summer daylight nor twenty-four hours of winter darkness. Snow accumulation is significant. As this snow gradually melts over the summer, it provides a constant supply of water to the vegetation (Biome Figure 8.5).

SIGNIFICANCE TO HUMANS

This biome represents some of the last large tracts of remaining wilderness. Some parts of the tundra contain significant oil and gas deposits, leading to conflicts over the development of these resources.

WHAT EVOLUTIONARY ADAPTATIONS ARE FOUND IN TUNDRA PLANTS?	
PLANT CHARACTERISTIC	**ADAPTIVE SIGNIFICANCE**
Hairy leaves and stems.	Reduce water loss.
Dark or red color.	Absorbs solar radiation and warms plant.
Rapid flowering and seed set.	Permits reproduction during brief summer.
Heliotropic flowers.	Warm flower and perhaps attracts pollinators.

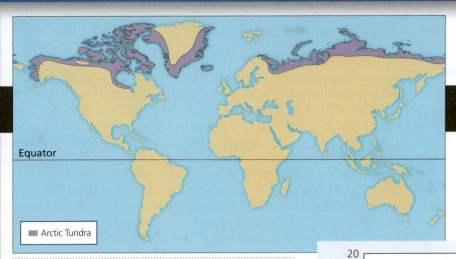

Equator

■ Arctic Tundra

Figure 8.2 Global Arctic Tundra

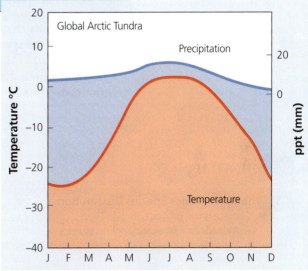

Global Arctic Tundra

Precipitation

Temperature

Figure 8.3 Climate Graph

Figure 8.4 Polygons in Arctic Tundra

CHARACTERISTIC ANIMALS

Figure 8.5 Alpine Tundra

LEMMING

CARIBOU

GOLDEN
PLOVER

SNOWY OWL

Figure 8.6 Animals of the Arctic Tundra

Putting It All Together

Early ecologists recognized that there are patterns of species distribution—that there are distinct community *types*. They also recognized that these community types occur in specific environments. Thus, Arctic tundra, savanna, and tropical rainforest each has a distinctive structure and a set of consistently co-occurring species. Moreover, we can describe precisely the kind of environment in which each occurs. This chapter summarizes the fundamental types of terrestrial communities. Perhaps the most remarkable feature of these systems is their diversity, ranging broadly from desert to tropical rainforest. However, three features are common to all of the communities we have considered. First, the species that occur there are adapted to the prevailing physical conditions. Second, many of the species comprising these communities are consistently found together. Finally, similar environments in different regions support similar communities.

We end this chapter without fully answering the question with which we began: How are species organized into communities? One reason is that the answer to this question depends on a full understanding of the ecological interactions among coexisting species. And of course this is the topic of much of the rest of this text.

Summary

4.1 What Determines Species Distributions?
- Climate is the product of
  - Altitude (via adiabatic cooling effects)
  - Latitude (via seasonality and angle of solar input)
  - Aspect
  - Proximity of large bodies of water
  - Rain shadows
  - Global air circulation patterns
- The substrate, especially soil, also determines plants' presence or absence.
  - Soil is the product of weathered parent material, external inputs, and organic matter.
- The combination of dispersal and species-specific tolerance limits determines a species' potential geographic range.

- Disturbance factors such as fire may affect the size of the geographic range.
- Biotic interactions such as pollination, grazing, and competition also determine the portion of the potential range that is actually occupied.

4.2 What Are the Fundamental Types of Terrestrial Communities?
- Biomes are the major types of terrestrial vegetation.
- Each biome is characterized by a specific type of vegetation that is determined by the regional climate.

Key Terms

adiabatic cooling p. 64
aspect p. 67
biogeographical realm p. 73
biome p. 73
climate diagram p. 74
continental climate p. 66

flora p. 73
geographic range p. 64
Krumholz growth form p. 68
life zone p. 63
maritime climate p. 66
mid-latitude desertification p. 67

parent material p. 67
rain shadow p. 67
tree line p. 68
vegetation p. 63

Review Questions

1. How do tolerance limits and dispersal interact to set a species' range limit?

2. What is the relationship between life zones and biomes?

3. How do biological interactions among species affect the range limit?

4. What is the role of historical factors in determining the range limit?

5. Why are moisture and temperature so important in the distributions of plants and animals?

6. What are the important differences between the Gleasonian and Clementsian concepts of plant communities?

7. For each biome, describe
   a. The main form of the variation
   b. The key limiting physical factors

Further Reading

Cuchman, S.A., et al. 2010. Toward Gleasonian landscape ecology: from communities to species, from patches to pixels. USDA Rocky Mountain Research Station. Research Paper RMRS-RP-84.

This is a sophisticated treatment of modern plant ecology based on Gleason's concept. Much of the paper is technical and assumes knowledge of information to be found later in this text. However, it is an important summary of the evolution of the Gleasonian community concept and its modern form.

Engelbrecht, B.M.J., et al. 2007. Drought sensitivity shapes species distribution patterns in tropical forests. *Nature* doi:10.1038/nature 05747.

The tropical climate is characterized by high rainfall and humidity and supports phenomenal species diversity. One might assume that moisture is not limiting in this system and therefore is of little importance in plant species distributions. This paper shows that subtle differences in species' moisture requirements play an important role in the local distribution and coexistence of many species.

MacArthur, R.H. 1972. Geographical ecology: patterns in the distribution of species. New York: Harper and Row.

This is a classic book by one of the most important ecologists of the twentieth century. Although some ideas are now dated or have been significantly modified, this book explores many of the topics of this chapter. It is historically important as the basis for much of modern geographic ecology.

Williams, S.E., et al. 2008. Toward an integrated framework for assessing the vulnerability of species to climate change. Public Library of Science 6 e25.

If climate is important to plant and animal distributions, it follows that climate change will alter the present geographic distributions of many species. This paper addresses this topic. Specifically, it explores the possibility that climate change will lead to novel climate patterns and thus novel plant communities with no analog in the current climate.

Chapter 5

Freshwater and Marine Communities

In 1977 Robert Ballard of the Woods Hole Oceanographic Institution conducted a search for deep-sea hot springs near the Galápagos Islands. His interest was primarily geological—scientists had predicted the existence of vents in the Earth's crust near the junctions of tectonic plates. He mounted his search at the boundary of the Pacific and Nazca Plates, the same rift that gave rise to the deep-sea volcanoes that became the Galápagos Islands. Ballard's team did indeed locate the predicted hydrothermal vents, but no one could have predicted the ecological community they discovered surrounding the hot springs. As the deep submersible, *Alvin*, moved over the vents, a rich, diverse, and bizarre group of species came into view. Giant tube worms (*Riftia*), some more than two meters long, were packed together near the hot-water vent. Huge clams (*Calyptogena*) were surrounded by shrimp, crabs, mussels, polychaete worms, and an array of fishes. Virtually all of these species were new to science. Moreover, they represented at least 16 new families of invertebrates. One group, the vestimentiferan worms, was deemed so different that it was originally placed in a new animal phylum (although today they are classified in the annelids). Certainly the community itself was novel—no one imagined that life so diverse and abundant could exist in that environment. After all, the vents lie more than 2,500 m below the surface, where no light penetrates and the pressure is crushing. Moreover, the water emanating from the vents contains high concentrations of heavy metals that are toxic to most organisms.

By 1977, most marine ecologists believed that the basic types of ocean communities were already known. The discovery of the Galápagos Rift hydrothermal vent communities harkened back to the time when much of the world was unexplored and we were discovering and categorizing its ecological systems. We saw in the last chapter that terrestrial explorers discovered the connection between species distributions and the physical environment. They also recognized groups of species that today we call communities. The Galápagos Rift community is but one example of an aquatic community. We explore the diversity of aquatic communities by addressing questions such as the following: What physical factors determine the composition of aquatic systems? How do they function? The answers to these questions help us answer the more general question: *What are the major types of aquatic communities?*

5.1 How Is Water Distributed Across the Earth?

The vast majority of the Earth's water is found in the ocean. In fact, from space, the Earth appears to be a blue planet. The total amount of water on Earth has been relatively constant for millennia. Water is found in seven major reservoirs: ice, oceans, lakes and ponds, rivers, groundwater, soil, and the atmosphere. Water moves among these reservoirs in a system known as the **hydrologic cycle** (Figure 5.1). Atmospheric water, in the form of water vapor, constitutes only 6 percent of all the Earth's water at any time, but precipitation from the atmosphere contributes to each of the other pools. Evaporation from the reservoirs, especially the oceans, lakes, and rivers, replenishes atmospheric water vapor. As air and the water vapor it contains move over the Earth's surface, precipitation distributes water from one region to another.

hydrologic cycle The movement of water among the living and nonliving components of the Earth.

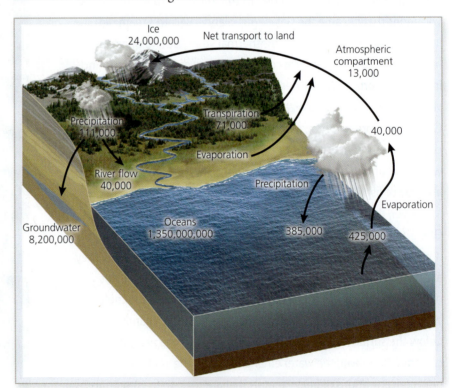

Figure 5.1 Hydrologic cycle. The Earth's hydrologic cycle: the reservoirs and pathways by which water moves. The amount of water in reservoirs is measured in km³; transfers are measured in km³/year.
Analyze: What components of the cycle do you think are vulnerable to the effects of climate change?

As water precipitates as rain or snow, it enters the other reservoirs by a number of pathways. Of course, as precipitation that falls on lakes, oceans, and rivers, water enters these reservoirs directly. Rain that falls on land may be intercepted directly by the vegetation. The rest is absorbed by litter and soil. From this reservoir, some water enters plants via the roots, some is held in soil, and some moves through the soil by gravity. Some of the water that enters plants returns to the atmosphere by **transpiration**, the evaporation of water from the internal surfaces of the plant, especially the stomata. The movement of water through the soil occurs by three main processes. **Infiltration** is the movement of water into the pores between soil particles. The nature of the soil (see Chapter 3) and the amount of water entering determine the rate of infiltration. When precipitation exceeds the capacity of the soil to absorb water, **sheet flow** carries water directly to bodies of water. Water that infiltrates soil gradually moves downward, or percolates, until it reaches a layer of rock or soil that is impervious to water. This water then flows laterally and eventually ends up in a pool of underground water known as an **aquifer**, or it may reach surface waters including lakes, streams, and the ocean.

transpiration The movement of water through a plant, including evaporation from the stem and leaves.

infiltration The movement of water into the pores between soil particles.

sheet flow The movement of water in broad sheets across the land.

aquifer An underground reservoir of water.

KEY CONCEPTS 5.1
- Water is found in the atmosphere, ice, oceans, lakes, rivers, groundwater, and soil.
- Water moves among these reservoirs in a series of processes that comprise the hydrologic cycle.
- The amounts and patterns of evaporation and precipitation determine the movement of water in the hydrologic cycle.

5.2 What Are the Important Physical Factors in Aquatic Systems?

One of the fundamental distinctions between aquatic and terrestrial systems is the degree to which oxygen is limiting in water. The partial pressure of oxygen is so high in the atmosphere that O_2 is rarely limiting for terrestrial species. In fact, this is one of the reasons that organisms colonized land in the first place.

The most basic dichotomy in aquatic ecology is the division between saltwater (marine) and freshwater systems. The structure, organization, and function of these two systems are so different that most ecologists analyze them independently. However, some factors are critically important in both. Aquatic photosynthesis, whether by plants or algae, requires light energy to drive the production of carbohydrate. Because water absorbs and scatters light, the light available to drive photosynthesis decreases with depth. In aquatic systems there is a depth, the light compensation point, at which photosynthesis exactly balances respiration. Below this depth, plants do not produce as much energy as they use.

Not only does the amount of light available change, the wavelengths available to plants change with depth as well. Short-wavelength light penetrates more deeply than long wavelengths. Also, water molecules scatter blue and violet light more than other portions of the spectrum. The result is that green light penetrates most deeply. One consequence is that different species of algae are adapted to exploit the light conditions at different depths. The green algae tend to be found in shallow water; their photosynthetic pigments are similar to those of land plants that use red and blue light. The red algae, which are found in deeper water, have pigments that convert green light to photosynthetic energy.

What Abiotic Factors and Processes Are Important in Freshwater Systems?

Freshwater systems are divided into flowing water, or **lotic systems**, and ponds and lakes, known as **lentic systems**. The rate and pattern of flow is perhaps the most important physical characteristic of rivers and streams. Flow is determined by the gradient, the angle of the underlying topography. Figure 5.2 shows the course of a river system from its origin at high elevation to its entry into a large lake or the ocean. Each branch of the system is assigned a rank, or **stream order**,

lotic systems Aquatic ecosystems composed of flowing water.

lentic systems Aquatic ecosystems composed of ponds and lakes.

stream order The hierarchical position of a stream in a lotic system based on the number of tributaries that form a stream or river.

depending on its position in the system. A first-order stream is one that has no other tributary to it. Thus, a high mountain stream beginning in a snowfield is a first-order stream. Two first-order streams that join form a second-order stream. The order increases as the water continues downhill and gathers new tributaries. Geographers define a **river** as a stream of at least 6th order. The Mississippi River is a 10th-order stream; the Amazon is a 12th-order stream. As the order increases, the gradient decreases and the volume increases.

Streams join to form larger stream.

river A stream of at least sixth order.

Figure 5.2 Streams. Small streams merge to form larger streams and rivers. **Analyze: What is the stream order of each stream in this figure?**

Photosynthetic production in streams is the result of a combination of rooted plants, phytoplankton in the water column, and attached algae known as **periphyton** or **aufwuchs**. The latter comprises attached diatoms, cyanobacteria, and green algae. The relative importance of each varies with the physical characteristics of the system. Phytoplankton is more important in high-order streams; periphyton is generally more important in the faster water of low-order streams.

periphyton (aufwuchs) Algae attached to the substrate in a lotic system.

Although we distinguish between terrestrial and aquatic ecosystems, they are connected in important ways. **Autochthonous sources** are the sources of energy derived from photosynthesis within the aquatic system. Organic materials that enter the water from terrestrial systems, known as **allochthonous sources**, may be important energy sources as well. Runoff from terrestrial systems carries materials into rivers and ultimately lakes or the ocean. Nutrients and organic material move from aquatic systems to land when rivers flood. Many of the world's agricultural areas derive a significant proportion of their fertility from past or continuing flooding. Biological processes also move energy and nutrients from aquatic to terrestrial systems. For example, salmon that return to river systems to spawn and die constitute an important input of materials to the surrounding terrestrial ecosystem. As the migrating and dying salmon are consumed by gulls, eagles, bears, and ravens, significant quantities of organic material and nutrients enter adjacent forest ecosystems (Quinn et al., 2009).

autochthonous sources Sources of organic energy derived from photosynthesis within the aquatic system.

allochthonous sources Organic energy that enters the water from terrestrial systems.

The ecological changes that occur from first-order to high-order streams are known as the **river continuum** (Figure 5.3). Near the headwaters, allochthonous sources provide most of the energy in the system. Coarse particulate organic matter (CPOM), such as leaves, that enters the stream from the terrestrial environment is the primary energy source for bacteria and fungi. As CPOM breaks down, fine particulate organic matter (FPOM) becomes available. Different groups of stream invertebrates feed on CPOM or FPOM. Where the stream widens and the gradient decreases, the main energy sources are FPOM from upstream and autochthonous production by algae, especially periphyton. FPOM continues to be important in rivers, but phytoplankton contribute organic matter where the water is slower and deeper. Each part of the continuum comprises a specific set of invertebrates adapted to shred CPOM, collect FPOM, or graze on algae. The fish species change with the temperature and oxygen concentration as well as their prey items.

river continuum Ecological and hydrological changes such as energy sources, gradient, and flow that occur along the course of a stream.

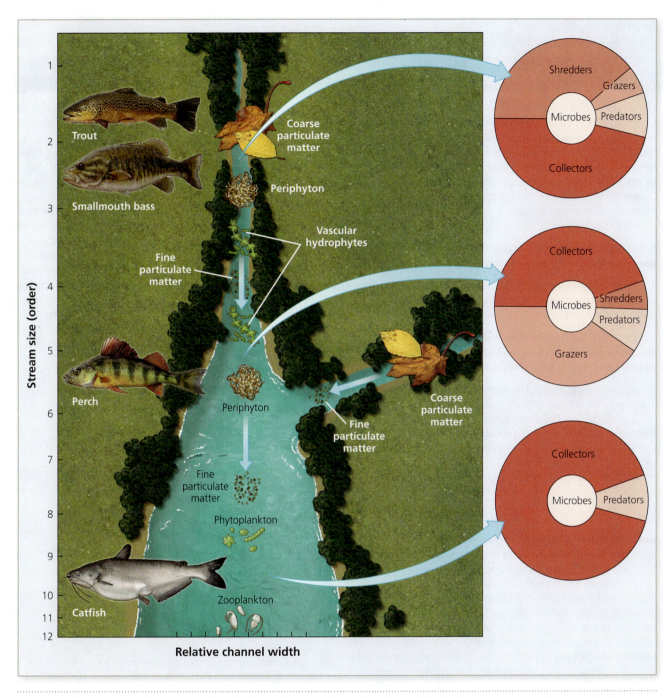

Figure 5.3 River continuum. Important ecological changes occur over the river continuum. Near the headwaters most of the energy in the stream comes from external sources of organic matter. This material supports communities of microbes and invertebrates that consume particulate matter. As the river widens and deepens, more of the energy input comes from plants and phytoplankton in the river. **Analyze: What factors do you think affect the oxygen concentration along the continuum?**

Two key chemical components of a river change with its order. First, small, high-gradient streams usually have a higher oxygen concentration. This occurs in part because in these shallow streams the surface area is large relative to the volume so a larger proportion of the stream is in contact with the air. Also, lower-order streams tend to be cooler and thus can hold more dissolved oxygen than warmer waters. Second, lower-order streams generally have lower nutrient content than higher-order streams that accumulate nutrients from a wide area. Runoff from the land is the major source of nutrients in flowing water. A mountain stream may flow through high, rocky terrestrial systems where soil is young

ON THE FRONTLINE

The Relationship Between the Environment and Dispersal in Aquatic Plant Communities

The interaction of dispersal and spatial variation in the physical environment has been a recurring theme in our analysis of both terrestrial and aquatic communities. The two processes are closely linked: the environment sorts species by their ability to tolerate the abiotic conditions they face; dispersal allows species to reach the places where conditions are suitable for them. Capers et al. (2009) examined the relationship between the environment and dispersal in aquatic plant communities in lakes in the northeastern United States.

HYPOTHESIS: Abiotic factors play a greater role in community composition when dispersal ability is high.

They reasoned that when dispersal ability makes all habitats available for colonization, abiotic factors then determine which species are present and which are absent.

PREDICTION: The proportion of the variation in aquatic community composition determined by abiotic conditions should be positively related to species' dispersal ability.

In other words, the distribution of mobile species should be explained primarily by environmental variables.

The study was based on surveys of the floating and submerged vegetation in 98 ponds and lakes across a 12,500 km^2 portion of Connecticut. The vegetation was sampled along transects across the bodies of water. They measured the similarity of communities using the Jaccard index of community similarity:

$$J_{SI} = \frac{N_{ij}}{N_i + N_j + N_{ij}}$$

where N_{ij} is the number of species in common between two lakes, N_i is the number of species found in lake i but not lake j, and N_j is the number of species found in lake j but not lake i.

A series of environmental variables was measured in each lake, including maximum depth, water clarity, water temperature, dissolved O_2, pH, alkalinity, total phosphorus, and conductivity. *Alkalinity* is the ability to neutralize acids due to the bases in solution. Conductivity is a measure of the ability of a solution to conduct electricity. It reflects the ion concentration of the solution. The lake surveys generated a presence/absence matrix of species for each lake. The researchers then statistically estimated the proportion of the variation in the community composition

that is associated with each of the abiotic variables. The role of dispersal was measured by analyzing the spatial distribution of the plants in relation to traits expected to increase their dispersal, such as sexual reproduction, production of many reproductive structures, production of vegetative fragments, and propagules that can be transported by birds or other vectors.

The abiotic factors they measured explained some of the distribution of species among lakes and ponds. However, dispersal turned out to be important as well. Specifically, 45 percent of the variation in plant composition among lakes was explained by the environmental variables, whereas dispersal accounted for 40 percent. There is also a clear effect of distance between lakes on community similarity. This is consistent with the importance of dispersal as a determinant of the aquatic plant community. Moreover, those species with functional traits that increase their dispersal ability were found across a wider spatial range. Thus, the distribution of these species is explained by physical factors in combination with dispersal ability.

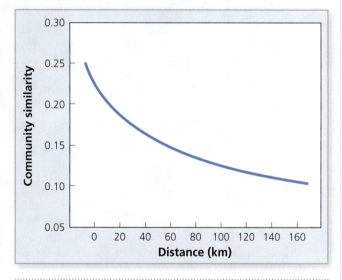

Figure 1 Community similarity. Community similarity of lakes as a function of distance from one another (from Capers et al., 2009). **Analyze: Do you think that there is a distance at which similarity no longer declines?**

■ **epilimnion** The warmer, well-mixed upper waters of a lake that lie above the thermocline.

■ **hypolimnion** The denser, colder water in a lake that lies below the thermocline.

■ **thermocline** A layer of a lake where the temperature and density of the water change rapidly.

and thin. They receive fewer nutrients than larger-order streams in lowlands with older, more nutrient-rich soils.

The chemistry and ecology of both lentic and lotic freshwater systems are dependent on dissolved gases, especially oxygen and carbon dioxide. Carbon dioxide enters aquatic systems at the water surface. CO_2 reacts with water to form carbonic acid:

$$CO_2 + H_2O \leftrightarrow H_2CO_3$$

The carbonic acid dissociates to form bicarbonate (HCO_3^-) and carbonate (CO_2^{-2}) ions, depending on the pH of the water.

$$H_2CO_3 \leftrightarrow HCO_3^- + H^+$$

$$HCO_3^- \leftrightarrow H^+ + CO_3^{-2}$$

Bicarbonate is ecologically important because it is the principal source of carbon for photosynthesis in aquatic ecosystems. Most of the alkalinity in lakes and rivers is from carbonate (CO_3^{-2}) and bicarbonate (HCO_3^-) ions. The addition of acid (hydrogen ions) shifts the second and third reactions above to the left. These processes determine the pH of the water by determining the degree of dissociation of H_2CO_3.

The physical characteristics of lakes are determined largely by size and depth. Perhaps the most important physical characteristic of deep lakes is the temperature gradient from the surface to the bottom. This gradient drives the distribution and abundance of oxygen and nutrients in a lake. The density of water is a function of its temperature—warm water is generally less dense than cold water. Moreover, waters of different density do not readily mix. Surface water warmed by the sun lies above deeper, colder water. If the gradient is sharp enough, two distinct layers develop: the **epilimnion** lies above the cold dense water of the **hypolimnion**. The **thermocline** is the boundary layer between the epilimnion and the hypolimnion where the temperature (and water density) changes rapidly.

During the summer months this stratification is stable (Figure 5.4). The functional significance of summer stratification is that oxygen and nutrients do not move readily between the two layers. Oxygen that dissolves at the surface remains in the epilimnion. Oxygen is consumed in the hypolimnion by the decomposition of organic material that sinks from the surface waters. Because this oxygen is not replaced, the oxygen concentration of the hypolimnion decreases and may lead to a condition known as **hypoxia**. The downward movement of organic material may also leave the epilimnion nutrient-poor. When the air temperature drops in the fall, the epilimnion cools, and eventually its temperature matches the hypolimnion. At this point the density of the lake water is uniform and mixing. This is known as the **fall turnover**. Nutrients and oxygen are mixed throughout the lake.

Water has an interesting and ecologically important physical property: its density increases as its temperature decreases but only until the water reaches 4°C. Below this temperature its density decreases. When the ice thaws in the spring and the surface water approaches 4°C, this denser water sinks and the waters of the lake mix again. This process, aided by spring winds, is known as the **spring turnover**.

What Abiotic Factors and Processes Are Important in Marine Systems?

Oceanographers define marine zones on the basis of depth, bottom profile, and the kinds of species they support (Figure 5.5). The distribution of these organisms reflects the variation in the abiotic factors with depth and proximity to the shore. Near-shore communities have diverse, conspicuous communities. Although the open ocean at great distance from shore appears superficially barren, staggering numbers of microorganisms are found in this water—a milliliter of ocean water contains approximately one billion organisms, many of them viruses. Even the aphotic zone, where no light penetrates, has a rich and diverse fauna thanks to the rain of dead organisms and organic matter from the photic zone above.

Salinity is the primary distinction between marine and freshwater systems. Sodium chloride is the primary salt in the ocean, but many other ions, including magnesium, sulfur, calcium, and potassium, are present as well. A portion of the

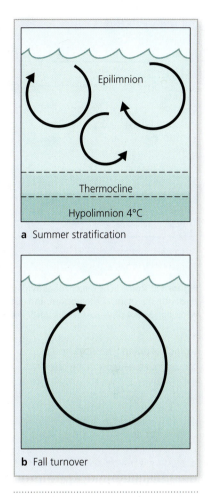

a Summer stratification

b Fall turnover

Figure 5.4 Summer stratification and fall turnover. Summer stratification and fall turnover in a lake deep enough to have a thermocline. In summer, circulation occurs in the epilimnion but not between the epilimnion and hypolimnion. In the fall, the thermocline breaks down and the entire lake mixes. **Analyze: What unique properties of water lead to these processes?**

THINKING ABOUT ECOLOGY:

The change in the density of water near the freezing point is an important property. How would the ecology of a lake change if the density of water did not increase near the freezing point?

hypoxia Low oxygen concentration.

fall turnover The mixing of the hypolimnion and epilimnion in the fall.

spring turnover Mixing of lake waters in the spring due to wind and warming lake temperature.

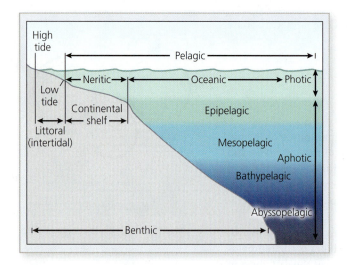

Figure 5.5 Zones. The major zones in marine systems are determined by depth, the shape of the bottom, and distance from shore. **Analyze: What important physical factors differ among these zones?**

■ **biological pump** The movement of carbon from the surface of the ocean to deeper waters due to living organisms and physical processes.

ocean's salinity was derived initially from runoff from terrestrial systems. Today ocean salinity is in dynamic equilibrium because salt is gained and lost via physical and chemical exchange with the underlying parent materials.

Salinity typically ranges between 3.1 percent and 3.8 percent in the open ocean. However, salinity is lower where large amounts of freshwater enter the ocean from rivers or melting glaciers. In places like the Red Sea, high temperature and evaporation in combination with minimal freshwater input lead to much higher salinity.

The input of carbon from the atmosphere and the chemistry of carbonic acid are similar to that of freshwater systems. Because of their huge volume and surface area, the oceans are a major sink for carbon. The carbon that enters the ocean moves from the surface to deep water by a phenomenon known as the **biological pump** (Figure 5.6). Carbon enters the ocean at the water-air interface. This carbon enters photosynthetic organisms and eventually the organisms that consume them in the euphotic zone. As these organisms die and sink, their carbon moves downward. In some places local currents also move dissolved carbon to deeper water. The result is that carbon is effectively pumped from the atmosphere and distributed throughout the ocean.

As atmospheric carbon dioxide levels rise, more CO_2 enters the ocean. Because the biological pump moves this carbon into deeper waters, the oceans potentially help mitigate the effect of increased atmospheric CO_2 and global warming. Although ecologists are still quantifying the rates of these processes, it seems unlikely that the oceans are capable of pumping enough CO_2 to the deep ocean to prevent climate change. There are negative impacts of the increased carbon content of the ocean as well. The amount of carbonic acid is increasing and thus the pH of marine systems is dropping. The acidification of the ocean represents a significant threat to many species of invertebrates whose bodies are composed of calcium carbonate. As the pH decreases, these species are at risk of reduced growth rates or dissolution (Raven et al., 2005).

The great depth of the oceans introduces a novel physical factor—pressure. In the deepest ocean trenches, the pressure exceeds 1,000 atmospheres. Some ocean vertebrates are relegated to the upper part of the water column by their limited ability to withstand the pressure at great depths. Many deep-water fish are unable to survive at the lower pressures found near the surface. For example, coelacanths, fish of the deep ocean, rupture if brought to the surface in nets. A few vertebrates, such as whales and seals, can withstand the high pressure of the deep ocean yet return to the surface to breathe.

As in terrestrial systems, ocean temperature varies with latitude. Polar waters may be below 0°C, whereas tropical seas often reach 30°C. However, ocean currents modify this latitudinal gradient by moving water great distances to or from the equator. A number of factors generate ocean currents. The rotation of the Earth generates

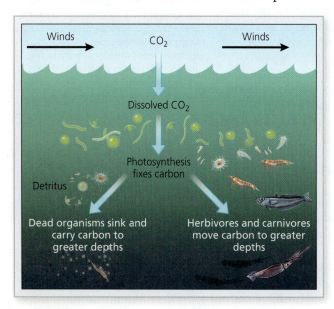

Figure 5.6 Biological pump. The biological pump moves carbon from the surface to deeper waters. The pump is initiated by the physical dissolution of CO_2 in water. Biological processes then move carbon to the depths. **Analyze: What role does the penetration of light play in this process?**

DO THE MATH

Linear Regression

Many physical and biological features of marine systems change with depth or latitude. For example, ocean temperature declines with latitude and light decreases with depth. Salinity increases with distance from the mouth of a river where it enters the ocean. In Chapter 2 we used measures of central tendency such as the mean or median to describe a data set. Here, our interest is not the average temperature, light, or salinity. Rather, it is the change in those variables with others such as depth or latitude. How can we quantify the relationship between two variables?

Imagine that you measure the ocean surface temperature at several points from south to north along a single longitudinal line. You obtain the data shown in Figure 1. The relationship between temperature and latitude appears to be linear. How can we quantify this relationship? One method used in ecology is *linear regression*. Linear regression determines the parameters of the line that best fit the data. These parameters, the slope and y-intercept, quantify the nature of the relationship between latitude and temperature and can be used to compare data sets.

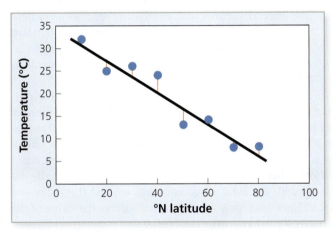

Figure 1 **Latitude and temperature.** A hypothetical relationship between latitude and temperature. **Analyze: Why do most points not fall exactly on the line?**

But how do we determine the line that best fits these data? The standard process is called *least squares regression*, the procedure used by most calculators and computer regression programs. The computations for least squares regression are tedious but the concept underlying them is straightforward. We define the "best line" through the data points as that line which lies closest to each of the data points. The computational procedure generates a line for which the sum of the squared distances between each point and the line is smallest—hence *least squares* regression. The procedure uses the squared distance between each point and the line because some points lie below the line; others lie above. If we simply used the absolute summed distances, the negative distances would cancel the positive ones and we could not be sure that the total distance from all points to the line is minimal.

We might also want to ask, "How strong is the association between temperature and latitude?" We measure this with the *correlation coefficient*. The most common correlation measure is the Pearson product-moment correlation, measured with the variable r. The value of the correlation coefficient is calculated as

$$r = \frac{\sum_{i=1}^{n}(X_i - \bar{X})(Y_i - \bar{Y})}{\sqrt{\sum_{i=1}^{n}(X_i - \bar{X})^2}\sqrt{\sum_{i=1}^{n}(Y_i - \bar{Y})^2}}$$

where x_1 is the *ith* value of one of the variables, y_i is the *ith* value of the other, and $\bar{X}$ and $\bar{Y}$ are the means of the two groups. The value of r ranges from -1.0 to $+1.0$. Negative values indicate that an increase in one variable is associated with a decrease in the other. Positive values mean that as one variable increases, the other does as well. A correlation coefficient of 0 means that the two variables are not associated; each changes randomly with respect to the other.

We can evaluate the strength of the correlation with the value r^2. The square of the correlation coefficient is a measure of the proportion of the variation in x that is explained by variation in y. The correlation coefficient between temperature and latitude in Figure 1 is -0.96. (Its square is 0.92.) Thus, 92 percent of the variation in temperature is explained by latitude. If the value of r^2 is low, other factors are probably important.

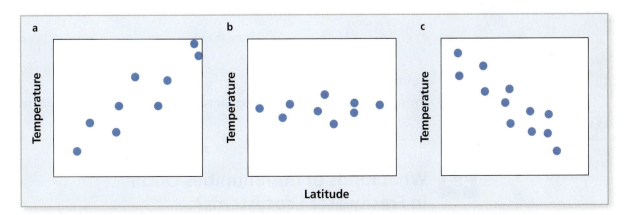

Figure 2 **Correlation.** Three possible correlations among temperature and latitude: (a) positive, (b) zero, and (c) negative. **Analyze: If temperature and latitude are negatively correlated, does this necessarily mean that latitude causes the temperature change?**

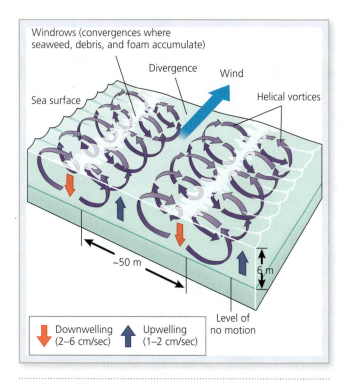

Figure 5.7 **Langmuir cell.** An ocean Langmuir cell is composed of alternating patterns of upwelling that occur parallel to the movement of wind across the surface. **Analyze: What forces might move water in these cells?**

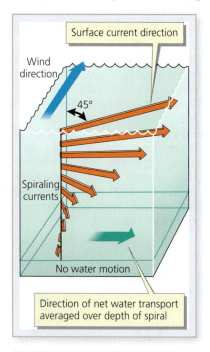

Figure 5.8 **Ekman spiral.** An Ekman spiral is the deflection of currents relative to surface winds. **Analyze: Why do the angles of deflection change with depth?**

Langmuir cell A pattern of circular water motion near the surface of bodies of water.

the Coriolis forces that initiate currents. Temperature and salinity gradients also produce currents. Warm, highly saline water is less dense than colder, less saline water. As waters of different density mix, the upwelling of warm saline water forms vertical currents. This leads to horizontal flow as water moves in to replace the water displaced near the bottom.

How Does Water Move in Marine Systems?

Superimposed on the broad patterns of ocean currents are other, local movements. **Langmuir cells** occur where winds generate circular rotation of the water just below the surface (Figure 5.7). Adjacent cells flow in opposite directions, which generates areas of convergence and divergence. Rows of accumulated seaweed and other materials are visible in the convergences of Langmuir cells. Wind also combines with the Coriolis forces to generate patterns of water flow known as **Ekman spirals**. When wind moves surface water, the Coriolis forces deflect the current to the right in the Northern Hemisphere and to the left in the Southern Hemisphere. Friction transfers the surface flow to deeper water, but the effect dissipates with depth. The result is a gradient of direction and current from the surface downward (Figure 5.8). Both these processes are important in moving and mixing nutrients, oxygen, and carbon dioxide from one region to another and by depth.

Tidal flow is a critically important factor in many near-shore marine environments. The gravitational attraction of the moon causes the surface water directly aligned with the moon to bulge upward. This bulge draws water from other regions and causes a high tide. Low tide occurs where the water has been pulled away. The tides change across the entire ocean, but their effect is most significant in shallow water near shore, where the tides have a much greater rise and fall relative to the bottom. In general, the tidal range is higher toward the poles and smaller near the equator. However, topographic features such as the shape of the bottom or a bay with a narrow inlet affect the tidal range.

KEY CONCEPTS 5.2

■ Many important physical factors, such as light, temperature, and oxygen, change with depth and latitude. Any of these factors that lie outside species' tolerance may be limiting.

■ The spatial patterns of the physical factors in aquatic systems are shaped by the movement of water.

QUESTION:

Which physical factors are most directly affected by depth? Which are least affected?

5.3 What Kinds of Communities Occur in Freshwater Ecosystems?

As we have seen, lentic and lotic systems differ in important physical characteristics. These differences also determine the differences in the ecological communities they support. **Wetlands** are semiaquatic habitats that are perpetually or periodically flooded. The depth of the water and the timing of inundation and desiccation, or **hydroperiod**, determine the dominant plant species. We identify wetland communities on the basis of the dominant vegetation type. **Marshes** are wetlands in which the dominant plants are herbaceous (Figure 5.9). They occur in depressions that accumulate and hold water, such as shallow basins along the edges of lakes and rivers. Prairie potholes occupy depressions gouged by the Pleistocene glaciers in the Great Plains of North America. These wetlands fill up and dry down with the cycles of drought and precipitation common in this region. **Swamps** are wetlands dominated by woody plants. Swamps occur in any deciduous forest system in which depressions hold water, but they are most common across the southern and southeastern states. Trees such as bald cypress (*Taxodium distichum*) and red maple are adapted to the saturated soils of swamps.

Bogs are wetland communities that typically occur in northern regions where glaciers created depressions with no outlet. These depressions, known as **kettle holes**, initially have very low nutrient content. As organic matter and nutrients slowly accumulate, sphagnum moss colonizes the bog. This species is adapted to low nutrient systems; it extracts nutrients with a cation exchange system. In the course of this exchange, hydrogen ions are released into the water, which gives the bog its characteristic low pH. The sphagnum gradually grows toward the center of the open water, eventually creating a floating mat on the surface. At the margins, organic matter and sediment accumulate, which allows new plants to colonize. Over time a bog develops a characteristic pattern of concentric rings of different types of vegetation: a floating sphagnum mat, rooted herbaceous species such as cotton grass (*Eriophorum polystachion*) and sedges (*Carex spp.*), and finally woody plants like tamaracks (*Larix larcina*).

Lakes and ponds contain zones defined by depth and light penetration (Figure 5.10). The **littoral zone** occurs in shallow water. The open lake in deep water is the **pelagic zone**. This region is further divided into two zones on the basis of light penetration: the **limnetic zone** occurs above the light compensation point and the **profundal zone** below. The **benthic zone** is the deepest water and substrate at the bottom of the lake.

Groups of higher plants occupy specific depths. **Emergents**, species that are rooted but protrude above the water surface, such as cattails (*Typha spp.*), occur near shore or in shallow water. **Floaters**, such as duck weed (*Lemna spp.*), are not rooted but float on the surface. In shallow water where light penetrates to the bottom, rooted **submergents** such as *Elodea* occur.

The surrounding landscape is an important determinant of lake ecology. Lakes accumulate material from adjacent terrestrial systems, including the regions that drain into the rivers that enter the lake. If both the nutrient content of the soils and runoff are high, nutrients accumulate in the lake, where they stimulate high rates of photosynthesis and plant production. These are known as **eutrophic** lakes. Most occur in basins with old, well-developed soils with high concentrations of organic material and nutrients. When humans augment natural sources of nutrients, the process of eutrophication may accelerate to the point that the lake becomes **hypertrophic**. Hypertrophic lakes are characterized by massive algal blooms, low oxygen concentration, and severely altered plant and animal communities. Low-nutrient, low-production lakes, such as many alpine lakes surrounded by barren granite, are called **oligotrophic**.

Although the vast majority of lakes are freshwater, there are a few saline lentic systems. The Great Salt Lake in Utah and Mono Lake in California exemplify these unusual lakes. Both are **terminal lakes**, basins that have no outlet. Because they are located in arid regions, evaporation concentrates salts and other substances. Great Salt Lake is the remnant of Lake Bonneville, a Pleistocene lake that was at least 10 times larger. As the Pleistocene waned,

Figure 5.9 Marshes. Marshes are characterized by fluctuating water levels and herbaceous vegetation.

Ekman spiral A pattern of water movement in which wind and the Coriolis force cause water to deflect differentially with depth.

wetland A semiaquatic habitat that may permanently or periodically contain water.

hydroperiod The timing and length of periods of the presence and absence of water in a wetland.

THINKING ABOUT ECOLOGY:

Chapter 4 outlines the basic types of terrestrial communities. What are the important similiarities and differences in the way we describe and classify aquatic versus terrestrial communities?

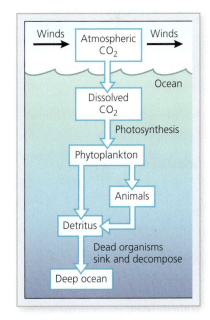

Figure 5.10 Zones. The major zones of a lake or pond are determined by light penetration and depth. **Analyze: Which of these zones would you expect to be most variable in size and extent?**

THE HUMAN IMPACT

Survival of the Everglades

The Everglades is a unique wetland community—no system like it exists anywhere in the world. This system once occupied most of southern Florida from Lake Okeechobee to Florida Bay. Today, only the southernmost portions of the Everglades remain, protected in Everglades National Park. This community is characterized by extensive areas of marsh dominated by sawgrass (*Cladium jamaicense*), a member of the sedge family. Islands of subtropical hardwood trees known as hammocks are scattered across the marsh. The area is known for the numbers and diversity of aquatic and wading birds, such as great egrets, roseate spoonbills, and white ibis. It is also home to the endemic Everglades kite, now an endangered species. The system is so unusual and rich in wildlife that it has been designated a World Heritage Site.

The Everglades illustrate an important characteristic of wetland communities: the flora and vegetation are intimately connected to the hydrology. The hardwood hammocks occur on sites just a meter or so higher than the surrounding marsh where conditions are dry enough to support these trees. Small depressions in the limestone substrate, where the water is slightly deeper, hold bald cypress trees, which are adapted to a longer hydroperiod. Historically, water entered Lake Okeechobee from the Kissimmee River. It flowed out of the southern margin of Lake Okeechobee not in a river but as a broad flow over the southern edge of the lake. This wide front of water flowed slowly south in a process known as sheet flow. Because the land drops only a few feet from Lake Okeechobee to the ocean, water crept slowly across the landscape.

Figure 1 The Everglades. The Everglades is a sawgrass marsh with scattered hardwood hammocks.

Today the hydrology of the Everglades has been radically altered, threatening the marsh and all the animals that depend on it. To reduce the impact of floods on the growing metropolitan Miami area, drainage ditches channeled water from the Kissimmee River and the northern Everglades directly to the Atlantic Ocean to the east. The effect was to cut off sheet flow through the marsh. Additional ditches drain the northern Everglades for

swamp A wetland in which the dominant vegetation is woody.

bog A small wetland that has no outlet, typically found in northern (glaciated) regions.

kettle hole A depression in the substrate created by glaciers.

littoral zone The near-shore, shallow region of a lake.

pelagic zone The open, deep water of a lake distant from the shore.

limnetic zone The region of a lake above the compensation point.

profundal zone The region of a lake below the compensation point.

marsh A wetland in which the dominant vegetation is herbaceous.

the climate of the Southwest became hotter and drier. The salts of ancient Lake Bonneville were concentrated in a decreasing volume of water. Consequently, the Great Salt Lake is almost 10 times as salty as the ocean. Their extreme salinity limits the number of species that can inhabit these lakes. The Great Salt Lake and Mono Lake are dominated by a few species of algae and astronomical numbers of brine shrimp and brine flies. These invertebrates provide critical nutrition for migrating waterfowl, gulls, and shorebirds. During migration the Great Salt Lake holds the greatest concentration of Wilson's phalarope in the world.

KEY CONCEPTS 5.3

- The fundamental dichotomy in freshwater systems is whether the water is flowing or standing.
- Lentic and lotic systems differ in the relative importance of their main photosynthetic species. Phytoplankton is more important in deeper, slower water; periphyton is more important in flowing water.
- Wetlands occur at the boundary between the aquatic and terrestrial environment. The specific type of wetland is determined primarily by the depth and hydroperiod.

QUESTION:

How are freshwater and terrestrial systems ecologically connected?

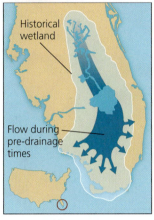

a Historical flow regime **b** Current flow regime

Figure 2 Water flow in the Everglades. Water flows in a broad sheet across the Everglades. This pattern has been significantly altered by a system of canals as well as roads, levees, and agricultural systems.

sugarcane plantations. Huge quantities of nitrogen fertilizer are applied to the cane fields, much of which makes its way into the remaining sheet flow and the sawgrass marsh. The accumulation of nutrients, and the excessive production by phytoplankton and rooted plants it stimulates, is known as *eutrophication*. Because so little water now reaches the southern margin of the marsh,

saltwater penetrates inland. Consequently, the native sawgrass is being squeezed by the invasion of other species from both the south and the north. Cattails, which tend to be nitrogen limited, are now expanding rapidly in the nitrogen-rich water in the north. Near Florida Bay, mangroves and other salt-tolerant species are moving inland just as rapidly.

The Comprehensive Everglades Restoration Plan (CERP) is designed to remedy many of these problems. Restoration of the historical hydrology is a central feature of the project. Engineers are re-creating the sinuous channel of the Kissimmee River, which will slow the flow of water into the remaining canals and Lake Okeechobee. More of the water leaving the lake will be allowed to reach the southern Everglades. Many of the sugar plantations will be purchased and retired. This will have two key effects. It will reduce the input of nitrogen and the rate of eutrophication. It will also provide additional marshland that can absorb excess water during floods, reducing the need for the drainage ditches that rob the Everglades of water. CERP is one of the largest restoration projects ever undertaken. If successful, it will restore one of the most unusual wetland systems in the world.

QUESTION:

The Comprehensive Everglades Restoration Plan emphasizes restoration of the abiotic environment. Explain why this is a logical priority.

5.4

What Kinds of Communities Occur in Marine Ecosystems?

The marine environment is composed of regions based on similar climate and current known as **oceanic domains** (Figure 5.11). Each domain consists of a series of divisions characterized by physical factors, especially currents and prevailing winds. Unlike terrestrial systems and most freshwater systems, there are no barriers to dispersal in the open ocean. Most pelagic species can theoretically move from one domain to another. Some, such as whales, move great distances over the course of their lifetime or in seasonal migrations. Others are geographically restricted by the environment or their dispersal ability.

Marine systems differ from terrestrial systems in that the ocean domains are not characterized by a distinct form or type of vegetation. Rather, in the open ocean, phytoplankton is responsible for most photosynthesis. Consequently, we categorize open-ocean communities by their physical and chemical attributes and the amount of photosynthetic production they support. Figure 5.12 shows the marine analogs of terrestrial biomes categorized on this basis. Each system has a unique combination of temperature, salinity, density, amount of photosynthetically active radiation, phosphate concentration, oxygen content, and photosynthetic production. These regions occur in latitudinal bands corresponding to the temperature gradient from the equator to the poles. However, the high nutrient content of the polar oceans also contributes to their distinction from more

benthic zone The deepest waters and substrate of a lake.

emergents Aquatic plants that are rooted but protrude above the water.

floaters Unrooted aquatic plants that float on the lake surface.

eutrophic Describes aquatic systems in which nutrients have accumulated and stimulated high rates of plant photosynthetic production.

submergents Aquatic plants that are rooted in the bottom but do not extend above the surface.

hypertrophic Highly eutrophic aquatic systems with massive photosynthetic production.

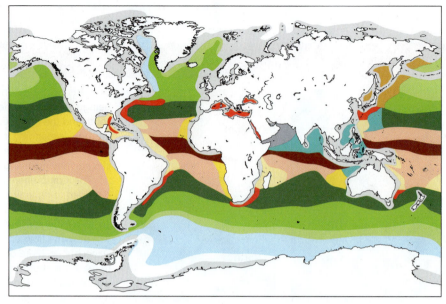

Figure 5.11 Ocean domains. This map depicts the major ocean domains. These are defined primarily by physical factors. **Analyze:** Why do you think these generally form latitudinal bands?

500 Polar domains	600 Temperate domains	700 Tropical domains
☐ 510 Inner polar division	610 Poleward westerlies division	710 Tropical monsoon division
520 Outer polar division	620 Equatorward westerlies div.	720 High salinity tropical monsoon div.
	630 Subtropical division	730 Poleward trades division
☐ S Shelf,	640 High salinity subtropical div.	740 Trades winds division
depth less than 200 m	650 Jet stream division	750 Equatorward trades division
	660 Poleward monsoon division	760 Equatorial countercurrent division

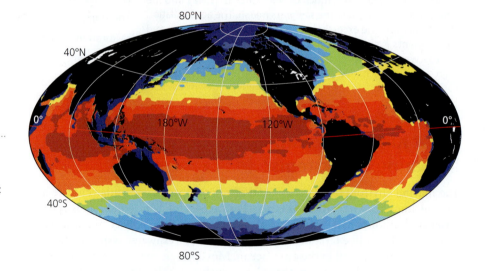

Figure 5.12 Marine biomes. Marine biomes are determined by the physical environment, including salinity, depth, and especially temperature. Colors depict ocean temperatures from warm (reds) to moderate (yellow and green) to cold (blues) (from Lewis et al., 2008). **Analyze: How do marine biomes differ from terrestrial biomes?**

▪ **oligotrophic** Describes aquatic ecosystems with low nutrient content and low rates of photosynthetic production.

▪ **terminal lake** A lake basin with no outlet.

▪ **oceanic domains** Marine regions characterized by similar climate and currents.

▪ **fringing reef** A coral reef that projects directly from shore and develops on rock.

southern zones. In both the Arctic and Antarctic, ocean current and the topography of the bottom cause an upwelling of nutrients, which stimulates photosynthesis. In contrast, much of the open ocean near the equator is nutrient limited and less productive.

In general, the most productive and diverse marine communities occur in shallow water near shore. Coral reefs are among the most spectacularly diverse marine communities—the aquatic rival of tropical rainforests. Reefs occur in subtropical regions where the water temperature does not fall below 18°C. Three basic types of reef occur, each associated with a specific geographic and topographic situation. **Fringing reefs** develop on rock substrate near islands and continents and project directly from the shore. The extensive coral reefs of the Florida Keys are fringing reefs. **Barrier reefs**, such as Australia's Great Barrier

Reef, develop parallel to the shore and form a lagoon between the reef and shore. Finally, **atolls** are island reefs, surrounded by deep open ocean. They develop as corals and sediments accumulate, gradually reaching the ocean surface. A typical atoll is a circular reef surrounding a shallow interior lagoon. Some formed as volcanoes rose or when the sea level dropped.

Colonial corals are the focal point of the reef. Large aggregations of corals build a living substrate that provides habitat for hundreds of other species. The corals contain symbiotic protozoans known as zooxanthellae. These photosynthetic organisms are extremely productive—they provide the coral with carbohydrate in return for protection and nutrients. The coral substrate and productive zooxanthellae form the base of a community comprising hundreds of herbivorous and predatory species of fishes and invertebrates (Figure 5.13).

Figure 5.13 Coral reefs. Coral reefs are remarkably productive and diverse.

A number of important marine communities occur at the boundary between marine and terrestrial habitats. One of the most diverse near-shore systems is the **rocky intertidal**. Such communities occur in many parts of the world, although they are most highly developed at high latitude where the tidal range is large. The rock substrate supports a diverse assemblage of sessile invertebrates, which in turn support mobile invertebrates and fish (Figure 5.14). The large tidal range and sloping rocky shore result in a gradient of exposure over the daily tidal cycle. The upper intertidal is exposed more often or longer than the lower. Consequently, there are zones of sessile species that differ in their tolerance of desiccation. Small indentations in the rocks hold water even at low tide. These **tide pools** provide a refuge for species that cannot tolerate desiccation.

Mangrove swamps occur along relatively flat tropical and subtropical shorelines with a soft bottom. Tides carry saltwater some distance inland, saturating the soil and depositing salt. This is a challenging physical environment for most plants. Consequently, the dominant vegetation consists of just a few species. Discrete bands, each dominated by a different species of mangrove, form along a gradient of soil saturation and salinity. In most New World mangrove swamps, the red mangrove (*Rhizophora mangle*), the species most tolerant of salt and saturated soil, occurs at the outer shoreline. Further inland it is replaced by the black mangrove (*Acicennia gerinans*), then the white mangrove (*Laguncularia racemosa*), and finally buttonwood (*Conocarpus erectus*).

The red mangrove pioneers the formation of a mangrove swamp. This species produces salt-tolerant seeds that have the unusual property of germinating while still attached to the tree. A small root forms before the seed is released. When the seed falls, it is carried by the currents, sometimes for great distances. If it reaches shallow water, it sticks in the mud, where it grows rapidly. The young mangrove produces many arching prop roots that anchor the plant and propagate it vegetatively. These roots slow the tidal flow, which allows sediment to accumulate, eventually forming a stable mud flat that other mangroves colonize.

Mangrove swamps provide important benefits and services to humans. The maze of prop roots and sediments is important breeding habitat for many economically important marine species such as shrimp and mangrove snappers. The swamp serves as a barrier to wave action and especially the storm surge of hurricanes, thus reducing the impact of these storms on coastal communities. These benefits are lost as coastal development intensifies and mangroves are destroyed to create waterfront property.

Estuaries occur where a river enters the ocean. An estuary is a partially enclosed body of water connected directly to the ocean. The key physical characteristic of this community is the salinity gradient that occurs where freshwater meets

barrier reef A coral reef that develops parallel to the shore and forms a lagoon between the reef and shore.

atoll A circular coral reef surrounded by open ocean.

rocky intertidal A community that develops on rocky shorelines between the high and low tide lines.

tide pool A depression in the rocky intertidal that holds water at low tide.

Figure 5.14 Rocky intertidal. The rocky intertidal occurs on steep shores with a large tidal range. Each zone in the intertidal experiences different physical conditions and supports a different set of organisms.

estuary A partially enclosed body of water connected directly to the ocean that occurs near the mouth of a river.

THE EVOLUTION CONNECTION

Not All Characteristics Are Adaptive

The origin of life occurred in water. Organisms remained exclusively aquatic for more than a billion years. The first colonists of land, perhaps 1.2 billion years ago, were microbes that splashed up on land and formed a mat near water. Plants and animals colonized terrestrial systems more than 400 million years ago. These first colonists faced a number of adaptive challenges, especially supporting their own weight and avoiding desiccation ("The Evolution Connection," Chapter 4).

The abundance and diversity of life on land today affirms the success of these early colonists. However, not all land plants and animals remained in the terrestrial environment. Animals that returned to the water are *secondarily aquatic*. Perhaps the most familiar are the cetaceans, which include the whales and dolphins. The transition back to water illustrates two important aspects of evolution.

First, the anatomy of these species retains evidence of their terrestrial origin. The fossil evidence suggests that cetaceans are derived from the artiodactyls, the even-toed ungulates such as deer and cattle. Whales maintain vestiges of the anatomy of land animals; they have small residual hind limbs that serve no purpose in the water. As the locomotion of whales adapted to the aquatic environment, the hind limbs gradually atrophied—not only did they have no benefit, they might have hindered the whale's locomotion. Vestigial structures such as these, with no adaptive value, constitute important evidence of the evolutionary process.

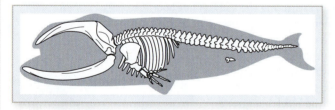

Figure 1 Vestigial structures. Whales have small, vestigial hind limbs. **Analyze: What function do the forelimbs perform in a whale?**

Second, some secondarily aquatic species illustrate the imperfections inherent in the process of evolution and adaptation. The mayflies in the order *Ephemeroptera* exemplify this phenomenon. The ancestors of this group are fully terrestrial. But the mayfly life cycle retains an aquatic stage. The nymph or naiad stage is aquatic—the creature feeds on algae and other material in the water for weeks or months, depending on the species. When it reaches a certain size and age, it rises to the water surface and transforms into a subadult known as a subimago. The subimago flies to the streamside vegetation, where it soon transforms into a sexually mature adult. The adults mate, the female lays her eggs on the water surface, and both sexes die. The adaptive imperfection is manifest in the transformation from naiad to subimago at the water's surface. This process is very inefficient, even clumsy. Not only must the subimago emerge from the nymphal case, it must dry its wings above the water before it can fly. Many subimagos become stuck in the nymphal case. Others drown when they can't break free of the surface tension of the

water. They are also extremely vulnerable to predation during this transition—trout key in on this stage of the life cycle.

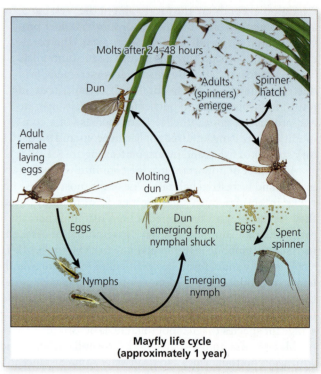

**Mayfly life cycle
(approximately 1 year)**

Figure 2 Life cycle with an aquatic stage. The life cycle of a mayfly includes both terrestrial and aquatic stages. The transition from the aquatic forms to the flying adults is awkward, and many individuals die or are eaten during emergence from the water. **Analyze: What are the advantages of an aquatic stage?**

This inefficient and high-risk transformation illustrates the imperfection inherent in the evolutionary process. If evolution is the result of random mutations and selection among the resulting phenotypes, we should expect that some adaptations will be superior to others. We marvel at the beautiful match between many organisms and their environment. But the mayflies (and many other examples) remind us that evolution does not always produce optimally adaptive features.

Mayflies are abundant and diverse—the clumsy transition from the aquatic to the terrestrial environment has obviously not doomed them to extinction. If a significant proportion of individuals die in transition, one adaptive solution is to produce so many that enough survive to reproduce. Mayflies channel huge amounts of energy into the production of many offspring. In the process, they compensate for their inefficient transition from the water with large numbers of offspring.

QUESTION:

Why do you suppose the vestigial pelvic girdle of whales has not been completely eliminated by natural selection?

the tidal flow of the ocean. Freshwater is less dense than saltwater, so it floats above the heavier saltwater. Consequently, a vertical salinity gradient stretches seaward until the freshwater and saltwater completely mix. The size and shape of the bay determine the pattern of mixing, the rate of evaporation, and the tidal range.

Salt marshes are estuarine communities that occur where a barrier island protects the bay from wave action. Most have a gradually sloping shoreline that allows the tide to spread across wide flats, where it drops sediments and nutrients. Because relatively few species can tolerate the high salinity, salt marshes are characterized by large numbers of a few emergent plant species such as salt grass (*Spartina spp.*) and pickleweed (*Salicornia spp*). Like mangrove swamps, salt marshes are important nurseries for economically important species. And like mangroves, they are at risk from coastal development and pollutants transported from far inland.

Estuaries are diverse and productive communities. The flow of freshwater seaward above the incoming saltwater creates a countercurrent circulation that traps nutrients in the bay, where they support high rates of photosynthesis. The wide range of habitat types, especially along the salinity gradient, results in gradients of species adapted to specific local conditions. However, this zonation is frequently disturbed because of the dynamic nature of this system. During drought, stream input may drop significantly. When this occurs, wind and storm surge move saltwater far upriver. Or flood conditions may carry a large volume of freshwater into the estuary. In either case, the disturbance to the normal salinity gradient disrupts the normal abundance and stratification of species.

THINKING ABOUT ECOLOGY:

Physical environments that vary over space or time pose special challenges for organisms (Chapter 3). Many of the communities in this section fall into this category, yet they are some of the most productive and diverse marine communities. How might you explain this apparent paradox?

KEY CONCEPTS 5.4

- The open ocean is subdivided according to physical-chemical properties that vary with distance from shore, depth, and latitude.
- The most variable marine communities occur near shore. Each is associated with a particular substrate, salinity gradient, and tidal regime.

QUESTION:

What factors make some communities, like coral reefs, so much more diverse than others?

Putting It All Together

Much of the open ocean has not been fully explored, especially the deepest basins. It is likely that new communities await discovery. Nevertheless, we know the major terrestrial and aquatic communities well enough that we understand the factors that determine their composition as well as the basic principles of their function. Chapters 4 and 5 summarize the basic information that constitutes the natural history of the communities of the biosphere.

Although there are profound differences among these communities, the composition of each is determined by common processes. The species that comprise a community are initially sorted by two factors: the prevailing physical conditions and the species' dispersal ability. This principle applies equally to the species found in a terrestrial desert, a high mountain lake, or a hydrothermal vent.

Of course, this is by no means the end of the story. Once species come together in a community, new ecological interactions among them develop. As these interactions unfold, new phenomena known as emergent properties arise. Predation, parasitism, mutualism, and competition are examples of biological interactions that arise in communities of plants and animals. We turn now to the analysis of these diverse ecological interactions.

Summary

5.1 How Is Water Distributed Across the Earth?

5.2 What Are the Important Physical Factors in Aquatic Systems?

- Aquatic systems are divided into freshwater and saltwater systems.
- Freshwater systems are composed of lentic (rivers and streams) and lotic (ponds and lakes).
- The absorption and scattering of light by water constrains the distribution and types of photosynthetic organisms.
  - Aquatic systems are composed of a number of zones determined in part by light penetration.
- The carbon used in photosynthesis enters aquatic systems from the atmosphere.
  - When CO_2 dissolves in water, it forms carbonic acid, which dissociates to form carbonate and bicarbonate.
  - In the ocean the biological pump moves carbon away from the surface waters and distributes it widely.
- Streams change in physical and biological characteristics with stream order along the river continuum.
- Deep lakes are characterized by temperature (and biological and chemical) stratification. Stratification, the distribution of materials, and the movement of water are due to the density changes in water as a function of temperature.

- Ocean salinity is a result of dynamic equilibrium with the underlying parent material.
- As in freshwater systems, light penetration defines important ecological zones in the ocean.
- Complex patterns of local and regional currents move materials vertically and horizontally in the ocean.

5.3 What Kinds of Communities Occur in Freshwater Ecosystems?

- Wetlands are important aquatic communities that occur at the interface between land and water.
  - The precise community is determined by the hydrology.
- Lotic systems differ markedly in chemistry and biology. For example, bogs and terminal lakes have vastly different origin, chemical composition, and organisms.

5.4 What Kinds of Communities Occur in Marine Ecosystems?

- Ocean communities are divided into a series of oceanic domains.
- Many of the most diverse and productive marine communities (such as coral reefs) occur in shallow water.
- Other important marine communities occur near the transition to land: for example, intertidal communities, estuaries, and mangrove swamps.

Key Terms

allochthonous sources p. 100
aquifer p. 99
atoll p. 111
aufwuchs p. 100
autochthonous sources p. 100
barrier reef p. 111
benthic zone p. 109
biological pump p. 104
bog p. 108
Ekman spiral p. 107
emergents p. 109
epilimnion p. 102
estuary p. 111
eutrophic p. 109
fall turnover p. 103
floaters p. 109
fringing reef p. 110

hydrologic cycle p. 98
hydroperiod p. 107
hypertrophic p. 109
hypolimnion p. 102
hypoxia p. 103
infiltration p. 99
kettle hole p. 108
Langmuir cell p. 106
lentic systems p. 99
light compensation point p. 99
limnetic zone p. 108
littoral zone p. 108
lotic systems p. 99
marsh p. 108
oceanic domain p. 110
oligotrophic p. 110
pelagic zone p. 108

percolates p. 99
periphyton p. 100
profundal zone p. 108
river p. 100
river continuum p. 100
rocky intertidal p. 111
sheet flow p. 99
spring turnover p. 103
stream order p. 99
submergents p. 109
swamp p. 108
terminal lake p. 110
thermocline p. 102
tide pool p. 111
transpiration p. 99
wetland p. 107

Review Questions

1. What are the major categories of aquatic systems?

2. What are the most important differences between lentic and lotic systems?

3. What physical factors depend on water flow?

4. What causes the patterns of mixing in deep lakes?

5. How do freshwater systems interact with terrestrial systems?

6. What physical factors are most important in marine systems?

7. What physical factors are important at the boundary of marine and terrestrial systems?

8. Describe the major zones of a lake.

9. How does photosynthesis differ between lentic and lotic systems?

10. Are there aquatic analogs of terrestrial biomes? Or are the systems too different for that kind of comparison?

Further Reading

Carpenter, K.E., et al. 2008. One-third of reef-building corals face elevated extinction risk from climate change and local impacts. *Science* 321:560.

This paper summarizes recent information on worldwide coral decline. The authors analyze the role of climate change in relation to other factors.

Douglas, M.S. 1947. *The Everglades: river of grass*. Fairhope, AL: Mockingbird Publishing.

This book is a classic in the history of the environmental movement. Marjorie Stoneman Douglas recognized the importance of the Everglades when few people appreciated wetlands. The book is both a description of the natural history of this system and a call for protection.

McIntyre, A.D. (ed.). 2010. *Life in the world's oceans: diversity, distribution and abundance*. Hoboken, NJ: Wiley-Blackwell.

This comprehensive book contains contributions from dozens of marine biologists. It summarizes all the major marine systems, including geographic variation within them. It also includes information on the ecology of the important animal species in each system.

Dodds, W., and M. Whiles. 2010. *Freshwater ecology: concepts and environmental applications of limnology*. Waltham, MA: Academic Press.

This book is a comprehensive summary of the ecology of freshwater systems, including the physical properties of water, the ecology of aquatic communities, and the key anthropogenic threats to these systems.

Chapter 6

Behavioral Ecology

In an alpine meadow, high in the Sierra Nevada, a female Belding's ground squirrel (*Urocitellus beldingi*) whistles an alarm call as a coyote sniffs along the edge of the meadow. All around her, squirrels dive for burrows at the sound, and a few moments later she enters her burrow as well. The coyote moves on through the meadow and out of sight. Gradually the ground squirrels emerge from their burrows and continue foraging.

At first glance, this seems a simple form of antipredator behavior. A vigilant individual detected a predator and warned the other members of the colony. The warning whistle elicited an immediate response in the nearby squirrels. However, research on this behavior reveals a remarkable aspect of the scenario: those individuals that give alarm calls are more likely to be preyed upon than the individuals they warn. The warning whistle calls the predator's attention to the caller, increasing its chances of being the one eaten. Thus, this system raises an interesting evolutionary question. If alarm-calling behavior has a genetic basis and callers are more likely to be preyed upon, why isn't it gradually eliminated from the population?

Belding's ground squirrels are interesting in other ways as well. They live in groups—colonies of individuals. Their preferred habitat ranges from high alpine meadows near or above tree line to lower-elevation sagebrush and bitterbrush flats. In this environment, summer is brief. Hibernation may last eight months or more at the highest elevations. The males emerge from hibernation first. When the females emerge, they quickly enter a short estrus period, usually just one day long. They are chased, courted, and mated by the waiting males; most females mate with more than one male. The gestation period is just 28 days. This allows the young to forage and gain sufficient weight during the short summer to survive the long hibernation. The young males disperse, usually to another colony, after weaning. Females remain near their mothers throughout their lives.

Much of the squirrel's natural history as described here involves species-specific behaviors in one way or another. Habitat selection is a behavioral choice. So, too, is living in a colony. The reproductive pattern, in which females mate with more than one male, depends on male courtship behavior and female choice of mates. Males engage in dispersal behavior; females do not. And most intriguing, some individuals increase their risk of predation by warning others of the presence of a predator. Moreover, each behavior is embedded in the squirrels' ecological context—their environment, including their predators.

How does this suite of behaviors arise? Is each of these behavioral traits adaptive? Are there different evolutionary processes at work here? What role does the environment play? Together, these questions lead us to the fundamental question of this chapter: *How does the ecology of the organism shape its behavior?*

6.1 How Does Behavior Evolve?

Darwin was well aware of the potential role his theory of evolution could play in explaining the behaviors exhibited by animals and humans. His 1872 treatise *The Expression of the Emotions in Man and Animals* laid out the foundation for the role of evolution and ecology in behavior. Today we understand that behavior is a phenotypic trait under some degree of genetic control. It follows then that we can address the adaptive nature of behavior as a response to the organism's ecology in much the same way we can analyze physiological or morphological traits.

The processes by which behavior evolves are fundamentally the same as those shaping other aspects of the phenotype. Behaviors, like morphological and physiological traits, contribute to fitness. And like other traits, behaviors that are genetically controlled are subject to natural selection. The evolution of behavior illustrates two key aspects of the adaptive process. Each behavior has both a fitness cost and benefit—each represents a trade-off. Selection favors those behaviors whose benefits outweigh the costs. For example, it may be advantageous for antelope to travel in herds because the group confers protection from predators. But there may be a cost to the individual in terms of the amount of food available in a large aggregation. Second, there may be a difference between the proximate and ultimate bases of behaviors. For example, in many species, females refuse to mate with related males. The proximate basis of this behavior is the recognition of kin via olfactory cues. The ultimate advantage is that she is likely to achieve higher lifetime reproductive success if she mates with non-kin and avoids the deleterious effects of inbreeding.

Other members of the same species constitute a significant component of an individual's environment. Thus, the behaviors elicited by other members of the population are a logical starting place for our analysis of the evolution of behavior. Of course, these behaviors—their ecology and evolution—are also a response to other components of the environment, including the abiotic conditions and other species.

What Are the Fundamental Mechanisms of Behavior Interaction?

The concept of the **stimulus-response** is central to animal behavior: a specific behavior is elicited by a specific stimulus. For example, a band of red coloration on the side of a male stickleback fish stimulates aggressive behavior in other

stimulus-response Specific behaviors elicited by a specific stimulus.

males. Red is the stimulus; aggression is the response. Aggressive males orient sideways to other males such that the red band is highly visible. When a female robin returns to the nest with an insect, the young rise up and open their bills wide. This stimulates the female to feed the offspring (Figure 6.1).

Often stimulus-response behaviors are highly stereotyped; that is, the behavioral response is invariant. Courtship displays exemplify this phenomenon. Many species engage in elaborate courtship rituals in which the male and female alternately perform very specific behaviors oriented toward one another. An example of courtship in the waved albatross (*Phoebastria irrorata*) is shown in Figure 6.2. Each action is invariant and elicits a precise response. In fact, if either individual does not perform the prescribed behavior, courtship breaks down and mating does not occur. This adaptation ensures that matings occur only between conspecifics—no eggs or sperm are wasted in interspecific mating.

Communication constitutes another important component of behavioral interactions among conspecifics. We define **communication** as any action on the part of one individual that alters the probability of a behavior in another individual. Thus, some stimulus-response systems, such as alarm calling, are forms of communication. Communication can also be much more complex and convey a range of information and elicit a range of responses.

The signals used in communication may take any form recipients are capable of receiving. Visual, auditory, olfactory, or tactile signals are commonly used. The optimal choice of the signal is determined by the nature of the communication and the physical properties of the environment. For example, whales maintain contact with other individuals over great distances in the ocean. Obviously, sounds, especially frequencies that travel well in water, make more sense for their kind of communication than, say, chemical signals that would disperse rapidly and travel slowly. In other species, such as primates, visual signals work well over short distances and facilitate rapid communication.

Communication may convey discrete or graded information. For example, the pheromones produced during estrus signal a discrete state: the female is physiologically and behaviorally ready to mate. Other signals convey a range of information. Black-capped chickadees (*Poecile atricapilla*) travel in groups of six to eight individuals. If a flying raptor is detected, the chickadee gives a high-frequency, low-amplitude "seet" call. If the predator is perched, they give a loud, broad-frequency "chick-a-dee" call. The latter elicits a mobbing response in the other birds: they gather around the perched predator and vocalize loudly (Figure 6.3). This often drives off the predator or at least causes it to abandon the area because the element of surprise is obviously gone. Their "chick-a-dee" alarm call is complex; graded information is conveyed by repetition, speed, and volume. Experiments in which chickadees were presented with live predators of different size showed that the birds change the nature and volume of the alarm call depending on the wingspan of the predator and its potential threat. Moreover, other chickadees respond to this information by mobbing larger and more threatening predators more vigorously (Templeton et al., 2005).

How Does Selection Differ for Males and Females?

Sexual dimorphism—differences in size, coloration, or morphology between males and females—is common in the animal kingdom (Figure 6.4). The prevalence of sexual dimorphism suggests that different selective forces operate on males and females. Darwin wrote extensively about this difference, describing the "struggle between the individuals of one sex, generally the males, for the possession of the other sex" (Darwin, 1871, p. 361). Darwin understood that one sex, typically the female, could potentially constitute an important selective force on the other.

Darwin's idea is particularly apt if one sex, or access to that sex, limits mating opportunities for the other. The sexes differ in this way because of fundamental

Figure 6.1 Stimulus-response. The wide, gaping bills in the nestlings stimulate feeding behavior in adult birds.

Figure 6.2 Courtship display. The mating dance of the waved albatross is a complex sequence of movements and responses by each member of the pair. Each member of the pair must perform the correct behavior at the correct time for mating to occur.

■ **communication** Any action by one individual that alters the probability of a behavior in another.

Figure 6.3 Mobbing. Small birds that are potential prey species or whose young are vulnerable often mob raptors, a behavior in which they chase and harass the predator.

■ **sexual dimorphism** Differences in size, coloration, or morphology between males and females.

THE EVOLUTION CONNECTION

Measuring the Contributions of Environment and Genetics to Behavior

Our approach to the study of behavior centers on behaviors as adaptations. We address not only the means by which they may have evolved but the central role of ecology in shaping their evolution. Thus, behavior falls under the rubric of the processes of adaptive evolution outlined in Chapter 2. However, a fundamental assumption underlies much of this analysis: that behavior has a genetic basis.

The obvious importance of learning in the social and behavioral interactions of many species immediately tells us that the environment and experience also play a role in behavior. For many years, the relative importance of genetics and the environment in the development and evolution of behavior, sometimes referred to as the nature/nurture debate, has been a contentious issue. One result of this controversy has been careful analysis of their relative roles in specific cases. This research reveals that few behaviors are purely genetically fixed and few are purely the result of the environment. Most are a mix of genetic tendency modified by experience and the environment. The ability to learn and modify behavior from experience is itself an adaptation.

Birdsong illustrates the interaction of the environment and genes in the development of a complex behavior. In the song sparrow (*Zonotrichia leucophrys*), a template with essential elements of the song is genetically encoded. Birds raised in isolation, who have never heard the adult song, sing this template song. However, experience modifies the template. If the birds do not hear adult sparrows sing, they do not develop the full adult song. Each region has a local song dialect, and this, also, fails to develop in the absence of examples heard by the young bird. Moreover, the birds must be able to hear themselves sing. If not, the song develops abnormally, indicating that birds practice and match their song to their experience (Baptista and Petrinovich, 1984; Marler, 1970). Many behaviors exemplify this kind of interaction between learning and genetically programmed traits.

How do we measure the relative contribution of the environment and genetics in behavior? A starting point is a measure of the heritable variation in a trait. In Chapter 2 we defined heritability as the proportion of the variation in a trait that can be attributed to genetic differences among individuals. Here we formalize the concept. We denote the total phenotypic variation in a trait that derived from the environment and genes as V_P. This variation has two bases: genetic variation among individuals (V_G) and environmental variation (V_E). *Heritability in the broad sense* is the proportion of the total variation that is due to genetic differences among individuals:

$$\text{Heritability} = \frac{V_G}{V_P} = \frac{V_G}{V_G + V_E}$$

We can refine this analysis by breaking the total genetic variance (V_G) into its component parts. A portion of the total variation (V_A) is due to additive genetic differences between individuals— that is, the sum of the contributions of their different genes. Some is due to the effect of dominance, the masking of recessive genes by dominant ones (V_D). We define the proportion of the total phenotypic variance that is due to additive genetic variance as *narrow-sense heritability (h_2)*. Algebraically,

$$h^2 = \frac{V_A}{V_P} = \frac{V_A}{V_A + V_D + V_E}$$

There are a number of ways to measure h^2. The slope of the regression of the average offspring phenotype as a function of the average of the parental phenotypes is a direct measure of h^2. Comparison of monozygotic and dizygotic twins also sheds light on heritability. If raised in the same environments, dizygotic twins share half their genes and their environment, whereas monozygotic twins share all their genes and their environment. If monozygotic twins are more similar to each other than dizygotic twins, we conclude that heritability is high. If monozygotic and dizygotic twins are similar, it suggests that h^2 is lower and the environment plays a significant role.

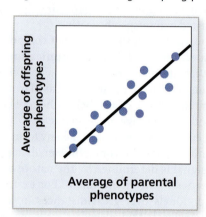

Figure 1 Heritability. Heritability is measured by the correlation between offspring and parental phenotypes. **Analyze: What is the significance of the variation in the data points around the line?**

Traits with low heritability do not respond to natural selection as rapidly as traits with high levels of additive genetic variation. Recall from Chapter 2 that the rate of response to selection is quantified by the equation

$$R = h^2 s$$

where *R* is the change in the trait, h^2 is heritability, and *s* is the selection coefficient. This equation shows that the response to selection depends on the combination of heritability and the intensity of selection. As heritability decreases, selection must be more intense in order to achieve the same response.

Still, environmental factors affect the response in ways not necessarily measured by the response equation. For example, in red deer (*Cervus elaphus*), antler size has high heritability ($h^2 = 0.33$) and selection for males with large antlers is intense ($s = 0.44$). However, over a 30-year study, antler size did not increase. Apparently, environmental factors, specifically male nutritional status, confound the effect of directional selection, reducing the response of antler size to selection (Kruuk et al., 2002).

QUESTION:

Is it possible that the combination of learned and genetic components of birdsong is itself adaptive?

Figure 6.4 Sexual dimorphism. In sexually dimorphic species such as mallards (a), elephant seals (b), and rhinoceros beetles (c), males and females differ in size, color, and morphology.

differences in their reproductive roles. Females bear the direct anatomical, physiological, and energetic costs of reproduction. Depending on the taxonomic group, they lay and perhaps incubate the eggs or carry them to term internally. In mammals, females are also responsible for lactation. As a result, there is a point at which female fitness cannot increase by producing more young. Rather, her fitness increases as a function of the quality of young she produces. Male reproduction is not limited in this way. The male contribution, sperm, can be produced in great numbers, and so male fitness increases with the number of females he inseminates.

These differences between males and females lead to **Bateman's principle**, the concept that males experience greater variation in reproductive success than females. The opportunity for selection to affect one sex increases with the degree of sexual dimorphism (Figure 6.5). There is strong selection among males for traits that increase their opportunity to mate or increase the number of matings they achieve. And females constitute a potent selective force on males: those males whose physical or behavioral attributes are preferred by females achieve higher reproductive success. The result is **sexual selection**: a form of natural selection in which traits that enhance the ability of one sex (usually males) to compete for or attract mates are favored.

Sexual selection operates in two basic ways. **Female choice** occurs when the female chooses among males on the basis of their physical or behavioral characteristics. Male phenotypic traits that improve their chances of being chosen may be greatly exaggerated by this process. The elaborate and gaudy tail of the bird of paradise or the brightly colored vocal sacs of grouse are useful for attracting females but little else (Figure 6.6).

Other bases of female choice are much more subtle. For example, female barn swallows choose mates on the basis of the symmetry of their forked tails. They choose males whose tails are most symmetrical (Figure 6.7).

The other basic mechanism of sexual selection is **male-male competition**, in which males vie for access to females. This competition may be direct, as for example in the battles among bighorn (*Ovis canadensis*) rams for females. Or it may be indirect: the males compete for an important resource, such as a high-quality territory, whose owner is attractive to females as a mate.

One of the most intriguing aspects of sexual selection is the degree to which male traits important to females are elaborated. Among the more extreme examples of the phenomenon are the antlers of the extinct Irish elk (*Megaloceros giganteus*), which weighed as much as 60 percent of the total body mass, an amount far larger than necessary or even useful for predator defense or male-male fighting. Apparently,

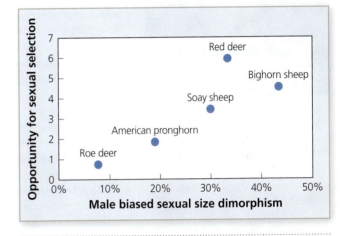

Figure 6.5 Sexual selection. The potential for sexual selection as a function of the degree of sexual size dimorphism—that is, the size of the male relative to the female (from Vanpé et al., 2007). **Analyze: Why might one species, such as the red deer, lie so far above the others?**

THINKING ABOUT ECOLOGY:

Explain the relative importance of the mean reproductive success and the variance in reproductive success for males and females.

■ **Bateman's principle** The idea that males experience greater variance in reproductive success than females.

■ **sexual selection** A form of natural selection in which traits that enhance the ability of one sex (usually males) to compete for or attract mates are favored.

■ **female choice** A mechanism of sexual selection in which females choose mates on the basis of their physical or behavioral characteristics.

■ **male-male competition** A form of sexual selection in which males compete with one another directly or indirectly for access to females.

Figure 6.6 Exaggerated traits. The tails of male birds of paradise (a) and the vocal sacks of sage grouse (b) are important in attracting mates.

Figure 6.7 Sexual selection in barn swallows. Female barn swallows choose males on the basis of the symmetry of the tail. Analyze: Why do you suppose this particular trait is used by females?

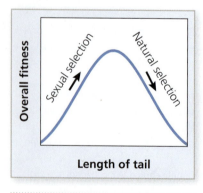

Figure 6.8 Short vs. long tails. The balance between natural selection and sexual selection on tail length in the bird of paradise. Both short and long tails are less fit than tails of intermediate length. Analyze: What would you expect to be the costs of short vs. long tails?

females chose males with the largest. At some point, intense directional selection of this type almost certainly results in lower overall fitness in the male—the elaborations females choose ultimately have fitness costs to the male. This raises the question: How does sexual selection interact with natural selection?

Two main hypotheses have been proposed to explain this relationship. The balance hypothesis suggests that male traits are exaggerated by female choice until their overall fitness cost is too high. In this view, natural and sexual selection gradually come into a balance in which the continuing selection imposed by the female reaches a limit due to natural selection (Figure 6.8). This hypothesis exemplifies the concept of evolutionary trade-offs—the notion that each trait has a cost and benefit. The balance hypothesis asserts that natural selection optimizes the costs and benefits of male traits.

A second hypothesis is based on the idea of "truth in advertising," that the elaborate phenotypic traits of males are an indicator of overall fitness. Recall that female reproductive fitness is constrained by the number of offspring she can produce. The female can produce only so many eggs, gestate so many young, or produce milk for only so many offspring. Given this quantitative limitation, she can increase her fitness only by increasing the quality of the young she produces. One means of achieving this is to mate with a genetically superior male. Male-male competition, in which males fight directly or fight for resources important to reproduction, such as a territory, is one way a female can evaluate the genetic quality of a potential mate. The male bighorn ram that defeats all others may be able to do so because of his genetic fitness. Similarly, the male that can acquire and defend the best territory not only may be superior himself, but he offers to the female resources that increase the quality of her offspring.

This hypothesis explains the elaboration of the traits important to females that otherwise seem irrelevant to fitness. For example, why should females choose male barn swallows on the basis of tail symmetry? The tail is genetically programmed to grow equally on both sides. Factors such as parasite load, disease, poor food resources, or genetic abnormalities may compromise this symmetry. The female uses tail symmetry as an index of overall male quality. Some highly exaggerated traits such as massive antlers or spectacular tail feathers are probably a burden to the males. A male that can support such a burden and still prosper is probably genetically superior to one who cannot. Thus, these elaborate structures signal male quality.

What Ecological Factors Influence the Mating System?

Sexual selection is a consequence of the different reproductive strategies of males and females. One consequence of this difference is the **mating system** of the species. The mating system is the set of relationships between males and females during reproduction. It consists of three components: (1) the number of mates an individual copulates with during the breeding season, (2) the relative contribution of males and females to parental care, and (3) how long the relationship between males and females lasts.

Monogamy is a mating system in which males mate with a single female. In **serial monogamy** this relationship lasts for just a single breeding season; in subsequent seasons the male mates with a different female. **Polygyny** is a mating system in which males mate with more than one female during a breeding season. In **polyandry** each female mates with more than one male.

Polygyny is the most common relationship between males and females. This system is clearly advantageous to males, whom we have said benefit from multiple mating. The value to females is less clear. Several factors associated with the advantages of being part of a group of females probably play a role. If there are survival benefits to membership in a group due to predator defense, the female's lifetime reproductive success may compensate for the cost of sharing a mate at each reproductive event. It may also be that the energetic cost of ensuring monogamy, either by fending off other females or garnering the male's fidelity, is too high. Selection on females probably favors any means by which they shift the balance toward monogamy.

The ecological context is a central factor in the development of the mating system. The potential for males to control and ensure access to more than one female is dependent on the spatial distribution of females. Aggregated females require far less energy to guard and maintain than widely scattered individuals. The spatial distribution of females is in turn determined by the distribution of ecologically important resources. Polygyny arising from aggregations of females guarded by males is known as **female defense polygyny**. This is a common pattern in ungulates and pinnipeds. For example, in the African impala (*Aepyceros melampus*), females congregate in prime habitat characterized by abundant food resources and cover that protects them from predation. This provides males with the opportunity to defend groups of females. Similarly, in pinnipeds such as the northern elephant seal (*Mirounga angustirostris*), females seasonally haul out onto land to give birth and mate. Because females are gregarious during this time and appropriate beaches are limited, males can defend harems of females.

The ecology of resource distribution may lead to polygyny via another pathway. If critical resources are distributed such that males can defend them, the males may obtain a territory that attracts multiple females. Females choose males that provide access to these resources, and the benefits of the resources outweigh the cost of polygyny to the females. For example, the orange-rumped honeyguide (*Indicator xanthonotus*) practices resource defense polygyny. The sole food source of this Asian species is wax from the hives of giant honeybees. Males defend beehives from other males. Females are allowed access to the hive and mate with the territorial male.

In some species, polygyny results from an unusual form of female mate choice. In **lek-mating species**, males defend neither resources nor females. Instead, they display together on traditional sites known as leks. Females observe the displays and choose the male with which to copulate. Sage grouse (*Centrocercus urophasianus*) and prairie chickens (*Tympanuchus cupido*), both birds of open habitats, have lek-mating systems. Displaying males vigorously defend their portion of the lek (Figure 6.9). Male reproductive success is highly variable; a few males obtain the majority of the copulations. Polygyny is a direct result of female choice as well as male-male aggression on the lek (Nooker and Sandercock, 2008).

Molecular techniques such as DNA fingerprinting have shown that pure monogamy is rare. Even in species in which a male and female form a pair bond, remain together, and both contribute to rearing the young, extra-pair mating occurs, sometimes for one sex and sometimes for both. For example, the upland sandpiper (*Bartramia longicauda*) is socially monogamous—the female bonds with a single male—but 15 percent of chicks and 30 percent of broods include extra-pair mating (Casey, 2011). There are a number of potential advantages accruing to females that mate promiscuously. Extra-pair mating may increase the genetic diversity of the offspring. It may also ensure that the female has the opportunity to mate with a number of males with high fitness or to increase the probability that at least one mate is fertile.

Even in species in which some extra-pair matings occur, the mating system may be skewed toward monogamy under certain ecological conditions. For example, if successful rearing of the young requires both parents, monogamy is advantageous for both sexes. Polygyny is of little value to the male if his offspring do not survive without his contribution to parental care. Some 90 percent of passerine birds are primarily monogamous. In many, male and female guarding and provisioning the young is required. If the seasonality and resource phenology lead to high reproductive synchrony in the population, monogamy is more likely because when most females are mated, a male would probably gain less from seeking additional copulations than remaining with his mate and providing parental care.

mating system The length of relationships between males and females, the relative contributions of males and females to parental care, and the number of mates an individual copulates with.

monogamy A mating system in which each male mates with a single female.

serial monogamy Monogamy that lasts for one breeding season.

polygyny A mating system in which males mate with more than one female.

polyandry A mating system in which each female mates with more than one male.

female defense polygyny A form of polygyny that occurs when groups of females are guarded by males.

lek-mating species A species that uses a form of polygyny and female choice in which males gather to display to females on traditional places.

Figure 6.9 Leks. Male prairie chickens gather on leks to display to females. The leks are traditional display grounds. Females gather there to choose among the displaying males.

THINKING ABOUT ECOLOGY:

Explain the costs and benefits of monogamous and polygynous mating systems in terms of evolutionary trade-offs.

Although rare, purely monogamous systems are known. Again, most are explained by the ecological context in which they occur. For example, in frogs, the size of the pool in which breeding occurs is an important determinant of the optimal mating system. High predation rates in large aquatic systems have led many species to breed in much smaller pools. In tropical rainforest, some species even breed in phytotelmata, water-filled leaf axils in the forest canopy. The trade-off is that pools this small may dry up before reproduction is complete. Additional parental care may offset this disadvantage (Figure 6.10). Some species accelerate the development of the larvae by trophic egg feeding, the production of unfertilized eggs as a food source for the other young. The Peruvian poison frog, *Ranitomeya imitator*, exemplifies this strategy. Both parents contribute to parental care, and genetic analysis confirms that they are entirely monogamous (Brown et al., 2010).

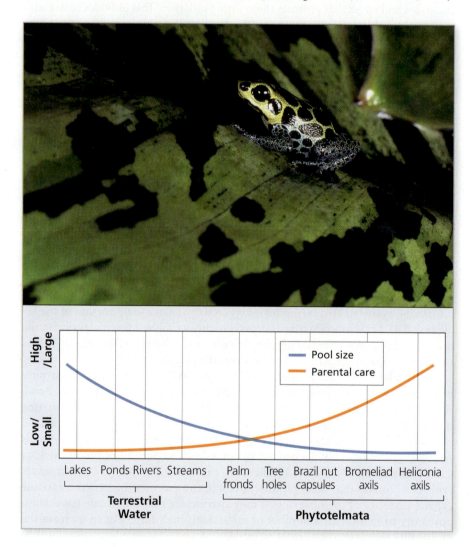

Figure 6.10 Mating pools. These photos show frogs breeding in phytotelmata. The amount of parental care increases as the size of the mating pool decreases (from Brown et al., 2010). **Analyze: What is the advantage of additional parental care in small pools?**

Polyandry is the rarest of the mating systems. It is known in some birds, such as the Galápagos hawk (*Buteo galapagoensis*) and the spotted sandpiper (*Actitis macularius*), and some insects. There are both direct and indirect fitness benefits that accrue to polyandrous females. Direct benefits include such things as nuptial gifts provided by male insects to their mates. These packages of energy or nutrient-rich resources increase the female's reproductive success. Indirect benefits include superior male genes, either from processes like sperm competition or simply increased genetic diversity. Among the insects, indirect benefits such as these increase the female's fitness as much as 70 percent (Arnqvist and Nilsson, 2000).

Other fitness benefits accrue to polyandrous females in the vertebrates. For example, polyandry guards against the fitness costs of choosing an infertile, closely related, or genetically inferior mate. By mating with multiple males, the female increases the probability of reproductive success and minimizes the risk of the catastrophic reproductive failure that might result from poor mate choice. In species in which the male chooses the nest site and constructs the nest, there can be considerable variation among nest sites, especially in extreme and unpredictable climates. For example, in the polyandrous Australian terrestrial toadlet (*Pseudophryne bibronii*), males construct nests along ephemeral watercourses. Many of the nests fail due to desiccation. The female's reproductive success increases with the number of males with which she mates because apparently multiple mating ensures that some nests succeed (Byrne and Keogh, 2009).

Although animal behavior is the central focus of this chapter, some aspects of behavioral ecology have analogs among plants. Plants, too, have mating systems defined by the relationships among the male and female floral components and the potential for self-fertilization. Plant mating systems take three basic forms: **outcrossing** (mating among different individuals), **autogamy** (self-fertilization), and **apomixis** (asexual reproduction). Plants vary enormously in the arrangement and distribution of male (pollen and anther) and female (egg and ovary) components of the reproductive system. Monoecious plants have both male and female flowers in the same individual. In dioecious species, some individuals are male; others are female. These patterns determine the potential for selfing. Hermaphroditic individuals and monoecious plants can at least potentially self-fertilize. In dioecious species, each flower must be pollinated by another plant. The variants such as andromonoecy and gynodioecy allow some selfing as well as some outcrossing.

Superimposed on these patterns is the second key component of the mating system: the degree of **self-incompatability**—the barriers to selfing. Genetic self-incompatability is based on a set of alleles, SI genes that, if identical in the pollen grain and the stigma, prevent growth of the pollen tube and fertilization. If the alleles do not match, fertilization can occur. In some species, physiological processes such as the developmental pattern of the inflorescence reduce the probability of selfing. For example, some species are **protogynous**—the female parts of the flower develop and mature before the male parts. The result is that a young flower has functional female parts before the anthers and pollen develop. The pollen deposited by an insect on a female flower comes from another individual with older flowers and male function. In **protandry** the anthers and pollen mature first; the female parts develop later.

Plant mating systems affect the degree of selfing, and this in turn affects the degree of inbreeding in the plant. And like animal mating systems, plant mating systems are shaped by the ecology of the species. For example, species that colonize new habitat after long-distance dispersal tend to be hermaphroditic and self-compatible. As a result, a single individual can initiate an entire population. Apomixis produces large numbers of genetically identical individuals. This is advantageous in species that rapidly exploit local resources or ephemeral favorable conditions.

THINKING ABOUT ECOLOGY:

What is the relationship between sexual selection and the tendency for most mating systems to include some degree of polygyny?

outcrossing A plant mating system in which mating occurs between different individuals.

autogamy A plant mating system in which individuals self-fertilize.

apomixis A plant mating system in which individuals reproduce asexually.

self-incompatibility A phenomenon in plants in which individuals cannot self-fertilize.

protogynous Describes a pattern of development in which female function develops first.

protandry A developmental phenomenon in plants in which the male parts of the flower mature before the female parts.

KEY CONCEPTS 6.1
- The stimulus-response system underlies much of animal behavior.
- Communication is a specialized form of stimulus and response that conveys information among conspecifics.
- Males and females differ in the means by which their reproductive success increases. Males gain by additional copulations. Females gain by increasing the quality of the male with which she mates.

- Because females tend to be the limiting sexual resource, female choice exerts selective pressure on males known as sexual selection.
- The mating system is a response to the nature of sexual selection in the species as well as the distribution of female and economic resources.

QUESTION:

Lek-mating systems include the phenomenon of sexual selection—females choose among the displaying males. How might this represent either the balance hypothesis or the truth-in-advertising hypothesis of sexual selection?

6.2 What Role Does the Ecosystem Play in the Evolution of Behavior?

We know that conspecifics—members of the same species—and certain components of the environment shape behavior. Here we extend this concept to explain the evolution of behavior and social systems as adaptations to the totality of the environment.

What Behaviors Are Important in Obtaining Resources?

Each species, plant and animal, has a preferred **habitat**, the physical and biological characteristics of the place it lives. For plants and many sessile animals, individuals cannot directly seek their preferred habitat. Instead, they distribute propagules—seeds or larvae—and those that end up in the proper environment persist. For mobile animals, habitat selection is an important behavior that is often genetically encoded. Newborn individuals choose the habitat type preferred by their species if offered an array of choices. Roe deer (*Capreolus capreolus*) choose habitats with high food availability, cover and protection from predation, and edge, the boundary between thickets and meadows. Lifetime reproductive success increases to the extent that each of these habitat variables is included in the habitat chosen by the individual (McCloughlin et al., 2007).

We use the term *habitat* in the broad sense to refer to the general features of the environment important to the animal. Thus, Arctic tundra is the preferred habitat of caribou (*Rangifer tarandus*). However, as we have seen in Chapter 4, tundra is a heterogeneous system. Caribou do not necessarily use all portions of this habitat. We refer to the specific portions of the habitat used by animals as their **microhabitat**. In winter, caribou tend to select microhabitats that have shallow, soft snow where foraging is easier (Tucker et al., 1991). In summer, they are likely to be found in higher, windy microhabitats, where they get relief from mosquitoes.

Microhabitat selection is important for many species as a means of maintaining homeostasis (Chapter 3). Aquatic organisms ranging from insects to fish choose pools, riffles, and other structural elements of a stream on the basis of temperature, oxygen concentration, flow rate, and other physical factors. As discussed in Chapter 3, microhabitat selection is an important aspect of behavioral thermoregulation in poikilotherms. Lizards choose sunny spots, perhaps on a dark substrate, to increase their body temperature in the early morning when the ambient temperature is low. They may retreat to the shade in the middle of the day to lower their core temperature.

The pattern of use and movement through the habitat, and even the microhabitat, is usually not random. The portion of the habitat used on a daily or seasonal basis is the **home range**. Figure 6.11 shows the home ranges of several white-footed mice (*Peromyscus leucopus*) in forest habitat. A home range may shift over time due to competition from other individuals or changes in the habitat.

- **habitat** The abiotic and biotic characteristics of the place where an organism lives.

- **microhabitat** The subset of the habitat that differs in important abiotic and biotic characteristics.

- **home range** The portion of the habitat used by an individual on a daily or seasonal basis.

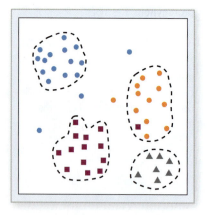

Figure 6.11 Home range. The home ranges of white-footed mice. Each symbol represents the radio-telemetry locations of a single individual. The dotted lines enclose the main home range of each individual. Note that some individuals are occasionally found outside their home range or in the home range of another individual. **Analyze: What factors might lead these home ranges to move or to increase or decrease in size?**

The Nature of Habitat Selection

Habitat selection is an important behavioral response to the physical and biological environment. For some species, abiotic factors determine where they can and cannot successfully settle and reproduce. For others, competitors, predators, or other biological constraints determine the optimal habitat or microhabitat. What cues do species use in the process of habitat choice?

Fletcher (2007) investigated this process in two species of potentially competing passerine birds, the least flycatcher (*Empidonax minimus*) and the American redstart (*Setophaga ruticilla*). The least flycatcher is more aggressive and dominates the redstart where they occur together. Fletcher assessed the role of song as a proximate cue for each species in the choice of breeding sites in the riparian habitat preferred by both species. Songs of potential competitors might reduce the probability of settlement. Alternatively, the songs of conspecifics might indicate suitable habitat and increase the possibility of settlement. He tested two specific hypotheses based on this logic:

Figure 1 Competition. The redstart (a) and least flycatcher (b) compete for nest sites in riparian zones.

HYPOTHESIS 1: Species avoid areas where another competitive species has settled.

HYPOTHESIS 2: Species are more likely to settle where conspecifics have already settled.

Two predictions regarding the role of song follow from these hypotheses:

PREDICTION 1: Both species should show conspecific attraction and heterospecific avoidance to song.

PREDICTION 2: Heterospecific avoidance should be stronger for the subordinate species, the American redstart.

The research was conducted at 25 riparian sites in Montana. Fletcher began by mapping habitat use of the two species for two "pre-treatment" years as a control. This was followed by two treatments: playback of calls of the flycatcher and of the redstart. The effect was measured by quantifying the settlement response of each species in the two treatments.

The control years showed that the subordinate redstart was much rarer than the dominant flycatcher in riparian habitat. The playback experiments affected both species. The flycatcher was more likely to settle in an area where it heard both flycatcher and

redstart song. In contrast, the redstart showed a negative response to flycatcher song. It responded positively to its own song, but the response was much weaker than the response of flycatchers to conspecific song. Fletcher concluded that the flycatcher uses both conspecific and heterospecific song to identify suitable habitat. The subordinate species, however, uses heterospecific song as a cue to avoid competition with flycatchers.

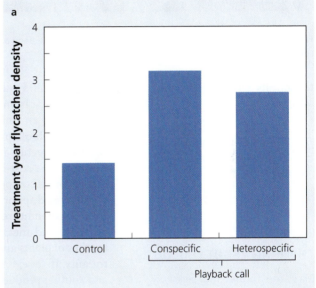

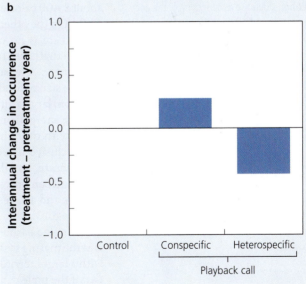

Figure 2 Heterospecific avoidance. The responses to playbacks of conspecific and heterospecific calls for (a) the least flycatcher and (b) the American redstart. In this experiment the controls are the territories in the years prior to the playback experiments (from Fletcher, 2007). **Analyze: What potential errors could arise because of these controls?**

Figure 6.12 Territoriality. Gannets nest in dense colonies and defend a small territory around the nest.

■ **territoriality** Exclusive use of a portion of the home range.

THINKING ABOUT ECOLOGY:

Draw a graph of the energy benefit of territorial defense as a function of the energetic cost of defense in the golden-winged sunbird (units not necessary). Show on the graph the point where the optimal strategy shifts.

For example, one effect of the reintroduction of wolves to Yellowstone National Park was a shift in the home range of elk. In the absence of wolves, elk spent much time in dense stands of willows and aspen. Subsequent to wolf reintroduction, they shifted their home ranges into open habitat, where they can detect wolves at greater distances (Creel et al., 2005).

Territoriality is exclusive use of a part or all of the home range. Intraspecific and sometimes interspecific aggression maintains exclusive access to the resources of the territory. The resources of importance and their spatial distribution determine the size of the territory. The defended area is determined by the trade-off between the energetic cost of defense and the gain due to exclusive access. For example, colonial nesting gulls and gannets defend just the immediate vicinity of the nest (Figure 6.12). Aggression from individuals near the nest reduces chick survival, so defense of the nest area is crucial. However, because they forage over the open ocean, defense of the food resource is energetically impossible. In contrast, for many passerine birds the territory encompasses the nest site as well as sufficient food resources for the pair and their offspring.

The energetic balance sheet of the golden-winged sunbird (*Nectarina reichenowi*) illustrates the cost-benefit relationship inherent in territoriality. This African bird forages on nectar produced in patches of mint. Rich patches of productive plants are defended; poorly producing patches are not. Detailed measurements of the energy expenditure in various behaviors, including perching, foraging, and aggression, and measurement of the variation in nectar production of different patches of plants are consistent with optimization of energy expenditure for territoriality. If, for example, a patch produces 1 microliter of nectar per flower per day and another produces twice that, the caloric benefit of defending the rich patch is about 2,400 calories per day. Thus, defense of the rich patch is energetically worthwhile. On the other hand, if a patch produces 4 microliters of nectar per flower per hour, there is no caloric saving from defense of the patch. Production is high enough that even if the resource is shared, the sunbird obtains so much energy that investing in defense provides no net energy benefit (Gill and Wolf, 1975).

The size of food-resource territories increases as a function of body mass. In addition, large animals have larger territories than predicted simply on the basis of their energetic needs. This is because large animals share more of their territories with other individuals than do small animals. In other words, they have less exclusive use of their territory and thus they require larger territories to compensate. Smaller animals such as mice cover their entire territory on a daily basis and can prevent intrusion by neighbors. Larger animals may not be able to do this, and as a result their territories are often less exclusive (Jetz et al., 2004) (Figure 6.13).

As noted earlier, territory size is also related to the process of sexual selection and the mating system. To the extent that territory size reflects resources, a male with a larger territory may attract more females (Figure 6.14). Also, we have seen that if the male can defend a resource important to the female, he may engage in resource defense polygyny.

How Does the Environment Affect Movement?

The external environment also determines the nature of movement behavior. Migration is the movement of individuals from one region to another and back. Some planktonic organisms undertake daily vertical migrations. Other migrations are seasonal. These movements permit the exploitation of resources during

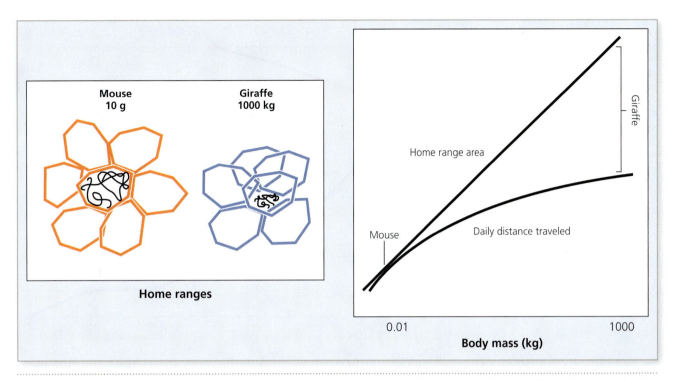

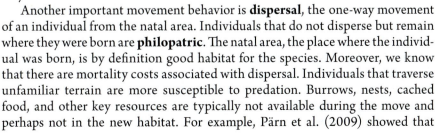

Figure 6.13 **Home range as a function of body size.** The home ranges of a mouse and giraffe as functions of body size. The mouse patrols its entire home range and has less overlap with neighbors; the giraffe does not cover its entire home range as frequently and intrusions occur. The result is that home ranges of large animals are bigger than predicted from their body mass and resource requirements (from Buskirk, 2004; Jetz et al., 2004). **Analyze: What factors other than the size of the animal might affect the portion of the home range used daily?**

productive periods but avoid harsh conditions in other seasons. Thus, many northern bird species migrate from the Arctic or temperate regions to avoid the physical conditions and lack of food there in the winter. Many marine mammals are migratory as well. Gray whales (*Eschrichtius robustus*) undertake the longest migration of any known mammal. They exploit the productive waters of Alaska in the summer and give birth to their calves in the warm waters of Baja California.

Migratory behavior has both genetic and environmental control. It is possible to select for increased migratory tendencies. Selection for individuals with greater unrest and mobility during a certain season results in lines with higher migratory tendency (Figure 6.15). On the other hand, some species modify their migration patterns to take advantage of new conditions and resources. Many species of waterfowl migrate only as far south as required to find food and open water. Thus, in some years they travel from the northern Great Plains all the way to the southern coasts or beyond; in others they may winter much farther north.

Another important movement behavior is **dispersal**, the one-way movement of an individual from the natal area. Individuals that do not disperse but remain where they were born are **philopatric**. The natal area, the place where the individual was born, is by definition good habitat for the species. Moreover, we know that there are mortality costs associated with dispersal. Individuals that traverse unfamiliar terrain are more susceptible to predation. Burrows, nests, cached food, and other key resources are typically not available during the move and perhaps not in the new habitat. For example, Pärn et al. (2009) showed that

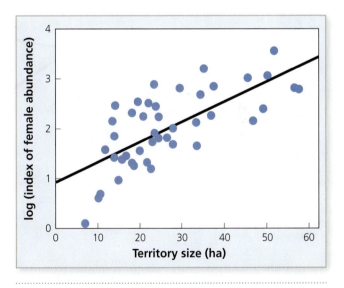

Figure 6.14 **Territory size and sexual selection.** Female abundance as a function of male territory size in roe deer (from Vanpé et al., 2008). **Analyze: Is this result dependent on the roe deer's harem social structure?**

dispersal The one-way movement of an individual from the natal area.

philopatric Describes individuals that do not disperse.

Figure 6.15 Migration. The effect of selection for increased and decreased migratory activity in the blackcap (*Sylvia atricapilla*). Selection could increase or decrease migratory activity (from Pudilo, 2007). **Analyze: Does this imply that there is naturally occurring genetic variation in migratory activity?**

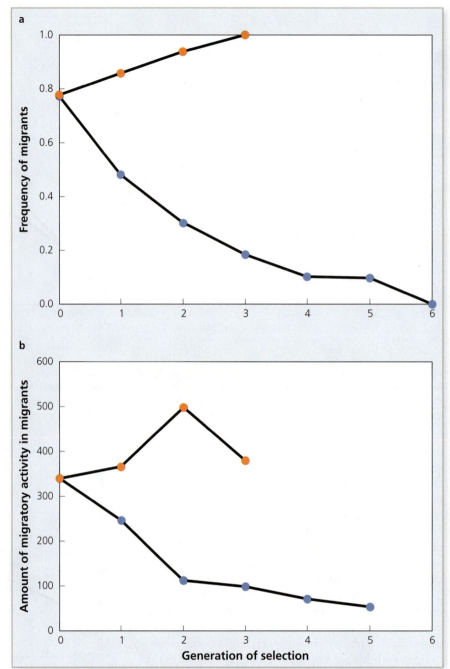

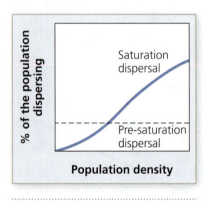

Figure 6.16 Saturation dispersers. Saturation and pre-saturation dispersal depicted as the proportion of the population dispersing as a function of density. **Analyze: Would you expect pre-saturation dispersal to ever increase with population density?**

■ **saturation dispersal** Dispersal of individuals that occurs when the habitat is filled or resources are limiting.

■ **pre-saturation dispersers** Individuals that disperse before the habitat is filled or resources are limiting.

dispersing house sparrows (*Passer domesticus*) have significantly lower lifetime reproductive success than residents. This cost is a consequence of higher mortality and lower mating success. Thus, we may ask why individuals leave one population to enter another.

In some cases dispersal occurs because economically important resources such as food or nest sites are in short supply. If the population density is high enough, competition precludes access to critical resources. Individuals that cannot gain access to these resources disperse despite the costs. Lidicker (1986) refers to individuals who disperse to obtain scarce economic resources as **saturation dispersers**. These individuals may have no choice but to seek better conditions elsewhere (Figure 6.16).

In some species, dispersal occurs even when economic resources are abundant. Lidicker defines **pre-saturation dispersal** (Figure 6.16) as the movement of individuals before resources become scarce at high density. Clearly some other

DO THE MATH

Using Hardy-Weinberg Analysis to Study Inbreeding

The potential deleterious effects of inbreeding are due to the increase in the frequency of homozygotes in the population. We can use the same Hardy-Weinberg analysis that we used in Chapter 2 to quantify this effect and to analyze the mechanisms of evolution and the rate of change in allele frequency under selection.

We begin with a variable to measure inbreeding, the *coefficient of inbreeding* (F). This value measures the probability that two alleles in an individual are *identical by descent*. A homozygote carries two copies of an allele. The two are identical by descent if they are derived from the same ancestral copy of the gene. We calculate the value of F from pedigree analysis. This figure shows the two ways the daughter of a half-sib mating can receive copies of a gene that are identical by descent. In DM-1a, one copy of the allele reaches the daughter by passing from the grandmother to both son and daughter and then from their mating to the daughter. Each transfer of the allele occurs with probability ½. Thus, the probability that the two alleles in the daughter are identical by descent is $½ \times ½ \times ½ \times ½ = 1/16$. The same is true for the other allele in the grandmother. Thus, there are two ways to obtain alleles identical by descent in the daughter, each of which has a probability of 1/16. The total probability, then, is $2 \times 1/16$ or 1/8.

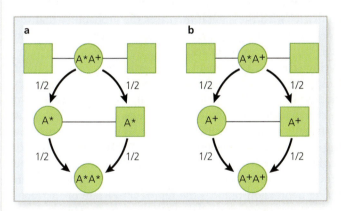

Figure 1 Allele transmission. Pathways of allele transmission in a half-sib mating that lead to an offspring with alleles that are identical by descent. In (a) the A* alleles combine; in (b) it is the A⁺ alleles that combine in the same individual. **Analyze: Why is the probability of each allele transfer ½?**

How does this process affect the frequencies of homozygotes and heterozygotes? The proportion of homozygous AA individuals is the sum of those homozygotes that are not identical by descent and those that are. If F is the probability that two alleles are identical by descent, the proportion of AA individuals is this fraction times the frequency of the A allele (*p*), or

$$p \times F$$

The remainder of the homozygous AA individuals, $1 - F$, is not identical by descent. Their frequency is given by the standard frequency of AA individuals (p^2) from Hardy-Weinberg:

$$p^2(-F)$$

The total frequency of homozygous AA individuals is the sum

$$p^2(1 - F) + pF$$

or

$$p^2 - p^2F + pF$$
$$= p^2 + pF(1 - p)$$
$$= p^2 + pqF$$

The process is the same for the aa homozygotes. Their frequency is

$$q^2 + pqF$$

The heterozygote frequency is calculated from the probability of an A and an a allele coming together. This can occur in two ways:

$$p \times q \text{ and } q \times p$$

or as in the original Hardy-Weinberg equation,

$$2pq$$

Since these alleles cannot be identical by descent, we multiply the frequencies by $1 - F$:

$$2pq(1 - F) \text{ or } 2pq - 2pqF$$

The frequencies of the three genotypes are thus given by the modification of the standard Hardy-Weinberg equation:

$$(p^2 + pqF) + (2pq - 2pqF) + (q^2 + pqF) = 1$$

Note that when inbreeding occurs, the frequency of both homozygotes increases relative to the equilibrium Hardy-Weinberg frequency by the amount pqF, and the frequency of heterozygotes decreases by the amount 2pqF.

The increase in frequency of homozygotes means that deleterious recessive alleles are more likely to appear in homozygous form under inbreeding. This is the basis of inbreeding depression, the decrease in fitness in inbred populations. Note, too, that these equations show that homozygosity increases each generation. As a result, inbreeding is a potent selective force that plays a significant role in behavior and social organization.

---

factor or factors must select for such behavior. One hint of the potential benefits of pre-saturation dispersal comes from the observation that in a number of vertebrate species, dispersal is sex-biased. In birds, most dispersers are females; in mammals, most dispersal is done by males. Greenwood (1983) hypothesized that

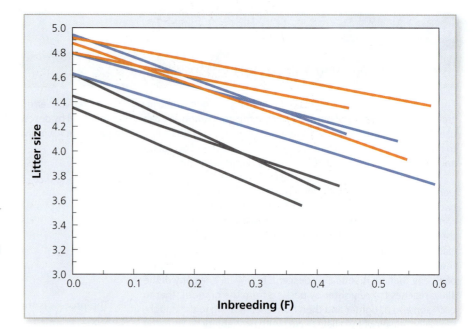

Figure 6.17 Inbreeding depression. The effect of inbreeding depression: litter size as a function of the intensity of inbreeding in families of *Peromyscus polionotus*. Each line represents a different family (from Lacy et al., 1996). **Analyze: What is the significance of the fact that these lines have almost exactly the same slope?**

inbreeding depression A decrease in fitness due to mating among related individuals.

THINKING ABOUT ECOLOGY:

You are studying a tropical bird that lives in small colonies associated with very specific microhabitat types. All the females disperse at fledging. What hypothesis might you devise to explain this behavior? What questions would be important to ask?

this pattern is the result of differences in bird and mammal mating systems. Most birds tend toward monogamy, and males defend territories to attract females. In contrast, mammals tend to be polygynous. Greenwood suggests that, as a result, males that cannot compete with other males for mates are better off dispersing in the hope of finding a population in which they can mate. In birds, males benefit from philopatry—familiarity with the habitat increases the probability that the male can obtain and defend a territory. Dispersing males would have little or no chance of securing a territory in a new population.

Other hypotheses emphasize genetic benefits to pre-saturation dispersers. For example, sex-biased dispersal reduces the rate of inbreeding for both dispersers and philopatric individuals. If most of the males in a population disperse at sexual maturity, they are unlikely to mate with relatives in their new habitat. The same is true of the philopatric females who mate with the unrelated males that enter the population. Inbreeding can have serious deleterious effects because deleterious recessive alleles are brought together in homozygous form in inbred individuals. The negative impact on fitness from this process is known as **inbreeding depression** (Figure 6.17). Inbreeding depression reduces the reproductive rate or increases mortality. If the population contains many deleterious alleles masked in heterozygotes by beneficial dominant alleles, there may be strong selection to avoid inbreeding. Thus, if inbreeding costs are high, the fitness advantage of inbreeding avoidance can outweigh the mortality costs of movement.

KEY CONCEPTS 6.2

- Species have genetically based habitat preferences—choice of the physical and biological factors important to their survival.
- Territorial behavior provides the individual with exclusive access to critical resources.
- Movements—migration and dispersal—are behaviors that ensure access to favorable conditions and critical resources.
- Territoriality, migration, and dispersal occur when the benefits of the behavior exceed its costs.

QUESTION:

How does the individual optimize its use of resources in the ecosystem?

6.3 How Does the Social System Evolve?

So far we have considered behaviors as independent adaptations to the physical environment, the opposite sex, and competition for resources. Of course, all these behaviors are integrated. Together they constitute a suite of social interactions known as the **social system**.

What Are the Components of the Social System?

The social system comprises three fundamental components: the group size and composition, the degree of cooperation among individuals, and the mating system. All three ultimately result from evolution of behaviors, as previously discussed. Here we examine each component in relation to the others.

Group size and composition vary widely and exemplify the principle of optimization discussed earlier. There are both costs and benefits to groups of different size, and an array of ecological factors determines the optimal number. Over time we expect selection to optimize the group size relative to the ecological situation and prey base. Some animals are purely solitary. They interact with conspecifics only rarely—at the time of mating and for the period the young remain with the female. Some carnivores, such as the mustelids, the family that includes wolverines (*Gulo bulo*), mink, and weasels (*Mustella spp.*), are entirely solitary. Two ecological factors probably underlie their solitary nature. First, their prey can be subdued by a single individual. Second, their prey is relatively less abundant, so that even the large home ranges of these species can support just a single individual. Short-term pair bonds form just for copulation. In fact, many of the mustelids are induced ovulators. This means that the female does not release eggs until copulation has occurred, an adaptation to infrequent and unpredictable encounters with males.

A range of types of group living occurs among animals. Some form loose aggregations that travel widely together. Many ungulates of the African savanna or bison of the North American prairie adopt this pattern. There are social groups within the large herds, in some cases harems defended by males. But the composition of the herd is relatively dynamic. Other group-living species occupy a specific area, and social interactions occur across the group. Belding's ground squirrels form colonies composed of many individuals congregated in a discrete patch of habitat. Their social unit is the matrilineal kin group—a female and her female kin. At the extreme end of the spectrum are species like honeybees that live in highly organized groups with significant division of labor. A typical honeybee hive consists of a single reproductive female, many reproductive males, and thousands of sterile workers that perform the jobs of foraging, hive maintenance, rearing the young, and defense.

Some of the group-living species also exhibit a degree of cooperation. In some, such as lions and wolves, this entails cooperative hunting. Cooperation sometimes includes other forms of helping behavior. Among the most interesting of these are **altruistic behaviors**. We define altruism as any behavior that increases the fitness of others but decreases the fitness of the altruist. Thus, the alarm-calling behavior of Belding's grounds squirrels is altruistic. The obvious question arises: How do such behaviors evolve? If the caller is more likely to be killed by a predator, genes for calling behavior should be eliminated. Noncallers should have higher fitness and come to predominate.

This was indeed a puzzle for evolutionary biologists until W. D. Hamilton solved the problem (Hamilton, 1964). He proposed that the definition of fitness be broadened to incorporate the concept that one can promote one's genes in future generations directly or indirectly. Specifically, an individual can derive fitness benefit from individuals who carry some of the same genes. Of course, relatives share more genes than random individuals in the population. Hamilton defined **inclusive fitness** as the relative ability to get one's genes, or copies of them, into the next generation. Thus, fitness is based on either personal reproductive success or that of individuals that carry copies of one's genes.

social system The organization of a group of individuals in terms of group size and composition, cooperation among individuals, and the mating system.

altruistic behavior Any behavior that increases the fitness of others at the expense of that altruist's fitness.

inclusive fitness A concept of fitness based on the relative ability of an individual to transmit its genes or copies of them to the next generation.

TABLE 6.1 Coefficients of relationship for various familial relationships (for diploid organisms)

RELATIONSHIP	R (AVERAGE PROPORTION OF GENES SHARED)
Parent–offspring	0.5
Full sibs	0.5
Grandparent–grandchild	0.25
Aunt/uncle–niece/nephew	0.25
First cousins	0.125

coefficient of relationship (r)
The proportion of genes shared by two individuals.

Relatives vary in the proportion of genes they share. This quantity is measured by the **coefficient of relationship** (*r*), the average proportion of genes shared by two individuals. The values of *r* for various relationships are shown in Table 6.1.

Hamilton proposed that genes for altruism can increase in frequency if

$$rB - C > 0$$

where *r* is the coefficient of relationship of the altruist, *B* is the fitness benefit to the recipient, and *C* is the fitness cost to the altruist. This simple equation explains how altruism can evolve. Imagine that the altruistic act results in the death of the altruist. Thus, all its genes are lost (*C*). However, if the act increases the success of close relatives (*B*) enough that more copies of the altruist's genes survive, the fitness of the altruist is actually higher. For example, if the altruist gives an alarm call and is killed by the predator, but that alarm increases the survival of a brother by a factor of 2.1, the equation becomes

$$(0.5)(2.1) > 1.0.$$

More of the altruist's genes are transmitted than if it had not given the call. This calculation led Hamilton to quip, "I would gladly lay down my life for two brothers or eight cousins."

An important question arises: How does the altruist know its behavior benefits kin? In other words, do individuals recognize kin? In some systems, direct kin recognition may not be necessary. Recall that in Belding's ground squirrels, males disperse from the colony before they breed. Thus, the colonies are composed of females and their female kin. Interestingly, males rarely give alarm calls; most are given by females. In this system, there is no real need for kin recognition. Because of the sex-biased dispersal, females who call are likely to be warning kin. Males, who are rarely related to their neighbors, do not have this altruistic behavior.

However, many species can in fact discriminate kin from non-kin. This occurs in a variety of ways, but for mammals, olfaction seems to be important. Mice are able to assess shared genes, such as components of the major histocompatability genes, by odor. It also turns out that kin discrimination is greater in species where the average relatedness in the group is lower (Cornwallis et al., 2009). Remarkably, there is even evidence that plants "behave" differently toward kin than non-kin. The annual plant *Cakile edutula* allocates more resources to competitive root growth if grown with non-kin than if it shares a pot with kin (Dudley and File, 2007).

The third component of the social system is the mating system. We have already discussed some of the ecological factors that underlie the mating system. These factors interact with the group size and structure and the frequency of altruism. The organization of the honeybee social system exemplifies this integration. As noted, the hive is a highly integrated group with significant division of labor. The mating system is polyandrous. The sterile female worker caste is especially interesting because they represent the epitome of altruism—they forego their own reproduction to care for the young of the queen. Kin selection helps explain this behavior. Honeybees, like most of the hymenopterous insects, are haplodiploid. This means that females are the diploid products of fertilized eggs; males are the haploid product of unfertilized eggs. Haplodiploidy changes the coefficients of relationships among sibs and parent-offspring. In most organisms, the value of *r* for both parents and their offspring and for sisters is ½ (Figure 6.18). In haplodiploids, the males have but one chromosome to contribute to offspring. Consequently, sisters are related by ¾, whereas parent-offspring *r* is still ½. This means that sisters share more genes with one another than they would with their own offspring. The sterile workers give up their own reproduction in order to care for the offspring of their queen sister.

How Are the Organism's Ecology and Social System Related?

We have examined the evolutionary ecology of the components of the social system. Each component is in some way related to the ecology of the species. Mating

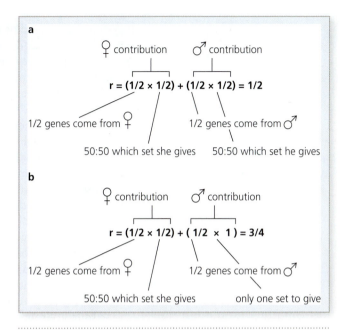

Figure 6.18 Coefficient of relationship. The values of the coefficient of relationship of sisters for (a) diploid species and (b) haplodiploid species. In haplodiploids, the male has only one set of genes to donate to the offspring; thus, sisters share more genes than in diploids. **Analyze:** Is the coefficient of relationship in haplodiploids affected by the number of mating males in the population?

THE HUMAN IMPACT

Conservation of Wild Populations

We have seen the adaptive importance of behavior and social systems. They also play a significant role in conservation biology, the science of maintaining sustainable populations of species. Management and conservation efforts must take into account relevant aspects of behavioral ecology. This is true for wild populations that we are attempting to preserve as well as for species whose populations declined so drastically that they required rescue by captive breeding.

The African wild dog (*Lyacon pictus*) represents an example of a species whose social organization and structure affects its vulnerability in the wild. This species is endangered over much of its range due to habitat loss as well as persecution by humans. African wild dogs were killed for many years to protect livestock and people. Although that effort has largely ended, the species is still at risk (Courchamp and MacDonald, 2001). Efforts to restore sustainable populations of this species are hampered by a central feature of their social system: obligate cooperative breeding. Helpers aid the rearing, protection, and feeding of young. As a result, pack size is particularly important. If pack size drops below a critical minimum, there may not be sufficient helpers to assure successful reproduction.

For other endangered species, captive breeding programs have been used to rescue populations that seemed doomed to extinction. California condors, whooping cranes, and the golden lion tamarin are examples of species rescued from extinction by captive breeding and subsequent release into the wild. For this approach to be successful, however, conservation biologists must account for the unique behavior and social system of each species. Captive-bred animals cannot simply be dumped into suitable habitat and be expected to survive, let alone restore the

species in the wild. Captive-bred individuals must sometimes be taught to forage. In the case of the golden lion tamarin, biologists prepared them for release by placing them in increasingly complex captive environments where they had to gradually learn to find shelter and food and to interact with other tamarins.

Although migratory behavior has a strong genetic basis in many birds, juveniles learn the details of the process by traveling with adults. This posed a significant problem for the release of captive-bred whooping cranes—the population in the wild was so low that there were no adults to help the naïve birds migrate. Extraordinary efforts, including the use of ultralight aircraft to lead cranes in migration, have been employed. There are now more than 100 whooping cranes in the eastern United States. Captive-bred young are placed with adult birds, who teach them the migration patterns.

Even animals born and raised in the wild cannot simply be released into new habitat. When wolves were reintroduced into Yellowstone National Park, the new packs were constituted from wild animals captured in Canada. Before release in the park, they were maintained together in large enclosures so they could develop the dominance patterns and social relations of a normal pack. This ensured that when they were released they would stay together as a functional pack rather than scatter to try to join an established pack.

QUESTION:

What behavioral characteristics would you expect to make conservation and restoration easier? Which would make it harder?

eusocial Describes a complex social system in which there is division of labor or castes, a high level of cooperation, and sometimes altruism.

systems are shaped by the pattern of resource distribution. Group size and composition as well as dispersal are products of the spatial distribution of resources. Predation pressure selects for behaviors such as alarm calling. For most species, the individual components of the social system are explicable by the kinds of ecological pressures we have described.

However, a few species have developed highly organized, complex social systems that require additional analysis. Social systems that include division of labor or castes and a high degree of cooperation or altruism are termed **eusocial**. Thus, the honeybee social system described earlier qualifies as eusocial. Ants and termites are also prominent eusocial insect species. Termites are of special interest because, unlike ants, bees, and wasps, termites are not haplodiploid, yet they show the same kind of reproductive altruism seen in the hymenoptera. Thus, haplodiploidy may facilitate altruism and eusociality, but it is not a requirement.

Only a few species of vertebrates qualify as eusocial. A number of bird species have high degrees of cooperation and altruism but do not have the sharp division of labor seen in the eusocial insects. For example, the acorn woodpecker (*Melanerpes formicivorus*) lives in cooperatively breeding groups. Their social structure centers on one or more granary trees. There the birds store thousands of acorns, their sole food source. Groups vigorously defend their granaries. Typically only a few members of the group breed; the rest actively help rear young that are not their own. Dispersal to or from a group is rare.

A similar social system, characterized by cooperative breeding, is found in a number of bird species, such as the Seychelles warbler, the Florida scrub jay (*Aphelocoma coerulescens*), and the groove-billed ani (*Crotophaga sulcirostris*). Initially, cooperative breeding was explained by kin selection: nonreproductive helpers might gain in inclusive fitness by helping kin raise their offspring. However, it now appears that the ecological situation, especially limited opportunities for entering a new group and high adult survival, are also important correlates of cooperative breeding (Stuchbury and Komadeur, 2009). When favorable home ranges, granary trees, or other important resources are limited, tightly controlled, and separated by inhospitable habitat, the dispersal and reproductive options of group members are limited. If individuals are long-lived, they may benefit from temporarily refraining from reproduction, helping other adults rear their young, and perhaps breeding themselves in the future. The result is philopatry and cooperative breeding.

The naked mole rat (*Heterocephalus glaber*) represents one of the few eusocial mammals (Sherman et al., 1991) (Figure 6.19). This species inhabits harsh deserts in East Africa, where they live in extensive underground burrow systems crucial to their survival. The colony digs these tunnels cooperatively. A long chain of individuals, each pushing dirt, excavates the system. A single female, the queen, breeds in each colony. She is significantly larger than other females. In addition to producing up to 27 pups each year, she also aggressively directs the work of the other members of the colony. There are two different morphs among the sterile females. The smaller one forages for food and helps care for the pups; the larger females defend the colony. Because of the composition of the colony and the single breeding female, the coefficient of relationship among members of a colony is as high as 81 percent.

How might this unusual system arise? Recent evidence from eusocial insects suggests that strict monogamy predisposes species to cooperative breeding. Of 267 species of eusocial bees, wasps, and ants, monogamy was the ancestral condition for all (Hughes et al., 2008). Moreover, the ecological conditions faced by mole rats are similar to those faced by many cooperatively breeding birds. First, the harsh environment and widely scattered resources make dispersal difficult. The fossorial (underground) habit is an adaptation that avoids the heat and desiccation of the desert. Group living is advantageous because the burrow system that affords this protection is impossible for a single individual to construct and is

Figure 6.19 Eusocial systems. The naked mole rat is a eusocial mammal that inhabits harsh deserts in Africa.

highly valuable once in place. Species such as honeybees, acorn woodpeckers, and naked mole rats have very sophisticated social systems. We can explain how they might arise and how they might prove adaptive in certain ecological situations. However, the ecological and evolutionary success of the millions of other species with a vast array of different social organizations argues against the conclusion that the eusocial species stand at the apex of some hierarchy of fitness.

KEY CONCEPTS 6.3

- Altruism directed toward kin may be adaptive; altruism directed toward non-kin is generally not.
- Cooperative breeding may result from kin selection or the improbability that an individual can leave the group and find a mate.
- The social system integrates a collection of behaviors that are adaptive in a specific environment.

QUESTION:

How do kin selection and harsh physical conditions interact in the evolution of the social system?

Putting It All Together

The evolutionary ecology of behavior helps explain the alarm-calling behavior of the Belding's ground squirrel. We began with the oddity of alarm calling—a seemingly disadvantageous behavior for the caller. However, the interaction of ecology and other aspects of the squirrels' behavior leads to a logical explanation. The squirrels are programmed to choose patches of open meadow habitat surrounded by forest. In their high-elevation habitat, there is strong selection to mate and bear young as soon as winter ends so the young have sufficient time to gain the weight that will allow them to survive a long winter in hibernation. Males disperse from the colony before they breed, a behavior whose main advantage probably lies in reducing inbreeding. This in turn sets the stage for the nature of their alarm-calling behavior. Females tend to be surrounded by kin—their alarm calls raise their inclusive fitness—and hence calling is almost exclusively a female behavior.

Of course, Belding's ground squirrels represent just one of millions of species-specific behavioral and social systems. This array of behavioral interactions begins to make sense when we analyze each as we did with the ground squirrels—in relation to the environment and ecology of the species. In fact, some behavioral ecologists refer to social systems as exploitation systems because they are so closely tied to the patterns of resource distribution and acquisition.

 Summary

6.1 How Does Behavior Evolve?

- Behavior is affected by natural selection in the same way as morphological and physiological traits.
- Like other adaptations, behavioral adaptations often involve trade-offs. Selection optimizes the cost-benefit ratio.
- There may be a difference between the proximate and ultimate basis of behavior.
- Stimulus-response systems and communication systems are central to behavioral interactions.
- In many species, females are a limiting resource. Consequently, females exert selection pressure on males.
- Bateman's principle states that males will have greater variance in reproductive success than females.

- Selection favors the female that chooses the male that will increase her reproductive success either by virtue of his traits or because he provides important resources.
- The mating system consists of three components:
  - The number of mates an individual has during the breeding season.
  - The relative contribution of males and females to parental care.
  - The duration of the relationship between males and females.
- The three main types of mating system are monogamy, polygyny, and polyandry.
- The nature of critical resources—their distribution and the potential that they can be defended—determines the mating system.
- Plant mating systems are defined by distribution of male and female function among flowers and plants and by the degree of selfing permitted.

6.2 What Role Does the Ecosystem Play in the Evolution of Behavior?
- Habitat selection behavior evolves to optimize access to the essential resources of the habitat and microhabitat.
- Territories are defended if the energy invested in defense is less than the energetic benefit of exclusive access to the resources. The trade-off is determined by the amount and distribution of resources.
- Migration occurs where seasonal changes in weather and resources vary significantly.
- Individual dispersal behavior is a response to local economic and genetic conditions.

6.3 How Does the Social System Evolve?
- Group size and composition, cooperation, and the mating system characterize the social system.
  - Group size is determined by an optimization process balancing the benefits of conspecifics for mating, cooperation, etc., with the individual's ability to acquire resources without competition.
  - Cooperation evolves when the benefits to the individual outweigh the costs.
  - Genetic relationships determine the degree of inclusive fitness. This determines the costs and benefits of altruistic behavior.
- Social species live in environments in which foraging and defense require groups of individuals and in which solitary individuals are at risk.
  - Eusociality is associated with altruism derived from high genetic relatedness and harsh environments in which defended resources are crucial to survival.

■ Key Terms

altruistic behaviors p. 133
apomixis p. 125
autogamy p. 125
Bateman's principle p. 121
coefficient of relationship (r) p. 134
communication p. 119
dispersal p. 129
eusocial p. 136
female choice p. 121
female defense polygyny p. 123
habitat p. 126
home range p. 126

inbreeding depression p. 132
inclusive fitness p. 133
lek-mating species p. 123
male-male competition p. 121
mating system p. 123
microhabitat p. 126
monogamy p. 123
outcrossing p. 125
philopatric p. 129
polyandry p. 123
polygyny p. 123
pre-saturation dispersal p. 1320

protandry p. 125
protogynous p. 125
saturation dispersers p. 130
self-incompatability p. 125
serial monogamy p. 123
sexual dimorphism p. 119
sexual selection p. 121
social system p. 133
stimulus-response p. 118
territoriality p. 128

■ Review Questions

1. What is the relationship between sexual dimorphism and sexual selection?

2. Explain how social systems can be explained as resource-exploitation systems; how is the social system related to the ecology of the organism?

3. What selective forces are important in the evolution of movement behaviors?

4. How do we explain the exaggerated traits produced by sexual selection?

5. What is the relationship between sexual selection and natural selection?

6. What ecological factors are associated with different mating systems?

7. How is kin selection an indirect form of natural selection?

8. What factors are important in the evolution of territory size and type?

9. A number of behaviors represent trade-offs with cost and benefits. Present examples from this chapter and explain the associated trade-offs.

10. What is the relationship between dispersal behavior and alarm-calling behavior in Belding's ground squirrels?

Further Reading

Darwin, C. 1871. *The descent of man and selection in relation to sex.* London: Murray.

Darwin had remarkable insights for his time regarding behavior and sexual selection. This book anticipates many modern topics in behavior.

Fletcher, J.A., and M. Zwick. 2006. Unifying theories of inclusive fitness and reciprocal altruism. *American Naturalist* 168:252–262.

This paper offers an alternative explanation for the evolution of altruism in some groups: reciprocal altruism. This concept is based on the idea that altruists benefit from an increased probability that they will receive altruistic acts from those they have benefited. It goes on to unify this theory with kin selection theory.

Stacey, P.B., and W.D. Koenig. 1990. *Cooperative breeding in birds.* Cambridge: Cambridge University Press.

Cooperative breeding has been a seminal topic in the study of behavior and social systems. This is an excellent review of the biology, ecology, and evolution of many species of cooperatively breeding birds. The author's comparative approach helps identify the principles important in the evolution of cooperation.

Wilson, E.O. 1975. *Sociobiology.* Cambridge, MA: Harvard University Press.

This classic, comprehensive work places behavior and social systems in an ecological context. In this book Wilson attempted to develop a unified theory of behavior and social evolution. Wilson's synthesis has been controversial, especially his views on the evolution of human behavior. Nevertheless, the book is a remarkable summary of information and data on the evolutionary ecology of behavior.

Chapter 7

The Ecology of Intraspecific Variation

Over the course of the summer, millions of salmon pass through Johnstone Strait in British Columbia as they home in on their natal streams. A succession of species, including king, sockeye, coho, and pink salmon, passes through the strait from May to September. Waiting for them are pods of orcas (*Orcinus orca*), some numbering dozens of animals. These pods are residents—they remain in the strait throughout the year. Their diet is almost exclusively fish and squid.

Smaller groups of orcas known as transients pass through the same waters. But these animals cruise along in groups of six or fewer. Their vocalizations are less frequent and much less complex than residents. More importantly, their prey is exclusively marine mammals such as seals, walrus, and sea lions. These two groups have radically different ecologies—they hunt different prey by different methods. Not only are they ecologically and socially distinct, but they differ morphologically as well. The dorsal fin of resident females is rounded with a sharp bend near the tip; that of transients is pointed and straight.

Presently all these killer whales are united in the single species *Orcinus orca*. However, some taxonomists believe that residents and transients are so different, and interbreeding between them so rare, that they should be classified as separate species. This raises a number of intriguing questions. How does variation like this arise? To what extent is it genetic? The answers to these questions further illuminate the close connection between ecology and evolution. We examined the adaptations to the abiotic environment (Chapter 3). In Chapters 4 and 5 we explored the role of environmental variation

■ **intraspecific variation** Genetic or phenotypic variation within a species.

in the distribution of species. In this chapter we focus on another component of the interaction between ecology and the physical environment: *How do populations respond to variation in the environment?*

What Are the Patterns of Local Environmental Variation?

If we are to fully understand the relationship between the ecology and evolution of populations, we must first analyze the nature of environmental variation. We begin with a fundamental ecological concept: the environment is not constant—it varies from place to place and over time.

How Does the Environment Vary Across Space?

Imagine that you travel from the crest of the Sierra Nevada down into the desert of Nevada. Figure 7.1 shows the diverse environments you would encounter—high alpine tundra, conifer forest, and desert. Each region of the Earth has its own pattern of spatial variation. And each of these environments has a unique set of abiotic and biotic characteristics that affect the ecology and adaptive evolution of the species inhabiting it.

Spatial variation in the environment is the result of discontinuities in physical and biological factors from place to place. Physical factors, such as temperature or soil salinity, vary—sometimes abruptly, sometimes gradually. Biological components of the environment, such as predators, grazers, pollinators, or parasites, also vary spatially. The natural world is a mosaic of patches that differ in these ways. There are three key aspects of these discontinuities that are important to the origin of **intraspecific variation**, the differences among phenotypes in a species:

1. The magnitude of the differences between patches
2. The degree to which the patches are isolated
3. The relative sizes

The magnitude of the environmental differences between patches determines the selective pressure exerted on the organism. When local environments differ markedly, the potential impact is exaggerated; when the differences are slight, adaptive shifts are less likely. For example, in desert environments, local substrates such as white gypsum sand, red rock, and lava differ markedly in color. In some places these different substrates are immediately adjacent to one another. For vertebrate animals living in these environments, matching the background is crucial to predator avoidance. The differences between adjoining habitats are significant: an animal camouflaged in one stands out in the others.

Ecotones are boundaries between habitat types. They reflect the shifting mosaic of spatial variation. The response to the mosaic depends on how sharp the differences are from place to place. The boundaries between substrates are abrupt. In other cases the changes occur more gradually. For example, at increasing elevation in the mountains, the environment becomes more and more hostile to trees until it no longer supports their growth and reproduction. Conifer forests give way to alpine tundra habitat at tree line (Figure 7.2). The shift is gradual—trees change in form and vigor gradually because the temperature and moisture conditions shift gradually.

The isolation and spatial scale of environmental variation also determine the potential adaptive response of the organism. Environmental variation in which the patches are large relative to the mobility of the organism is known as **coarse-grained variation** (Figure 7.3). In coarse-grained environments the organism tends to experience just one or perhaps a few different environments because its

Figure 7.1 Spatial variation. A transect from the crest of the Sierra Nevada into the desert reveals a number of different environments, each associated with a different altitude. These include alpine tundra (a), conifer forest (b), and sagebrush (c).

Figure 7.2 Tree line. Near tree line, the vigor and form of trees change.

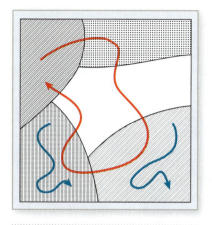
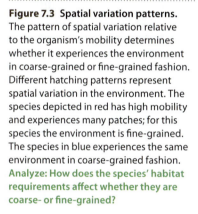
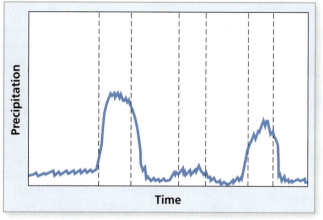

Figure 7.3 Spatial variation patterns.
The pattern of spatial variation relative to the organism's mobility determines whether it experiences the environment in coarse-grained or fine-grained fashion. Different hatching patterns represent spatial variation in the environment. The species depicted in red has high mobility and experiences many patches; for this species the environment is fine-grained. The species in blue experiences the same environment in coarse-grained fashion. **Analyze: How does the species' habitat requirements affect whether they are coarse- or fine-grained?**

Figure 7.4 Temporal change. Unpredictable precipitation in a desert environment. There are regular seasons during which rainfall can occur. However, the variation in rainfall among seasons is high. **Analyze: What is the difference between "variation" and "unpredictability"?**

movement is small relative to the scale of the spatial variation in the environment. In contrast, in **fine-grained** environments the scale of the environmental differences is small relative to the mobility of the organism. Consequently, individuals will experience many environments.

How Does the Environment Vary in Time?

The environment in any one place is rarely constant (Figure 7.4). Temporal change occurs over many time scales. We are all familiar, for example, with the fact that light and temperature change over the course of the day and across a longer seasonal scale. Many of these temporal changes are ecologically significant. For example, in the intertidal zone some sessile species experience both exposure to air and submersion twice each day and thus two radically different physical environments. Some temporal changes, such as droughts or El Niño events, occur irregularly. Others, such as global climate change, either natural or human-induced, constitute long-term directional shifts.

Although the pattern of change is important, the speed of change is also a key factor. The sudden and rapid occurrence of frost is far more detrimental to many plants than a gradual temperature decrease, even if both ultimately reach the same low temperature. We also see this phenomenon in a broad comparison of the temperature regime in the Arctic and the tropics. The mean annual temperature is lower in the Arctic, and this is certainly important. However, the huge seasonal change, the difference between summer and winter, is perhaps even more important.

For some organisms, unpredictability poses a greater challenge than a constant environment. When day length begins to shorten in the Arctic, both animals and plants can predict that the stresses of winter are imminent. Animals migrate, hibernate, or accumulate fat for the winter. Plants become dormant or produce resistant seeds. In contrast, species in unpredictable environments may

coarse-grained variation Spatial variation in the environment that is large relative to the mobility of the organism.

fine-grained variation Spatial variation in the environment that is small relative to the mobility of the organism.

Figure 7.5 Vernal pools. Vernal pools are characterized by concentric rings of vegetation. These occur because as the pool dries up, the soil moisture changes with distance from the center of the pool.

THINKING ABOUT ECOLOGY:

Imagine you are charged with predicting where the effects of global climate change will be most severe. Discuss the relative challenges that climate change poses in (a) an unpredictable environment such as a desert, (b) a seasonal extreme environment like the Arctic, and (c) a relatively stable, benign environment like the tropics.

- **subspecies** A local, phenotypically distinct form. *Race* and *subspecies* are synonymous.

- **phenotypic plasticity** The ability of an organism to produce different phenotypes in different environments.

not have the chance to prepare. Their environment changes without warning—sometimes favorably, sometimes unfavorably. This makes it more difficult to take advantage of favorable conditions or to avoid unfavorable ones. For example, although rains are seasonal in many deserts, whether they will occur in any year, or how much rain will fall in any one locale, is unpredictable. Animals and plants whose activity and reproduction are limited by moisture must respond quickly and appropriately to the changes they encounter.

Of course, spatial variation interacts with temporal variation. For example, in many dry grassland habitats, seasonal rains create small pools of surface water known as vernal pools. As the ponds gradually evaporate, the moisture conditions change sharply from the outer edge of the drying pond to the deeper water in the center. The result is a temporal and spatial gradient of soil moisture. Plants, especially annuals, respond to these patterns with rapid growth where and when they find optimal moisture conditions. Because different species have different moisture requirements, vernal pools are characterized by concentric rings of different species of plants. And as the pool dries, these rings shift as each species grows and dies with the changing environment (Figure 7.5).

KEY CONCEPTS 7.1

- Spatial and temporal variation in the environment constitutes an important selective force on organisms.
- The key characteristics of spatial variation are the intensity of selection among patches, the isolation of patches, and the spatial scale over which they occur.
- The key characteristics of temporal variation are the mean conditions, the speed of changes, and the predictability of the changes.

QUESTION:

How do rapid changes and unpredictable changes differ in the adaptive challenges they pose?

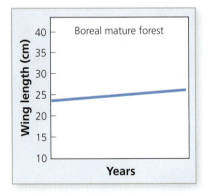

Figure 7.6 Adaptation. There has been a subtle but significant increase in wing length in conifer forest birds over the past 100 years (from Desrochers, 2010). **Analyze: Would you expect this trend in migratory as well as permanent resident species?**

7.2 What Are the Adaptive Responses to Local Environmental Variation?

Rapid anthropogenic changes in the environment show that adaptation is an ongoing process. For example, logging and habitat destruction have decreased the size and increased the isolation of conifer forest ecosystems. The result is that forest birds inhabit increasingly open forests with more open space between forest patches. Desrochers (2010) has documented that the wings of forest birds have become more pointed over the last 100 years. Pointed wings are an adaptation that increases mobility and the efficiency of sustained flight (Figure 7.6).

One of the most thoroughly studied examples of intraspecific variation is the land snail *Cepaea nemoralis* in Britain (Cain and Sheppard, 1954). The shells of these snails vary in the presence or absence of banding as well as in color (Figure 7.7). Cain and Sheppard showed that the color patterns in museum collections correlate strongly with the vegetation background in different regions: the banding pattern of each variant camouflages the snail in its native habitat. Bird predation seems to drive these changes: Cain and Sheppard showed that snails that don't match the background are eaten at a higher rate.

In sexually reproducing organisms, we define a *species* as a group of populations that can interbreed and produce fertile offspring. All the color morphs of *Cepaea* can interbreed with one another and thus are assigned to a single species. Populations of the same species, living in different geographic regions and differing in important biological characteristics, are assigned to different **subspecies**, or races. Figure 7.8 shows a set of subspecies of the salamander *Ensatina eschscholtzii* that inhabit moist woodlands in the mountains of California and Oregon. Each race is found in a specific geographic region. Although each is morphologically distinct, neighboring groups can interbreed with one another.

The essential feature of intraspecific variation is that the phenotype varies with the patterns of environmental variation. There are two basic ways that the phenotype varies with the environment. Genetically distinct populations adapted to the local environmental conditions are known as ecotypes. Nongenetic changes that arise as a developmental response to a particular environment constitute **phenotypic plasticity**.

Figure 7.7 Intraspecific variation. The snail *Cepaea* shows intraspecific variation in color patterns.

How Do Ecotypes Arise?

The variation among *Ensatina* subspecies occurs over a large geographic scale. Local habitat variation can lead populations to adapt to the specific conditions they encounter. Many plant ecotypes inhabiting localized, unusual soil types have been described. The serpentine soils described in Chapter 3 exemplify this phenomenon. Recall that the soil that develops on serpentine rock is inhospitable to plants. Some nutrients, such as magnesium, occur in levels so high they are toxic, whereas the concentration of others, such as nitrogen, phosphorus, and potassium, are extremely low. Serpentine also has poor water-holding capacity, resulting in much more xeric conditions than adjacent nonserpentine soil. Consequently, serpentine outcrops are often nearly devoid of plants. However, a few serpentine ecotypes, populations physiologically adapted to these conditions, do inhabit these sites. Many are morphologically identical to nonserpentine populations of the same species located just a few meters away on other soil types. The same phenomenon is well known in plant populations inhabiting mine tailings, which often have high concentrations of heavy metals.

These adaptations arise by natural selection operating on the species' genetic variation. Directional, disruptive, and stabilizing selection eliminate some variants and favor others. This suggests that the amount of variation on which these processes can act is important to the process. Specifically, the amount of genetic variation in the population may limit the adaptive process. As noted in Chapter 2, R. A. Fisher (1930) formalized this idea in what has become known as **Fisher's Fundamental Theorem**: the rate at which the mean fitness of a population increases by natural selection is exactly equal to the additive genetic variation in fitness. Additive genetic variation is the summed phenotypic effects of the variation within the gene pool. In other words, additive genetic variation is the combined effects of all the different genes that affect a trait.

The intensity of selection also determines whether a local ecotype will arise. This in turn is the direct result of the ecological conditions. For example, in rainforest streams in Trinidad, populations of guppies (*Poecilia reticulata*) vary markedly in color pattern. Upstream, above a series of falls and rapids, guppies are sexually

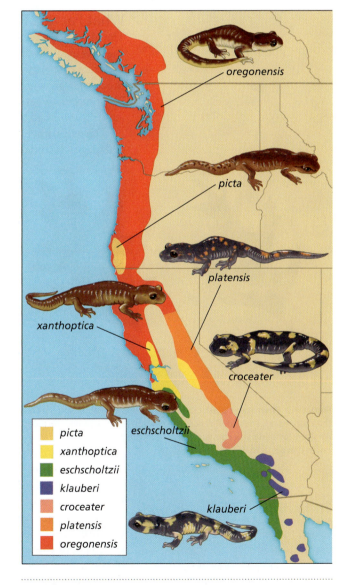

Figure 7.8 Subspecies. The subspecies of *Ensatina escholtzii* are found in the Coast Range and Sierra Nevada. Each race is morphologically distinct and found in a particular portion of the species' range, yet all can interbreed. **Analyze: What evolutionary forces do you think cause the color differences among subspecies?**

■ **Fisher's Fundamental Theorem** The principle that the rate of change in fitness by natural selection is equal to the additive genetic variance in the population.

dimorphic—males are brightly colored and females are drab. Downstream both males and females are drab. The spatial pattern of variation is the direct result of differences in predation pressure and sexual selection in different parts of the stream. These streams tumble down from high elevation in a series of rapids and falls that prevent predatory fish from moving upstream. In upstream populations where predators are absent, females mate with the most brightly colored males. Downstream, predation on brightly colored males constitutes strong selection that results in drab, camouflaged males (Endler, 1983).

Finally, ecotypic differentiation is more likely to occur if the spatial variation in the environment also includes barriers to gene flow among populations. Gene flow tends to oppose local natural selection (Chapter 2). Immigrants from other regions bring genes that may not be locally adaptive. Their effect is to homogenize allele frequencies, which significantly reduces the effect of local selection. In the guppy example earlier, the falls and rapids reduce upstream and downstream gene flow, thus facilitating local differentiation by natural selection.

What Factors Favor Phenotypic Plasticity?

The alternative to adaptation to the local environment is phenotypic plasticity, the development of different phenotypes in different environments. For example, oak trees that grow in deep shade produce wider leaves with smaller indentations; those that grow in full sunlight have narrow, highly lobed leaves (Figure 7.9). There are no genetic differences between these trees. Rather, the development of leaves is modified by the light regime the individual experiences.

Darwin's finches and the marine iguana (*Amblyrhynchus cristatus*) of the Galápagos Islands exemplify genetic adaptation and phenotypic plasticity, respectively. Both species are sometimes affected by extreme environmental conditions. During droughts, small seeds become scarce and the finches must rely on large, hard seeds. Directional selection favors those individuals with larger bills, and genes for large bills increase in frequency. In contrast, marine iguanas are phenotypically plastic. During El Niño events, the cool water and upwelling nutrients near the islands are replaced by warmer, less productive water. This in turn causes a significant decrease in the red and green algae that constitute the primary food source for iguanas. Under starvation conditions, smaller iguanas fare better because they use energy more efficiently. During an El Niño, the mean size of individuals declines. However, the change is not genetic; the bones and connective tissues of individuals actually shrink. Some individuals have been found to reduce their body size by as much as 20 percent, then regrow when the algae return (Wikelski and Thom, 2002).

Sun leaf   Shade leaf

Figure 7.9 Phenotypic plasticity. Shade leaves have larger surface area than sun leaves. These differences can even arise in the same individual. **Analyze: What do you conclude about this adaptation if the differences can occur in the same individual?**

Conifers near tree line on a mountain have a horizontal growth form. This might be an adaptive phenotypic change that reduces dehydration from cold wind and snow damage to the branches. Or it might simply be the negative impact of a harsh environment. How might you distinguish between these alternatives?

Not all phenotypic responses are adaptive. A pine tree, for instance, growing in wet, low-nutrient soil, may have slow, stunted growth. In this case, the phenotypic changes are not adaptive—they are simply the detrimental effects of poor habitat. We can distinguish between phenotypic differences, which arise due to the direct detrimental effect of the environment, and phenotypic plasticity, in which the phenotype shifts in adaptive ways in different environments. Adaptive phenotypic plasticity requires a mechanism that not only detects environmental change but alters the developmental pathways so as to produce the optimal phenotype. This characteristic is both the advantage and disadvantage of phenotypic plasticity: the organism can respond to its environment, but the developmental complexity and energetic requirements for phenotypic changes may be a burden.

ON THE FRONTLINE

Melanism and Phenotypic Plasticity in Beetles

Phenotypic variation can be an important strategy for tracking changes in the environment. This approach and the alternative, genetically programmed adaptation, are not mutually exclusive. Given the advantages and disadvantages of each approach, it is not surprising that mixed strategies occur. Michie et al. (2010) have elucidated an example in the harlequin ladybird beetle (*Harmonia axyridis*), a species endemic to central and eastern Asia.

This species varies enormously in color and spotting. Some races are melanistic—they have darker bodies with more and larger dark spots. For poikilotherms, such as these beetles, dark pigment increases heat absorption and thus provides thermal advantages in cold environments.

Figure 1 Color variation. Variation in the spotting patterns of the harlequin ladybird beetle (*Harmonia axyridis*) (from Michie et al., 2010).

This study was possible because the genetics of the color morphs of the harlequin ladybird beetle are well known. Color and spotting are controlled primarily by four alleles. Among natural populations, nonmelanistic forms are more common in hotter climates; melanistic forms increase in frequency in colder climates. Also, within a population, the frequency of melanism changes over the course of the year—melanistic forms predominate in colder months, nonmelanistic forms in the summer. These shifts are known to be genetic; that is, the melanistic alleles increase in frequency with latitude and during the colder months within a population. Mitchie and his associates were curious if there is also phenotypic plasticity for melanism.

The researchers addressed the question of whether the beetles employ a mixed phenotypic and genetic strategy.

HYPOTHESIS: Melanism is the result of both genetic differences among populations and phenotypic responses to temperature.

If this hypothesis is true, we expect to find genetic differences among some populations. In others, the degree of melanism should be determined by the temperature at which development occurs. The latter possibility leads to the key prediction of their hypothesis.

PREDICTION: Eggs reared in colder temperatures should produce more melanistic adults with larger black spots.

Their experiment compared four major spotting genotypes ranging from ones with low frequency of melanism to one that is largely black in color. The eggs of each of the genotypes were reared at 14°F, 21°F, and 28°F. The researchers quantified the degree of melanism and the size of black spots in individuals from each of these treatments.

Consistent with their prediction, the nonmelanistic form showed a significant increase in the number and size of black spots in cold temperatures. In the melanistic form, spot size and number also changed, but the effect was much smaller. They concluded that there is an interaction between phenotypic plasticity and genetically programmed melanism; different morphs have different degrees of plasticity and genetically programmed melanism. Interestingly, the nonmelanistic form occurs in more temperate regions, where it may experience both cold and hot temperatures. In that environment, phenotypic plasticity would be a great advantage. In contrast, the native range of the melanistic form is Siberia, where the temperature regime is cold and much less variable. A genetically based melanism is adaptive there because there is rarely the need to develop a different color morph.

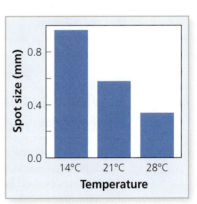

Figure 2 Melanism. Melanistic spot size increased in individuals from eggs reared at lower temperature (from Mitchie et al., 2010). **Analyze: Explain the importance of melanism in temperature regulation.**

What determines whether the adaptive response to the environment is genetic or phenotypic? Ultimately this is determined by the temporal and spatial pattern of variation in the environment. In environments in which variation is fine-grained, the organism typically experiences many ecological conditions.

In this case, phenotypic plasticity may be the optimal response—a single genotype can develop more than one phenotype depending on the environment it encounters. Plasticity is also advantageous if the pattern of temporal variation includes rapid shifts or highly unpredictable conditions, for which a genetic response would be too slow. On the other hand, if the pattern and timing of spatial and temporal variation in the environment is such that the organism experiences a single environment, a genetically fixed adaptation may be the superior response. There is no adaptive value to phenotypic plasticity—the organism encounters only one environment. Compare two fish—a tuna and a steelhead. The steelhead spends part of its life in the ocean but breeds in freshwater rivers. It must modify its physiology to survive in both saltwater and freshwater. But the tuna lives its entire life in saltwater. There is no adaptive value in the ability to shift its phenotype to tolerate freshwater.

KEY CONCEPTS 7.2

- Ecotypes are most likely to arise when (a) there is sufficient genetic variation on which selection can act, (b) local selection pressure is intense, and (c) gene flow from regions with different selection pressure is minimal.
- Phenotypic plasticity allows the organism to respond appropriately to the spatial or temporal variation it encounters.
- Ultimately it is the pattern, intensity, and spatial or temporal variation in the environment that determines whether a genetic or phenotypic response evolves. If the patterns are such that the organism is likely to face a single set of conditions, a genetic adaptive response is more likely; if the pattern is temporally or spatially variable such that the organism encounters a changing environment, phenotypic plasticity is advantageous.

QUESTION:

In phenotypically plastic organisms, how are the developmental and phenotypic response times of the organism related to the patterns and timing of changes in the environment?

7.3   What Factors Determine the Amount of Intraspecific Variation?

Fisher's Fundamental Theorem affirms the importance of genetic variation in the adaptive response to the environment. The variation within and among populations represents the interaction of evolutionary processes unfolding in each species' unique ecological context.

What Mechanisms Increase Genetic Variation?

Ultimately, all genetic variation derives from mutations. Once new mutants arise, a number of mechanisms increase, decrease, or maintain that variation. Sexual reproduction is certainly among the most important of these. Because genes from more than one parent contribute to the genotype of the offspring, sex increases the number of new combinations of genes. The meiotic process itself also contributes variation when crossing over among chromosomes recombines genes. Dominant and recessive genes also contribute to genetic variation. In diploid species, dominance masks the phenotypic effects of recessives and can allow deleterious recessive alleles to persist in the population.

The importance of sexual reproduction as a source of new gene combinations is epitomized by examples in which environmental stress triggers massive sexual

THE EVOLUTION CONNECTION

The Evolution of Sex

Sexual reproduction generates genetic variation in populations by producing new combinations of alleles. The obvious advantage of sexual reproduction is the potential adaptive value of these new genetic combinations. You might then assume that the evolutionary origin of sex is based simply on these advantages. However, a problem arises that illustrates an important evolutionary principle. Sexual reproduction is advantageous only when it is established in the population. When sex is rare, as would have been the case when it first arose in an asexual species, it is not adaptive. We say that sex is not an *evolutionarily stable strategy (ESS)*. To be an ESS, an adaptation must be able to increase in frequency when rare.

Thus, the origin of sex remains a puzzle for evolutionary biologists. Imagine a world in which all reproduction is asexual. Any mutations that lead to sexual reproduction would initially face a considerable disadvantage. Each offspring of sexual reproduction carries one half the genes of each parent. In contrast, each offspring of asexual reproduction carries all the genes of the parent. Evolutionary fitness is based on the relative ability of individuals to pass their genes on to the next generation. The asexual type has twice the fitness of its sexual counterpart because twice

as many of its genes are conveyed to its offspring. In direct competition, asexual reproduction will quickly outcompete sexual reproduction. How, then, did sexual reproduction arise?

One important hypothesis for the origin of sexual reproduction centers on parasite-host systems. Imagine a eukaryotic host infected by a prokaryotic parasite. The generation time of the parasite is much shorter than its host's, and thus the parasite can evolve much more rapidly than the host. The host's defense systems will always be a step behind. Theoretical biologists term this the *Red Queen hypothesis*, from *Alice in Wonderland*, in which the Red Queen had to run faster and faster to stay in place. Sexual reproduction could be an ESS if it increased host resistance (Lively, 1996), because the gene combinations compensate for the rapid evolution of the parasite. Of course, once established, sexual reproduction provides all the advantages of recombination, and its broader fitness value can be realized.

QUESTION:

What are some of the costs and benefits of sexual versus asexual reproduction?

---

reproduction. Some plants—oaks, for example—produce relatively low numbers of flowers and seeds in most years. They may even skip reproduction in some years. However, an extreme environmental stress such as a drought sometimes triggers massive flowering and sexual reproduction. The many new recombinants produced may include some that are adaptive in the new conditions.

The mechanisms of natural selection also play a role in maintaining genetic variation. Disruptive selection maintains genetic variation in the population because more than one phenotype is adaptive and selection ensures that each persists. **Frequency-dependent selection**, in which the fitness of a genotype varies with its frequency in the population, also maintains variation by favoring rare genes (Figure 7.10). The elderflower orchid (*Dactylorhiza sambucina*) illustrates this phenomenon. Populations of this orchid contain two color morphs, yellow and purple (Figure 7.11). The orchid provides no nectar reward for insect visitors. The behavior of its main pollinator, a bumblebee, explains how selection favors the rare genotype. Bees, like many insects, sample flowers and learn which produce rewards. If a bee visits a flower of a particular color and receives no reward, it shifts to a different flower type. Now, imagine that yellow flowers are rare and purple flowers are common. Bees that enter the population are likely to sample the more frequent purple flower. Because they receive no reward, they next sample a yellow flower. Bees may switch back to purple and to yellow again before giving up and leaving the population. However, this behavior ensures that both yellow and purple flowers are pollinated and produce seed. In fact, visits to yellow occur more frequently than if visits to the two morphs were random—each purple visit is followed by a yellow visit. Thus, the rare yellow morph receives a proportionally larger number of visits and it persists despite its rarity. No matter which morph becomes rare, the proportional increase in its pollination success ensures it is not eliminated.

frequency-dependent selection A form of natural selection in which the fitness of a gene is determined by its frequency in the population.

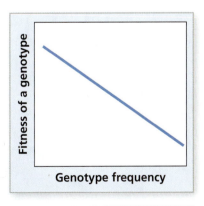

Figure 7.10 Frequency-dependent selection. Frequency-dependent selection occurs when the fitness of a genotype varies with its frequency in the population. **Analyze: What evolutionary forces balance or limit this shift?**

Figure 7.11 Elderflower orchid. Populations of the elderflower orchid contain both purple and yellow morphs.

■ **common garden experiment** An experimental design that distinguishes between genetic variation and phenotypic plasticity. Phenotypically different plants from different environments are grown in a common garden. Differences that persist are genetic; similar phenotypes in the common garden indicate phenotypic plasticity.

What Mechanisms Decrease Variation?

A number of evolutionary processes work in the opposite direction—to decrease intraspecific genetic variation. One form of natural selection, stabilizing selection, eliminates extreme phenotypes and their genotypes, resulting in reduced variation in the population. Directional selection favors phenotypes in one tail of the bell curve and eliminates those at the other extreme. Inbreeding, the result of mating among related individuals, also decreases genetic variation. One effect of inbreeding is to increase the frequency of the homozygotes and decrease the frequency of heterozygotes. Variation is thus lost in two ways. Heterozygotes, which by definition contain two different alleles, are themselves a component of genetic variation. Also, the increase in homozygosity exposes recessive alleles to natural selection. Variation is reduced as selection eliminates recessive alleles that otherwise would be protected in heterozygous form. Genetic drift reduces genetic variation as well. As allele frequencies shift stochastically, some are fixed ($p = 1.0$) and others are lost ($p = 0$). Each time this happens, the total genetic variation in the population is reduced.

Virtually all of the mechanisms that increase or decrease genetic variation are affected directly by the ecology of the organism. We have seen that extreme environmental stress sometimes leads to massive sexual reproduction. In other cases, such as when drought forces Galápagos finches to consume larger seeds, directional selection for large bills reduces the overall genetic variation. The ecological processes that determine population size are particularly important because the size of the population determines the relative importance of inbreeding and genetic drift. Interactions with other species also contribute to the connection between ecology and population genetics, as illustrated by pollinator behavior in elderflower orchids and predation on beach mice.

How Do We Measure Genetic Variation in Populations?

We know that the phenotype is the product of complex developmental processes directed by the genes and modified by the environment. If we are to assess genetic variation, we must first distinguish between genetic variation and phenotypic plasticity.

Superficially, genetic and phenotypic responses are similar—distinct phenotypes are associated with specific environments. One traditional approach to distinguish among the possibilities in plants is to employ a **common garden experiment**, in which individuals with different phenotypes in the field are grown under similar conditions—in a common garden—to see if the field phenotypes persist when the plants are grown under identical conditions. If they do, we know that the differences among populations are genetically controlled. If the phenotypes that arise in the common garden are similar, we conclude that the variation in the field is due to phenotypic plasticity.

Common yarrow (*Achillea lanulosa*) is found across a gradient of habitats in California, where it occurs on the coast, in the Central Valley, and in the high Sierra. Plant height varies widely across this range of environments (Figure 7.12). One of the first and still classic applications of the common garden technique was used by Klausen et al. (1948) to determine which of these phenotypic differences have a genetic basis. Seeds from the coast, the Sierra foothills, and the Sierra crest were collected and planted in common gardens at all three sites. The results are shown in Table 7.1.

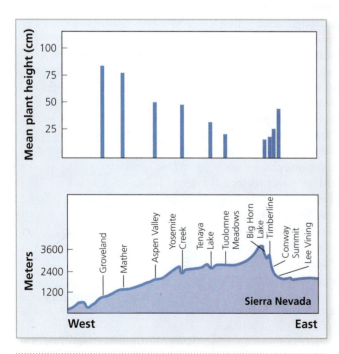

Figure 7.12 Plant height. Variation in plant height in *Achillea lanulosa* across a transect from the coast to the Sierra Nevada (from Klausen et al., 1948). **Analyze: What might be the adaptive advantage of small stature in high-elevation populations?**

TABLE 7.1 Plant heights of yarrow (*Achillea lanulosa*) from different populations grown in common gardens (from Klausen et al., 1948)

GARDEN	POPULATION ORIGIN		
	Groveland	Mather	Timberline
Stanford	83.6	58.2	15.5
Mather	79.6	82.4	34.3
Timberline	21.2	31.6	23.7

There is evidence of both genetic differentiation of the populations and phenotypic plasticity. Note that the small plants from the high-elevation population are small regardless of where they grow; this morph seems to be genetically programmed for small stature. On the other hand, the coastal plants are tallest in the garden on the coast but are dwarfed when grown at high elevation, suggesting that for this population, high-altitude dwarfing is a phenotypic response to the environment.

Statistical methods can also distinguish between genetically based responses and phenotypic plasticity. For example, long-term data on the breeding date of the great tit (*Parus major*) in England show that it is breeding earlier than it did in the past (Figure 7.13). One effect of climate change is that both warm temperatures and insect emergence now occur earlier in the spring. The shift in breeding in the tit could be the result of a genetic change in breeding or a plastic response to each year's conditions. Charmantier et al. (2008) used data from birds that bred over several years and experienced different temperature regimes to show that the shift is a phenotypic response. Figure 7.13 shows that the same individuals breed early or late depending on the temperature that year. Specifically, the mean difference in laying dates within an individual was correlated with the year-to-year difference in temperature. If the population shift to earlier dates were genetically programmed, each individual should breed at the same time regardless of temperature.

Prior to the advent of modern molecular technology, biologists interested in intraspecific variation were limited to the analysis of observable phenotypes. However, individuals vary in many genes whose phenotypic effects may not be obvious or quantifiable. Today an array of molecular techniques allows us to quantify a larger proportion of the genome in comparisons of the genetic differences within and among populations. DNA fingerprinting is perhaps best known

THINKING ABOUT ECOLOGY:

The high-altitude morph of yarrow is genetically programmed to grow to small stature. What adaptive advantages might there be to small stature in high-elevation populations? How could you test your hypotheses?

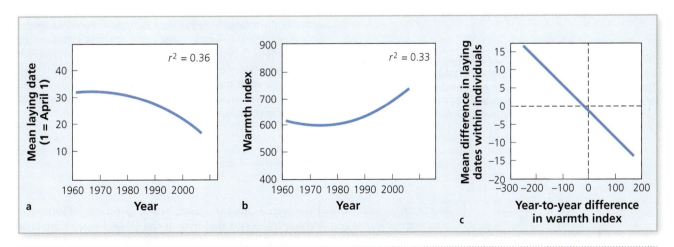

Figure 7.13 Breeding date changes. Laying dates for great tits (*Parus major*) over time (a) correlate with an increase in an index of sun and temperature during the breeding period (b) (from Charmantier et al., 2008). **Analyze: Do these data prove that temperature is the critical factor?**

DO THE MATH

Measuring Genetic Variation

Imagine a series of discrete populations separated in space. To fully understand the genetic variation in this system, we must understand the distribution of variation within and between populations. One way to quantify hierarchical genetic variation employs Wright's F-statistics (Wright, 1935). The analysis is based on the concept of partitioning variation into its components. In the set of populations in Figure 1, we measure the allele frequencies of one or more genes in each population. The total genetic variation in this group of populations is composed of two constituent parts: the variation *within* populations and the variation *among* populations.

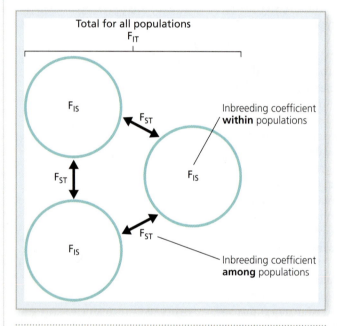

Figure 1 Variation within and among populations. Schematic representation of the partitioning of genetic variation into the portion of the total variation (F_{IT}) due to variation within populations (F_{IS}) and the portion due to variation among populations (F_{ST}). **Analyze: What do you conclude if FIS is a larger proportion of FIT than FST?**

The basis of this analysis is the inbreeding coefficient, F, the probability that two alleles in a population are identical by descent; that is, the two alleles are derived from the same ancestral copy of the gene. Mating of related individuals is the chief cause of identity by descent; thus, a large value of F signifies a high level of inbreeding. The variable F also measures the correlation among alleles in a population—their probability of co-occurrence. Alleles within and among populations can be similar (correlated) due to inbreeding, genetic drift, natural selection, or gene flow. Another way of thinking of the variable F is the genetic similarity of two entities (individuals or populations). The more similar they are genetically, the higher the correlation between them.

In Wright's method, the total variation (F_{IT}) is due to the summed effects of variation within populations (F_{IS}) and that between populations (F_{ST}). How are these variables related mathematically? We begin by considering the probability that there is no correlation among alleles: $1 - F$. Thus, for the entire set of subpopulations, the probability that two allele frequencies are *not* correlated is given by the equation

$$(1 - F_{IT}) = (1 - F_{ST})(1 - F_{IS})$$

This equation is based on the product rule of probability: the probability of two independently occurring events happening simultaneously is the product of their independent probabilities. Thus, the probability that allele frequencies in the set of populations is not correlated ($1 - F_{IT}$) is the product of the probabilities that they are not correlated within (F_{IS}) or among (F_{ST}) groups.

Now we can algebraically derive the relationship between the three components of variation

$$1 - F_{IT} = 1 - F_{IS} - F_{ST} + F_{ST}F_{IS}$$

or

$$F_{IT} = F_{IS} + F_{ST} - F_{ST}F_{IT}$$

In effect, we have partitioned the total variation (F_{IT}) into two components, the amount derived from among populations (F_{ST}) and that derived from within populations (F_{IS}).

F_{ST}, the variation among populations, is particularly important. In terms of the other components of variation,

$$F_{ST} = (F_{IT} - F_{IS})/1 - F_{IS}$$

This is an important measure of the degree of genetic differentiation of populations: the higher the value of F_{ST}, the larger the proportion of the total genetic variation is due to differences *among* populations; the smaller the value of F_{ST}, the less differentiated are populations. Ecological factors, such as barriers to movement among populations (reducing gene flow), radically different environments for each population (strong selection pressure), and the size of the population (and hence the importance of drift), contribute to the variable F_{ST}.

as a forensic technique. However, it can also be applied to questions of intraspecific genetic variation. Although the vast majority of the DNA in individuals of the same species is identical, there are regions in which variation occurs. DNA fingerprinting uses a portion of the genome that contains short repeated sequences of a few bases. The most commonly used type contains short tandem repeats, sets of four bases repeated a variable number of times. Individuals differ in the number of repeats; intraspecific variation is measured by the differences within or among populations in these genotypes.

Direct sequencing of DNA also provides a measure of genetic differences among individuals. As sequencing has become more sensitive, inexpensive, and rapid, sequence information has been brought to bear more often for this purpose. Sequence data allows us to measure directly the proportion of the genome that differs between individuals and populations.

The great advantage of molecular techniques is that they measure variation in a much larger proportion of the genome than specific traits such as the shell color patterns in *Cepaea*. Some molecular markers, such as the tandem repeats used in DNA fingerprinting, probably have no phenotypic effect. These markers are selectively neutral; that is, none confers a particular advantage or disadvantage on the individual. Thus, their frequencies do not change by natural selection. Instead, any differences among individuals and groups arise by genetic drift. Consequently, they provide an index of genetic differentiation independent of the environment.

There is also selectively neutral variation among alleles in some genetic loci that code for proteins. Many protein enzymes occur in allelic forms that have no physiological effect. Even some genes for important morphological traits may be selectively neutral. For example, there is considerable variation in the striping pattern of tigers (Figure 7.14) within and among populations (Kitchner, 1999). However, it is unlikely that this variation has important fitness effects.

Regardless of the technique employed, the goal is to measure the number of allelic differences among individuals. This information can then be applied to the pattern of spatial or temporal variation among individuals. We quantify genetic differences among populations with measures of **genetic similarity** (Nei, 1972). If we can determine allele frequencies for a number of genetic loci, we can quantify the genetic similarity of two populations as

$$I_N = \frac{\sum_{i=1}^m p_{ix}\ p_{iy}}{\sqrt{\left(\sum_{i=1}^m p_{ix}^2\right)\left(\sum_{i=1}^m p_{iy}^2\right)}}$$

where p_{ix} is the frequency of allele I in population x, p_{iy} is the frequency of the same allele in population y, and m is the number of genetic loci. Notice that when p_{ix} and p_{iy} are very different, the numerator becomes small relative to the denominator and the index becomes small. In contrast, if the frequencies of the two alleles are similar, the numerator and denominator will also be similar and the index will be large. For example, if $p_{ix} = p_{iy} = 0.5$, their product is 0.25. But if $p_{ix} = 0.9$ and $p_{iy} = 0.1$, their product is just 0.09 and the value of the index, the genetic similarity of the two populations, declines.

The great advantage of molecular measures of genetic variation is that they sample a large proportion of the genome, which allows us to make broad comparisons of the variation among species or analyze intraspecific changes in variation over space and time. However, these methods do not necessarily measure the amount of *adaptive* variation among populations. Further analysis is required to determine if genetic differences are adaptive. For example, the beach mouse (*Peromyscus polionotus*) is a classic case of color matching. This species, found in light sand and dune habitats, is much lighter in color than its relatives that live on darker substrates. Moreover, we know that coat color is determined genetically—these differences represent ecotypic differentiation. However, a comparison of the genomes of mice living on light and dark substrates using molecular techniques revealed no statistical differences between them (Mullen et al., 2009). The adaptive color morphs are the result of a small number of genes under strong selection. These differences are not manifest in the rest of the genome that molecular techniques sample.

Figure 7.14 Striping pattern of tigers. Variation in the striping patterns in tigers from different regions (from Kirchner, 1999).

genetic similarity The measure of the proportion of alleles shared by two populations.

THINKING ABOUT ECOLOGY:

Do subspecies necessarily represent adaptation to regional conditions?

THE HUMAN IMPACT

Conservation Genetics in Small Populations

Small populations are important in the process of evolution. Of course, one of the most significant human impacts on natural populations is to reduce their size and increase their isolation. The field of conservation genetics deals with these effects and ways we might mitigate them. The genetics of small populations is dominated by the increased importance of chance effects. Moreover, many of these chance effects reduce the genetic variation in the population. Because it is variation that allows the organism to respond to environmental change, the long-term fitness of the species may be compromised when the population declines.

Genetic drift operates in all populations, but its effects are proportionally large in small populations. As the total population, N, decreases, N_e necessarily declines as well. Thus, human impacts that reduce the total and effective population sizes increase the relative importance of genetic drift. And as we have seen, one long-term effect of drift is the fixation or loss of alleles, ultimately reducing genetic variation. For example, grizzlies persist in several small populations in North America. In some, N_e is only 25 percent of N (Harris and Allendorf, 1989). Several populations have demonstrably low levels of genetic diversity, especially in small remnant populations in the Northwest Territories of Canada and Yellowstone (Paetkau et al., 2008). The second important consequence of small population size is the increased probability of the deleterious consequences of inbreeding. In small populations there are fewer potential mate choices and thus higher rates of inbreeding. We do not yet know if inbreeding affects remnant grizzly populations.

Even if a small population later expands again, the effects of drift and inbreeding may persist. This phenomenon—reduction and subsequent expansion—is known as a *population bottleneck*. Lions exemplify the persistent consequences of this phenomenon. Populations that passed through a bottleneck and later recovered are still burdened by deleterious alleles that produce abnormal sperm and lower testosterone levels (Wildt et al., 1987).

Recovery programs generally seek to restore both population size and genetic variation. The latter may require import of new individuals and new genotypes in an attempt to restore genetic variability where it has been lost. This, too, may be problematic because of the phenomenon known as *outbreeding depression*. If two populations are adapted to different local environments, mixing them may result in lower fitness because locally adapted gene complexes may be broken up. For example, hybrids of disjunct populations of the larkspur (*Delphinium nelsonii*) have 48 percent smaller size as well as negative impact on pollen production (Waser and Price, 1994).

All these factors are further complicated by the fact that each species and each population is genetically, historically, and ecologically unique. For example, many plant species are self-compatible—that is, able to self-fertilize. Consequently, they have naturally high levels of inbreeding and often high levels of homozygosity. However, over time, selection may have purged the population of deleterious alleles. If so, inbreeding and drift will not have the negative consequences we expect.

Unfortunately, the genetic consequences of small population size are idiosyncratic. Although we understand the important phenomena in general, we are in the difficult position of having to know the details of each species' biology in order to predict the effects of human impacts as well as to develop management and recovery strategies.

QUESTION:

Why might you expect that management practices that simply increase the size of a population may be insufficient to mitigate the genetic problems it faces?

KEY CONCEPTS 7.3

- According to Fisher's Fundamental Theorem, adaptive evolution depends on the genetic variation on which selection can act. Mutation is the ultimate source of this variation.
- Sexual reproduction, diploidy and dominance, disruptive selection, and frequency-dependent selection all maintain or increase genetic variation in populations.
- Inbreeding, genetic drift, and stabilizing selection are important mechanisms that reduce genetic variation.

QUESTION:

Compare and contrast the speed of adaptive evolution in small compared to large populations.

Putting It All Together

The patterns of spatial and temporal variation in the environment are fundamentally important to ecology and evolution. The adaptive responses of the organism must keep pace with the ecological changes it experiences. Strong selection pressure, barriers to gene flow, and predictable conditions all lead to the evolution of local ecotypes. When the conditions faced by the organism are less predictable, because they change over time or because their spatial pattern is small relative to the mobility of the organism (fine-grained environments), phenotypic plasticity is often the result.

Each of these broad categories of response depends ultimately on the amount of genetic variation available for evolutionary forces to work on. The amount of genetic variation in turn depends on the balance between opposing mechanisms. Processes such as mutation, sexual reproduction, and frequency-dependent selection increase and preserve genetic variation; processes such as genetic drift and population bottlenecks decrease it.

These principles apply to the origin and persistence of resident and transient orcas. Variation inherent in the environment, specifically the presence of two distinct food sources, salmon and marine mammals, ultimately led to the origin of two morphs. Ancestral orcas that followed either of these two distinct prey bases gradually adapted their behavior and hunting and ultimately produced a specialized form. The tactics and group sizes most adaptive to feed on salmon and seals differ markedly—a mixed strategy would probably not be successful for either.

We know that adaptations this specialized require intense selection pressure and restricted gene flow. And because marine mammals are dispersed widely in the ocean environment, the transients must keep moving to locate their food source. One result is less contact between transients and residents and thus reduced gene flow between the groups.

The orcas also exemplify another principle of intraspecific genetic variation: not all genetic variation is tied to fitness. For example, the differences among orcas in the location and size of white patches have no known adaptive value. This variation probably arose by the random processes of mutation and genetic drift. And like other nonselected variation, it might become important in the future—perhaps for sexual selection or as markers that facilitate individual recognition within the group.

Summary

7.1 What Are the Patterns of Local Environmental Variation?

- Ecological conditions vary spatially in the intensity of the selective pressure they exert.
- Spatial environmental variation can be coarse-grained (the scale of the variation is large relative to the mobility of the species) or fine-grained (the scale of the variation is small relative to the mobility of the species).
- Some environments change abruptly, others gradually.
- Shifts in ecological conditions may be gradual or rapid.
- Unpredictable changes pose significant adaptive challenges.

7.2 What Are the Adaptive Responses to Local Environmental Variation?

- Ecotypes are most likely to arise when selection is strong, boundaries between environments are sharp, and gene flow is reduced.
- Phenotypic plasticity is advantageous when temporal environmental change is unpredictable or spatial environmental variation is fine-grained.

7.3 What Factors Determine the Amount of Intraspecific Variation?

- Sexual reproduction, disruptive selection, and frequency-dependent selection tend to increase local genetic variation.

- Inbreeding and genetic drift are important processes that decrease variation.
- Population bottlenecks, during which population size is reduced, contribute to those processes.
- Fisher's Fundamental Theorem states that adaptive evolution is limited by the additive genetic variance in the population.

- Phenotypic plasticity and ecotypic differentiation can be distinguished by common garden experiments.
- Molecular techniques such as DNA sequencing and DNA fingerprinting measure the variability in the genome.

■ Key Terms

coarse-grained variation p. 143
common garden experiment p. 150
fine-grained variation p. 143

Fisher's Fundamental Theorem p. 146
frequency-dependent selection p. 149
genetic similarity p. 153

intraspecific variation p. 142
phenotypic plasticity p. 144
subspecies p. 144

■ Review Questions

1. What are the key features of spatial variation in the environment?

2. What are the key features of temporal variation in the environment?

3. What aspects of environmental variation are most difficult for organisms to adapt to?

4. What is the difference between subspecies and ecotypes?

5. What factors increase the development of ecotypes? What factors retard their development?

6. What are the relative advantages and disadvantages of genetic adaptation and phenotypic plasticity?

7. Explain the main mechanisms that increase or maintain genetic variation in populations.

8. Explain the main mechanisms that decrease genetic variation in populations.

9. How do we experimentally distinguish between genetic adaptation and phenotypic plasticity?

10. Explain the advantages and disadvantages of using molecular measures to quantify intraspecific variation.

■ Further Reading

Falk, D.A., and K.E. Holsinger. 1991. *Genetics and conservation of rare plants.* New York: Oxford University Press.

This book is a classic treatment of the effects of rarity on the genetics of plants and the genetic consequences of rarity. It begins with a thorough discussion of the biology and genetics of rarity. It then progresses to discussions of management and conservation strategies for small, genetically impacted populations.

Mullen, L., et al. 2009. Adaptive basis of geographic variation: genetic, phenotypic and environmental differences among beach mouse populations. *Philosophical Transactions of the Royal Society B* 276:3809–3818.

This paper is a clear and well-documented analysis of the relationship between adaptive variation in coat color and molecular measures of differences among populations. It merges classical and molecular analyses of intraspecific variation.

Harrison, S., and N. Rajakauna (eds.). 2011. *Serpentine: the evolution and ecology of a model system.* Oakland, CA: University of California Press.

Serpentine outcrops are classic systems for the analysis of physiology and local adaptation. This book summarizes and synthesizes the details of ecotypic differentiation in serpentine systems.

Pigliucci, M. 2001. *Phenotypic plasticity: beyond nature and nurture.* Baltimore: Johns Hopkins University Press.

Phenotypic plasticity is an important component of intraspecific variation. Because plasticity itself can evolve, it bridges the gap between purely genetic and purely environmental variation. This book summarizes the state of our understanding of the evolution and ecology of phenotypic plasticity.

Part 2

Populations

| **8** Demography | **9** Population Regulation | **10** Life History Strategies |

Steelhead trout hatch from eggs laid in freshwater rivers. The juveniles migrate to the sea, grow, and eventually return to their natal streams to mate. The next three chapters explore the ecological mechanisms that determine the characteristics and size of a population. These mechanisms are related to the physical environment and the presence of other species, such as predators and parasites. The mechanisms are also determined by the life history of the species itself—its reproduction and life cycle.

Chapter 8

Demography

Imagine a cloud nearly 12 miles long and 6 miles wide, dense enough to block out the sun, its shadow rolling across the landscape. Now imagine that this cloud is composed of insects numbering in the billions. When the cloud descends, wheat, grass, shrubs—any living plant—is consumed. The vegetation decimated, the cloud moves on and eventually dissipates. The land may be visited again by such a plague the next season or perhaps not for a generation.

Locust swarms or "plagues" are probably beyond comprehension for those who have not witnessed one. But the statistical description of them is impressive enough. In 2004, a swarm of locusts 3.7 miles long was recorded in Australia. In 1954, a series of 50 locust swarms invaded Kenya. Aerial measurements of the swarms estimated their total numbers at 5×10^{10} and a mass of 100,000 tons. The appetite of a locust swarm has devastating consequences for farmers in Australia, Africa, and Asia. Accounts of locust plagues are recorded deep in the history of these regions. For much of this history, the appearance of locust swarms was a mystery. They simply appeared, seemingly at random, and then just as abruptly disappeared.

The most common species of swarming locust is the desert locust (*Schistocerca gregaria*). We now understand that populations of this species are typically composed of small numbers of solitary individuals. But these solitary individuals sometimes undergo a remarkable transformation. The adults shift from brown with green marks to brown with yellow patterns. More importantly, they become highly gregarious and come together in huge swarms. The appetite of these swarms triggers their migratory search for new food sources. The shift from the solitary to the gregarious form is so radical that the two were considered separate species until 1921. We now understand the physiology of the transformation. Under favorable conditions of weather and food supply, the population increases locally by reproduction. If these favorable conditions change such that habitat and food supply become limited, large numbers of individuals may be forced into a relatively small space. Physical contact among individuals triggers a hormonal change, which in turn alters their morphology and behavior. The aggregation grows, strips the local food supply, and begins its migratory phase.

Population ecologists are interested in these phenomena for obvious practical reasons. Predicting swarm formation and behavior and controlling or reducing their impact depends on understanding their root causes. These locust population eruptions are important more generally as well. They represent an extreme form of processes many populations exhibit: population growth, variation in the population size across the landscape, movement in search of key resources, and a strong connection between these processes and climate. As such, they exemplify the key question of this chapter: *How do populations change and grow?* The answer to this question comprises the subject of **demography**, the quantitative description of a population and its characteristics. The demographic attributes of populations include measures such as the number of individuals, the proportions of males and females, and the patterns of survival and reproduction of individuals of different ages. As we will see, populations and their demographic parameters vary in both space and time. By quantifying both, we can analyze and predict how populations grow and decline. This is the first step in a full ecological understanding of population dynamics.

■ **demography** The quantitative description of a population and its characteristics.

Figure 8.1 **Isolated population.** Isle Royale in Lake Superior is home to an isolated population of moose.

■ **population** A group of individuals of a single species inhabiting a particular area.

■ **connectivity** The link between two or more populations by dispersal of individuals.

■ **genet** A genetically distinct individual in a plant population. Genets may comprise many individuals, especially when reproduction is by cloning.

8.1 What Is a Population?

Ecologists define a **population** as a group of individuals of a single species inhabiting a particular area. For example, we can identify the population of moose on Isle Royale in Lake Superior (Figure 8.1). On this island there is a distinct boundary to the group's range and thus to the population. For the moose inhabiting the mainland, the boundaries are not necessarily so obvious. The key component of the definition, "a particular area," is not so easily defined. Is there one large integrated population on the mainland? Or are there many distinct but interacting populations?

What Defines the Boundary of a Population?

This question arises because individuals are not homogeneously distributed across the landscape. Differences in demographic parameters, such as the number of individuals, their spacing in the habitat, and their reproductive rate, distinguish local groups. Figure 8.2 shows the differences in the demographic attributes of white-footed mice (*Peromyscus leucopus*) across an array of geographically connected habitat types. In practice, ecologists define the boundaries of a population on the basis of these kinds of discontinuities—that is, sharp spatial changes in demography.

Population boundaries are also largely defined by the movement of individuals. **Connectivity** is the degree to which populations are connected by dispersal. Populations with high connectivity tend to differ less than those with low connectivity. The movement of individuals among a set of populations makes them more similar or homogeneous. The demography of groups that are isolated from other groups by habitat barriers or great distances is more likely to exhibit distinct patterns, the kinds of discontinuities that form the practical boundaries of a population. The movement of individuals to or from the Isle Royale moose population is sufficiently rare that its demography is distinct from populations on the mainland. However, where the exchange of individuals is more frequent, ecological discontinuities are less likely to arise and persist. And of course the degree of movement, and thus integration, varies along a spectrum, which complicates assigning population boundaries.

Researchers must define the boundaries of the populations they study. The nature of the ecological question being addressed determines the boundaries they assign. For example, the Great Lakes are interconnected. Fish such as walleyes (*Sander vitreus*) can move among the lakes. A researcher might pose questions about the population of walleye in the Great Lakes. However, Lake Superior (like all the Great Lakes) has unique ecological attributes, such as temperature, currents, pollution levels, and the set of fish species that coexist there. Thus, a researcher might also compare the populations of walleye in Lake Superior and Lake Erie. In that case each lake is considered a separate population. In essence, defining a population is also a question of scale: What spatial scale—that is, what size area—is appropriate for the ecological question we are addressing? In this case, is the population of interest Lake Superior or the entire Great Lakes?

Spatial scaling is even more important to populations of sessile organisms. Individuals are exchanged between regions only by propagules such as seeds, in the case of plants, or larvae, as in some marine invertebrates. For organisms that reproduce by cloning, spatial scale is even more important, because we distinguish between two types of "individuals." Clones are genetically distinct from other such groups. In plants, we refer to these genetically distinct groups as **genets**. Physiologically distinct individuals, whether genetically related or not, are called **ramets**. For example, may apples (*Podophyllum peltatum*), a common herb in eastern deciduous forests, reproduce by cloning. Local patches of genetically identical individuals are formed by clonal expansion of one genotype (Figure 8.3). Each individual stem in such a clone is a ramet. What is the population in this system? Is it the genet or the group of ramets? The isolated patches of ramets separated by open habitat from other such groups could reasonably be considered populations. However, a set of genets might also be considered a population. Once again, the appropriate definition of the population in this example depends on the particular question we are asking.

Another distinction between kinds of organisms is important to scale. **Unitary organisms** exist as separate individuals. Most animals are unitary. They develop as separate individuals from the zygote according to strict and sometimes irreversible patterns of development. In contrast, **modular organisms** are those that develop and increase in size by repetitive patterns of growth and development. In plants, modular growth is achieved by the growth of meristems that lead to shoot elongation and occasional differentiation into new structures such as leaves, new meristems and shoots, and flowers. Colonial animals such as corals and sponges develop in the same way. Much of demographic analysis is based on the numbers, survival, and reproduction of the individual. This is simpler in unitary organisms than modular species, in which individuals persist for long periods but the individuals change continuously in size and structure.

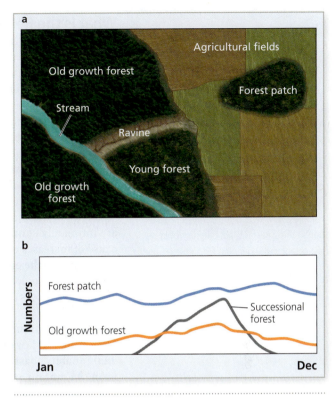

Figure 8.2 Population boundaries. The white-footed mouse occupies a complex of habitat types (a), each of which has unique demographic parameters and population dynamics (b). **Analyze: For the mouse, is this a coarse-grained or fine-grained habitat?**

ramet A physiologically distinct individual in a plant population.

Figure 8.3 Local clones. Many apples reproduce vegetatively, producing local clones of individuals.

unitary organisms Organisms that exist as separate and distinct individuals.

modular organism An organism that develops by repetitive patterns of growth of body parts.

THINKING ABOUT ECOLOGY:

Many coral reef systems in the world are declining. Imagine that you are studying the population dynamics of corals on a reef. Corals are colonies composed of thousands of polyps. Scattered along the reef are many large corals separated from other colonies. How would you define a population in this system? What criteria would be important in the definition?

◾ **deme** The evolutionary population unit; a group of randomly mating individuals.

What Is the Relationship Between the Ecological and Evolutionary Population?

As defined above, a population is a group of individuals of a species that occupy a defined area. The boundaries are determined by an ecologically relevant change in the environment. Evolution also operates within populations. However, evolutionary biologists define a population somewhat differently. The genetic change in a population that occurs by evolution results from some combination of four processes: natural selection, genetic drift, gene flow, and mutation pressure. Each of these processes operates on the gene pool, the summed genetic constitution of the population. Evolutionary biologists use the term **deme** to denote a local group of individuals that mate at random. It is within this randomly mating group that evolution occurs. Thus, for evolutionary biologists, the deme is the relevant population concept. However, the deme and the ecological population are intimately connected. The genetic change that defines evolution occurs because of increases or decreases in certain genotypes. And genetic change is intimately tied to the environment and thus to the ecological population. The size of the deme is also determined by ecological factors acting on the population. Any ecological process that reduces the population size is also likely to reduce the size of the deme.

The discontinuities that define the ecological and evolutionary populations often coincide. For example, the small annual plant rock cress (*Arabidopsis thaliana*) inhabits coastal and montane habitats in its native range in the Mediterranean. Mountain populations face longer, more severe winters, but moisture from snowmelt is available all summer. Coastal populations experience more benign winters but must contend with summer drought when moisture becomes critically limiting. Montesinos et al. (2009) showed that there are both genetic and ecological differences among coastal and montane populations of this species. Montane populations are characterized by higher mortality of seeds and young plants, and later flowering than coastal populations. The populations are

THE EVOLUTION CONNECTION

The Interaction of Evolution and Population Ecology

Natural selection operates through the differences in survival and reproduction of individuals according to their genotypes. Consequently, additions and losses from populations have demographic effects. Coulson et al. (2006) describe this connection quantitatively by analyzing the process of natural selection on the basis of the effect of an individual on future growth of the population. Their technique measures the change in population size with and without certain individuals (genotypes) present. We define the fitness of an individual in terms of its contribution to population growth. The value $w_{t(-i)}$ is the growth of the population with individual i removed. If N is the population size and $\xi_{t(i)}$ is the reproductive contribution of individual i at time t, the growth of the population with individual i removed is

$$w_{t(-i)} = (N_{t+1} - \xi_{t(i)})/N_t - 1.$$

The growth of the population between time t and $t + 1$ is the population size at time $t + 1$ divided by the population size at time t. If we remove individual i at time t, we must reduce the population N_t by 1. Thus, the denominator is $N_t - 1$. We remove the performance of this individual in the future by subtracting $\xi_{t(i)}$ from the

population size N_{t+1}. We denote the contribution of individual i to population growth as $p_{t(i)}$. It is calculated by subtracting the population growth without i ($w_{t(-i)}$) from population growth with i present (w_t):

$$p_{ti} = w_t - (N_{t+1} - \xi_{t(i)})/N_t - 1$$

or, by rearrangement,

$$p_{ti} = (\xi_{t(i)} - w_t)/N_t - 1.$$

In other words, the average contribution of an individual is the per capita difference between the contribution of an individual in the future and the rate of population growth without the individual.

QUESTION:

Does this analysis mean that intense selection can cause the population to decline?

also genetically distinct: no genotypes are shared among coastal and mountain populations. These genetic differences probably result from selection imposed by the environmental differences among the populations as well as from genetic drift within each population. In this system the ecological and genetic populations overlap; that is, the plants differ ecologically and genetically in the two environments.

KEY CONCEPTS 8.1

- We define the population on the basis of demographic discontinuities: demographic features that distinguish one group from another.
- The evolutionary population, or deme, is the unit within which evolution occurs. A deme is a randomly mating group of individuals.

QUESTION:

You find that plants of the same species in two different environments differ in features like flowering time, life span, and seed production. However, there are no genetic differences between the populations. How do you interpret this information? What does it mean for the concepts of the ecological and evolutionary population?

8.2 What Are the Key Quantitative Characteristics of a Population?

The demographic characteristics of a population provide a snapshot of the characteristics of the population at a moment in time. There are three fundamental demographic parameters of a population: the number of individuals, their distribution in space, and the characteristics of the individuals in the population.

How Do We Quantify the Number of Individuals in a Population?

The most basic demographic aspect of a population is the number of individuals it contains. We represent the total number of individuals in the population at a particular time with the symbol N_t. We represent the number of individuals in terms of **population density**, the number of individuals per unit area (or per unit volume, for suspended aquatic organisms). For example, Admiralty Island in Alaska supports one of the highest-density populations of grizzly bears (*Ursus arctos*) known—approximately one animal per square mile. Population density has important ecological consequences. For example, individuals in dense populations may face more intense competition for resources and thus may interact more frequently.

For some organisms, measuring population size or density is simple. If the organism is readily observable, direct counts may be possible. The moose population on Isle Royale is measured by aerial surveillance in the winter when moose are visible against the snow. Smaller, cryptic, or nocturnal species would not be amenable to this kind of technique. The population sizes of such species must be measured using different techniques.

One such technique, the **Lincoln index**, estimates the number of individuals from a set of samples of the population. The procedure is based on capturing and marking a portion of the population, releasing them, and then obtaining another sample of the population at a later time. The relative numbers of marked and unmarked individuals in this second sample can be used to calculate the total population size. Specifically, we capture and mark a sample of the population, n_m,

population density The number of individuals per unit area or volume.

Lincoln index A method for determining population size by marking and recapturing portions of a population.

index of relative abundance
A quantitative measure of the relative size of a population using indirect evidence of the presence or absence of individuals.

transect A sampling method of measuring the presence or absence of individuals along systematic paths through the habitat.

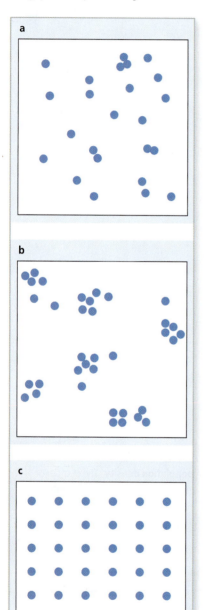

Figure 8.4 Dispersion patterns. There are three possible dispersion patterns: (a) random, (b) clumped, and (c) regular.
Analyze: What ecological factors could lead to a random, clumped, or regular distribution?

at time t_1. We know that we have not captured the entire population; thus, n_m represents a proportion, n_m/N, of the total population. The marked individuals are released into the population, where they intermix with the unmarked individuals. At a later time, t_2, we capture another sample of the population. This second sample will contain some of the individuals previously captured and marked (n_m), as well as some unmarked individuals (n_{um}). We now have all the information needed to estimate the total population size, N. At time t_1 we captured a proportion of the total population (n_{m1}/N). If the marked and unmarked individuals thoroughly mixed, the second sample should contain the same proportion of marked and unmarked animals. This value is represented by the proportion n_{m2}/n_{um}. Thus, we can set the two proportions equal to each other:

$$n_{m1}/N = n_{m2}/n_{um}.$$

Because we know the quantities n_M, n_m, and n_{um}, we can calculate the total population size N:

$$N = n_{m1}\,(n_{um}/n_{m2}).$$

Some ecological questions do not require that we know the exact number of individuals. It may be sufficient to know if there are more in one place than in another or more at one time or another. For example, we might wish to evaluate a management program for a population of game animals or measure the efficacy of control activities for an insect pest. In either case it may be sufficient to determine if the population is increasing or decreasing and by what proportion. If so, we can use an **index of relative abundance** to quantify the population size. An index of relative abundance uses a quantitative measure of indirect evidence of the presence or absence of individuals as a comparative measure of abundance. A wide array of evidence can be used as an index of abundance. In deer populations, the number of fecal pellets in an area reflects the total number of deer present. For some species, the number of individuals encountered on a set of systematic paths through the habitat, or **transects**, correlates with abundance. For example, waterfowl biologists fly transects across the northern plains, counting ducks on the ponds. Their counts do not quantify the actual number of ducks in the area. But the numbers encountered on a transect are an index of relative abundance from which comparisons across habitat or time can be made. Indices of relative abundance can also provide population information back in time, where direct counts are obviously impossible. For example, the activity of the larch bud moth (*Zeiraphera diniana*) retards the growth of larch trees (*Larix spp.*) in distinctive patterns. Thus, the growth rings of larches record the relative abundance of this insect back in time.

How Do We Measure the Spatial Distribution of Individuals in a Population?

The population density is an average—the mean number of individuals per unit area in the population. However, the individuals may or may not be distributed evenly across the habitat. The distribution of pikas is determined by factors such as elevation, the type of lava formation, and the vegetation. Their distribution is not uniform across the landscape; some combinations of habitat attributes, particularly high elevation and a type of lava known as pahoehoe, have the highest pika density (Rodhouse et al., 2010).

The pattern of spatial distribution of specific individuals is the population **dispersion**. There are three potential dispersion patterns in populations (Figure 8.4). The individuals of the population may be randomly distributed in the habitat. Or they may have a clumped dispersion pattern: groups of individuals clustered together, separated from other such groups. The third possibility is that they are evenly distributed at regular intervals.

In Figure 8.4 the three dispersion patterns are readily distinguished. However, in practice, categorizing the dispersion pattern is not always so easy, because the three patterns grade into one another. Thus, it is important to be able to quantitatively assign a dispersion pattern to a population. Figure 8.5 shows the dispersion of cholla cactus (*Cylindropuntia fulgida*) in a desert habitat. Are the cacti randomly distributed across space?

One important method for answering this question is based on the statistical distribution pattern known as the **Poisson distribution**. The Poisson distribution is a random pattern that applies to relatively rare events. We compare the actual pattern of distribution to this theoretical random dispersion. If they match, the dispersion pattern is random. If not, the dispersion pattern is either clumped or regular.

To use this method, a grid is laid over the dispersion pattern of the population such that the mean number of individuals in each grid square is small. In Figure 8.6 the grid results in a mean of 0.51 individuals per square. As you see in the figure, some have 0 individuals, some have 1, some have 2, and so on. We calculate the expected number of grid squares with certain numbers of individuals under the Poisson (random) distribution.

The expected proportion of squares with x individuals is calculated with the formula

$$P_x = a^x e^{-a}/x!$$

where x is the number of occurrences, a is the mean number of occurrences per grid square, and e is the base of the natural logarithms. The value $x!$ is known as "x factorial." This means that we multiply the value x times the values $x - 1$, $x - 2$, $x - 3$, and so on, until the remaining term is 1. For example, $5! = 5 \times 4 \times 3 \times 2 \times 1 = 120$. The expected number of squares with x individuals is then calculated by multiplying the expected proportion of squares by the total number of squares.

For example, the expected proportion of squares with 2 individuals is

$$P_2 = 0.51^2 e^{-0.51}/2! = 0.078.$$

We can now calculate the expected number of squares containing 2 individuals by multiplying the expected proportion by the total number of squares, in this example 100:

$$0.078 \times 100 = 7.8.$$

Thus, we expect 7.8 squares to contain 2 individuals if the dispersion is random. We do similar calculations for each expected number up to the maximum number observed. These calculations are presented in Table 8.1.

You see that the observed number is not exactly the predicted number in each case. How do we interpret these data? Are the differences between the observed and expected values small enough that we attribute them to chance? We determine this by asking if the observed distribution differs significantly from the predictions of the Poisson using the chi square test. The chi square test is a statistical test (see Chapter 4, "Do the Math") that measures the difference between frequency distributions. In this case the χ^2 value is highly significant ($\chi^2 = 738.9 \ 6df$; $p < .001$), indicating the dispersion of cholla in the desert is not random.

If the distribution of cholla is nonrandom, how do we determine if it is clumped or regular? Another feature of the Poisson distribution is useful in this regard. For distributions that follow the Poisson, the mean occurrence and the variance in occurrence are equal (see Chapter 3, "Do the Math"). Thus, if the mean/variance ratio equals 1, the dispersion is random. If we reject the Poisson distribution, we must then determine if the dispersion is clumped or regular.

THINKING ABOUT ECOLOGY:

You are studying deer populations in two different habitats. In one, the deer are eating primarily herbaceous plant material. In the other, the diet is mostly woody material such as bark and twigs. To quantify the relative numbers of deer, you plan to use pellet counts as an index of relative abundance. What assumption inherent in this process should you test before applying the procedure?

dispersion The pattern of spatial distribution of species in the habitat.

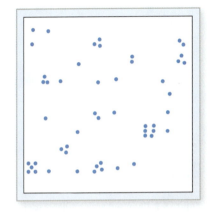

Figure 8.5 Cholla cactus. The dispersion pattern of cholla cactus in the desert. Dots represent individual cacti. **Analyze: What ecological factors might determine this dispersion pattern?**

Poisson distribution A statistical distribution in which rare events occur at random. It can be used to analyze dispersion.

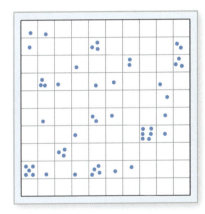

Figure 8.6 Poisson distribution. The dispersion patterns from Figure 8.4 overlaid with a grid for comparison to the predictions of the Poisson distribution. **Analyze: What are the key characteristics of the Poisson distribution?**

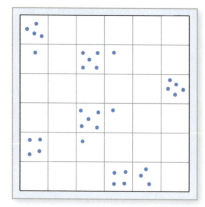

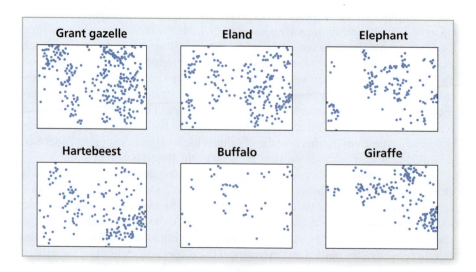

Figure 8.7 Clumped and regular dispersion patterns. A clumped dispersion (top) and a regular dispersion (bottom) with grid lines overlaid. **Analyze:** What is the effect of the grid size chosen for the analysis?

TABLE 8.1 The observed and expected numbers of individuals in each grid square according to the Poisson distribution applied to Figure 8.6

NUMBER OF INDIVIDUALS	EXPECTED PROPORTION	EXPECTED NUMBER OF SQUARES	OBSERVED NUMBER OF SQUARES
0	0.60	60.00	72
1	0.306	30.60	18
2	0.039	3.90	3
3	0.013	1.30	4
4	0.0017	0.17	1
5	0.00017	0.017	1
6	0.000015	0.0015	1

We can distinguish between these alternatives because they differ in the relationship between the mean and variance. Figure 8.7 shows clumped and regular dispersion patterns with a grid laid over them. In the clumped dispersion, the grid squares differ markedly in the number of individuals they contain—some have none whereas others have many. In other words, the variance in the number per square is high. This contrasts with the situation in the regular dispersion pattern, in which each square holds about the same number of individuals. In this case the variation is small. A mean/variance ratio greater than 1.0 occurs if the variation among squares is small, indicating a regular dispersion pattern. In contrast, a mean variance ratio less than 1.0 indicates a clumped distribution. This occurs when some squares contain many individuals whereas others contain few—high variance in occurrence. In the case of the cholla, the mean/variance ratio = 0.43, indicating that the plants are clumped.

We encounter all three dispersion patterns in natural populations. Figure 8.8 shows the distribution of individuals of six species of large mammals across the same habitat in Kenya. The dispersion patterns differ markedly in terms of the portions of the habitat most heavily used by each species and in the dispersion pattern within the area of use. Many ecological factors determine the pattern for any one species. Critical resources such as water holes or vegetation types may lead to clumped dispersion patterns. Competition or territorial defense may lead to regular patterns. Moreover, the dispersion pattern may change when analyzed on different spatial scales.

Figure 8.8 Savanna dispersion patterns. The dispersion patterns of several species across the savanna habitat in Kenya (from Stein and Georgiadis, 2008). **Analyze:** Which would you visually characterize as random?

For sessile organisms, the movement of seeds or propagules and their interaction with the environment affects dispersion as well. Tropical trees tend to be widely scattered: most show a random or regular dispersion with large distances between adults. The Janzen-Connell hypothesis (Janzen, 1970; Connell, 1971) proposes that seed predators and herbivores are responsible for this dispersion pattern (Figure 8.9). They reason that adult trees attract seed predators because the greatest concentration of seeds is found near an adult tree. As a result, virtually all the seeds in the immediate vicinity of the adult are consumed. Seeds and young individuals have the greatest probability of success at great distance from the parent tree, hence the widely scattered dispersion pattern.

Data from Chinese rainforests support the Janzen-Connell hypothesis. Lan et al. (2010) compared the dispersion patterns of saplings and adult trees of 131 species. The vast majority (83.2 percent) of saplings are clumped near the parents. However, 96.2 percent of adults are randomly distributed, suggesting that mortality of the high-density saplings shifts the final dispersion pattern toward the more regular pattern seen in adults.

What Are the Key Demographic Characteristics of a Population?

Just as populations are not uniformly distributed in space, not all individuals in a population are identical. They vary in age, sex, reproductive rate, and survival. Moreover, reproduction and survival vary with age and sex. These quantitative demographic parameters tell us much about the ecology of a population. They also allow us to analyze and even predict the growth or decline of the population.

The **age structure** of the population is the distribution of individuals by age groups. We can learn much about the ecology of the organism by these patterns. For example, Figure 8.10 shows the proportion of regional forest samples in which new trees have established in populations of Douglas fir (*Pseudotsuga menziesii*) in different forests in Oregon. These graphs were generated by measuring the age structures of each population. We can infer that new individuals are not establishing if the population is composed only of older age classes. Note that the population in Figure 8.10 established beginning in the early 1500s. Almost no new establishment has occurred since. In contrast, in the forest shown in Figure 8.10, two peaks of establishment have occurred during the same time span. Poage et al. (2009) demonstrate that these patterns reflect the fire history experienced by the population. The population has not burned since the fire that led to the establishment of the current forest in the late 1500s. A new wave of establishment occurred in the forests in Figure 8.10 after a second fire occurred in the late nineteenth century.

We represent age structure information in standardized form in a **life table**. A life table shows the numbers of individuals in each age class as well as a set of age-specific parameters such as mortality, reproduction, and life expectancy. Life table information is generally obtained in one of two ways. A **cohort life table** follows a single group of individuals born at the same time (a cohort) through time until the last individual dies. A **static life table** is based on a sample of the population and the ages of individuals at one moment in time.

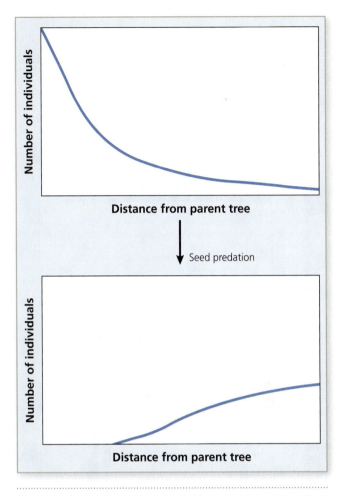

Figure 8.9 Janzen-Connell hypothesis. The Janzen-Connell hypothesis for the dispersion of tropical forest trees. Saplings near the parent tree experience high mortality, resulting in widely scattered trees at great distance from the parent tree (Janzen, 1970; Connell, 1971). **Analyze: What would be the effect of different seed dispersal mechanisms on this hypothesis?**

age structure The distribution of individuals according to their ages.

life table A table showing the numbers of individuals of different ages and their age-specific mortality and reproductive rates.

cohort life table A life table based on following a single cohort from birth to the death of the last individual.

static life table A life table based on a sample of the population and the distribution of individuals of different age.

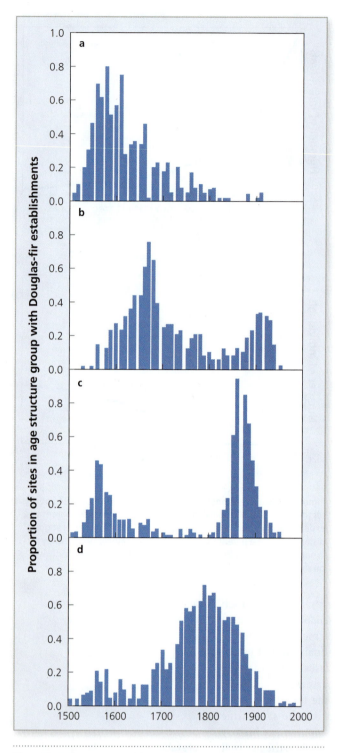

Figure 8.10 Age structure. The age distribution of Douglas firs from different sites in Oregon reflects the fire history of each forest (from Poage et al., 2009). **Analyze:** What specific fire history would lead to each of these age structures?

Table 8.2 shows a life table for the population of Dall sheep (*Ovis dalli*) at Denali National Park in Alaska.

These data were obtained from a sample of 608 sets of horns from Dall sheep collected in the park. Annual growth rings in the horns allow each individual to be aged. That is, we know the age of the individual when it died. Because the horns were collected over a period of time, they generate a static life table. The column n_x represents the number of individuals living at the start of each age category. A total of 608 horns were found; thus, 608 individuals were born in this sample: $n_0 = 608$. Of that initial group, 487 lived to be 1 or more years old. Thus, $n_1 = 487$. The third column in this life table contains the age-specific survival, l_x. We represent this information as the proportion of newborn individuals surviving to age *x*. The value of l_x is calculated by dividing n_x by n_0. Thus, for example, $l_3 = 480/608 = 0.79$. This means that approximately 80 percent of newborn sheep live to be 3 years old. The number of individuals in each age class also allows us to calculate the number of individuals dying in each age category, d_x. Of the 608 newborn sheep, 487 died in the first year of life. Thus, $d_0 = n_0 - n_1 = 608 - 478 = 121$.

A **survivorship curve** graphically represents the pattern of age-specific survival. Figure 8.11 shows a survivorship curve based on the Dall sheep data in Table 8.2. The values of l_x are depicted on a log scale as a function of age. Note in this graph that young sheep experience high mortality. Then survival is high between the ages of 2 and 10, at which point mortality increases significantly.

Survivorship curves vary widely among species. However, three broad patterns emerge, as shown in Figure 8.12. Type I survivorship curves are characterized by low survival in young ages, then high survival until old age, when mortality increases rapidly. Dall sheep as well as many other mammals, including humans, have Type I survivorship curves. In Type II survivorship curves, survivorship is constant across ages, leading to a linear relationship between l_x and age. This pattern is common among some birds, such as the starling (*Sturnus vulgaris*) and the American robin (*Turdus migratorius*). Finally, in a Type III survivorship curve, early mortality is very high but decreases in older ages. Many invertebrates have curves of this form. For example, barnacles produce huge numbers of offspring that drift in the water column, where they suffer high mortality. When they settle onto the substrate and metamorphose into adults, survival increases dramatically. There is much variation among plants in the form of their survivorship curves. In general, annuals typically have Type I curves. Perennials that reproduce just once (monocarpic perennials) usually have Type II curves. Polycarpic perennials, long-lived plants that reproduce repeatedly, show Type III curves.

Survivorship curves can also provide useful comparative information. For example, in polygynous mammals, the age-specific survivorship of males and females often differs (Figure 8.13). Male-male competition leads to higher age-specific

■ **survivorship curve** A plot of the log lx as a function of age.

TABLE 8.2 A life table for Dall sheep at Denali National Park, Alaska (data from Murie, 1944)

AGE IN YEARS (X)	NUMBER ALIVE AT START OF AGE (N_x)	PROPORTION SURVIVING AS FRACTION OF NEWBORN (L_x)	NUMBER DYING IN AGE INTERVAL (D_x)
0	608	1.000	121
1	487	0.801	7
2	480	0.789	8
3	472	0.776	7
4	465	0.764	18
5	447	0.735	28
6	419	0.689	29
7	390	0.641	42
8	348	0.572	80
9	268	0.441	114
10	154	0.253	95
11	59	0.097	55
12	4	0.0065	2
13	2	0.003	2
14	0	0.000	0

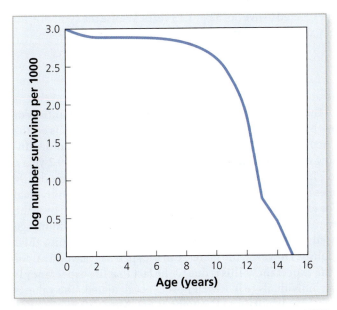

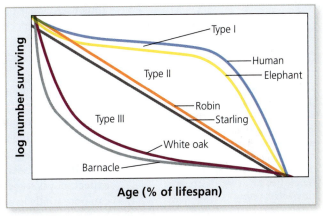

Figure 8.12 Types of survivorship curves. The three types of survivorship curves and examples of species with each type. **Analyze: Why do we use % of lifespan for the x-axis in this graph?**

Figure 8.11 Survivorship curve. A survivorship curve for the Dall sheep data presented in Table 8.2 plots the log number surviving versus age (data from Murie, 1947). **Analyze: What sources of mortality would you expect to increase with age?**

mortality for males. Differences in habitat quality may be reflected in the survivorship curve as well.

The age-specific mortality rate can be used to calculate the age-specific **life expectancy (e_x)**. This value represents the mean amount of time an individual of age x is expected to live. Table 8.3 contains a life table for a perennial plant.

To calculate e_x, we first calculate a new column, L_x, which represents the average number of individuals alive in an age category: $L_x = (n_x + n_{x+1})/2$. For

life expectancy (e_x) The mean expectation of further life for an individual of age x.

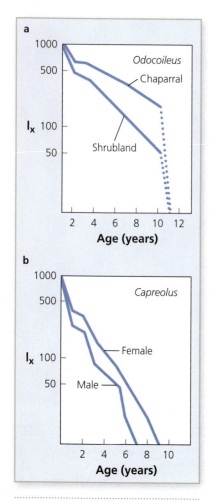

Figure 8.13 Male and female survivorship curves. (a) A survivorship curve for black-tailed deer (*Odocoileus hemionus*) in chaparral and shrub habitat. (b) A survivorship curve for male and female roebuck (*Capreolus capreolus*) (adapted from Hutchinson, 1978). **Analyze: Why should habitat or sex affect the survivorship curve?**

■ **life span** The maximum age to which an individual can live.

■ **net reproductive rate (R$_0$)** The average number of individuals produced by a female in her life span.

TABLE 8.3 A life table for a perennial plant

AGE (X)	N$_x$	L$_x$	L$_x$	T$_x$	E$_x$
0	125	1.000	91.5	172.5	1.38
1	58	0.464	45.0	81.0	1.40
2	32	0.256	24.0	36.0	1.13
3	16	0.128	10.0	12.0	0.75
4	4	0.032	2.0	2.0	0.50
5	0	0	0	0	0.00

example, in Table 8.3, $L_2 = (32 + 16)/2 = 24$. The column T_x depicts the total number of "plant-years" to be lived by all individuals of age x. A "plant-year" is one individual living for one year, or two individuals living for one half year, etc. We use the L_x column to calculate T_x

$$T_x = \sum_{x}^{\infty} L_x.$$

So T_2 is the sum $L_2 + L_3 + L_4 + L_5 = 36$ plant-years. This means that the individuals alive at age 2 live on average 36 additional plant-years. We then calculate the mean number of years to be lived for a single individual at age 2 by dividing the number of plant-years to be lived by the number of individuals alive at age 2. There are 32 individuals alive at age 2. This group of 32 individuals will live a total of 36 plant-years. Thus, each 2-year-old is expected to live $36/32 = 1.13$ years. We use the symbol e_x for the mean life expectancy at age x. The general form for its calculation is

$$e_x = T_x/n_x.$$

Life expectancy is an important human demographic variable, one that is commonly reported in the media, especially in comparisons among countries or when changes occur within a country. The statistic commonly reported is e_0, the expectation of life at birth. In the United States, life expectancy continues to increase. At the beginning of the twentieth century, the value was only 49.2 years. By 2006 it had risen to 77.5 years. This statistic is commonly misunderstood. Life expectancy is sometimes confused with the **life span**, the total time an individual can potentially live. For example, the human life span has also increased over the last century. Thanks to modern medicine and hygiene, the average maximum age of humans has increased. In other words, more individuals are living into their 80s and 90s than previously. However, the increase in life span has played a minor role in the increase in life expectancy. This is due to the fact that life expectancy is calculated for newborn individuals (e_0). Recall how e_0 is calculated. The value is affected by the age-specific mortality rate throughout life. Mean life expectancy increases when age-specific survival (l_x) increases. Changes in l_x at young ages have a proportionally large effect on e_0. In fact, much of the increase in human life expectancy is the result of improvements in infant survival; relatively little is the result of older individuals living longer.

The final variable important in a life table is the mean number of offspring produced by an individual of age x. Table 8.4 shows the age-specific seed production for a population of the annual plant *Phlox drummondii*.

When reproductive information is available, we can use it to calculate the **net reproductive rate (R_0)**. The net reproductive rate is the average number of offspring produced by an individual over the life span. Table 8.4 shows how it is calculated. Note there is a new column in this life table, $l_x b_x$. Whereas b_x is the expected reproductive output of an individual *that has reached age x*, $l_x b_x$

TABLE 8.4 A life table for the annual plant *Phlox drummondii* (data from Leverich and Levin, 1979)

X (DAYS)	N_x	B_x	L_x	L_xB_x
0–299	996	0	1.000	0
299–306	158	0.34	0.159	0.0532
306–313	154	0.80	0.155	0.1231
313–320	151	2.40	0.152	0.3638
320–327	147	3.19	0.148	0.4589
327–334	136	2.54	0.137	0.3470
334–341	105	3.16	0.105	0.3330
341–348	74	8.66	0.074	0.6436
348–355	22	4.30	0.022	0.0951
355–362	0	0	0	0
				Sum = 2.4177

measures the *realized* production by including the probability that an individual survives to that age. In effect, we discount b_x by the chance that an individual does not live to age to reproduce. The net reproductive rate is the sum of the l_xm_x column. It represents the sum of the realized reproductive output at each age over the entire life span. R_0 for the phlox population is 2.4177, which means that the average individual produces 2.4177 offspring before it dies. We will return to this important variable when we consider the mathematics of population growth.

The b_x column requires that we specify precisely how the life table is developed. For the phlox table, we count all individuals in the population because each individual contains both male and female flower parts and function. However, if the sexes are separate, we must accommodate this in the life table in one of two ways. We can generate a life table for males and females combined. If so, the value of b_x (and thus l_xb_x and R_0) is expressed per mated pair. In other words, b_x represents the average number of offspring produced per pair in the population. Or we can simply produce a life table with data for females only. In this case, b_x represents the age-specific birth rate of a female, and R_0 measures the average number of offspring a single female is expected to produce in her life. In species with polygynous mating systems (see Chapter 5), males typically mate with more than one female. A female life table more accurately depicts the demographic parameters of such a population.

Sex ratio, the proportion of the population that is female or male, constitutes another important demographic parameter. The sex ratio may differ among populations of the same species. For species in which a male can inseminate more than one female, sex ratio plays an important role in the potential growth of the population. For example, elk are polygynous and a bull can inseminate many cows. Consider two populations of the same total size, one with an equal number of males and females and one with a sex ratio skewed toward females. The skewed population can grow faster because there are more cows reproducing.

The sex ratio can also change within a population for groups of individuals of different ages. One important reason is the difference between male and female survivorship curves, shown in Figure 8.13. Thus, demographers refer to the sex ratio at specific points in the life history:

1° sex ratio	sex ratio at fertilization
2° sex ratio	sex ratio at birth or hatching
3° sex ratio	sex ratio at sexual maturity
4° sex ratio	sex ratio of the adult population

sex ratio The proportions of males and females in a population.

In the boat-tailed grackle (*Quiscalus mexicanus*), the secondary sex ratio is 1:1 males to females. However, the 3° sex ratio is biased toward females by a ratio of 1.42:1 (Selander, 1965). The shift occurs due to the effect of sexual selection on the males: females choose males on the basis of the size of the tail, but large tails incur a mortality cost that shifts the sex ratio as males reach sexual maturity.

The concept of sex ratio in plants is complicated somewhat by the arrangement of male and female flower parts within and among individuals. Some plant species are **monoecious**, meaning that they have both male and female function in the same individual. In some cases, each flower contains both stamens and

■ **monoecious** Describes a plant in which male and female function occur in the same individual.

ON THE FRONTLINE

Herbivores and Trillium Sex Ratio

Dense populations of herbivores significantly affect the density of their forage plants. In much of the eastern United States, white-tailed deer (*Odocoileus virginianus*) are at record high population densities. As a result, grazing and browsing by deer are significantly altering the diversity of forest herbs. We might expect that the preference of deer for certain plants and avoidance of others might shift the balance of plant diversity on the forest floor.

Heckel et al. (2010) investigated the impact of deer foraging on palatable and unpalatable forest herbs. Jack-in-the-pulpit (*Arisaema triphyllum*) is relatively unpalatable to deer; they avoid this plant when possible. In contrast, giant trillium (*Trillium grandiflorum*) is highly palatable to deer and is readily eaten by them.

HYPOTHESIS: Deer foraging selects against trillium and favors jack-in-the-pulpit.

The researchers studied the dynamics of these two species in Pennsylvania forests with high deer density (4–18 deer/km²). They established a series of seven paired plots: one open to deer foraging, the other fenced to exclude deer. The population dynamics, seed production, and plant size of both herbs were compared in the deer access and deer exclosure plots.

PREDICTION: When deer are excluded, both jack-in-the-pulpit and trillium should

thrive. **When deer are present, trillium should decline, but there should be no difference between fenced and unfenced populations of jack-in-the-pulpit.**

As expected, deer foraging resulted in a significant negative impact on the palatable trillium. Surprisingly, however, plant size and seed production of the unpalatable jack-in-the-pulpit was positively correlated with damage to trillium. That is, where trillium damage was highest, the negative impacts on jack-in-the-pulpit were also highest.

A shift in jack-in-the-pulpit sex ratio hinted at the mechanism behind the indirect deer effect. The sex ratio of this plant was skewed toward males in populations in which trillium was heavily browsed. Heckel et al. suggest that high deer density decreases soil quality because the activity of deer compacts the soil and reduces the amount of leaf litter, an important source of nutrients. Consequently, all plant growth is adversely affected. We know that when jack-in-the-pulpit faces poor growing conditions, plants switch to male function due to the lower energetic cost of being male. Thus, where deer have reduced soil quality, the population of jack-in-the-pulpit shifts to a male-biased sex ratio.

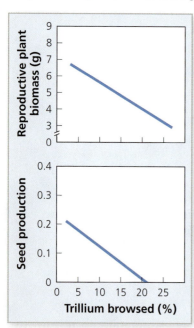

Figure 1   Seed production of forage plants. The biomass of reproductive structures and the seed production of trillium as a function of the percentage of the plants browsed by deer (from Heckel et al., 2010). **Analyze:** What would you expect these graphs to look like for jack-in-the-pulpit?

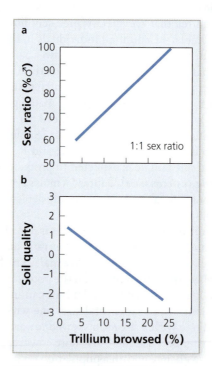

Figure 2   Effects of herbivory. The impact of deer browsing on trillium on the sex ratio of jack-in-the-pulpit (a) and soil quality (b) (from Heckel et al., 2010). **Analyze:** How is the sex ratio of jack-in-the-pulpit affected if the deer tend to avoid this species?

pistils. In other species, the plant produces separate female and male flowers. For monoecious species, the traditional concept of the sex ratio does not apply. In **dioecious** species, such as cottonwoods and ginkos, there are separate male and female plants. In these species, sex ratio is expressed as in animals.

dioecious Describes a plant in which male and female function occur in separate individuals.

The primary sex ratio may be determined purely genetically or may be influenced by environmental effects. In some species of reptiles, such as alligators and turtles, the 2° sex ratio is determined by the temperature at which the eggs are incubated. The temperature varies in different parts of the nest. Warmer eggs produce males; cooler eggs develop into females. These environmental effects can also lead to variation in sex ratio among clutches if nests differ in temperature.

Interestingly, some dioecious plant species can shift their reproductive function from male to female or female to male. This provides the opportunity for adaptive shifts in the plant's gender. For example, the forest herb known as jack-in-the-pulpit (*Arisaema triphyllum*) can switch from male to female. Female function is up to 3.6 times as energetically costly as male function. The sex ratio of populations inhabiting high-light environments differs significantly from that of populations in dense shade. Where light energy is abundant, the plants shift to the more energy-demanding female function; where light is limiting, most plants function as males, skewing the sex ratio in that direction.

KEY CONCEPTS 8.2

- The numbers of individuals in some populations are measured directly; in others, indirect measures are required.
- The dispersion pattern reflects important ecological forces acting on the spatial distribution of individuals.
- The age structure of a population reveals key patterns of mortality and reproduction.
- The sex ratio of a population reflects sex-specific mortality patterns and the potential reproductive rate of the population.

QUESTION:

What are the key advantages and disadvantages of static and cohort life tables?

8.3 How Do Populations Change in Time and Space?

We know that individuals are not distributed evenly across the landscape. And dispersion analysis reveals that some populations are clumped—abundant in some places and rare in others. A population in one place at one time has a specific density, age structure, dispersion pattern, and sex ratio. These demographic parameters as we have discussed them so far represent a snapshot of the population at a moment in time. However, populations are not static—key demographic features change over time. The locust swarms described at the beginning of this chapter represent an extreme form of this kind of variation: the population grows, morphs into the gregarious migratory form, and gathers into a vast swarm that moves across the landscape.

A simple set of factors determines the dynamics of a population. The change in the number of individuals in a population is a result of four processes:

1. Immigration to the population (I)
2. Emigration from the population (E)
3. Additions to the population via births (B)
4. Losses from the population via deaths (D)

The change in population size (N) from time t to $t + 1$ is thus determined by the following equation:

$$N_{t+1} = N_t + B - D + I - E.$$

Two of these factors are measures of the movement of individuals to and from the population. The other two, B and D, the total birth and death rates in the population, depend on processes that operate within the population, specifically the per capita birth and death rates, b_x and l_x columns of a life table. If we understand the evolutionary ecology of these four factors, we can understand the dynamics of populations.

What Determines the Movement of Individuals Within and Among Populations?

Animals and plants undergo an array of movement patterns. For example, many species undertake migrations. The movement requires energy and exposes individuals to mortality from predation and other risks inherent in travel, and this ultimately affects the number of individuals in the population. Dispersal, the one-directional movement of individuals to or from a population, can have important impacts on demography.

In plants and sessile animals, dispersal occurs via the movement of seeds or larvae. Many kinds of structural adaptations facilitate seed dispersal (Figure 8.14). Animals transport plant seeds internally or externally. Externally dispersed plants have seeds with structures that increase the probability that a seed will be inadvertently picked up by a passing animal and transported. The seeds of some species are transported in the course of an animal's feeding. Seeds gathered by an animal for food may be buried or cached. If forgotten or not consumed, these seeds may germinate some distance from the parent plant. Some plant fruits or seeds are consumed and the seed is later defecated in a new habitat. Mutualistic interactions between plants and their animal dispersal agents abound. For example, in New World tropical forests, strangler figs (*Ficus spp.*) produce large nutritious fruits on their highest branches that attract vertebrates, including monkeys, parrots, and bats. These animals consume the fruits and, as they move through the forest canopy, defecate the tiny seeds contained in the fruit high on the crowns of other trees. The seeds of the strangler fig germinate not on the ground but in the canopy. The plant grows down the trunk of the host tree, using the host for support.

A number of adaptations are associated with wind-dispersed seeds. Because wind disperses seeds at random, there is no guarantee that a seed will be deposited in suitable habitat. Thus, wind-dispersed plants produce large numbers of seeds because only a fraction will succeed. Structural elements either keep the seed aloft or cause it to flutter to the ground some distance from the parent plant. The parent plant and the fruiting structures may also have adaptations that facilitate wind dispersal of their seeds (Soons and Bullock, 2008). Wind-dispersed seeds tend to be produced higher on the plant, where wind is stronger. The heather plants *Calluna vulgaris* and *Erica cinerea* have wind-speed thresholds below which the seeds are not released. This adaptation ensures dispersal over longer distances.

In essence, dispersal allows individuals to exploit spatial variation in key resources. These may be economic resources such as food or open habitat, reproductive resources such as mates or territories, or genetic resources in the form of nonrelatives with which to mate. And because these resources are themselves dynamic, most populations experience net immigration or net emigration at different times.

The demographic impact of dispersal is both local and regional depending on the spatial scale of the movements. Dispersal sometimes leads individuals to move far beyond the normal range of the species. For example, when new habitats

Figure 8.14 Seed dispersal. Structural adaptations that facilitate seed dispersal by animals include hooks or barbs that attach to animal fur.

arise, it is dispersers that colonize them. This type of long-distance dispersal was responsible for the colonization of South America, Antarctica, and eventually Australia by marsupial mammals. Long-range dispersal also underlies invasions by species introduced by humans. For example, the Eurasian collared dove (*Streptopelia decaocto*) was introduced to the Bahamas in the 1970s. From there it quickly colonized the United States, where its range rapidly expanded and its numbers are increasing (Figure 8.15).

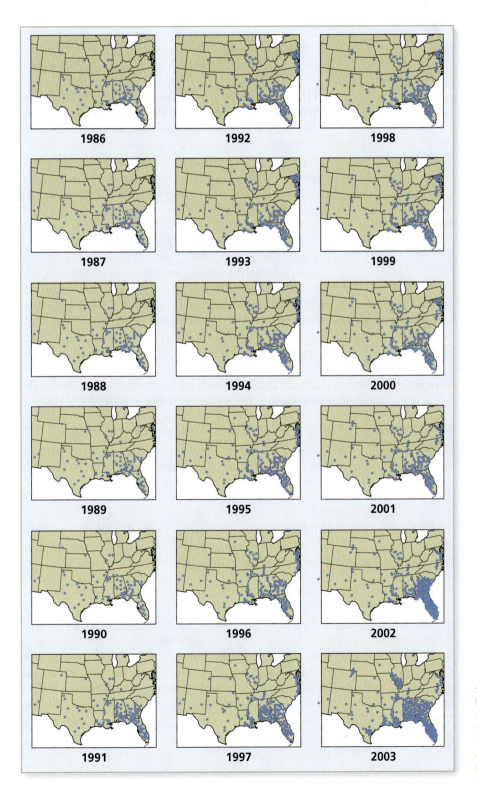

Figure 8.15 Range expansion. The range expansion of the introduced Eurasian collared dove, *Streptopelia decaocto* (from Hooten and Wikle, 2008). **Analyze:** What ecological factors might limit the maximum expansion of this species?

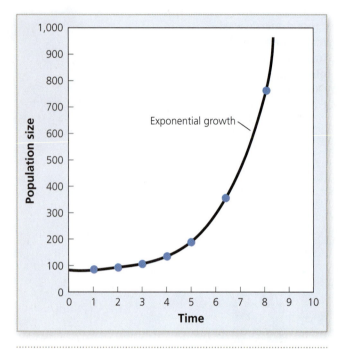

Figure 8.16 Exponential growth. When a population is introduced into new, suitable habitat, it often grows exponentially. **Analyze: What would this graph look like if the log population size were plotted on the y-axis?**

Dispersal maintains connections between distinct populations. Population connectivity determines the degree to which the demography or genetics of one population is affected by another. The populations of white-footed mice in Figure 8.2 are connected by dispersal. Distance and obstacles to movement reduce dispersal and the connectivity of populations. For passively dispersed organisms like sessile marine invertebrates, a number of factors determine the connectivity of populations. Reproductive output, the physical factors of ocean currents, larval behavior, and predation determine how many individuals move among populations.

How Do Birth and Death Rates Determine the Growth Rate of a Population?

The birth and death rates are the two key internal factors in the dynamics of a population. We have already seen how these parameters change as a function of the age of the individual. We turn our attention now to the mathematics of population growth as determined by the birth and death rate.

When dispersing individuals occupy new and suitable habitat, the population grows exponentially. A small population in exponential growth may increase slowly, but over time the rate of growth increases (Figure 8.16).

We mathematically characterize exponential growth in two ways depending on the reproductive pattern of the species. Some organisms have discrete generations that do not exist simultaneously. For example, annual plants germinate, grow, flower, and set seed in one year. All the adults die and the population continues from seed the next year. There is no overlap of generations. Other organisms breed more than once per season, and some individuals survive to reproduce in another season. These species have overlapping generations. The mathematics of population growth in these two kinds of species differs.

How Do Populations with Discrete Generations Grow?

Consider a population of annual plants that doubles each year. The population in the next year, time $t + 1$, is calculated from the population at time t as follows:

$$N_{t+1} = \lambda N_t.$$

multiplicative growth rate The factor by which the current population size increases in each time period.

The variable λ is the **multiplicative growth rate**, the factor by which the population increases in each time period. For this population, $\lambda = 2.0$. We calculate the population at any time t from λ and the starting population, N_0:

$$N_t = \lambda^t N_0.$$

We can also measure population growth for populations with discrete generations using the net reproductive rate, R_0, from a life table. R_0 measures the mean number of offspring a female is expected to produce in her lifetime. If $R_0 > 1.0$, each female is more than replacing herself; the population is growing. If $R_0 < 1.0$, the population is in decline. If we know the value of R_0, we can calculate the population size at any future time. For organisms with discrete generations, the population size at any time t can be calculated if we know the starting population size (N_0) and the net reproductive rate:

$$N_t = R_0^t N_0.$$

Human Reproductive Rates

Human populations are affected by the same demographic variables and growth parameters as other populations. We refer to the *total fertility rate (TFR)* as the number of children a female is expected to have in her lifetime. Of course, this is the same as the variable, R_0, calculated from the values l_x and b_x. There are significant differences among countries in growth rate. For example, in many African nations, the TFR exceeds 6.0; in Singapore and Taiwan, the value is just slightly larger than 1.0.

The age structure of the population plays a central role in the value of TFR. The effect of age structure is determined by the total number of individuals in the population of childbearing age. We can represent the age structure of a population as an age pyramid. Note the difference in the age structure of industrial and developing countries. In the latter, a large proportion of the population is in the youngest age classes, which will lead to tremendous potential population growth. In industrial countries, a large proportion of the individuals have moved past reproductive age; their populations will thus grow more slowly.

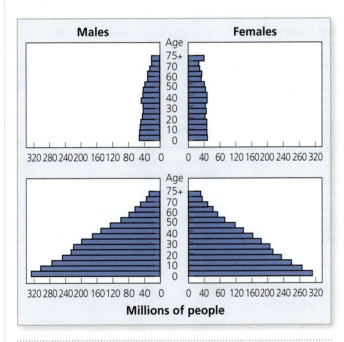

Figure 1 Age-structure pyramids. Age-structure pyramids for developing and developed countries. Note the skew toward younger reproductive age classes in the developing world (United Nations Department of Economic and Social Affairs, 2012). **Analyze: What factors increase or decrease the symmetry of the male and female age structures in humans?**

In the early twentieth century, the demographer Warren Thompson proposed that as countries develop, their populations undergo a decline in the birth and death rates that ultimately slows the growth rate. He termed this set of changes the *demographic transition*. We recognize four stages in this transition. In preindustrial societies, both birth and death rates are high. Birth control and family planning are not priorities, partly because children are

important contributors to the family economy and because child mortality is so high. The growth or decline of the population in this stage is heavily dependent on outside forces, such as drought or changes in the food supply, that affect the birth and death rates.

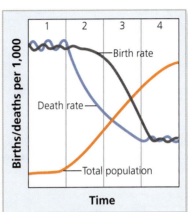

Figure 2 Demographic transition. The demographic transition is a set of changes in the birth and death rates during the course of development from an agrarian to an industrial society. **Analyze: What factors increase the differences between birth and death rates?**

In the second stage of this process, development of the country leads to more productive agriculture and improvements in public health. Consequently, mortality, especially of children, declines dramatically. However, birth rates remain high, leading to an age pyramid heavily skewed toward the younger reproductive age classes. The result is rapid population growth during this stage.

Stage 3 is characterized by a decline in the birth rate. A number of factors contribute to this change. As education and opportunities for women increase, the desire for large numbers of children declines. With affluence and education, birth control and family planning are more available and desirable. The population may still be increasing due to the carryover of the skew in age structure, but the growth rate has slowed. Eventually the population reaches Stage 4, in which the birth and death rates are both low and relatively stable. The total population remains high, but the increase has slowed or stopped. A few countries have moved into a stage in which the age structure has shifted toward older ages and the birth rate has declined to the point that the total population is actually declining.

This pattern was observed in a number of Western countries during the Industrial Revolution. However, it is not clear whether this model applies to all countries, or whether it applies today, when so many cultural and technological changes affect populations in developing countries. Obviously, the key stage for rapid population growth is Stage 3, when the death rate has declined but the birth rate has not yet declined to match it. To the extent that countries remain in Stage 3 for long periods, their populations grow exponentially. The world population recently exceeded 6 billion. How large it becomes in the next century will be determined in large part by this demographic transition.

QUESTION:

What cultural or technological factors do you think might alter this process in developing countries today?

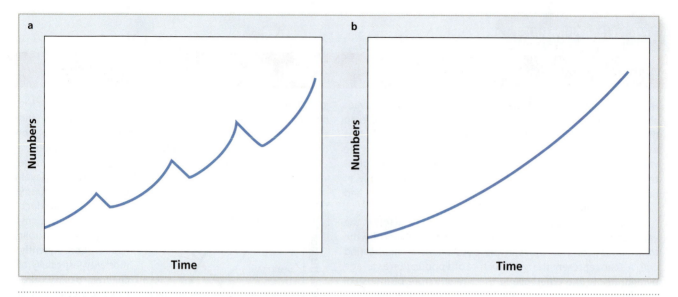

Figure 8.17 Discrete generations. The population trends for species with (a) discrete generations and (b) overlapping generations. **Analyze:** What causes the periodic dip in the population with discrete generations?

THINKING ABOUT ECOLOGY:

In this section we have generated two equations for population growth,

$$N_t = \lambda^t N_0 \text{ and } N_t = R_0{}^t N_0,$$

with essentially the same form. What is the difference between them? Does $\lambda = R_0$, or are these quantities in fact different?

The growth curve in Figure 8.17 shows continuous increase in the population. For populations with discrete generations, it is likely that some mortality during the nonreproductive season causes a decline in the population size before the next reproductive event. The result is a sawtooth pattern of change, as shown in Figure 8.17.

How Do Populations with Overlapping Generations Grow?

The situation is slightly more complex for organisms with overlapping generations. In any year or season, reproduction will be accomplished by young first-time reproducers as well as by older individuals who reproduced previously. The result is the presence of multiple generations together in the population. To mathematically describe this kind of growth, we return to the fundamental equation from population change:

$$N_{t+1} = N_t + B - D + I - E.$$

To simplify the mathematics, we will assume that emigration and immigration are equal and cancel one another. Thus,

$$N_{t+1} = N_t + B - D.$$

The change in the population, ΔN, is calculated

$$\Delta N = N_{t+1} - N_t.$$

If $N_{t+1} = N_t + B - D$, we can calculate ΔN by subtracting N_t from both sides of the equation,

$$N_{t+1} - N_t = N_t - N_t + B - D.$$

Then

$$\Delta N = B - D.$$

The values B and D are the total births and deaths. Because the values of B and D will be larger in larger populations, we need per capita measures of reproduction (b) and death (d). Thus,

$$B = bN \text{ and } D = dN.$$

The change in the population becomes

$$\Delta N = bN - dN = (b - d)N.$$

Because birth and death are occurring constantly, we express growth as the instantaneous change in the population using the differential equation

$$dN/dt = (b - d)N.$$

We replace $(b - d)$ with a new variable, r:

$$dN/dt = rN.$$

The variable r is the **instantaneous growth rate**, or the **intrinsic rate of increase**. It is a measure of the potential growth rate of the population based on the difference between the per capita birth and death rates. Traditionally the value of r is measured when the population is small (signified by the 0 in the subscripts for b and d).

$$r = b_0 - d_0.$$

As in the case of discrete generations, we can derive a general equation to calculate the population size at time t if we know the initial population size and the reproductive rate (in this case r, the instantaneous rate of growth):

$$N_t = N_0 e^{rt}$$

where e is the base of the natural logarithms. As is the case for R_0 in populations with discrete generations, the value of r determines how rapidly the population grows. Figure 8.18 depicts exponential growth of populations with different values of r.

It is also possible to quantify population growth directly from the age-specific birth and survivorship rates in a life table. Each age or stage in the life history has a specific survival and reproductive rate. A **transition matrix**, or **Leslie matrix**, is a technique for measuring population growth using the probabilities of transition from one age to another and the corresponding reproductive output that accumulates in the process. The process is shown schematically in Figure 8.19. N_0 is the number of newborn individuals, N_1 is the number in age class 1, etc. Each age group has a specific birth and mortality rate. The total population size, N_t, is the sum of the numbers of individuals in all the age classes.

From this information, we can calculate N_{t+1}, the number of individuals in the population at the next time interval. New individuals of age class 0 enter the population only by reproduction (we are ignoring dispersal in this example). Thus, N_0 at time $t + 1$ is calculated solely by the number of individuals born at time t to individuals of different ages at time t. This is the contribution from b_0, b_1, b_2, and so on. For all other age classes, the number of individuals at time $t + 1$ is the result of survivorship from previous ages. We represent the transitions shown in Figure 8.19 by a set of linear recurrence equations:

$$N_{0(t+1)} = (N_{0t} \times b_0) + (N_{1t} \times b_1) + (N_{2t} \times b_2) + (N_3 \times b_3)$$
$$N_{1(t+1)} = (N_{0t} \times l_0)$$
$$N_{2(t+1)} = (N_{2t} \times l_2).$$

instantaneous growth rate (also, intrinsic rate of increase) The potential growth rate of a population based on the difference between the per capita birth and death rates, measured at small population size.

transition (Leslie) matrix A method of predicting population growth from the probabilities of transition from each age class to the next and the age-specific reproductive rate.

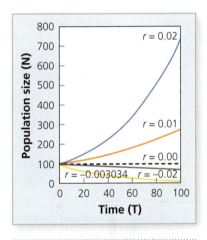

Figure 8.18 Intrinsic rate of growth. The growth and decline of populations with different values of the intrinsic rate of growth, r (from Gotelli, 1995). **Analyze: What biological factors contribute to the value of r?**

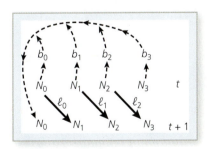

Figure 8.19 Transition matrix. A transition matrix showing the paths of change from one age or stage to another and the effect on the population size. N_x represents the population at age x; l_x represents the age-specific survivorship; b_x represents the age-specific birth rate. **Analyze: What is the total population at time t?**

DO THE MATH

Matrix Algebra

The transition matrix in Figure 8.19 can be analyzed using matrices. A matrix is a group of numbers known as elements, arranged in rows and columns. Matrices are denoted by boldface letters. For example, a matrix (M) might contain the following values:

$$\begin{matrix} 2 & 4 & 6 \\ 10 & 12 & 14 \\ 18 & 20 & 22 \end{matrix} = M$$

Another, A, might be constructed thus:

$$\begin{matrix} 0.2 \\ 0.4 \\ 0.6 \end{matrix} = A$$

A matrix containing a single column of numbers is referred to as a *column vector*. To multiply a column vector and a matrix, we first multiply each element in the column vector by each element in the first row of the matrix. We then add all these products and place the resulting sum in the first row of a new column vector. Next, the values in the column vector are multiplied by the values in the second row of the matrix, summed, and placed in the second row of the new column vector. This process is completed for all rows in the column vector. If the new column vector is labeled X, the calculations for $M \times A = X$ are as follows:

$$\begin{aligned} X = (2 \times 0.2) + (4 \times 0.4) + (6 \times 0.6) &= 5.6 \\ (10 \times 0.2) + (12 \times 0.4) + (14 \times 0.6) &= 15.2 \\ (18 \times 0.2) + (20 \times 0.4) + (22 \times 0.6) &= 24.8 \end{aligned}$$

In multiplying matrices, $M \times A \neq A \times M$. We distinguish among the patterns of multiplication by designating pre- and postmultiplication. In $M \times A$, we say that M is postmultiplied by A.

We can use these procedures to arrange demographic information in matrix form, then multiply the matrices to calculate the population size at any time in the future. To do this for the population in Table 8.5, we devise a square matrix that consists of the age-specific birth and survivorship rates. We place zeroes in the matrix so that when we multiply by the numbers of individuals in the age classes, the results are the numbers of individuals in those age classes in the next time period:

$$\begin{matrix} 0 & 2 & 6 & 8 \\ 0.2 & 0 & 0 & 0 \\ 0 & 0.5 & 0 & 0 \\ 0 & 0 & 0.2 & 0 \end{matrix} \times \begin{matrix} 500 \\ 100 \\ 50 \\ 10 \end{matrix} = \begin{matrix} 580 \\ 100 \\ 50 \\ 10 \end{matrix}$$

The general form is that the matrix containing the age-specific reproduction and survival data (A) is multiplied by the matrix containing the numbers of individuals in each age class (N_t) to create a new matrix for time $t + 1$ (N_{t+1}):

$$A \times N_t = N_{t+1}$$

Or

$$\begin{matrix} b_0 & b_1 & b_2 \ldots \ldots b_{n-1} & b_2 & 0 \\ l_0 & 0 & 0 \ldots \ldots 0 & 0 & 0 \\ 0 & l_1 & 0 \ldots \ldots 0 & 0 & 0 \\ 0 & 0. & l_2 \ldots \ldots 0 & 0 & 0 \\ . & . & . & . & 0 \\ . & . & . & . & 0 \\ 0 & 0 & 0 & l_{n-1} & 0 \end{matrix} \times \begin{matrix} N_{0t} \\ N_{1t} \\ N_{2t} \\ . \\ . \\ . \\ N_{nt} \end{matrix} = \begin{matrix} N_{0t+1} \\ N_{1t+1} \\ N_{2t+1} \\ . \\ . \\ . \\ N_{nt+1} \end{matrix}$$

One of the advantages of this kind of analysis is that matrices can be analyzed by computer, enabling us to carry out the predictions of population size many generations into the future.

Table 8.5 shows a life table for a population of 660 individuals at time t. From this information and the recurrence equations, we can calculate N_{t+1}, the number of individuals at time $t + 1$, as follows:

$$\begin{aligned} N_{0(t+1)} &= (500 \times 0) + (100 \times 2) + (50 \times 6) + (10 \times 8) = 580 \\ N_{1(t+1)} &= (500 \times 0.2) = 100 \\ N_{2(t+1)} &= (100 \times 0.5) = 50 \\ N_{3(t+1)} &= (50 \times 0.2) = 10. \end{aligned}$$

TABLE 8.5   A life table for a population with overlapping generations

X (AGE)	N_x	B_x	L_x
0	500	0	0.2
1	100	2	0.5
2	50	6	0.3
3	10	8	0.0

The population size in the next time interval (N_{t+1}) is the sum of the number of individuals in these age classes at time $t+1$:

$$N_{t+1} = N_{0(t+1)} + N_{1(t+1)} + N_{2(t+1)} + N_{3(t+1)} = 580 + 100 + 50 + 10 = 740.$$

Thus, the population has grown from 660 to 740 in that time interval.

KEY CONCEPTS 8.3

- The growth of a population is determined by the number of births, immigrants, deaths, and emigrants.
- Individuals disperse from one population to another to secure economic resources (e.g., food), reproductive resources (e.g., mates), or genetic resources (e.g., nonrelatives with which to mate).
- Dispersal leads to population connectivity. Connected populations are less demographically distinct.

QUESTION:

Why is it necessary to use different mathematics for populations with discrete and overlapping generations?

Putting It All Together

The patterns of increase, aggregation, dispersal, and eventual decline of a locust population can all be quantified using the demographic analyses developed in this chapter. The spectacular increase in their numbers is the result of exponential growth during favorable periods and the aggregation of individuals that have transformed into the swarming morph. The movement of the swarm is really a dispersal event, as the population destroys the local food resources and must emigrate to find new sources. Eventually the locust swarm declines, sometimes nearly as dramatically as it increased.

Recent studies have clarified the factors that lead to outbreaks. Stige et al. (2007) studied the pattern of locust outbreaks over a 1,000-year time span in the Yangtze Delta in China (Figure 8.20). By comparing the outbreaks and declines

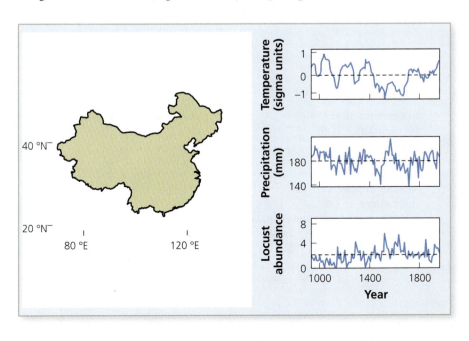

Figure 8.20 Locust outbreaks. The historical record of locust plagues in China (from Lima, 2007). **Analyze:** How do we measure the interaction of locust abundance and either precipitation or temperature?

with historical weather patterns determined from ice cores, tree-ring growth patterns, analysis of pollen types in lake sediments, and historical records, the researchers showed that locust outbreaks are associated with weather patterns at two different time scales. They found that locust outbreaks are more common during decades characterized by cold, wet weather. Flooding was more common under these climatic conditions, and this apparently leads to an increase in suitable locust habitat. However, outbreaks tend to occur in the warmest year in such a decade, when the rate of locust development and survival are highest.

Locusts are among the most prolific and locally abundant animals on earth. Nevertheless, as abundant as they sometimes are, their populations eventually decline. Thus, locusts also illustrate a fundamental principle of population ecology: no population can grow indefinitely. Something limits its growth and eventually initiates a decline. This leads to another fundamental question, one we will address in the next chapter: What regulates populations?

Summary

8.1 What Is a Population?
- A population is defined as a group of individuals inhabiting a particular area.
- The boundary of a population is defined by a sharp change in demographic parameters: density, age structure, and so forth. The boundaries of some populations, such as on islands, are relatively easy to define. Others are less discrete.
- Evolution is a population phenomenon. It occurs in the deme, a population of randomly mating individuals.

8.2 What Are the Key Quantitative Characteristics of a Population?
- The number of individuals in the population and population density are key population parameters. They can be measured directly or indirectly.

- Other important parameters are dispersion, age structure, age-specific mortality and reproduction, and sex ratio.
- The age structure of a population is depicted in a life table and survivorship curve. It reveals the key patterns of mortality and reproduction.

8.3 How Do Populations Change in Time and Space?
- Four variables determine population size: births, deaths, immigrants, and emigrants.
- Dispersal connects populations. The relative numbers of immigrants and emigrants determine the effect on population size and the similarity of two populations.
- Some populations have discrete generations. Their growth is modeled by the equation $N_t = \lambda^t N_0$.
- Other populations have overlapping generations. Their growth is modeled by the equation $N_t = N_0 e^{rt}$.

Key Terms

age structure p. 167
cohort life table p. 167
connectivity p. 160
deme p. 162
demography p. 160
dioecious p. 173
dispersion p. 165
genet p. 160
index of relative abundance p. 164
instantaneous growth rate; intrinsic
    rate of increase (*r*) p. 179

Leslie matrix p. 179
life expectancy p. 169
life span p. 170
life table p. 167
Lincoln index p. 163
modular organism p. 161
monoecious p. 172
multiplicative growth rate (λ) p. 176
net reproductive rate (R_0) p. 170
Poisson distribution p. 165
population p. 160

population density p. 163
ramet p. 161
sex ratio p. 171
static life table p. 167
survivorship curve p. 168
transect p. 164
transition matrix p. 179
unitary organisms p. 161

Review Questions

1. Why is it sometimes difficult to define the boundaries of a population?

2. Why is the deme sometimes different from the ecological population?

3. How is the Poisson distribution used to identify the dispersion pattern in a population?

4. How are static and cohort life tables different? The same?

5. Why can we use data for just females in some life tables?

6. What role does dispersal play in population regulation?

7. What is the relationship between R_0, λ, and r?

8. What is the difference between e_x and life span?

9. What role does the sex ratio play in population growth?

10. What role does survivorship play in determining the sex ratio?

Further Reading

Alford, R., et al. 2009. Comparisons through time and space suggest rapid evolution of dispersal behaviour in invasive species. *Wildlife Research* 36:23–28.

This paper explores the evolution of dispersal behavior in species in the process of range expansion. The authors examine the rate of change in dispersal behavior in a population of toads in Australia.

Fisher, R.A. 1930. *The genetical theory of natural selection.* New York: Oxford University Press.

This classic book discusses the basis of modern sex-ratio theory. Much current work in this area still connects to Fisher's theory.

Gotelli, N.J. 2008. *A primer of ecology.* Sunderland, MA: Sinauer Associates.

This book expands on the mathematics of demographic analysis and population growth in this chapter.

Yaukey, D., et al. 2007. *Demography: the study of human population.* Long Grove, IL: Waveland Press.

This book surveys the population dynamics and analysis of human populations. It extends the concepts presented in this chapter.

Chapter 9

Population Regulation

Isle Royale National Park is a spectacularly beautiful island in Lake Superior in Michigan's Upper Peninsula. Covered with a lush conifer forest, it is home to two icons of wild country—the moose, *Alces alces*, and its main predator, the wolf, *Canis lupus*. Today, visitors are likely to see moose and perhaps glimpse a wolf as well. But how different the island was only a hundred years ago, when neither moose nor wolf could be found there.

Historically, moose and wolves were not known on the island. Then, early in the twentieth century, a major fire swept through the mature conifer forests of the island. The result was abundant new growth of woody vegetation. Moose were first recorded on the island in 1914. We do not know how many initially colonized or how they arrived. However, when they did, they found ideal habitat, thanks to the fire. Moose are browsers—they eat woody vegetation—and the flush of postfire vegetation meant that food was abundant. Not surprisingly, under these ideal conditions, the moose population grew exponentially, peaking at about 3,000 by about 1930. Around the same time, the postfire conifer forest was maturing and the shrubby browse that had been prevalent was disappearing due to overconsumption. The large moose population had literally eaten itself out of house and home. When a major fire consumed still more of the food base, starvation set in and their numbers plummeted.

Then, in the especially severe winter of 1947, Lake Superior froze over and a small group of wolves crossed the ice to the island. In the decades since, the Isle Royale moose-wolf system has been studied intensively. There have been periods of relative stability, punctuated by significant fluctuations in their populations. Sometimes the moose population increases or decreases radically and independently of the wolf population. Similarly, the wolf population seems tied to moose numbers at some times but has also decreased on its own at other times.

The story of these two intimately connected species parallels the development of our understanding of population growth and its limits. Why do these populations fluctuate so much? We can envision a number of factors that affect the population size of each:

- The condition of the vegetation
- Winter conditions, especially snow
- The diseases and parasites of each species
- Changes in the vegetation caused by fire
- Competitive interactions with other herbivores or predators

How do these forces interact to determine the number of moose and wolves on Isle Royale? What is the maximum number of moose the island can support? How many wolves can the moose support? Will the population ever stabilize, or will it continue to fluctuate? Each of these questions reflects a fundamental element of moose and wolf population dynamics. Together they lead to the question central to this chapter: *What limits population growth?*

9.1   What Limits Population Growth?

The concept that population growth is limited in some way is a fundamental component of ecology. The roots of this idea go all the way back to Darwin. As we saw in Chapter 2, Darwin's theory of natural selection was based on a logical argument:

1. Each individual in a population has a high potential reproductive rate. In the absence of other forces, populations increase rapidly.
2. Eventually, food (or perhaps other resources) limits the population. This was the essence of Malthus's argument that human populations increase geometrically whereas their food supply grows arithmetically. Ultimately, the population is limited by food.
3. When competition for resources becomes intense, the inherent differences among individuals become important: some individuals will succeed; some will fail.

Selection favors those variants that can obtain resources, especially when conditions are harsh. Darwin was interested in the limits of population growth because they inevitably lead to competition and ultimately to natural selection. Vital resources may become limiting, especially at high density. If so, competition for them becomes a regulatory factor. But an array of other processes, including physical factors, affects population size. Storms, fire, drought, in effect any abiotic factor that changes mortality or reproduction, can determine the size of a population. Today, ecologists study populations to elucidate the relative importance of, and interaction among, these factors.

What Determines the Population Growth Rate?

In one sense, this is an easy question to answer. The mathematics of population growth presented in the last chapter highlight the main set of regulatory variables. Populations grow according to the equation

$$dN/dt = rN$$

top-down control A process in which organisms higher in the food chain control the density and diversity of lower trophic levels.

bottom-up control A process in which organisms lower in the food chain control the density and diversity of higher trophic levels.

where *r* is the intrinsic rate of growth. Recall that

$$r = (b + i) - (d + e)$$

where *b* is the per capita birth rate, *i* is the per capita immigration rate, *d* is the per capita death rate, and *e* is the per capita emigration rate. In this respect, the answer to our question is easy—only four variables determine the value of *r*. However, the rich complexity of population biology, in fact the reason it fascinates population ecologists, is the array of biological factors encompassed by these four mathematical variables and the interactions among them. We can organize these factors according to two important (and nonexclusive) characteristics: (1) the relationship of biotic and abiotic factors and (2) the effect of density on limiting factors.

How Do Biotic and Abiotic Factors Affect Populations?

Biotic factors are any processes that ultimately depend on the actions of living organisms. These may be factors external to the population of interest, such as predation or the food supply (Figure 9.1). Competitors, parasites, and pathogens also fall into this category. If the species participates in symbiotic relationships with other species, their presence or absence can have profound effects on the population, especially if the symbiosis is obligate. Abiotic factors include any physical factors that cause the population to increase or decrease. Among these are storms, fire, and drought—in essence, any physical factor that can affect mortality or reproduction.

Biotic regulatory factors can operate in a **top-down** or **bottom-up** fashion (Figure 9.2). These terms refer to the direction of regulation through the **food chain**—the hierarchy of plants, the animals that consume them, the predators that eat the consumers, and so on. The set of species that feeds at a particular place in a food chain is known as a **trophic level**. Top-down factors operate from higher to lower trophic levels. For example, in Yellowstone National Park, wolves affect the numbers of elk in a top-down fashion. When wolves were removed from the park in the early twentieth century, the elk population exploded due to the release from top-down control (Coughenour and Singer, 1996).

Bottom-up factors, as the name implies, act on the population via limitations imposed by lower trophic levels. Lush growth of aquatic plants and algae in the waters surrounding the Galápagos Islands supports large populations of consumers. One reason for this large food base is that the cold Humboldt Current carries nutrient-rich water from the Antarctic to the Galápagos, where it supports large populations of marine invertebrates and fish. This food base is exploited by a population of the endemic Galápagos penguin (Figure 9.3). El Niño occurs when the normal pattern of ocean currents and flow in the Pacific Ocean reverses. During an El Niño, warm water from the western Pacific flows eastward toward South America and the Galápagos. The cold, nutrient-rich water is replaced with warmer, less nutrient-rich water. When this occurs, the marine food base collapses in the Galápagos and the penguin population declines radically (Figure 9.4) and may take years to recover (Valle and Coulter, 1987).

How Do Density-Dependent and Density-Independent Factors Affect Populations?

Some population processes are affected by density. These **density-dependent factors** increase or decrease in intensity as the population increases or decreases (Figure 9.5). The competition Darwin envisioned at high population density falls into this category. In this case, as the population increases, the amount of food available to each individual declines. The result is that either the birth rate declines or the mortality rate increases, or both. The net effect is to decrease the value of *r* and eventually the size of the population.

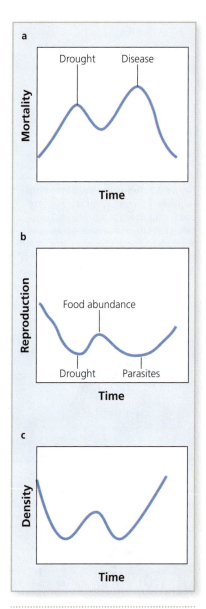

Figure 9.1 Factors that regulate population. Populations are regulated by a combination of factors acting on mortality (a) and reproduction (b). These factors can be abiotic, such as weather, or biotic, such as intraspecific competition for food. Their summed positive and negative effects (c) determine the population density. **Analyze:** How might food abundance affect mortality from parasites?

food chain The hierarchical pattern of energy transfer from primary producers to consumers.

trophic level A feeding level; one level in a food chain.

density-dependent factor Population processes that change with the population density.

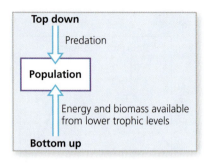

Figure 9.2 Top-down and bottom-up regulatory factors. Regulatory factors can be top-down or bottom-up. These terms refer to mechanisms whose actions are from higher trophic levels (top-down) or from lower trophic levels (bottom-up). **Analyze:** Are top-down and bottom-up factors mutually exclusive?

Figure 9.3 Galápagos penguin. The Galápagos penguin (*Spheniscus mendiculus*) is endemic to the Galápagos. Its population is affected by El Niño events.

The density-dependent effects on food supply are bottom-up processes. Top-down density-dependent effects also occur, primarily through the action of predators and parasites. For example, the red grouse (*Lagopus lagopus*) is known for the cyclic increases and declines in its populations in Great Britain. Long-term population records show that the cycle lasts between four and eight years. During the crash of the population, the birds are heavily infected by a parasitic nematode (*Trichostrongylus tenuis*). However, for some time it was not clear if the parasite is the cause of the decline or simply increases in frequency because the birds are in poor condition from other factors. Now we understand that the parasite is a key density-dependent cause of the cycles. When the parasite burden of grouse was experimentally reduced, the population crash did not occur (Hudson et al., 1998; Figure 9.6).

We model density-dependent effects with a simple algebraic manipulation of the basic equation for population growth. Recall that for populations with overlapping generations, we use the equation

$$dN/dt = rN$$

or the integral form,

$$N_t = N_0 e^{rt}.$$

In these equations, r is the intrinsic rate of growth of the population. Under favorable conditions, the population grows exponentially. If we include the Malthusian argument that resources will eventually limit population growth, this exponential growth cannot continue indefinitely. Eventually growth must slow, stop, or even decline. We modify the growth equation to include this phenomenon with this addition:

$$dN/dt = rN\,(K - N)/K$$

or, as it is sometimes presented,

$$dN/dt = rN\,(1 - N/K)$$

THE HUMAN IMPACT

Biological Control of Pests

An array of pests affects agricultural systems, forests, and rangeland. The cost of these pests in lost production and the effort to eradicate them runs into the tens of billions of dollars annually. When synthetic insecticides and herbicides became available in the 1950s, large-scale pest control appeared feasible. However, it eventually became clear that chemical pest control has serious disadvantages. Pests resistant to chemicals quickly evolved, reducing the effectiveness of many agents or requiring larger and larger doses. In addition, many of these pesticides were found to have significant negative environmental impacts. DDT is perhaps the classic example: although it was highly effective in the control of insects, it resulted in mortality of nontarget species, especially birds.

These problems prompted increased interest in biological control agents. Biological control uses natural enemies of the pest—predators, herbivores, pathogens, or parasites—to reduce its population size and impact. However, the choice and development of a biological control agent is sometimes difficult. Some 600 species of insects have been introduced into the United States as potential biological control species. Only about 55 percent of these species have succeeded in controlling their intended pest. The remainder either failed as control agents or could not successfully establish a population. We now recognize that potential biological control species have the following characteristics:

1. Narrow host range. The most successful control species rely exclusively on the target species. This increases their effect on the pest and minimizes its impact on nontarget species.
2. Climatic compatibility. The species must be able to reproduce in the climate of the new habitat.
3. Life-cycle synchrony. The species must be present and active at the same time as the target species.
4. High reproductive potential. The control species must be able to increase its population size rapidly enough to keep pace with the target.

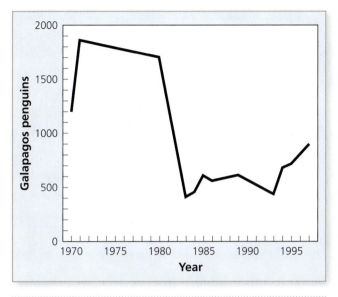

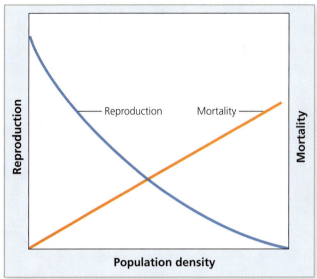

Figure 9.4 El Niño. The crash of the Galápagos penguin population as a result of an El Niño event (from Valle and Coulter, 1987). **Analyze: Can you tell from this graph if the decline is a result of mortality or reproductive failure?**

Figure 9.5 Density-dependent regulatory factors. Density-dependent regulatory factors change as a function of population density. The changes may be linear or curvilinear. **Analyze: Is it possible for mortality to decrease as a function of density?**

where K is the **carrying capacity**, the maximum population size sustainable by the resources available. Notice how this equation behaves. If $N > K$: dN/dt becomes negative; that is, the population declines. If $N < K$: dN/dt is positive, the population grows. If $N = K$: $dN/dt = 0$, the population is stable. Our equation now includes a density-dependent effect: at high density, the growth rate is negative; at low density, the growth rate is positive (Figure 9.7).

Density-dependent factors act as negative feedback systems on populations. The factor being regulated, the population size, acts to turn on or turn off the factors that change it (Figure 9.8). You may have encountered other such systems in

carrying capacity The number of individuals that can be supported by the resources available to a population.

5. Efficiency. The ideal control agent must be able to locate its host even when it is rare. However, it must not be so efficient that it reduces its host to such low levels that its own population dies out.

As you can see, the development and implementation of a successful biological control system is no simple task. Nevertheless, a number of remarkable successes have been achieved. For example, the cottony cushion scale (*Icerya purchase* Maskell) arrived in the citrus groves of California in 1868 with devastating effect. The insect feeds on sap in the phloem of young branches. A team of entomologists discovered that in its native Australia, the vedalia ladybird beetle feeds on the scale insect. Some 500 beetles were released in California in 1888. By 1890, cottony cushion scale had been virtually eliminated.

Klamath weed (*Hypericum perforatum* Linnaeus) was introduced into North America in the nineteenth century from Europe and Africa. Some introductions were accidental—seeds were

carried here with bags of crop seed stocks. Others were intentional because the plant was believed to have medicinal value. By the 1940s, the species covered 400,000 acres in California, where it caused gastrointestinal disease in cattle and sheep and crowded out useful plant species. Australia was facing the same challenges from invasions to that continent. Researchers there found that a beetle from Klamath weed's native range, *Chrysolina quadrigemina*, met the five criteria listed above. When the beetle was introduced into California, the range of Klamath weed declined to 1 percent of its precontrol area in just 10 years.

QUESTION:

What potential negative effects of biological control agents might be important to measure?

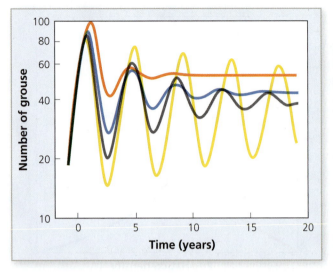

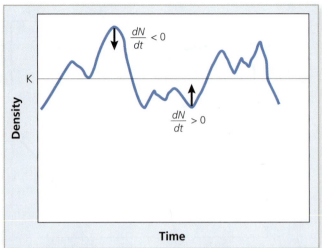

Figure 9.6 Fluctuating population. The effect of experimentally reducing the parasite burden in grouse on the population size. No treatment, yellow line; 5% treated, gray line; 10% treated, blue line; 20% treated orange line (from Hudson et al., 1998). **Analyze: Why do the fluctuations of these populations decrease over time?**

Figure 9.7 Carrying capacity. When the population is above the carrying capacity, density-dependent factors result in dN/dt < 0. If it is below K, dN/dt is positive. **Analyze: Is it likely that the value K is constant?**

biology. For example, a number of physiological properties, such as blood glucose, body temperature, and water balance, are regulated by negative feedback systems. For example, if body temperature exceeds a certain value, physiological cooling mechanisms such as sweating switch on.

DO THE MATH

Logistic Growth with Dynamic Resources

The logistic growth equation

$$dN/dt = rN\,(1 - N/K)$$

accounts for the change in population growth in relation to the resource level, K. Although the equation models the behavior of a population below K, at K, and above K, it does not account for changes in resources that affect the value of K. How might we adjust the equation to account for this phenomenon?

The simplest case is a renewable resource (R) independent of its use by the consumer. For example, the algae in a pond use the sunlight that penetrates the water and nutrients carried to the pond from the land. The rate at which these resources become available is given by the equation dR/dt. From the point of view of the algae, the resource is supplied at a constant rate (k_R):

$$dR/dt = k_R.$$

The growth of the population using this resource is a function of its own density and the rate at which the resource is supplied:

$$dN/dt = f(k_R, N).$$

Individuals require some level of resource (a) to maintain their normal function. Any resources beyond that amount can be used

for reproduction, thus increasing N. In other words, the maintenance use of resources is aN. The difference between this amount and the rate of resource supply is the amount available for reproduction:

$$k_R - aN.$$

Then the rate of population growth is a function of the birth rate (b), the population size (N), and the rate of resource supply beyond maintenance ($k_R - aN$):

$$dN/dt = bN(k_R - aN).$$

If we rearrange this equation by factoring k_R from the second term, we obtain

$$dN/dt = bk_R N\,(1 - aN/k_R).$$

Notice that this equation has the same form as the standard logistic equation. In place of the carrying capacity (K), we have the rate of resource supply per individual (k_R/a). We can use this equation to identify the population size, resource use, and resource supplies at which the population is constant, growing, or declining.

The equations for population growth simply state that above K, dN/dt is negative, whereas below K, dN/dt is positive. As we know, negative population growth is the result of decreased reproduction or increased mortality, whereas positive population growth occurs when mortality is low and reproduction is high. Thus, we identify density-dependent effects by the correlation between mortality or reproduction and population density. This seems a straightforward task. However, in practice, a number of factors complicate such analyses. For example, there may be a delay between the change in population density and its effect on reproduction. A variety of biological factors can underlie this delay, such as the time interval between the density increase and the next seasonal breeding season. This kind of effect leads to a time lag in the correlation between density and reproduction. For example, in the cotton rat (*Sigmodon hispidus*), reproduction decreases at high density (Reed and Slade, 2008). However, the probability that a female is pregnant is negatively correlated with the population density one to two months before (Figure 9.9).

Density-independent factors, as the name implies, do not change with population size (Figure 9.10). Abiotic factors, such as weather, fire, and floods, act in a density-independent fashion. Because the occurrence and nature of abiotic factors are often unpredictable, they add a random dimension to population dynamics. For example, in populations of bobwhite quail (*Colinus virginianus*) in the Southeast, the unpredictably occurring heavy spring rains negatively affect reproduction and thus population size (Rosene, 1969). This kind of mortality is independent of density. If 10 percent of chicks are killed by heavy rains, it does not matter if there are 100 chicks or 1,000 in the population. One in ten will die in both populations.

Some population ecologists use the term **population regulation** to refer to density-dependent mechanisms that determine population size, as described by the equation

$$dN/dt = rN(K - N)/K.$$

In this view, density-independent factors occasionally disturb the population such that it moves away from K. These factors represent noise—that is, random variation that density-dependent factors quickly compensate for. This kind of equilibrium dynamics dominated population ecology for many years. We now understand the importance of abiotic disturbances—fire, flood, drought, storms, and excessive heat or cold are all recurring physical factors that represent the norm, not the exception, for most populations. Long-term studies have shown equilibrium to be the exception rather than the rule—it is only a matter of time before disturbance of some kind destroys the equilibrium.

Although many abiotic factors decimate the population, others enable exponential growth. For example, one effect of climate change and warmer mean temperature is that the natural population outbreaks of bark beetles in the conifer forest of the western United States have recently been much more significant (Raffa et al., 2008). Moreover, these abiotic factors ultimately have the same effect as the resource limitation Darwin emphasized: they determine whether populations increase or decrease. They simply do this in a manner that is both random and independent of density.

Two species of Darwin's finches (*Geospiza fortis* and *G. scandens*) on the island of Daphne in the Galápagos illustrate how important density-independent factors

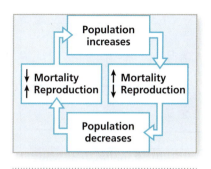

Figure 9.8 Negative feedback system. In a negative feedback system, the component being regulated causes changes in key parameters that affect it. In this case, population size affects mortality and reproduction, which in turn affect population size. **Analyze:** Explain how this mechanism could lead to (a) a constant population size or (b) population cycles.

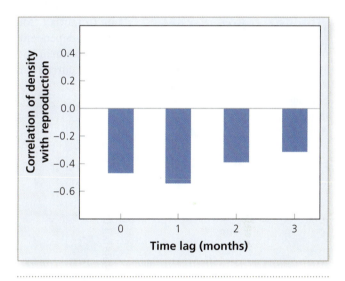

Figure 9.9 Density and reproduction. The correlation between density and reproduction in the cotton rat (*Sigmodon hispidus*). Note that there is a significant negative correlation between reproduction and the density one and two months before (from Reed and Slade, 2008). **Analyze:** How can there be significant correlations in the past but not in the most recent time period?

density-independent factor Population processes that are not affected by density.

population regulation Mechanisms that determine population size by density-dependent processes.

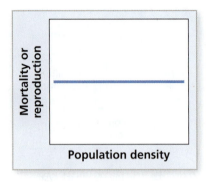

Figure 9.10 Density-independent factors. Density-independent factors, as the name suggests, do not change as the population size changes. **Analyze:** What factors are most likely to be density independent?

THINKING ABOUT ECOLOGY:

A number of paired concepts are outlined in this section: density dependence/density independence, top-down/bottom-up, and biotic/abiotic. We describe each pair of terms independently of the others. But some clearly overlap. For example, some top-down factors are also density independent. Which of these factors would you expect to overlap? Which cannot?

Figure 9.11 Heavy rainfall. The extraordinary increase in rainfall in the Galápagos during the 1982–1983 El Niño event (bars) and the correlated increase in the populations of *Geospiza fortis* and *G. Scandens* (points and line) (data from Gibbs and Grant, 1987). **Analyze: Does rainfall affect the finches directly?**

can be. In the dry conditions that normally prevail on Daphne, finches occur at consistently low density, limited by seed availability (Figure 9.11). El Niño events result in unusually heavy rain in the Galápagos that leads to prolific seed production, which in turn leads to dramatic increases in the finch population. The El Niño event of 1982–1983 was particularly strong—rainfall increased tenfold over previous rainy-season totals. Both species of finch responded with significant population increases (Gibbs and Grant, 1987). Note that the impact of this climatic event was indirect: rainfall altered seed production, a biotic factor that limits finch reproduction and survival, and hence density. Clearly, persistent dry conditions result in stable but low densities most of the time. It is the rare El Niño event that reveals the role that predominantly dry conditions play in most years.

Populations of the desert bighorn sheep (*Ovis canadensis*) exemplify the interaction between density-independent and density-dependent factors (Marshall et al., 2009). In the arid mountains of the Southwest, the vegetation on which sheep depend varies with precipitation: when rainfall is high, there is abundant forage; when rainfall is low, the food supply declines. However, the *variability* in precipitation changes across this landscape: low mean precipitation is associated with high variation in rainfall (Figure 9.12). Marshall et al. compared the population growth rate to rainfall (forage) and sheep abundance (density-dependent competition). Density-dependent effects were more likely in populations inhabiting regions of low variation in rainfall. Populations strongly influenced by the amount of precipitation occurred in regions with high variation in rainfall.

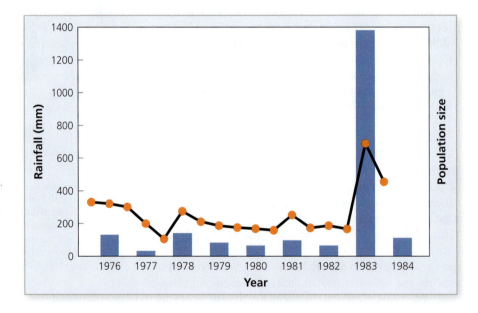

KEY CONCEPTS 9.1

- No population can grow indefinitely. Some factor will ultimately limit further growth.
- Several important regulatory factors comprise paired mechanisms that limit growth.
  - Biotic and abiotic factors
  - Top-down and bottom-up factors
  - Density-dependent and density-independent factors

QUESTION:

Explain the effect of each of these paired factors on the four variables that determine population growth.

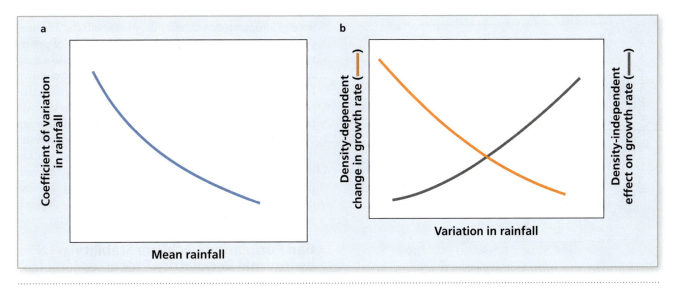

Figure 9.12 Variable rainfall. (a) Variation in rainfall increases as mean rainfall decreases in the range of the desert bighorn sheep (*Ovis canadensis*). (b) Density-dependent population growth is more pronounced in regions with little variation in precipitation. Where rainfall is highly variable, density-independent factors increase in importance (from Marshall et al., 2009). **Analyze: Why should the variation in rainfall affect the nature of population growth?**

9.2 Are Populations Ever Stable?

Some populations achieve relative stability and only fluctuate within narrow limits. As noted above, if a set of density-dependent factors consistently maintain the population near *K*, it is an equilibrium population. We reserve the term *equilibrium* for the special case described in Figure 9.7. However, a population may be stable even if it is not regulated at an equilibrium value. **Population stability** simply implies that the population fluctuates within relatively narrow limits. Stability may be achieved by the summed effects of all the regulatory factors operating on the population—both density dependent and density independent. And we have seen how abiotic factors, even though they operate in density-independent fashion, may lead to relative population stability. Density-independent factors operating at irregular, random intervals check the reproductive potential of the population, effectively determining the upper boundary to population size (Figure 9.13). If so, the population may be stable but not at equilibrium.

population stability The state of a population that fluctuates within narrow limits.

What Is the Relationship Between Density Dependence and Stability?

We might expect that when density-dependent factors predominate, the population will be relatively stable. Certainly, populations that operate in this way are known, especially from short-term studies. However, the more populations we study over longer periods of time, the more we recognize that density dependence and stability are not necessarily linked.

One of the classic data sets illustrating this comes from long-term records of fur sales by the Hudson's Bay Company in Canada (Figure 9.14). These records show that the snowshoe hare and lynx populations fluctuate in a regular, cyclic pattern in which each population repeatedly increases to high levels, then declines. In this system, the period of the cycle is about 10 years. The cycles are slightly out of phase with one another (Keith, 1963). In other words, the peak of the hare population occurs slightly before the peak lynx population. The decline in the hare population precedes the decline in lynx numbers. Each

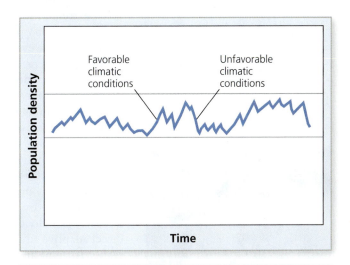

Figure 9.13 Population stability. Density-independent factors can result in stability if they combine to limit the range of population fluctuation. **Analyze: What is the significance of the scale on the y-axis in evaluating population fluctuation?**

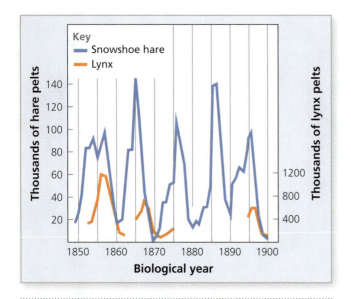

Figure 9.14 Hare and lynx population cycles. Long-term fur records from the Hudson's Bay Company in Canada reveal the dynamics of snowshoe hares and lynx. Both populations cycle with a period of about 10 years (from Keith, 1963). **Analyze: Is the hare population driving the lynx population or is the lynx population driving the hare population?**

self-regulation Internal density-dependent processes that operate independently of outside biotic or abiotic forces.

species is regulated by density-dependent effects. The hare population is regulated by the top-down effect of predation: hare mortality increases with density due to an increase in the lynx population. Lynx predation drives down hare numbers. The collapse of the lynx prey base constitutes a bottom-up force on the lynx population. As the lynx population declines in the absence of hares, the hares can once again increase and provide bottom-up support for the lynx population. Notice, however, that neither population is stable. Neither lynx nor hare ever achieves an equilibrium population size, despite the important role that density-dependent factors play for each. Clearly, density dependence is not sufficient to achieve population stability.

Can Populations Achieve Stability by Internal Regulatory Processes?

The factors we have discussed so far are external forces that affect the population size. **Self-regulation** refers to regulatory mechanisms within the organism itself that operate in a density-dependent fashion. Self-regulation requires that internal mechanisms affect the components of the basic equation

$$r = (b + i) - (d + e)$$

in a density-dependent fashion. For example, territorial behavior is an adaptation in which a male or a breeding pair obtains exclusive use of the resources in a habitat. If access to these resources is a prerequisite for successful reproduction, the value of b may be negatively related to density. Similarly, if, at high density, resources become limiting and emigration increases as individuals seek better conditions or a territory elsewhere, the value of e will be positively associated with density. Both represent behaviors inherent in the population that operate in a density-dependent regulatory fashion. Of course, neither is independent of the status of resources external to the population.

We might imagine that populations that can self-regulate would be favored by natural selection. A population that swings violently up and down may, by chance, reach extremely low densities, even local extinction. Internal mechanisms that maintain stability perhaps avoid this potential catastrophe. However, we should be careful of the evolutionary logic of self-regulation.

In northern regions, especially the Arctic, lemmings (Figure 9.15) are known for their unusual population dynamics. Their populations cycle from low to high density every four years. One legend about them is that during the high phase of the cycle, lemmings rush into the sea to commit mass suicide. Behind this legend is the idea that mass suicide is a behavior that limits the population when it is too high. That is, mass suicide is adaptive because in its absence the population would continue to grow, vastly outstripping its resources. Now consider this behavior in evolutionary terms. We can appreciate the advantage of such a system for the *population*—the population is ultimately regulated by such behavior. But what about the adaptive value for the *individual*? Recall that selection operates most intensely on individuals. Suicide, especially if it limits future reproduction, has no direct adaptive advantage. This is the reason the lemming legend is not based in reality. Imagine an ancestral population in which suicide did not occur. Then imagine that a mutation arises that leads to suicide at high population density. The fate of that mutant individual is obvious—death—and hence the loss of the new mutation. Consequently, population regulation by suicide is not an evolutionarily stable strategy. The lemming legend is just that, a legend. It probably arose when, during the high phase of the cycle, a few individuals were found in rivers or ponds.

Figure 9.15 Lemmings. Lemmings are well known for their cyclic pattern of population fluctuation.

THE EVOLUTION CONNECTION

Group Selection and Evolutionarily Stable Strategies

The elements of population regulation illustrate two key principles that connect Darwinian evolution and ecology. First, an essential component of Darwin's theory of natural selection is the fact that no population can grow forever—something will eventually prevent further growth. This idea underlies much of this chapter. Second, natural selection operates primarily on the individual.

The second principle is based on the idea that a new adaptation can spread throughout a population only if it benefits the individual. Only in this way can a new, rare mutation increase in frequency. We saw this principle, the concept of the evolutionarily stable strategy, or ESS, in previous discussions of the evolution of sex. An adaptation is an ESS only if it can increase in frequency in the population when it is rare.

The concept of the ESS has important implications for population dynamics. Any mechanism of population regulation that relies on an advantage that accrues to a *group* rather than the individual may not be an ESS. The myth that lemmings commit suicide to stabilize the population is a specific example of the general phenomenon known as *group selection*. Consider a set of isolated populations. Some are regulated and persist; others do not regulate and go extinct. Now imagine that in this system, selection operates at the level of the population: those populations with self-regulation that prevents overpopulation persist. This might arise by reproductive altruism—that is, by an individual reducing its reproductive rate. It might arise by mechanisms that increase mortality, as in the lemming myth. In this scenario, populations without self-regulation eventually go extinct; those that regulate persist. Over time, all populations self-regulate.

Although the logic of this example is analogous to natural selection operating on the individual, group selection fails due to the principle of the ESS. The advantage of reproductive restraint is only manifest when it is common in the population; it cannot increase from rarity. Consequently, explanations that rely on a group selection scenario generally fail.

QUESTION:

How does group selection differ from kin selection (Chapter 6)?

These were animals that accidentally drowned. At high lemming densities, the number of individuals and the crowding is astounding. By chance, a few end up in the water, and the notion of population regulation by suicide was born.

In some species, high density leads to crowding, competition, and increased aggressive interactions among individuals. This in turn can effect physiological changes in the individual, such as stimulating the production of stress hormones. Under stressful conditions, physiological changes can potentially increase mortality or decrease reproduction. If so, internal density-dependent mechanisms decrease the population size. At lower density, these effects are reversed. There is some evidence of this kind of process in vertebrate populations that occasionally reach very high density. This kind of self-regulation is not constrained by the principle of the ESS, because, in essence, high density causes pathological changes that ultimately decrease the population size.

KEY CONCEPTS 9.2

- All populations fluctuate to varying degrees. Population stability is usually short-lived.
- Density-dependent factors tend to limit the degree of population fluctuation. However, they do not necessarily lead to stability.

QUESTION:

Under what conditions do density-dependent factors lead to stability?

9.3 How Do Regulatory Factors Interact?

There is a long history of experimental analysis of the factors that regulate populations. For many years, this was a contentious field in which proponents of one kind of regulatory factor or another devised experiments to demonstrate the action of one form of regulation and perhaps reject the importance of another.

Consequently, there is a rich literature that demonstrates the activity of an array of factors in population regulation. For each of the population processes outlined above, there is a set of empirical studies that demonstrate it in action.

The debate about the supremacy of one or another regulatory factor has shifted to address a more sophisticated question: How do the many types of regulatory factors interact in nature (Krebs, 2002; Lawton, 1992)? This question represents a major advance in the field—the recognition and analysis of the complexity of the regulatory process.

One reason the understanding of population regulation deepened is that ecologists undertook long-term studies of populations. In short-term studies, lasting perhaps three or four years, it is often possible to identify one factor that clearly regulates the population. The incorrect assumption following such an empirical study is that this factor is always the key regulatory mechanism. As long-term data sets accumulated, it became clear that the key regulatory factor may shift over time. Not only are populations dynamic in terms of their numbers, the regulatory mechanisms may be as well.

The long-term studies of the moose population on Isle Royale illustrate the dynamic nature of the regulatory process (Figure 9.16; Delgiudice et al., 1997; Peterson, 1999). As the wolf pack increased in size, the moose population declined due to predation and the density-dependent effects of the food shortage. A short period of relative stability, during which the moose population fluctuated within fairly narrow limits, occurred roughly in the period 1960 until the late 1970s. At the time, some ecologists suggested that because of top-down control by wolves, the moose population had achieved equilibrium. But eventually the equilibrium (or period of stability) broke down as a number of other factors changed the dynamics of both moose and wolves. Note in Figure 9.16 that these factors included a severe infestation of winter ticks, canine parvovirus that devastated the wolf population, deep winter snows, and so forth.

THINKING ABOUT ECOLOGY:
....
How does the nonequilibrium nature of populations make it more difficult to discern the factors regulating a population?

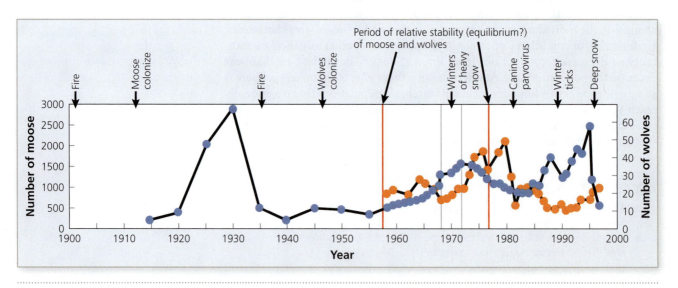

Figure 9.16 Moose and wolf populations. The pattern of fluctuation of moose and wolves on Isle Royale and the suite of factors that play a role in their dynamics. Note the short period of relative stability in both species. Eventually other factors disrupted this stability (from Peterson, 1999). **Analyze:** What future trends would you predict for these two populations?

As we look at this long history, we see no evidence of a true equilibrium maintained within narrow limits for long periods by density-dependent factors. Instead, we see evidence of a nonequilibrium population whose dynamics are constantly changing as a multitude of external factors change. Postfire forest development, wolf predation, winter snow, parasites, and disease affecting the wolf

Long-Term Studies of Song Sparrow Population Dynamics

Long-term studies are difficult to design. Often the researcher has a specific set of questions and hypotheses in mind. However, it takes great foresight to imagine just what questions and methods are going to be useful over a long period. If the methodology changes over time, comparison of results from different periods may be difficult or impossible. Some methods and measurements are best performed by the same individual each time. The longer a study proceeds, the more difficult this may be.

An important long-term study of the population dynamics of song sparrows (*Melospiza melodia*) illustrates these points. Chase et al. (2005) asked a fundamental question regarding population regulation: How do density-independent and density-dependent forces interact to determine the size of a song sparrow population?

HYPOTHESIS: Abiotic factors are the main factor determining population size.

Their study site was located in Point Reyes National Seashore, 20 km north of San Francisco in California. California has a Mediterranean climate characterized by wet winters and dry, essentially rainless summers. In this region, song sparrows are found primarily in moister, coastal habitats such as Point Reyes.

PREDICTION: Reproduction should correlate positively with precipitation; mortality should decrease with precipitation.

If abiotic factors are the primary factor underlying population dynamics, we can make a second prediction:

PREDICTION: There should be no evidence of density dependence in the population dynamics of song sparrows.

Chase et al. captured and marked more than 90 percent of the sparrows on their 36 ha study site each year for more than 20 years. They also mapped every song sparrow territory on the study. Nests provided data on the number of young produced and their survival. The number of territories and nests are measures of population density—when the population is high, the habitat is packed with many territories and nests.

The first prediction was supported. In the Mediterranean climate of Point Reyes, precipitation was found to influence reproductive success. Specifically, the number of fledglings per female was positively correlated with a factor they termed "bioyear precipitation." Bioyear precipitation is the total rainfall in the period July to the following June. It was defined in this way simply to match the seasonality of rainfall in their system rather than the calendar year. In addition, daily nest survival increased as bioyear precipitation increased. These data support their hypothesis that abiotic factors drive the population dynamics of song sparrows.

At first glance, density-dependent effects were absent. For example, the reproductive rate did not change with density. However, the long-term data set allowed them to tease out an important interaction between density-independent and density-dependent effects that might have been missed in a shorter study. The population density in any given year is related to the number of young fledged per female in the previous year. This is a delayed density-dependent effect. But the population density is a function of the amount of precipitation in that bioyear—a density-independent effect. The population density in year y is highly correlated with the bioyear precipitation in the previous year, $y - 1$. Thus, density-independent and density-dependent factors interact.

QUESTION:

What assumption must be correct in order to use the number of territories as a measure of population size?

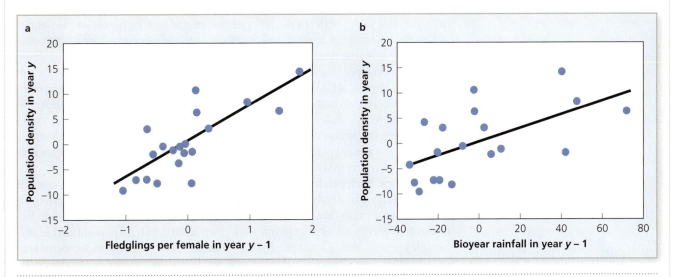

Figure 1 **Interaction between density-dependent and density-independent factors.** Time lags in the population dynamics of song sparrows. In both graphs, the y-axis is the population size in the present year. In (a), the x-axis is the reproductive output the year before; in (b), the x-axis is an index of the precipitation in the previous year (from Chase et al., 2005). **Analyze:** How might precipitation in the current year affect the population?

population all interact to determine the number of moose on Isle Royale. At any one moment in time, we might identify one of these factors as dominant. But none persists as *the* overriding factor because the entire system is so dynamic.

KEY CONCEPTS 9.3

- Species participate in an array of different interactions, many of which limit population growth. Consequently, most populations are regulated by a number of interacting factors.
- Over time, the importance of regulatory factors may change. This leads to two important ideas:
  - Long-term studies are important for understanding the process of population regulation.
  - It is not likely that any single factor will regulate a population.

QUESTION:

Explain the role that chance plays in the long-term dynamics of a population.

How Do Populations Interact?

The Isle Royale moose population history illustrates the importance of long-term studies that reveal how regulatory factors change over time. Variation over *space* is another important component of the interaction of regulatory mechanisms. In our previous treatments of the processes that regulate populations, the focus has been on one population in one particular place. For most species, however, individual populations are connected as part of a larger complex of populations. Thus, we need to consider each population in a broader landscape perspective. This is the essence of the concept of a **metapopulation** (Figure 9.17), a group of populations or a "population of populations," if you will, embedded in a landscape composed of habitat of varying quality and suitability. Some populations within the metapopulation inhabit discrete regions of high-quality habitat. Other parts of the landscape may be less suitable and hold smaller or more ephemeral populations. Still other parts of the landscape may be completely uninhabitable. Dispersal connects the populations to varying degrees. However, the potential for dispersal also varies with the mobility of the species and the nature of the habitat across the landscape. Linear segments of appropriate habitat may constitute dispersal corridors that connect populations.

The checkerspot butterfly, *Euphydryas editha*, exemplifies the metapopulation concept in its coastal California habitat (Figure 9.18). This species is restricted to grasslands growing on patches of serpentine soil (Harrison et al., 1988). As we saw in Chapter 7, endemic serpentine plant ecotypes have evolved on these patches. One of the main host plants of the checkerspot butterfly, on which it lays its eggs, is the annual plantain (*Plantago erecta*), a serpentine endemic. Also, several of the plants from which this butterfly obtains nectar are restricted to serpentine outcrops. Each patch of serpentine habitat represents a discrete population of the checkerspot butterfly. Weather, especially drought, is the prime source of mortality in a patch due to its negative effect on the life history synchrony between the insect and its host plant. Any given habitat patch (population) is vulnerable to extinction. Vacant patches are recolonized by long-distance movements of butterflies. The larger the habitat patch, the lower the frequency of extinction; the more distant the patch, the lower the probability of recolonization.

A complex suite of factors—density dependent and density independent, biotic and abiotic—operates across a metapopulation, differentially affecting each local population. The net result is that each local population has its

metapopulation A group of populations connected to varying degrees by dispersal.

THINKING ABOUT ECOLOGY:

How do the spatial size and the arrangement of populations in a metapopulation affect the concepts of stability and equilibrium?

own pattern of scarcity and abundance. None achieves equilibrium or even stability, but the metapopulation is comparatively stable because it is composed of a set of populations that fluctuate independently. If some populations are in decline, others may be increasing. The net effect is that stability occurs at the landscape level even though individual populations fluctuate.

Metapopulation structure can vary over time as well. In Antarctica, the Adelie penguin (*Pygoscelis adeliae*) occupies near-shore ice in a metapopulation structure in which colony size varies by an order of magnitude. Shifting ice conditions and the movement of large icebergs change the size, isolation, and resource conditions of individual populations. When two large icebergs ran aground in the 1990s, the population of penguins on each was effectively isolated from the other, thus altering the structure of the metapopulation, especially the rate of dispersal and local habitat quality (Dugger et al., 2010).

KEY CONCEPTS 9.4

- Regulatory factors vary over space. Thus, populations are spatial mosaics of different regulatory processes.
- The metapopulation concept incorporates this spatial variation. The populations that comprise a metapopulation fluctuate and are regulated independently. However, they are potentially connected by dispersal among populations.

QUESTION:

Explain how, in a metapopulation, population processes in separate populations are both independent and connected.

Putting It All Together

This chapter began with a history of the Isle Royale moose and wolf populations. Each of the potential factors discussed at the beginning of the chapter does indeed affect population dynamics of this system. On Isle Royale or in any other population, we can identify no single overriding factor that regulates the population at all times. Rather, a suite of factors interact over time such that each is necessary to explain regulation but none is sufficient. Figure 9.19 shows a general conceptual model of population regulation that can apply to a wide range of populations and species—from Isle Royale moose to tropical insects.

The complexity of this model, and the fact that key mechanisms vary over time, makes it unlikely that any population will achieve long-term equilibrium. Now we understand why true equilibrium at K, like that depicted in Figure 9.7, is relatively rare in nature. Too many factors combine to allow density-dependent

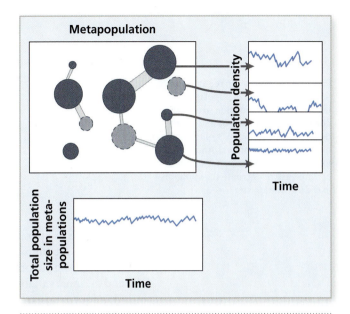

Figure 9.17 Metapopulation. The concept of a metapopulation, a *population of populations*. Subpopulations (shown as circles), connected to varying degrees by dispersal (shown by the width of the band connecting populations), are embedded in a heterogeneous matrix of habitat of varying quality. Populations inhabiting high-quality habitat are shown with solid lines; populations in marginal habitat are shown with dotted lines. Examples of the dynamics of four individual populations are shown in the graphs to the right. Note the variation in mean density, the amplitude of the fluctuations, and the fact that one population in marginal habitat regularly goes extinct. The population of the entire metapopulation (shown in the graph below the metapopulation) varies within rather narrow limits despite the wide fluctuations of individual populations. **Analyze: What is the impact of local extinction on the entire metapopulation?**

Figure 9.18 Checkerspot butterfly. The checkerspot butterfly (*Euphydryas editha*) exhibits a metapopulation structure on outcrops of serpentine soil in the Coast Range of California. Note that the landscape is composed of large areas of serpentine separated from smaller outcrops by inhospitable habitat (Harrison et al., 2008). **Analyze: What is the importance of chains of small islands of habitat near one another?**

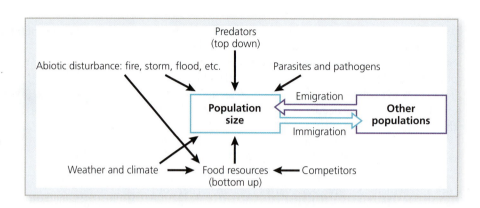

Figure 9.19 Population regulation. A conceptual model of population regulation. The population size at any one time is determined by a suite of density-dependent/density-independent, biotic, and abiotic factors. **Analyze:** Explain how each factor in this figure can increase or decrease the variation in the population size.

regulation to operate long enough to establish and maintain equilibrium. Long-term data sets show that when equilibrium does occur, it is ephemeral.

The more factors regulate the population, the more complex their interactions and the more variable, and even unpredictable, the long-term population dynamics. This does not mean, however, that we are incapable of understanding the regulation of any one population. As complex as the process seems to be, we have identified the key factors that combine to determine population density. We have the tools to predict population changes. This information also allows us to manage populations of species importance to us, and preserve species at risk of extinction.

Summary

9.1 What Limits Population Growth?

- Darwin's theory of natural selection was based on the notion that the growth of populations is limited.
- The basic equation for the intrinsic rate of growth depicts the key factors that regulate population size. Additions occur via reproduction and immigration, and losses occur by mortality and emigration.
- Population regulation occurs via
  - Top-down and bottom-up factors
  - Density-dependent and density-independent factors

9.2 Are Populations Ever Stable?

- All populations fluctuate, although some vary more widely than others.
- True equilibrium maintained solely by density-dependent mechanisms is probably rare or short-lived. Some populations that fluctuate widely are controlled by density-dependent mechanisms.
- Mechanisms internal to the population can regulate numbers, but such mechanisms must evolve through benefits to individuals, not the entire population at the expense of the individual.

9.3 How Do Regulatory Factors Interact?

- Most populations are regulated by a complex suite of factors.
- Over time these factors change in importance. Short-term studies can be misleading because they may identify a single overriding factor in that time period.
- Long-term studies reveal the complex interaction of factors at any one moment in time and how they change in importance over time.

9.4 How Do Populations Interact?

- Populations are spread across a complex landscape of habitats of varying quality that can support different population sizes.
- The metapopulation is a "population of populations" connected by dispersal.
- Any population within a metapopulation can increase, decrease, or even go extinct. Connections among the populations maintain relative stability within the metapopulation even if the local populations fluctuate.

Key Terms

bottom-up control p. 186
carrying capacity p. 189
density-dependent factor p. 187
density-independent factor p. 191

food chain p. 187
metapopulation p. 198
population regulation p. 191
population stability p. 193

self-regulation p. 194
top-down control p. 186
trophic level p. 187

Review Questions

1. Distinguish between equilibrium and stability.

2. What are the major categories (types) of regulatory factors? Explain the relationships among them.

3. How do random factors support the nonequilibrium view of population regulation?

4. How do regulation and stability differ between individual populations and the metapopulation?

5. How can random abiotic factors be regulatory?

6. Does each population have its own unique mechanism of regulation?

7. Explain the difference between density-independent and density-dependent forces.

8. How might abiotic factors act on a population through biotic factors?

9. Why are random processes so important in small populations?

10. Explain why it would be difficult for suicidal behavior to evolve as a regulatory mechanism.

11. The theory of natural selection is based on the relative survival and reproduction of individuals. Explain why the intrinsic rate of increase, r, is a useful measure of evolutionary fitness.

Further Reading

Boag, P.T., and P.R. Grant. 1984. Darwin's finches (*Geospiza*) on Isla Daphne Major, Galapagos: breeding and feeding ecology in a climatically variable environment. *Ecological Monographs* 54:463–489.

This classic paper documents the role of weather and climate on the populations of finches on the Galápagos. It explores the interaction of abiotic factors and the food supply, and their impact on the population dynamics of the finches.

Krebs, C.J. 2002. Two complementary paradigms for analyzing population dynamics. *Philosophical Transactions of the Royal Society B* 357:1211–1219.

Krebs discusses two paradigms for population regulation: (1) the older regulation by density-dependent factors and (2) the more modern, mechanistic approach emphasizing the forces that affect reproduction, survival, and dispersal.

Lawton, J.H. 1992. There are not ten million kinds of population dynamics. *Oikos* 63:337–338.

In this short, interesting paper, Lawton discusses the wide array of regulatory factors and their interaction. He discusses the issue of a general theory of regulation in the context of the species-specific differences in patterns.

Losos, J., et al. 2010. *The theory of island biogeography revisited.* Princeton, NJ: Princeton University Press.

This important new book discusses the patterns of population regulation and metapopulation dynamics in the context of MacArthur and Wilson's theory of island biogeography. Island biogeography is relevant to this discussion because it focuses on spatial patterns and isolation of populations and their connection via dispersal.

Peterson, R.O. 1999. Wolf-moose interaction on Isle Royale: the end of natural regulation? *Ecological Applications* 9:10–16.

This paper summarizes the long-term studies of the moose-wolf interaction on Isle Royale. It discusses the lack of density-dependent regulation of the moose population and its significance for a nonequilibrium view of population regulation.

Chapter 10

Life History Strategies

An adult sockeye salmon (*Oncorhynchus nerka*) feeds voraciously on small fish in the Bering Sea off the coast of Russia. It has been at sea for two years, following schools of bait fish and growing rapidly. Its foraging pattern shifts to the east and south so that by midsummer it is cruising into Bristol Bay in Southwest Alaska. The scent of just one particular stream is fixed in its memory, and it is that river it seeks out of the hundreds that enter the Bering Sea and Bristol Bay. Orcas flash through the schools of salmon near the mouth of the Mulchatna River. Those not taken by orcas make their way upstream. Bears line the river, fishing for the incoming salmon. The salmon's kidneys, adapted to saltwater, begin to fail as the fish moves further into freshwater. The fish no longer eats; it digests its own muscles for the energy to push upstream. It pauses in slow pools below each set of rapids, and then pushes on through. Eventually it enters the lake in which it was spawned. The female lays hundreds of eggs in a depression in the gravel known as a *redd*. As she moves off, the male releases his sperm over the eggs. Both adults die within hours of mating.

Like other species of salmon, sockeyes reproduce in freshwater. The young fish feed on algae, and as they grow, their diet shifts to animals of increasing size. The juveniles spend one to five years in freshwater, depending on the population. Eventually they migrate downstream. As they near the ocean, their color pattern changes to a camouflage more appropriate for open water. Their kidneys are reorganized to process saltwater. Once in the open ocean, they travel thousands of miles, exploiting the rich food sources there.

Why make the shift from freshwater to saltwater and back again? Why reproduce only in the stream from which they came? Why reproduce just once, then die? Why do they not provide any parental care for the offspring? The salmon life cycle is just one of thousands that have evolved. Other species take radically different approaches. Whales such as the humpback (*Megaptera novaeangliae*) share the Bering Sea with salmon. But rather than reproduce just once, they give birth to a calf every two to three years. The calf stays with the mother, who provides parental care in the form of milk and protection for up to a year. Still, each of these species is successful and has been so for thousands of generations. And of course even though they share parts of the same habitat, their ecologies differ in important ways. Each strategy has been shaped by the selective forces of the species' ecology. If we expand our survey of life history strategies to all organisms and all habitats we find an astonishing array of life cycles and patterns of life span, development, and reproduction.

This diversity leads to the fundamental question of this chapter: *How does the life history strategy evolve?* To answer this question, we begin with the components that make up the life history. We can then examine the ecological forces that shape the spectrum of life histories adopted by plants and animals.

10.1 What Is a Life History Strategy?

The **life history** of an organism consists of three main elements: the pattern of development and growth, the life span, and the timing and quantity of reproduction. As we shall see, each of these components is a response to species-specific ecological conditions. Together, they constitute an adaptive **life history strategy**. We use the term "strategy" because the three components of the life history are integrated—each relates to and affects the others. Of course, this is not a strategy in any conscious sense (see "The Evolution Connection"). Rather, it is a suite of characteristics shaped by natural selection in response to selection pressure from the environment. We will examine each of these three life history components in a separate section. However, it will be important to bear in mind that none is independent of the others.

The integration of these three components is the result of the finite energy available to the organism. Every organism, whether it is a photosynthesizing plant, a carnivorous predator, an herbivore, or a parasite, obtains a finite amount of energy from its environment. This energy allows the organism to accomplish all the life tasks required for survival and reproduction. And because the energy input is limited, the organism faces a series of trade-offs: to invest energy in one life history component limits the investment in other areas (Figure 10.1). Each energy use has a benefit to the organism as well as a cost. The cost is often an opportunity cost—possible uses of energy that are not chosen. This means that there is an optimal investment in each stage of the life history. Investment beyond that optimum results in lower fitness because energy is unavailable for other important functions (Figure 10.2).

What Is Optimized in the Evolution of the Life History Strategy?

The life history strategy is a complex suite of processes and patterns. The concept of trade-offs means that it is the overall strategy rather than a single life history component that determines fitness. The components of the life history strategy evolve as a unit, an integrated suite of patterns and processes. Organisms face an array of ecological conditions. Some inhabit stable, productive environments in

life history The combination of development and growth, the life span, and the timing and quantity of reproduction.

life history strategy The suite of life history traits adaptive in a particular ecological context.

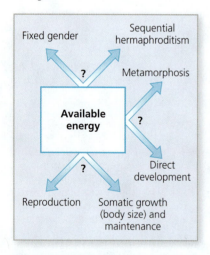

Figure 10.1 Allocation strategies. The organism has finite energy to allocate to many different life functions. Energy used for one is unavailable for others. Consequently, selection favors an allocation strategy that maximizes fitness.
Analyze: How would you define the optimal allocation strategy?

THE EVOLUTION CONNECTION

Teleology and Evolutionary Strategies

Adaptive evolution by natural selection has been an important component of our discussion of ecology thus far. Selection and adaptation are especially important in the ecology of life history strategies. This is a topic in which precision in the language we use and the connotations of that language are especially important.

When we refer to a life history "strategy," we do not mean to imply that the organism makes any conscious decision about how to apportion energy to various tasks. In effect, we use the term "strategy" as shorthand for a more involved explanation. Specifically, strategy refers to the process by which the life history is organized. If there is genetically based variation among individuals in a population in the ways they allocate energy, and there are fitness consequences to these allocations, natural selection will come into play.

The problem of teleology is a more general manifestation of the care we need to take in the language we use. Teleology is the idea that purpose exists in evolution in the same sense that it does for human intention. So, for example, the statement "meiosis evolved in order to reduce the chromosome number by half" is teleological. Such statements are philosophically invalid. They imply purpose and design and impart an intent that evolutionary biologists reject.

This chapter is replete with statements that can be misconstrued to imply that an adaptation evolved to serve a purpose. Just as "strategy" is shorthand for a more complex correct logic, these superficially teleological statements are a convenient way of expressing our understanding of the evolution of the life history. Despite our shorthand language, biologists recognize that adaptations are ultimately the result of random mutations that change the phenotype. The environment sorts those phenotypes by the process of natural selection.

QUESTION:

What biological processes underlie the random nature of evolution?

---

which competition is intense. Others exploit rich but ephemeral resources. In some, predation is intense; in others, competition is the key ecological factor. For each ecological situation, there is a different optimal solution to the apportionment of energy among the life history components. Thus, the myriad life history strategies we observe in nature result from the wide array of ecological conditions under which organisms live.

The concept of **reproductive value (V_x)** is a useful framework with which to understand the optimization process. The reproductive value of an organism is the expected reproductive contribution of an individual of age x to the next generation. Thus, reproductive value is closely tied to the organism's fitness. Mathematically we define reproductive value

$$V_x = \sum_{t=x}^{t=tmax} (l_t/l_x)\,(b_x)$$

where l_x and b_x are the age-specific survivorship and birth rates, respectively. The value x is the current age; t represents future ages from x on to the end of the life span, t_{max}. The value (l_t/l_x) represents the probability of surviving from age x to age t in the future. When age x is achieved, b_x young are produced. Thus, the sum of $(l_t/l_x)\,(b_x)$ for all future ages is the expected number of offspring the individual will produce over the rest of the life span. The logic of this equation is that the relative contribution of an individual is the result of its age-specific survival and birth rate.

The reproductive value changes over the course of the life span because the age-specific survival and birth rates determine the number of offspring expected in the future. So, for example, if there is high juvenile mortality, V_x will be lower for young individuals—they must survive the early ages to contribute births. However, once an individual survives that early mortality, its reproductive value increases. Thus, V_x often increases with age to a maximum just as the organism

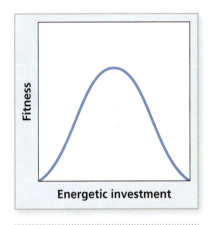

Figure 10.2 Optimal investment. Fitness as a function of the energetic investment in a component of the life history strategy such as clutch size. The optimum energy allocation to any life history function is not necessarily the maximum investment. At some point, additional energy devoted to one task reduces fitness, because that energy is not available for some other task. **Analyze: Does this curve have to be bell shaped?**

■ **reproductive value (V_x)** The expected reproductive contribution to the next generation of an individual of age x.

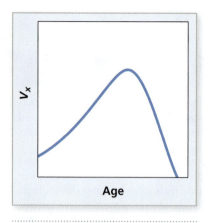

Figure 10.3 Reproductive value. The reproductive value, the organism's age-specific expected reproductive contribution to the next generation, changes with age. **Analyze:** At what age is the effect of natural selection greatest?

THINKING ABOUT ECOLOGY:

What is the difference between the reproductive value at a particular age and fitness?

enters the reproductive years. As the birth rate declines with age and survivorship begins to drop, V_x declines as well (Figure 10.3).

All the components of the life history—development, life span, and reproduction—contribute to the value of V_x. Developmental and reproductive patterns have mortality costs. In this way, reproductive value integrates the components of the life history: a change in one component affects others. If there is variation among individuals in the apportionment of energy to various aspects of the life history, the effect will be manifest in the individual's reproductive value.

However, selection operates on the life history only if the differences among strategies have a genetic basis. The life history may also vary among individuals due to environmental effects. We refer to nongenetic variation of this type as phenotypic plasticity. For example, some amphibians alter the timing of metamorphosis to increase body size when predators are present. This effect is apparently triggered by chemical cues that signal the presence of predators (Lardner, 2000). For example, killifish (*Rivulus hartii*), potential predators of guppies (*Poecilia reticulata*), are able to capture and consume prey only of relatively small size. The size component of the guppy life history strategy is phenotypically plastic: guppies grow larger when killifish are present (Gosline and Rodd, 2008). Phenotypic plasticity permits the organism to respond appropriately to some crucial aspect of its environment. In this example, the size change itself is not genetically based but the ability to alter the phenotype may be. In other words, the phenotypic plasticity may itself be a genetically coded adaptive response to a variable environment.

KEY CONCEPTS 10.1

- The life history strategy is the product of evolution.
- Each organism has a finite amount of energy to devote to life functions such as development, maintenance, and reproduction. Therefore, it must allocate that energy in ways that maximize fitness.
- The reproductive value measures the expected contribution of an individual age *x* to the next generation. It incorporates key elements of the life history.
- Some aspects of the life history are not genetically programmed (phenotypic plasticity). However, the nature of the plastic response may itself be an adaptation.

QUESTION:

Is the maximum value of V_x in a V_x-versus-age curve a measure of fitness?

resting stage A developmental stage in which the organism is dormant, inactive, and often resistant to harsh environmental conditions.

10.2 Why Do Some Species Have Complex Life Cycles?

The life cycle is one of the three components of a life history strategy. The life cycle itself comprises three key developmental features: the process by which an embryo becomes an adult, the presence of dormant stages during development, and the development and constancy of the organism's sex. In a simple life cycle, juveniles develop from the fertilized egg, grow into adults whose gender is determined genetically, live out their lives as active adults, and eventually die. Humans exemplify this simple life cycle.

A complex life cycle includes changes in the body plan, including resting stages and change in the individual's gender. For example, the egg may hatch into a larval form quite different from the adult. At some point, the larva metamorphoses into the adult form. Or at some point in the life cycle, the individual enters a dormant **resting stage**, such as the long-lived seed of a plant. In some species, individuals change gender during the life span, either in a genetically programmed manner or as a plastic response to the environment.

What Are the Advantages of Metamorphosis?

Developmental pathways can be divided into two broad types (Figure 10.4). In **direct development** the adult develops directly from the fertilized egg without a larval stage. For example, mammals and birds develop directly—the juvenile is a smaller version of the adult and there is no intermediate larval stage. **Metamorphic development** entails a larval stage that is often radically different from the adult individual. Most invertebrates have metamorphic development. Some vertebrates, such as amphibians, do as well.

direct development A process in which the adult develops from the fertilized egg without a larval stage.

metamorphic development A developmental pattern that includes a larval stage.

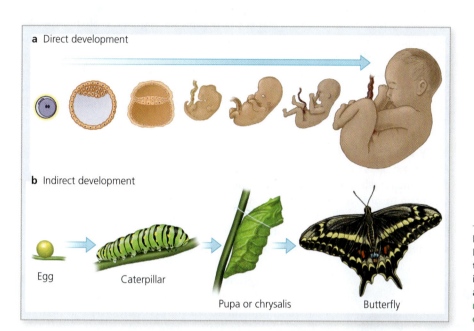

a Direct development

b Indirect development

Egg   Caterpillar   Pupa or chrysalis   Butterfly

Figure 10.4 Developmental pathways. Development can be direct (a), in which the adult develops directly from the fertilized egg, or indirect (b), in which there is a larval stage. **Analyze: What are the relative advantages and disadvantages of these two life histories?**

There are many significant costs to metamorphosis. Perhaps the most important is that the complete transformation of the body plan requires enormous energy. The genetic systems that organize the body plans and the timing and developmental process of remodeling the organism are complex. Also, the individual may be vulnerable to predation at various stages. For example, mayfly larvae live in rivers and streams, where they forage on plant material and detritus. When it is time to metamorphose, the larva rises through the water column and then transforms into the subadult form at the surface. The winged subadult must emerge from the larval body case, dry its wings, and take off. During this transformation at the surface, they are highly vulnerable to predation by fish.

The prevalence of metamorphosis suggests that its advantages are sufficient to outweigh these costs. Metamorphosis allows the organism to take advantage of highly specialized body plans for different functions over the life span. For example, many sessile marine invertebrates produce pelagic larvae that drift in the water and settle great distances from the adult. These larval forms disperse passively in ocean currents, something the sessile adults cannot accomplish. In other groups, such as the mayflies discussed above, the larva is specialized to feed and accumulate energy. The winged adult does not eat; it is adapted only for mobility from the water and for sexual reproduction.

Metamorphosis also adapts the organism to exploit resources otherwise unavailable to it. Some aquatic ecosystems are more productive than the surrounding terrestrial system. Aquatic larvae, such as the tadpoles of frogs, can exploit this energy for growth. At the appropriate size, they metamorphose into the adult form. The frog is adapted to feed on insects or other animals in its semiterrestrial habitat. The combined energy available in the two habitats supports this

neoteny The attainment of functional sexuality by larval forms that do not undergo metamorphosis.

Figure 10.5 Neoteny. Some populations of the spotted salamander are neotenic—the larval form becomes sexually mature.

THINKING ABOUT ECOLOGY:

It has been hypothesized that at high elevation, temperature is too low to support metamorphosis in poikilotherms. What field experiments would allow you to test this hypothesis?

Figure 10.6 Desert annuals. Desert annuals compress the life cycle into short periods of moisture availability. When conditions are favorable, most of the population reproduces in a short period.

Figure 10.7 Diapause. The reproductive cycle of the red kangaroo includes diapause, a period during which there is an unimplanted embryo in the uterus.

life history. Moreover, separating the young and adults in this way reduces competition between them. In some environments, the energy source is ephemeral. In that case, the larval stage can exploit this temporarily available energy. Thus, some amphibians lay their eggs in temporary spring ponds. The eggs hatch and the larvae use the food resources there. Metamorphosis relieves the adult of complete dependence on a short-lived aquatic habitat.

Some metamorphic species have populations that exhibit **neoteny**, the development of sexual larval forms that no longer metamorphose into adults. For example, some populations of the spotted salamander, *Ambystoma maculatum*, are neotenic (Figure 10.5). In fact, the sexually reproducing larvae of this species were originally classified as a separate species. The selective forces leading to neoteny are not fully understood. However, the phenomenon seems to be more common in extreme environments, such as at high altitude, where the temperature regime in the terrestrial environment is too low for poikilothermy (Sprules, 1974). Neoteny also seems to occur more frequently where the aquatic environment is unusually productive (Alford and Harris, 1988).

Why Have Resting Stages?

A resting stage is a developmental stage in which the organism is dormant, inactive, and perhaps resistant to harsh environmental conditions. Resting stages allow the organism to survive conditions that would be physiologically taxing. They take a variety of forms in different organisms. Deciduous trees drop their leaves and become physiologically dormant during the winter. Mammals such as ground squirrels, bats, and bears hibernate. Bacteria and protozoa form spores that are highly resistant to environmental extremes of temperature, desiccation, chemicals, and even radiation. Plant seeds fulfill this function as well. Seeds may lie dormant in the ground for decades until conditions for growth are favorable. Many invertebrate parasites form cysts that protect the individual during harsh conditions.

Extreme environmental conditions fall into two broad categories. Some, like winter or a dry season, are predictable. Cues such as changing photoperiod or decreasing soil moisture trigger the physiological and structural changes that initiate the dormant stage. Or the developmental pattern includes time lags that avoid unfavorable periods. For example, some seeds require exposure to low temperature for a certain amount of time before they germinate. This is an adaptation that prevents seeds produced in the summer from germinating just as winter begins.

Unpredictable harsh conditions pose a special problem for the organism: it must quickly respond to changes or face the consequences. There may be cues that precede difficult times, but it cannot rely on regular patterns, such as photoperiod, to trigger the response. Its adaptations require a more flexible, plastic response. Desert precipitation is often highly unpredictable—there may be years with little or no rain, then suddenly a series of heavy rains. Annual plants adapt to unpredictable moisture conditions by lengthening the period of seed resistance and shortening the vegetative and flowering cycle. Seeds may lie dormant in the desert for decades. When rains do come, the seeds germinate rapidly; the plant flowers and sets seed during the brief window of moisture availability (Figure 10.6).

This adaptation is not restricted to plants. The red kangaroo (*Megaleia rufa*) inhabits some of the harshest deserts in Australia (Figure 10.7). Its reproductive physiology includes a resting stage that adapts it to the unpredictability of its desert habitat. Like most marsupials, young are born relatively undeveloped after a short gestation period. They enter the pouch, where they suckle and develop. Soon after giving birth, the female mates again. However, this embryo does not implant and develop. Rather, it enters a resting period known as **diapause**, in which it floats in the uterus without developing. It remains there for 204 days and implants just before the first young leaves the pouch. After 31 days, this second young is born and enters the pouch. Mating occurs again, and this embryo, too, remains in diapause for 204 days. The result is a reproductive system and resting state that adapts the kangaroo to the unpredictable conditions of its desert habitat. In effect, she maintains three young at various stages of development. If conditions are favorable, the female carries each young to independence. If, however, the resources disappear partway through development, the female simply abandons the young in the pouch and does not implant the embryo in diapause. Her investment in these undeveloped young is small—she has wasted little resources on an offspring with poor survival prospects.

> **diapause** An embryonic stage in which the embryo does not implant in the uterus and suspends development.

Why Change Sex?

In a simple life cycle, the sex of the individual is determined early in development and remains constant throughout life. Some complex life cycles include shifts in sex over the course of the life span. We refer to this phenomenon as **sequential hermaphroditism**. In some, gender changes more than once. A life cycle that includes gender change has significant energetic costs. Nevertheless, such changes are not uncommon; they occur in many plants and animals. The adaptive value of the shift must exceed its cost.

> **sequential hermaphroditism** A complex life cycle in which the sex of the individual changes over the course of the life span.

In some species, the individual switches gender depending on the environmental conditions it faces. We have seen this phenomenon in Chapter 8 in the case of the forest herb jack-in-the-pulpit (*Arisaema triphyllum*). This species is dioecious—there are separate male and female plants. Moreover, any individual can switch sex. Female function is much more costly than male function. Thus, in high-light environments, there is sufficient energy to support female function, and most individuals are female. When light and nutrients are scarce, however, individuals switch to the less energetically costly male function.

In jack-in-the-pulpit, individuals can change sex multiple times depending on the conditions. In other species, the shift occurs at a specific point in development but not multiple times. If the individual is male first and develops female function later, it is **protandrous**. If female function precedes male function, it is **protogynous**. These changes may be developmentally programmed and always occur at a particular age. Others are triggered by environmental or social conditions. For example, in the rainbow wrasse (*Thalassoma lucasanum*), a coral reef fish of the eastern Pacific, the social group determines whether sex change occurs. Large, brightly colored males do most of the mating. If there is no such male in a social group, one or more females change sex and breed the females in the group (Warner, 1982).

> **protandrous** Describes a pattern of development in which male function develops first.

> **protogynous** A pattern of sequential hermaphroditism in which female function develops first.

Other factors may select for sequential hermaphroditism. For example, if the social system limits male reproductive success to a small number of older or larger male fish, protogyny may be adaptive. Individuals have female function when small because as males they would have little reproductive success. As they mature and grow, they switch to male function to take advantage of the reproductive success of large males. In other species, female **fecundity**, the number of offspring produced, is size dependent: large females lay many more eggs than small females. Protandry is adaptive in such species, particularly if female reproduction has a high survival cost. Individuals reproduce as males when small, achieving reproductive success without the mortality costs of females. Then, when they are large, they switch to female function and take advantage of the size effect on egg production.

> **fecundity** The number of offspring produced. It can be measured for a single reproductive event or over the life span.

10.3 What Determines the Life Span?

The variation among species in life span has long intrigued ecologists. What accounts for the fact that some organisms live for just a few hours or days whereas others live for centuries? Among unitary organisms, life spans range from invertebrates such as *Drosophila*, whose life span is measured in days, to plants such as the bristlecone pine (*Pinus longaeva*), that may live for 5,000 years. It is more difficult to define the life span for species that reproduce by fission, and for modular organisms. If a bacterium divides after 20 minutes, two individuals carry on but no death has occurred. At the other end of the spectrum, some modular organisms are truly ancient. The desert shrub creosote bush (*Larrea tridentata*) has modular growth in which the plant grows as an expanding ring. The shoots in the center of the ring die, but the same genetic individual expands outward (Figure 10.8). Carbon14 dating of the dead wood in the center of creosote bush rings shows the "individual" to be as much as 10,000 years old.

We define **senescence** as degenerative changes that increase the mortality rate as the individual ages. Basic metabolic processes release a number of molecules containing oxygen, known as **reactive oxygen species (ROS)**. Various oxygen ions and peroxides are important types of ROS. Many are produced in the pathways associated with energy production. Some ROS are even used as signaling molecules by the cell. They are important in senescence because they are toxic to cells. Cells have mechanisms to reduce exposure to these molecules, but those mechanisms decline with age, resulting in cumulative damage to cells by ROS.

Figure 10.8 Creosote bush. Creosote bush grows as an expanding ring. The inner tissue may be thousands of years old.

- **senescence** Degenerative changes that increase the probability as age increases.

- **reactive oxygen species (ROS)** Oxygen ions and peroxides produced in normal metabolic activity that are toxic to the cell.

Can Life Span Evolve?

For many years it was assumed that senescence and life span are the products of these kinds of physiological insults and age-related declines. This view was supported by the finding that life span correlates positively with body size. Larger species tend to live longer than smaller species in the same taxonomic group. This was thought to be a consequence of small species' higher metabolism and heart rate—that they simply "wore out" more rapidly. However, Ricklefs (2008) showed that captive birds live about as long as a wild individual, which suggests that mortality is not entirely due to external factors. Moreover, experiments with *Drosophila* and the nematode *Caenorhabditis elegans* showed that it is possible to select for longer life spans. Finch (1990) developed lines of *Drosophila* with a 30 percent longer life span than wild-type flies. Kenyon et al. (1993) doubled the life span of the nematode by selection for the longest-lived individuals in each generation.

Curran and Ruvkun (2007) screened 2,700 genes important to development in *C. elegans* and found 64 that extend the life span when activated after development is complete. Interestingly, these genes are highly conserved in evolution—they are also found in organisms ranging from yeast to humans. These studies suggest that the life span is under genetic control and subject to natural selection. They raise an obvious question: Why does selection not simply result in the maximum possible life span in every species? In other words, why is senescence adaptive?

Three important hypotheses have been proposed to explain the evolution of senescence (Monoghan, 2008). The concept of reproductive value is central to these ideas. First, it has been suggested that the intensity of natural selection declines with age. This would occur because of the shape of the V_x curve. Genes whose main effects occur after the peak in V_x are not subject to the same intensity of selection as those that occur when V_x is high or increasing. Specifically, deleterious mutations that affect the individual after reproduction, or when reproductive value reaches zero, are not affected by natural selection. These mutations then accumulate and cause senescence. A second, related hypothesis emphasizes **pleiotropic effects**, the action of a single gene that affects several phenotypic traits. Genes that benefit younger individuals whose reproductive value is high will be selected for even if they have deleterious pleiotropic effects that occur in old age.

The third hypothesis is sometimes referred to as the **limiting soma theory**: limiting resources devoted to reproduction result in decreased somatic maintenance. This in turn leads to cell death and senescence. This is another manifestation of the trade-offs that occur in the evolution of life history strategies. And like the other two hypotheses, the effects of the trade-off are affected by the point in the V_x-age curve at which the costs occur.

pleiotropic effects The actions of a single gene that affect more than one phenotypic trait.

limiting soma theory The hypothesis that senescence is caused because resources devoted to reproduction are not available to maintain somatic tissue.

THINKING ABOUT ECOLOGY:

What is the potential flaw in the hypothesis that individuals reach senescence and die *for the good of the population*?

KEY CONCEPTS 10.3

- Senescence results from the combination of external and internal factors that increase mortality as a function of age.
- Life span also has a genetic component. Selection experiments can increase the life span.
- Deleterious mutations that act as V_x declines after its peak value are not subject to the same intensity of selection as those that act prior to the peak. Late-acting deleterious mutations may be responsible for senescent changes in the organism.

QUESTION:

Is V_x always maximal at the point of reproductive maturity?

10.4 What Determines the Optimal Reproductive Pattern?

Plants and animals employ a wide array of reproductive modes. The most fundamental dichotomy is between asexual and sexual reproduction. **Asexual reproduction** preceded sexual reproduction in evolutionary history. The first organisms probably reproduced by simple fission. Over time, many variants of asexual reproduction arose. Some species, such as sponges, reproduce asexually by budding off new individuals from an adult. Asexual reproduction in plants is referred to as vegetative reproduction. The adult individual sends out a leaf, stem, or root that generates a new and independent individual. Among algae and fungi, asexually produced spores develop into new individuals. Sexual reproduction is most prevalent in multicellular organisms, although it does occur in some single-celled organisms, such as algae. Sexual reproduction employs meiosis to halve the genetic material in the formation of gametes or after fertilization.

Each individual in an asexual population contributes 100 percent of its genes to its offspring; in sexual species, the genetic contribution is only 50 percent. This

asexual reproduction A mode of reproduction in which offspring arise from a single parent and inherit its genes directly.

cost of meiosis The reduction in genetic contribution to the offspring in sexual species.

reduction in genetic contribution is known as the **cost of meiosis**. Fitness is directly related to the individual's relative ability to get its genes into the next generation. Thus, all other things being equal, asexuals have higher fitness than sexuals because of their larger proportional genetic contribution through their offspring. Obviously, the presence of sexual species means that other factors do come into play. For example, species that inhabit varying environments benefit from the recombination that occurs during sexual reproduction and the variety of genotypes it produces. For organisms that inhabit relatively constant environments in which it is advantageous to produce large numbers of offspring, asexual reproduction has significant advantages. Not only is each offspring genetically identical to the adult, a successful genotype can be reproduced in huge numbers. Also, the sheer number of offspring is usually higher among asexual species. Some species employ a combination of sexual and asexual reproduction and thus realize the advantages of each. Fungi, many algae, and invertebrate parasites have complex life cycles in which there are phases of both sexual and asexual reproduction.

Is Reproduction Always Maximized?

Given the importance of reproduction to evolutionary fitness, you might expect that selection simply maximizes reproductive output within the physiological constraints of the organism. However, there are both theoretical and empirical reasons to reject this hypothesis. The two key fitness components of a life table are l_x and b_x. So far we have treated these as essentially independent entities. But because reproduction requires energy that is then unavailable for other tasks, including somatic maintenance, predator avoidance, and so forth, reproduction may also decrease l_x (Figure 10.9). If the cost of reproduction is high, selection may favor reproductive rates lower than the physiologically maximum.

Empirical work supports this idea. For example, many shorebirds have an invariant clutch size. Females lay four eggs with little or no variation. Szabolcs et al. (2009) experimentally enlarged clutches in the pied avocet (*Recurvirostra avosetta*) by adding eggs to naturally occurring clutches. These artificially large clutches hatched more young, but pairs with enlarged clutches occupied poorer territories and had significantly higher chick mortality (Figure 10.10), thus decreasing their net reproductive success. Small lizards known as anoles (*Anolis sagrei*) lay a single egg at intervals during the reproductive season. Cox and Calsbeek (2009) surgically removed eggs from female anoles. Females whose eggs were removed had 56 percent higher survival to the next reproductive event and their survival to the next breeding season doubled (Figure 10.11). This occurs because the presence of an egg reduces the female's stamina, sprint speed, and growth. These data suggest that there is a limit to the fitness benefits of high reproductive output.

One might argue that artificially increasing reproductive output does not definitively support the concept of a cost to increased reproduction. If reproduction is already at the physiological maximum, enlarged clutches will necessarily be detrimental. However, costs

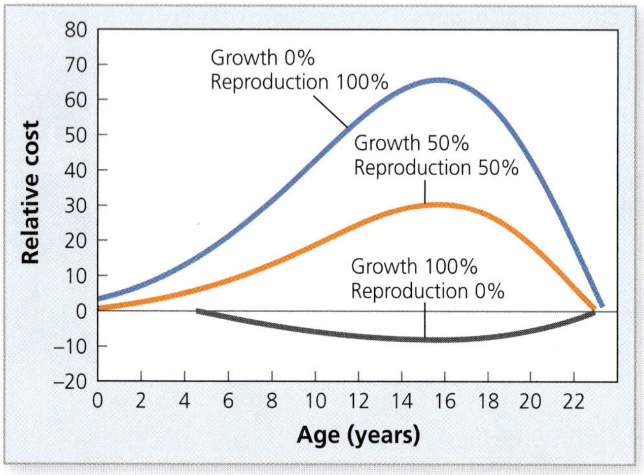

Figure 10.9 Cost of reproduction. The relative cost to l_x of different investments in growth and reproduction as a function of age. **Analyze: Why is the relative cost of investing 100% in growth negative?**

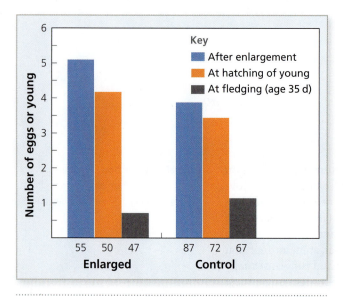

Figure 10.10 Experimentally enlarged clutches. The effect of experimentally increasing clutch size in the pied avocet on the number of offspring at different stages (from Szabolcs et al., 2009). **Analyze: What factors might contribute to the lower number of fledged young in experimentally increased clutches?**

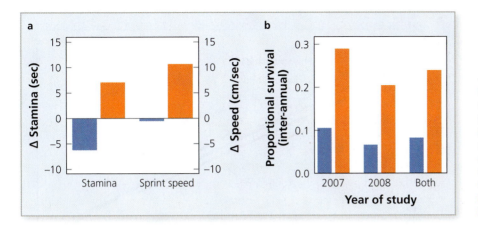

Figure 10.11 Egg removal. The effects of surgical addition or removal of eggs in anoles (orange bars: eggs removed; blue bars: control). (a) Stamina and sprint speed of females increased when eggs were removed. (b) Female survival also improved when eggs were removed (from Cox and Calsbeek, 2009). **Analyze: What is the purpose of "sham" operations in which no eggs were removed?**

are also apparent in naturally occurring variation. For example, the weaning weight of individuals in large litters of the collared lemming (*Dicrostonyx groenlandicus*) is less than that in small litters (Figure 10.12). In the desert wood rat (*Neotoma lepida*), the mean birth weight is smaller in large litters. Although the young from large and small litters are similar in weight at weaning, it takes the female 2.5 times as long to wean young from a large litter (Cameron, 1973).

Thus, there may be adaptive reasons to produce fewer offspring than physiologically possible at any one time. Here is another manifestation of the trade-offs inherent in the investment of energy in various life history components: energy invested in reproductive output at one point in the life history results in costs elsewhere.

What Are the Trade-Offs Among Reproductive Strategies?

The total parental expenditure of energy on offspring through the numbers and size of offspring, their care, feeding, and defense is the **parental investment**. This investment requires energy and thus incurs some cost to the parent or offspring. Large, well-provisioned eggs, large numbers of eggs, and parental care and protection are all forms of parental investment. The optimal reproductive strategy apportions this parental care such that fitness is maximized. The fact that species vary widely in reproductive output means that different ecological situations select for different allocations of reproductive investment.

One of the most fundamental reproductive trade-offs is between the number of offspring and their size. All other factors being equal, large clutches can be produced only if the offspring are smaller or receive less parental investment. Figure 10.13 shows this relationship across an array of lizard species (Warne and Charnov, 2008). The size of the offspring is important to its survival and competitive ability. Of course, if insufficient numbers are produced, fitness declines, too. The result is selection to optimize the facets of this trade-off.

Another key trade-off is the number of reproductive events in the female's lifetime. Here, too, the age-specific reproductive rate and the age-specific survival rate are intimately connected. Should a female reduce the size of the reproductive investment in order to have more reproductive events in the life span? Or should she invest in a single, maximal reproductive effort? Females that reproduce multiple times are known as **iteroparous**; those that reproduce just once are **semelparous**. The latter is sometimes called **big bang reproduction**, because semelparity often results in a large reproductive effort.

Salmon exemplify the kind of ecological situation that selects for semelparity. Salmon make the journey back to their natal freshwater streams to reproduce. This has tremendous energetic and physiological costs. Consequently, it may be impossible to make this trip multiple times and the result is a single, massive reproductive event.

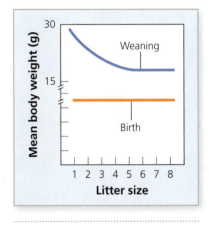

Figure 10.12 Birth and weaning weights. Birth and weaning weights of collared lemmings (*Dicrostonyx groenlandicus*) as functions of litter size (from Hasler and Banks, 1975). **Analyze: Why do lemmings ever produce clutches of more than four?**

parental investment The total parental expenditure of energy on offspring through the numbers and size of offspring, and their care, feeding, and defense.

iteroparous Describes a reproductive strategy in which females reproduce more than once over the life span.

semelparous (big bang reproduction) A reproductive strategy in which a female reproduces once in the life span.

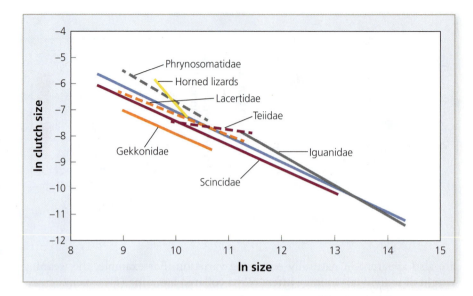

Figure 10.13 Clutch size vs. offspring size. The trade-off between clutch size and the size of offspring in lizards. Each line represents a different family (from Warne and Charnov, 2008). **Analyze:** How do horned lizards and the Teiidae differ from other lizards?

Iteroparity is a strategy that allows the female to spread her reproductive effort over a longer period. Organisms with Type II or Type III survivorship curves (Chapter 8) are more likely to survive from one reproductive event to the next. If so, the iteroparous female spreads her reproductive effort over a longer period. Potential advantages include increasing the probability that conditions will be favorable for at least some of the offspring. If a semelparous female happens to reproduce under unfavorable conditions, her entire reproductive effort is lost. Also, if iteroparity is associated with smaller clutch or litter sizes, each offspring may be larger or receive additional parental investment that increases its chances of success.

The equation for reproductive value can be modified to illustrate this trade-off. We begin with the basic equation

$$V_x = \sum_{t=x}^{t=tmax} (l_t / l_x)(b_x).$$

We modify it to separate present reproduction from future reproduction as follows:

$$V_x = b_x + \sum_{t=x+1}^{t=tmax} (l_t / l_x)(b_t).$$

residual reproductive value The remaining reproductive value of an individual after the current reproductive effort.

The value b_x represents current reproduction. The remainder of the equation is known as the **residual reproductive value**. In other words, we have separated reproduction in the present (b_x) and the expected reproduction over the rest of the life span (the right side of the equation). This separation quantifies the trade-off implicit in the difference between semelparity and iteroparity. For semelparous species, residual reproductive value is 0. Iteroparous species reduce current reproduction in order to maximize residual reproductive value.

A third reproductive trade-off is the age at which reproduction should begin—that is, the age of sexual maturity. One might think that all organisms should be sexually mature at the earliest possible age. This would increase the probability that the individual survives long enough to achieve reproduction. Also, in iteroparous species, the sooner reproduction begins, the more reproductive events can potentially fit into the life span. However, if the survivorship curve permits it, there are some advantages to delaying sexual maturity, particularly if each reproductive effort has a high cost. If there are social advantages to age or size that

DO THE MATH

The Difference in Production Between Semelparity and Iteroparity

Semelparity and iteroparity are significantly different reproductive strategies. However, mathematical analysis of the difference yields a surprising result: the difference between the two strategies is small. We show this by calculating the production of an annual and a perennial species using the growth equations of Chapter 8. If we hold all other factors equal, the population size of an annual species can be calculated by multiplying the current population size by the number of offspring in the average litter:

$$N_{t+1} = b_a N_t$$

where b_a is the litter size of the annual species. The current generation dies; the population in the next time period is solely the result of reproduction. We do the same calculation for a perennial species, but we include the assumption that survival from one year to the next is 100 percent:

$$N_{t+1} = b_p N_t + N_t = N_t (b_p + 1).$$

That is, the population at $t + 1$ is the sum of the offspring produced by the N_t individuals plus the N_t individuals that survive to $t + 1$. The annual and perennial species have the same growth rate if

$$b_a = b_p + 1.$$

This result is known as Cole's paradox: the difference between the annual and perennial is remarkably small. Cole wrote, "For an annual species, the absolute gain in intrinsic population growth which could be achieved by changing to the perennial habit would be exactly equivalent to adding one individual to the average litter size" (Cole, 1954, p. 126).

Of course, the assumption of immortality (100 percent survival) in the perennial is impossible. We can relax that assumption by including variables for adult and juvenile mortality (Charnov and Schaeffer, 1973):

$$N_{t+1} = b_p N_t (p_a/p_j)$$

where p_a and p_j are adult and juvenile mortality, respectively. If the annual and perennial grow at the same rate,

$$b_a = b_p + p_a/p_j.$$

If $p_a/p_j = 1.0$, Cole's initial calculation is correct. If $p_a/p_j \neq 1.0$, the difference between the annual and perennial increases or decreases depending on whether adult or juvenile survival is larger.

---

increase the probability of obtaining a mate or a territory, delayed sexual maturity will be favored. If fecundity is size related, females may benefit from delaying reproduction until they are larger.

What Are the Ecological Correlates of Reproductive Trade-Offs?

The association between the reproductive strategy and the ecology of the organism provides insight into the selective advantages associated with reproductive trade-offs. Some consistent patterns have emerged from this analysis.

Clutch size increases with latitude, both within species and among related species (Brown and Lomolino, 1998; Jetz et al., 2008). This trend is particularly strong in birds (Figure 10.14). A number of hypotheses have been offered to explain this pattern. Lack (1956) hypothesized that the longer days during the breeding season at high latitude permit the parents to forage longer and thus to

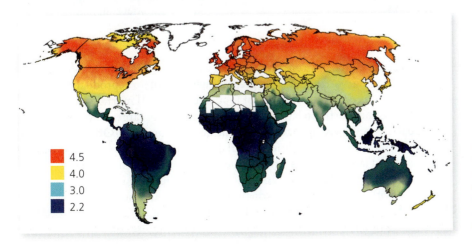

4.5
4.0
3.0
2.2

Figure 10.14 Clutch size and latitude. Clutch size in birds increases with latitude (from Jetz et al., 2008). **Analyze:** Why do you suppose clutch size does not increase as much in the southern continents as it does in the Northern Hemisphere?

The Importance of Predation to Clutch Size

One possible explanation for the latitudinal increase in clutch size is based on the intensity of predation in the tropics. Biancucci and Martin (2010) offered reasoned arguments that predation and nest size are connected.

HYPOTHESIS: High predation rates in the tropics select for smaller nests and thus smaller clutch sizes.
They tested this hypothesis in a study of nest predation in Venezuelan cloud forest. The nest-size hypothesis makes several testable predictions:

PREDICTION 1: Nest predation rates increase with nest size.

PREDICTION 2: Clutch size increases with nest size.

PREDICTION 3: Nests in the tropics are smaller than nests at higher latitudes.

The second and third predictions require that nest size be corrected for the body size of the bird. In other words, nest sizes should be larger in larger clutches for birds of similar size, and nests should be smaller in the tropics relative to the bird's body size.

Biancucci and Martin used two techniques to test these predictions. First, they performed nest-swap experiments in tropical and temperate sites. In these experiments, two nests were placed near a natural nest in the environment. One was a nest belonging to the same species as the original nest at the site; the other was a larger or smaller nest of a different species. Nests were baited with quail eggs and checked daily for predation. Daily predation rate was smaller in smaller nests in both the temperate and tropical site. This confirms the first prediction of the nest-size hypothesis.

Prediction 2 was not supported by their data. When the nest size was adjusted for body size, there was no increase in clutch size in large nests. Similarly, prediction 3 was not supported. Tropical nests were not smaller relative to body size in the tropics than at higher latitude.

What should we conclude from the mixed support for the nest-size hypothesis? The researchers reject the nest-size hypothesis. Two important predictions of the hypothesis were not supported. Nevertheless, they did detect a strong effect of nest size on predation rate. They suggest that nest predation may be an important factor in the latitudinal trend in clutch size for other reasons. Because body size, and thus nest size, is related to other life history traits, nest predation may favor smaller clutches in the tropics by other mechanisms.

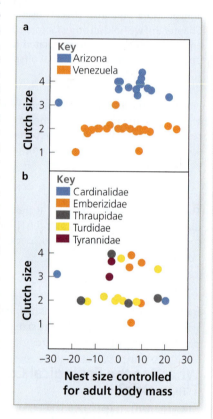

Figure 2 Clutch size and nest size. Clutch sizes as a function of nest size (controlled for body size): (a) data for species in temperate and tropical sites and (b) the same plot for families with members in both field sites. Each symbol type represents a different family (from Biancucci and Martin, 2010). **Analyze:** Why was it necessary to control for body size?

Figure 1 Predation rate and nest size. The predation rate as a function of nest size (from Biancucci and Martin, 2010). **Analyze:** What other characteristics of the nest might affect the predation rate?

provision larger clutches. In effect, this energy enables the parents to devote more energy to foraging, which leads to a greater energy pool for reproduction. Other hypotheses focus on the seasonality of food abundance. If mortality is higher in the more seasonal northern latitudes, selection will favor larger clutches because the probability of surviving to reproduce the next season is low. Others argue that high adult mortality results in lower population density and thus a larger resource base to support larger clutches. These hypotheses are not mutually exclusive, and, indeed, recent work (Jetz et al., 2008) shows that the most important variables associated with this trend are seasonality and high adult mortality. They found that clutch size increases with latitude to compensate for the high adult mortality, either in situ or during migration, which seasonality imposes on adults.

The latitudinal trend in clutch size perhaps represents an example of two more general correlates of reproductive strategies: variability in conditions and resources and external mortality factors. These two factors represent key selective forces. If conditions are harsh, mortality increases and this in turn affects the optimal reproductive strategy in ways described above. If conditions are variable, the organism faces another selective force: it must arrange its reproductive pattern to take advantage of favorable conditions and avoid times when the probability of success is low.

These effects are particularly well documented for plants. Grime (1977) categorized plants on the basis of the ecological pressures they face. He argued that plants are limited by two main factors: (1) stress, any external physical factor that inhibits growth, and (2) disturbance, any biological or physical factor that reduces the plant's biomass. There are four possible combinations of these two key factors. One, however, high stress and high disturbance, essentially prohibits plant growth. Thus, Grime developed a three-axis model of plant ecological situations (Figure 10.15). When plants experience low stress and low disturbance, they are more limited by competition between individuals than external factors. These he termed **competitive plants**. A forest tree species found in stable, old-growth forests, such as sugar maple (*Acer saccharum*) in eastern deciduous forest, would be an example. When stress is high and disturbance is low, **stress-tolerant plants** are favored. Stress-tolerant plants, such as the high-altitude white-bark pine (*Pinus albicaulis*), inhabit physically demanding habitats. The third environment is low stress and high disturbance. The plants adapted to these conditions are called **ruderals**. They tend to be plants we sometimes consider weeds—plants adapted to disturbed sites, such as dandelions (*Taraxacum officinale*).

Each of these three types of plants employs a different life history strategy appropriate for its ecological situation. Competitive plants are more likely to produce few well-provisioned seeds, because the ability of young plants to grow and compete is crucial. Stress-tolerant plants face herbivory and poor resource conditions, both of which lead to very slow growth. The life history of these plants emphasizes energy allocation to iteroparity, evergreen foliage, efficient use of water and nutrients, and herbivore deterrence. Ruderals are adapted to rapidly exploit ephemeral favorable conditions between disturbances. Thus, their life history strategy emphasizes rapid growth, a short life span, semelparity, and sometimes long dormant stages. One ruderal, the dandelion, has even secondarily evolved asexual reproduction; although the plant flowers, seeds are produced asexually in large numbers.

The theory of *r*-selection and *K*-selection was an important attempt to develop an overarching ecological explanation for the variation among reproductive life histories (MacArthur and Wilson, 1967). According to this theory, organisms can be divided into two broad groups: those that face intense competition for limited resources and those that exploit disturbed habitats in which resources are abundant, competition is low, but mortality from disturbance is high. These two habitat types select for different reproductive strategies (Figure 10.16). In low-disturbance, high-density situations near the carrying capacity (*K*; see Chapter 6), a reproductive strategy that emphasizes offspring *quality* is paramount. Reproductive strategies that improve offspring quality include late age at sexual maturity, iteroparity, small clutches, and high parental investment and care. In contrast, if disturbance is frequent, the organism inhabits an environment in which resources are abundant and competition is low. However, the next disturbance and its attendant mortality are always imminent, so the reproductive strategy emphasizes *quantity*—it maximizes the intrinsic rate of growth, *r*, while conditions are favorable. Large clutches, young age at maturity, little parental care, and semelparity characterize the *r*-selected reproductive strategy. Of course, the *r*/*K* dichotomy is actually a continuum; few species exhibit pure *r*- or *K*-selected strategies.

Figure 10.15 Ecological axes. The three ecological axes of Grime's model for plant reproductive strategies. Each species faces a specific level of competition, disturbance, and stress (from Grime, 1977). **Analyze: Are these three axes completely independent of one another?**

■ **competitive plant** In Grime's classification of plant strategies, a plant that is limited more by competition than other factors.

■ **stress-tolerant plant** In Grime's classification of plant strategies, a plant that inhabits physically demanding environments.

■ **ruderal** In Grime's classification of plant strategies, a plant that experiences low stress but high levels of disturbance of the habitat.

■ ***r*-selection and *K*-selection** The division of life history strategies into species that experience competition, whose strategy emphasizes offspring quality, and those that experience disturbance but abundant resources, whose strategy emphasizes offspring quantity.

THE HUMAN IMPACT

The Effect of Life History on the Risk of Extinction

The number of species threatened with extinction is increasing rapidly. The Endangered Species Act classifies species as endangered—that is, in imminent threat of extinction—or threatened, which means there is a high probability the species will become endangered. The act also requires federal agencies such as the US Fish and Wildlife Service, National Park Service, and Bureau of Land Management to develop recovery plans for endangered and threatened species. This task is daunting because of the sheer numbers of species that are listed and the fact that each has unique ecology, threats, and population dynamics. Thus, mechanisms that identify the demographic or life history parameters most crucial to the species' survival are of great value.

Heppel (1998) has developed a method that uses life table information to ascertain the key elements of the life history that will have the most impact on the population of a threatened or endangered species. The method employs elasticity analysis. This mathematical procedure uses life table and Leslie matrix information to address a key question about the relative importance of life history traits: What is the effect of a proportional change in survival, fecundity, or growth on the population growth rate, λ? For example, we might ask if a 5 percent increase in juvenile survival produces a change in λ greater than 5 percent. Or we might ask if a 5 percent increase in juvenile survival results in more or less than a 5 percent increase in adult survival.

At least 33 species of reptiles are listed as threatened or endangered by the federal government. Relatively little is known about the ecology and demography of many of these species. Consequently, any analysis that uses basic life table information to generate predictions about the potential benefit of different management plans is of great value. For freshwater turtles, elasticity analysis showed that adult survival has the greatest impact on λ; fecundity elasticity was consistently low. In contrast, for two other threatened species, the desert tortoise (*Gopherus agassizii*) and the loggerhead sea turtle (*Caretta caretta*), juvenile and subadult elasticity was much higher. Note that for the desert tortoise, adult elasticity is still the highest. Management plans that focus on increasing subadult survival of loggerheads and adult and subadult survival of desert tortoises will have greater impact on recovery of the populations than enhancement of fecundity.

In fact, Heppel's method allows us to predict quantitatively the relative importance of enhancing different life history stages. We can calculate this impact:

Proportional increase in
$\lambda \approx$ proportional increase in survival $\times$ elasticity.

Consider a population that is declining 5 percent per year (proportional increase = 0.95). Management plan 1 (MP1) increases adult survival by 5 percent. Management plan 2 (MP2) increases subadult survival by 25 percent. In this example, the elasticity of adult and subadult survival is 0.6 and 0.2, respectively. Thus, for MP1 the proportional change in proportional increase would be

$$0.05 \times 0.6 = 0.03.$$

Thus, λ would increase 3 percent, or $0.95 \times 1.03 = 0.9785$. For MP2 the proportional change in λ would be

$$0.25 \times 0.2 = 0.05.$$

This 5 percent increase would mean that λ would be $0.95 \times 1.05 = 0.9975$. Under both plans the population would decline, but less so under MP2.

Just as in life histories themselves, managers have finite resources to devote to recovery plans. Information such as this allows them to make the optimal allocation of those resources.

QUESTION:

What factors make juvenile and adult survival rates different?

What factors make them similar?

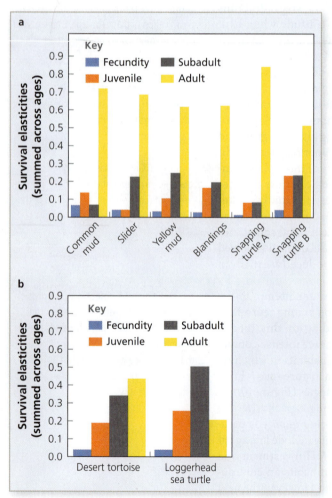

Figure 1 Elasticity. Elasticity of different life history stages in (a) freshwater turtles and (b) the desert tortoise (left) and loggerhead sea turtle (right). (From Heppel, 2006.) **Analyze:** What is meant by the concept of "elasticity"?

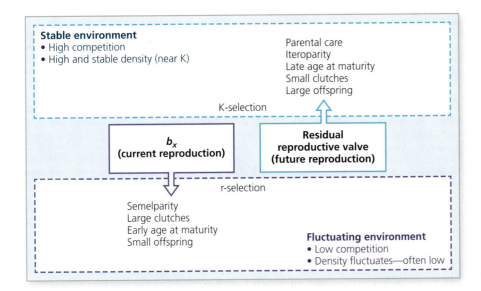

Figure 10.16 *R*-selection and *K*-selection. A schematic representation of the key environmental differences between *r*-selected and *K*-selected species, and their adaptive responses. **Analyze: Must these two strategies be genetically encoded?**

This theory was a useful model to characterize the interaction of the ecological situation and the reproductive strategy. However, many exceptions to the suites of predicted reproductive patterns have been found. For example, the marine gastropod (*Littorina rudis*) occupies two distinct habitats—vertical rock faces on which wave action and bird predation are significant mortality factors, and gently sloping boulder fields. The two populations have distinctly different life histories (Table 10.1).

However, they do not appear to match the *r*-*K* continuum: each is a compound strategy consisting of both *r*- and *K*-selected features (Emson and Faller-Fritsch, 1976; Rafaeli and Hugues, 1978; Hart and Begon, 1982). Because of the prevalence of compound strategies such as these, we no longer take this theory as strictly correct (Reznick et al., 2002). Nevertheless, ecologists still refer to *r*- or *K*-selected strategies as broad categories that emphasize either the quality or the quantity of offspring.

A third theory that correlates the life history strategy with the ecological context is **bet-hedging theory**. A bet-hedging reproductive strategy emphasizes reduced *variation* in fitness even if that occurs at the expense of *mean* fitness. The impetus for this idea comes from prevalence of unpredictable environments. Unpredictable conditions also result from fire, flood, drought, storms—all manner of physical disturbance—in all kinds of habitats.

The concept of a bet-hedging strategy, first proposed by Cohen (1966), is based on the idea that in unpredictable environments, there is great potential for catastrophic loss of an entire reproductive effort. If there is no reliable environmental cue to the occurrence or length of favorable reproductive periods, the organism is vulnerable to a complete reproductive failure if its reproductive effort is ill-timed. The bet-hedging strategy reduces the magnitude of each reproductive event, thus spreading the risk over multiple events (Figure 10.17). The organism

bet-hedging theory The idea that organisms that experience unpredictable environments should reduce variation in fitness by spreading out the risk of reproductive failure.

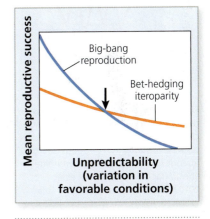

Figure 10.17 Optimal strategy. The optimal strategy, measured by reproductive success, differs with the unpredictability of favorable conditions. **Analyze: What is the significance of the point where the two lines cross?**

TABLE 10.1 The life history traits of *Littorina* in rock-face and boulder habitats; (r) and (K) refer to *r*-selected or *K*-selected components of the life history (from Hart and Begon, 1982)

ROCK FACE	BOULDER
Small size (r)	Large size (K)
Early age at maturity (r)	Delayed maturity (K)
High reproductive effort (r)	Low reproductive effort (K)
Few, large offspring (K)	Many, small offspring (r)

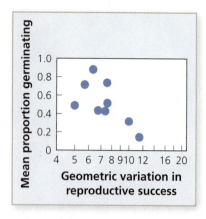

Figure 10.18 Bet-hedging. The relationship between the mean proportion of seeds germinating in a season and variation in reproductive success (adapted from Childs et al., 2010). **Analyze: Why is the proportion of seeds germinating the dependent variable in this graph?**

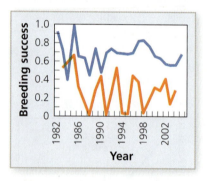

Figure 10.19 Black-browed albatross reproductive success. The breeding success of the black-browed albatross (*Thalassarche melanophrys*) on South Georgia (orange line) and the Kerguelens (blue line). (Adapted from Nevoux et al., 2010.) **Analyze: What is it about this graph that suggests that a bet-hedging strategy might be advantageous on South Georgia?**

forgoes the potential for the major reproductive success it would achieve if it reproduced massively when conditions are good in favor of reducing the risk of the catastrophic reproductive failure it would experience if it were to reproduce massively during an unfavorable period.

Any aspect of the life history can be modified by bet-hedging. For example, desert annuals produce seeds that are extremely long-lived. Moreover, subsets of seeds germinate in different years, thus "hedging the plant's bet" about when a favorable year will occur (Childs et al., 2010). Slight differences in seed mass and extreme sensitivity to microhabitat conditions result in temporal variation in germination. Among desert annuals, the proportion of seeds that germinates in any year is negatively correlated with variation in reproductive success (Figure 10.18). Other desert species modify other aspects of the life history, such as life span and iteroparity, to ameliorate variation in reproductive potential. In these species, delayed germination does not occur.

Bet-hedging is observed in animals as well. Nevoux et al. (2010) studied two populations of the black-browed albatross (*Thalassarche melanophrys*): one in the highly variable environment of South Georgia Island in the southern Atlantic, the other in the more stable environment of the Kerguelen Islands in the Indian Ocean. The population at South Georgia experiences much larger variation in reproductive success (Figure 10.19). The two populations differ significantly in their life histories in the way predicted by bet-hedging theory. On South Georgia, annual reproduction is lower, but one consequence is greater interannual survival and longer life span. Thus, in an unpredictable environment, the birds trade mean reproductive output for lower variation in success. This case is important because it also shows that life histories can vary within species in the same ways they do between species.

Each of these three theories—Grime's classification of plants, the theory of *r*- and *K*-selection, and bet-hedging theory—emphasizes the importance of disturbance and the nonequilibrium nature of many ecological systems. Each theory emphasizes the nonequilibrium ecology that constitutes an important selective force on many species. Each recognizes that species occupy different positions on a spectrum from mature, stable communities to dynamic, disturbed systems. And thus each theory accounts for both equilibrium and nonequilibrium ecology in the evolution of life history strategies.

KEY CONCEPTS 10.4

■ The optimal reproductive output is not necessarily the physiologically maximum possible.

■ Energy can be devoted to reproduction in many ways. The reproductive life history represents a set of trade-offs—there are costs and benefits to each potential type of energetic allocation.

■ The age-specific reproductive rate (b_x) and the age-specific survival rate (l_x) are reciprocally connected: reproduction has a mortality cost, and the reproductive pattern is a response to the mortality schedule.

■ The reproductive life history is a response to the species' unique ecology. A number of conceptual frameworks, such as *r/K* selection, Grime's triangle, and bet-hedging theory, organize these interactions according to the key ecological forces acting on reproduction. Important selective forces imposed by ecology are quantity versus quality offspring and the mean reproductive output versus the variation in reproductive output.

QUESTION:

How would you expect the V_x-age curve to differ for an *r*-selected and a *K*-selected species?

Putting It All Together

The life history of the sockeye salmon seems odd—the switch from freshwater to saltwater and back to freshwater, long periods at sea without reproducing, an all-consuming and ultimately lethal trip to the natal water, and a single massive reproductive effort followed immediately by death. It epitomizes the underlying question of the chapter: How does the life history evolve?

Now that we have worked our way through the intimate connection between the ecology of an organism and the evolution of its life history, this example should make more sense. We begin with the fundamental tenet of life history evolution: the organism has a finite amount of energy to invest in all the components of the life history. The salmon life history represents one successful response to the ecological conditions of the North Pacific and the freshwater tributaries that flow into it.

We can infer that one advantage of spawning in freshwater systems is the lack of predation potential relative to the open ocean, possibly a crucial advantage for offspring that receive no parental care. The same barrenness of these streams that decreases predation risk also limits the growth potential of the young salmon. Their growth potential is far higher in the ocean, where the prey base is orders of magnitude larger than in freshwater streams. Because the journey back to freshwater to spawn is so arduous and physiologically challenging, salmon can make the trip just once. The result is a relatively long-lived species with a semelparous reproductive strategy. A single massive reproductive event is supported by the energy diverted from somatic maintenance into egg production.

The fact that whales also take advantage of the high ocean productivity in the North Pacific but employ a radically different life history emphasizes two key aspects of life history evolution:

1. There may be different adaptive solutions to similar selective forces.
2. The components of the life history strategy are interconnected such that a change in one may result in significant changes in others.

For whales, which are homeotherms, large body size is important in the cold ocean waters. This changes the predation risk, especially for the young. This in turn affects the reproductive strategy: single calves, iteroparity, and significant parental care over a long period.

Summary

10.1 What Is a Life History Strategy?

- The life history strategy is the product of evolution.
- Each organism has a finite amount of energy; consequently, the life history strategy involves trade-offs.
- The reproductive value measures the expected contribution of an individual age x to the next generation. It incorporates key elements of the life history.
- Aspects of the life history may be phenotypically plastic.

10.2 Why Do Some Species Have Complex Life Cycles?

- The life cycle is composed of (1) the development of the individual, (2) the presence or absence of resting stages, and (3) the constancy of the individual's gender.
- Complex life cycles include one or more of the following: metamorphosis, a resting stage, and a change in gender over the course of the life span.
- Metamorphosis allows organisms to exploit energy in more than one habitat. It also reduces competition between juveniles and adults and allows for specialization for specific tasks such as dispersal.
- Resting stages are adaptations to harsh environments. They allow the organism to avoid predictable or unpredictable periods of unfavorable conditions.
- Because male and female functions have different energetic requirements and different behavioral requirements, selection may favor shifts in gender over the life span.

10.3 What Determines the Life Span?
- The accumulation of cellular damage over time limits the life span.
- Senescence may evolve as well. Artificial selection can increase the life span.
- Genes with deleterious effects that arise after reproduction accumulate, thereby affecting life span.

10.4 What Determines the Optimal Reproductive Pattern?
- Natural selection does not simply favor the maximum potential reproductive output.

- Finite energy resources for reproduction result in a set of trade-offs among reproductive strategies:
  - Size and number of offspring
  - Parental care and number of offspring
  - Iteroparity and semelparity
  - Early or late age at sexual maturity
- The ecology of the organism determines the optimal reproductive strategy. Disturbance, competition, the physical environment, and environmental predictability are important ecological factors that select for different reproductive patterns.

Key Terms

asexual reproduction p. 211	life history p. 204	reactive oxygen species (ROS) p. 210
bet-hedging theory p. 219	life history strategy p. 204	reproductive value (V_x) p. 205
big bang reproduction p. 213	limiting soma theory p. 211	residual reproductive value p. 214
competitive plant p. 217	metamorphic development p. 207	resting stage p. 206
cost of meiosis p. 212	neoteny p. 208	ruderals p. 217
diapause p. 209	parental investment p. 213	semelparous p. 213
direct development p. 207	pleiotropic effects p. 211	senescence p. 210
fecundity p. 209	protandrous p. 209	sequential hermaphroditism p. 209
iteroparous p. 213	protogynous p. 209	stress-tolerant plant p. 217
K-selection p. 217	*r*-selection p. 217	

Review Questions

1. How does the concept of reproductive value reflect the concept of trade-offs in life history strategies?

2. For each of the following, describe the associated trade-off:
   a. Iteroparity/semelparity
   b. Metamorphosis/direct development
   c. Sexual/asexual reproduction
   d. Current reproduction/residual reproduction

3. How do Grime's categories of plants relate to the *r/K* dichotomy?

4. Why are unpredictable conditions often more difficult for organisms than consistently harsh conditions?

5. What selective forces might extend the life span beyond the last reproductive event?

6. Some species can shift from asexual to sexual reproduction. Under what ecological conditions would you expect this to occur?

7. How do we know that the optimal fecundity is not necessarily the physiologically maximum fecundity?

8. Explain the kinds of life history adaptations that adapt the organism for an unpredictable environment.

9. Explain the ecological and social situations that select for sequential hermaphroditism.

10. What is the relationship between semelparity, iteroparity, the l_x curve, and residual reproductive value?

Further Reading

Roff, D. 2002. *Life history evolution.* Sunderland, MA: Sinauer.

This is an excellent summary of the evolutionary ecology of life history evolution.

Stearns, S.C. 1992. *The evolution of life histories: theory and analysis.* New York: Oxford University Press.

Like the Roff book, this book summarizes and synthesizes life history theory.

Stearns, S.C. 1977. The evolution of life history traits: a critique of the theory and a review of the data. *Annual Review of Ecology and Systematics* 8:145–171.

This classic paper analyzes the logic underlying much of life history theory. Although many new ideas and hypotheses have emerged since its publication, its critiques of life history theory are still relevant.

Vuorisalo, T.O., and P.K. Mutikainen (eds.). 2001. *Life history evolution in plants.* Amsterdam: Kluwer Academic Publishers.

The approach to life history theory by botanists has historically been somewhat distinct from that of other ecologists. This volume bridges that gap.

Eagles prey on migrating salmon. Part 3 examines the ecology and evolution of many kinds of interactions between species. Some interactions are mutually beneficial, such as that between flowers and their insect pollinators. Other interactions can be detrimental to one or both species, as in competition or predation. These interspecific interactions determine how communities are organized or structured and how they respond if disturbed.

Chapter 11

Competition

Life is hard in the Galápagos. These volcanic islands are subject to extremes of weather including El Niño events that bring torrential rains, as well as extended droughts. In drought years, plant flowering and seed set may completely fail. No species are more directly connected to their plant resources than Darwin's finches, the group of seed-eating birds that inhabit the islands. Chapter 2 illustrated the selective impact of drought on the medium ground finch, *Geospiza fortis*, on Daphne Island. In the intense drought of 1973, the small seeds easiest for the finches to crack open were eventually depleted, leaving only the largest, hardest seeds. Directional selection shifted the bill size of the medium ground finch upwards—only those individuals with the largest bills could open the remaining seeds, and only they lived to reproduce. Then, an intense El Niño event in 1983 resulted in abundant rainfall and copious seed production, which relaxed the selection pressure on the finches.

A second drought hit Daphne in 2004. We might have expected another shift to larger bill size in *Geospiza fortis* as small seeds were once again depleted. However, this time no such shift occurred—in fact, the bill sizes of *G. fortis* decreased in size. What was different this time?

In 2004 *G. fortis* faced not just drought but a new biological challenge as well. During the 1983 El Niño event, a new finch, the large ground finch (*Geospiza magnirostris*), established a population on the island. When the 2004 drought occurred, *G. fortis* faced two problems: vastly reduced food resources and a competitor for the seeds it once had all to itself. The large ground finch has a more massive bill than the medium ground finch (Figure 11.1). As a result, it is more adept at opening the largest, hardest seeds on the island. Its efficiency effectively eliminates large seeds as a potential food resource for the medium ground finch.

Figure 11.1 Darwin's finches. The massive bill of the large ground finch (*Geospiza magnirostris*) can open large, hard seeds (from Grant and Grant, 2009).

In the 2004 drought, medium ground finches with the largest bills were at a selective disadvantage because they competed directly with the large ground finch. There were two important consequences of this new situation. During the drought, the bill size of the medium ground finch *decreased* as selection pushed it away from competition with the larger species. And this left the finch with only small seeds, which were rapidly depleted. Second, its population declined radically.

This example raises many questions. Is this always the effect of competition? Are other results possible? How do the finches actually interact? Can these species coexist? These are the questions we address in this chapter. As we work through them, we will ultimately be able to answer the fundamental question: *What are the effects of competition for resources?*

11.1 What Is Competition?

The interaction of the large and medium ground finches on the Galápagos contains all the key elements of **interspecific competition**: the interaction among two or more species over a limiting resource that results in a decrease in the population size of at least one of the species. Central to this definition is the fact that during the drought, seeds became a *limiting resource* for both populations. Although both require oxygen, it is not limiting, and thus they are not in competition for it. Also important in this example is the fact that the abundance of the limiting resource changes over time. Consequently, the degree to which it limits both species changes—there are times when competition for seeds is especially intense, others when it is relaxed.

Many kinds of resources are limiting. For sessile organisms like plants or barnacles, space on the substrate may be the resource in short supply. Although oxygen is usually not limiting for terrestrial animals, it can be in aquatic ecosystems, where strong gradients in oxygen concentration occur. Sometimes the essential factor is the means of access to another resource. For example, rainbow trout compete with Atlantic salmon when they inhabit the same streams (Blanchet et al., 2008). When trout are present, the salmon activity pattern shifts from mostly nocturnal to diurnal (daytime) activity (Figure 11.2). However, activity time itself is not the resource; at stake is access to rich feeding areas in the stream. Because the trout exclude the salmon from such sites, the salmon shift their activity patterns to times when trout are less active.

Intraspecific competition occurs when individuals of the same species are limited by the abundance of a key resource. In this case the effect is primarily on the individual. Those individuals better able to obtain and use the resource survive and reproduce at a higher rate than those who cannot. Intraspecific competition underlies many of the density-dependent effects, discussed in Chapter 9, that limit population growth.

For plants, space is essential, because it ensures access to nutrients and water. At high intraspecific density, the population size declines but the remaining

interspecific competition The interaction among two or more species over a limiting resource that results in a decrease in the population size of at least one of the species.

intraspecific competition Competition among members of the same species.

individuals grow larger, a phenomenon known as **self-thinning** (Figure 11.3). The decrease in density over time occurs with a characteristic slope of −3/2.

Many species exhibit a decline in population density under intraspecific competition. The human parasite that causes African sleeping sickness (*Trypanosoma brucei*) occurs in a number of genetically different strains. It is common for an individual to be infected by more than one strain at a time, which opens the possibility of intraspecific competition among the trypanosomes in their human host. Bulmer et al. (2009) have shown that when multiple infections occur, competition among the strains reduces the population sizes attained by each, effectively diminishing the virulence of the infection.

How Do We Demonstrate That Competition Is Occurring?

We assume that competition is important to many species. Still, we need a precise mechanism for demonstrating that it is in fact operating. The definition of interspecific competition provides the key to empirical tests of competition: we need to determine that a resource is limiting and that diminished access to that resource negatively impacts the putative competitors. This appears to be a straightforward procedure. However, ecological interactions are so complex that it can be difficult to disentangle the effects of other factors that negatively impact a species, such as predation, parasitism, and abiotic factors.

A study of competition in barnacles (Connell, 1961) remains one of the clearest demonstrations of competition ever documented. In the rocky intertidal, two species of barnacles, *Semibalanus balanoides* and *Chthamalus stellatus*, occur in different zones. The adults are sessile filter feeders—they filter detritus or microorganisms from the water as it washes over them. *Chthamalus* is found higher in the intertidal zone, where it is exposed to air longer and to wave action; *Semibalanus* is found below the mean tide line. The larvae of both species are released into the water column and drift for some time before settling randomly onto the substrate, where they develop into the sessile adult form. Both species settle throughout the intertidal, yet their adult distributions differ. Connell asked the question: Is the adult distribution the result of competition or the different physical conditions high and low in the intertidal?

Virtually all the rock substrate in this system is covered with barnacles, suggesting that space is potentially limiting. Connell addressed the question of competition with two key experiments. First, he moved rocks bearing *Semibalanus* into the upper intertidal to see if they could survive the conditions there. He also did the reciprocal experiment—he moved rocks bearing *Chthamalus* from the upper to the lower intertidal. Second, he performed removal experiments to assess the absence of one species on its potential competitor. He removed *Chthamalus* from rocks in the upper intertidal and followed the fate of *Semibalanus* there. In the reciprocal experiment he removed *Semibalanus* from rocks in the lower intertidal and followed the fate of *Chthamalus*. He found that although *Chthamalus* can survive in the lower intertidal, *Semibalanus* cannot tolerate the conditions in the upper zone. Thus, *Semibalanus* is limited by the physical conditions. Experimental removal of *Semibalanus* in the lower zone resulted in *Chthamalus* establishing there (Figure 11.4). Connell took a series of photographs of rocks in the lower intertidal that showed *Semibalanus* slowly growing under *Chthamalus* and prying the individuals from the rock. Thus, the distribution of *Chthamalus* is limited by competition from *Semibalanus*.

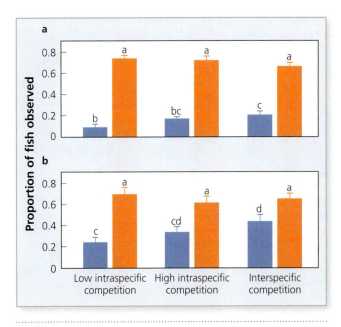

Figure 11.2 Atlantic salmon activity patterns. The mean proportion (+/− Standard Error) of Atlantic salmon active during daytime (blue bars) and twilight (orange bars) when maintained at low and high intraspecific density and with rainbow trout in (a) 2005 and (b) 2006 (from Blanchet et al., 2008). **Analyze: Is this an example of direct or indirect competition for a resource?**

self-thinning A phenomenon in plants in which individuals at high density have smaller population size but larger individuals.

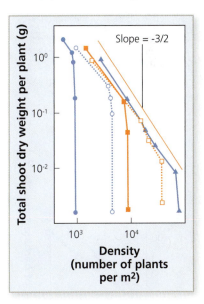

Figure 11.3 Self-thinning. The process of self-thinning in the grass *Lolium*. Each line represents a trajectory of weight per plant in a population over time. The straight line of the shift in density over time for these populations has a slope of −3/2 (from Lonsdale and Watkinson, 1983). **Analyze: What would this graph look like if intraspecific competition were not important in these plants?**

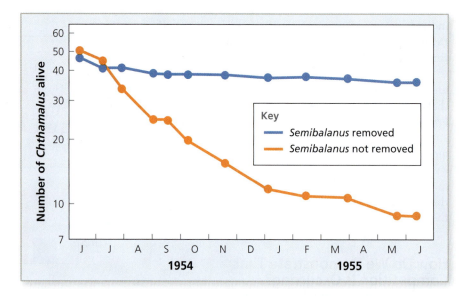

Figure 11.4 Intertidal competition.
When *Semibalanus* is removed from rocks in the lower intertidal, *Chthamalus* persists, indicating that competition from *Semibalanus* limits *Chthamalus* to the upper intertidal (from Connell, 1961).
Analyze: If there were no other evidence in this study, would this graph unequivocally demonstrate competition?

THINKING ABOUT ECOLOGY:

Imagine that in Connell's experiment *Semibalanus* and *Chthamalus* were each restricted to a portion of the intertidal by their tolerance of the physical factors. That is, neither species can survive in the other's preferred zone. Is it possible that competition is still ultimately responsible for their distribution?

Demonstrating competition can be even more difficult if both intraspecific competition and interspecific competition are affecting the species. We expect that members of the same species are more likely to be limited by the same limiting resource. In other words, because they are more likely to use the same resource, we might expect intraspecific competition to be more intense than interspecific competition. Of course, the more similar the competing species, the more likely it is that they, too, compete. How can we assess the relative importance of intraspecific and interspecific competition?

For plants, which are immobile and can be grown under controlled conditions, the analysis is relatively straightforward. For example, two species of oats, *Avena fatua* and *A. barbata*, inhabit the same grasslands in California. Marshall and Jain (1969) used an experimental design known as a DeWitt replacement series to measure the relative importance of intraspecific and interspecific competition in these two species. In a DeWitt replacement series, the two species are sown in a set of pots at constant total density. However, across the series, the relative numbers of the two species vary. A series of pots was established, each of which contained a total of 128 plants. The relative numbers of the two species varied from 0 *A. fatua* and 128 *A. barbata* at one extreme to 128 *A. fatua* and 0 *A. barbata* at the other.

Thus, each end of the spectrum represented pure intraspecific competition. At one extreme, *A. barbata* was competing only with itself; at the other, *A. fatua* was competing only with itself. Intermediate combinations represented varying degrees of intraspecific and interspecific competition. Interspecific competition was most intense in the pots with 64 individuals of each species.

Marshall and Jain measured the survival and reproduction of each species. They generated a null hypothesis in this analysis: that competition within species is exactly equivalent to competition between species. For example, if *A. fatua* produces *x* seeds per plant when competing only with itself, under the null hypothesis this same number of seeds should be produced if some of the competing individuals are *A. barbata*. This is shown by the dashed lines in Figure 11.5. They then measured the reproductive output of each individual in each pot and plotted it on the same graph. Note in Figure 11.5 that *A. fatua* produced more seeds than predicted by the null hypothesis, especially at the most intense interspecific competition (64 individuals of each species). Thus, it

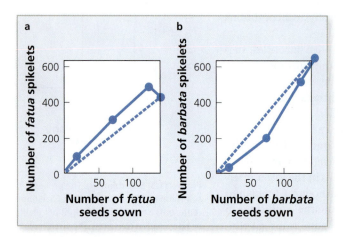

Figure 11.5 Replacement experiments. DeWitt replacement experiments between *Avena fatua* and *A. barbata*. The dashed line represents the null hypothesis—the seed production of *A. fatua* (left) and *A. barbata* (right) in the absence of interspecific competitors. The actual performance of each species is shown by the solid line (from Marshall and Jain, 1969).
Analyze: Why does the dotted line represent the null hypothesis?

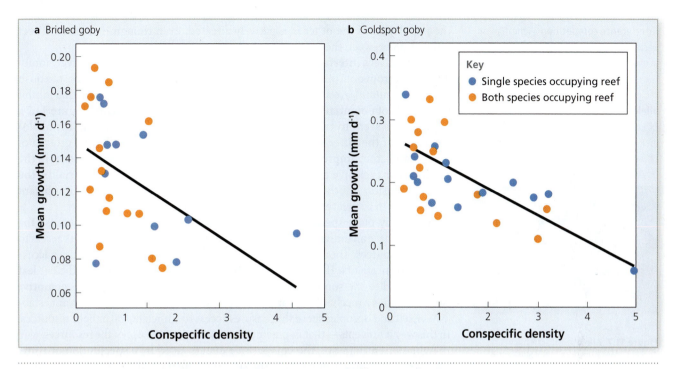

Figure 11.6 Gobie growth rates. The growth rate of bridled gobies (a) and goldspot gobies (b) as functions of density. Blue circles are the results from single reef populations; orange circles are for reefs occupied by both species (from Forrester et al., 2006). **Analyze: Why do these data suggest that intraspecific competition is more important to these species than interspecific competition?**

performed better in interspecific competition than in intraspecific competition: it was limited more by members of its own species than another. In contrast, *A. barbata* performed worst when interspecific competition was most intense. Thus, *A. barbata* was limited more by interspecific than intraspecific competition.

Demonstrating intraspecific and interspecific competition is more complex for mobile species. Some systems, however, are amenable to experimental manipulation. Forrester et al. (2006) studied the magnitude of competition in two coral reef fish, the bridled goby (*Coryphopterus glaucofraneum*) and the goldspot goby (*Gnatholepis thompsoni*). These species inhabit small patch reefs that represent isolated island habitats. The researchers manipulated goby densities by stocking reefs with either one species or equal numbers of the two species. The growth rate of tagged individuals was used to measure the impact of conspecifics on the presence of the other species. Both gobies' growth rates were significantly affected by the presence of their own species (Figure 11.6), indicating intense intraspecific competition. Growth rates also declined as a function of the population size of the other species, but only about half as much. Thus, these fish are affected by both intraspecific and interspecific competition, but interactions with their own species have the greater effect.

What Are the Mechanisms of Interspecific Competition?

Access to a resource in short supply is central to the process of competition. Any adaptations that improve the species' ability to exploit a limiting resource confer a selective advantage. In general, the mechanisms that accomplish this fall into two broad categories. In **exploitation competition**, one species reduces the amount or availability of the limiting resource. This often occurs by adaptations that improve the efficiency with which the resource is used. The two competitors may never even encounter each other; one species simply exploits the resource so efficiently that the other species is compromised. In the deserts of the southwestern United States, the seeds of annual plants remain in the soil for years awaiting sufficient rainfall for germination. Two very different taxa exploit this seed resource: granivorous rodents such as kangaroo rats, and ants (Brown and Davidson, 1977). Each depends on this resource. To the extent that either does that more efficiently,

exploitation competition A mechanism of competition in which one species reduces the amount or availability of the limiting resource.

interference competition A mechanism of competition in which one species actively inhibits another from obtaining the resource.

allelopathic Describes a plant that produces and releases chemicals that inhibit the growth of nearby individuals of the same or another species.

Figure 11.7 Allelopathic chemicals. The creosote bush produces allelopathic chemicals that inhibit the growth of other plants.

preemptive competition A mechanism of competition in which a plant establishes access to resources by establishing itself and occupying space.

diffuse competition The summed effects of all competitors.

the population of the other is negatively affected, even though there is no direct interaction between them.

Alternatively, **interference competition** occurs when one species actively inhibits another from obtaining the resource. Some desert plants, such as creosote bush (*Larrea tridentata*), are **allelopathic**. They secrete chemicals into the soil that inhibit the growth of potential competitors for scarce water and nutrients (Figure 11.7). Some plants seem to take this competitive advantage a step farther. The allelochemicals of the bitter vine (*Mikania micrantha*) not only inhibit the growth of other plants, they improve the availability of any mineral nutrients present in the soil for their exclusive benefit (Chen et al., 2009). Interspecific territoriality, the active exclusion of other species from a portion of the habitat, is an important mechanism of interference competition in animals.

For many plant species, space is a critically important resource, because the amount of space an individual occupies in the habitat directly determines its access to essential resources. The amount of soil space occupied by the roots determines access to water or nutrients. If light is the limiting resource, there must be sufficient leaf area exposed to the sun for effective photosynthesis. Thus, for plants, **preemptive competition** is a mechanism that ensures access to these resources. The plant establishes itself and occupies space that ensures access to sufficient soil or light resources. In this way it prevents—that is, preempts—its competitors' use of the resources.

More often than not, the competitive effects faced by a species do not come from a single competitor species. The summed effect of all competitors is known as **diffuse competition**. It may be that no single competitor is important; rather, it is the combined effect of two or more competitors that is significant. For example, Brown and Davidson (1977) demonstrated diffuse competition in the ant–rodent desert system by excluding rodents from experimental plots and measuring the resulting increase in ant density. However, their removal experiments eliminated not just a single rodent species but a group of species whose summed competitive effect was significant to the ants. A tree seedling in a mature forest competes for light not only with a group of other seedling species but with the adults of a number of species. Adult trees absorb light before it reaches the seedlings. Thus each adult fractionally reduces the light available to the seedlings.

KEY CONCEPTS 11.1

- Limited resources lead to competition between species as well as within species.
- Competition is empirically demonstrated by showing that (a) the resource is limiting and (b) the interaction between the species has a negative impact on one or both of them. Competition for limited resources may be between different species, members of the same species, or a combination of the two.

QUESTION:

Why would you expect interspecific competition between vastly different taxa such as ants and rodents to be relatively rare?

11.2 What Determines the Intensity of Competition?

Competition occurs when two species exploit the same resource. Thus, a crucial factor is the degree to which they overlap in their use and reliance on a particular resource. For example, in tundra habitat in Alaska, plant production is limited primarily by soil nitrogen, which occurs in a variety of chemical forms, such as ammonium nitrate and as part of free amino acids such as glycine. Moreover, the temporal and spatial availability of these nitrogen compounds varies. However, the intensity of competition for nitrogen is mediated by the specific form of nitrogen the plants are adapted to utilize (McKane et al., 2002). The more similar the species' use of nitrogen forms, the more their production is reduced by competition (Figure 11.8). This suggests that it is important to be able to quantify the specific limiting resource.

How Do We Quantify the Use of Resources?

Early in this history of ecology, Joseph Grinnell (1917) described the **functional niche** as the ecological role that the organism plays in the community. We have alluded to a number of functional niches in this chapter. For example, Darwin's finches and ants are granivores that inhabit variable environments. And the functional niche of a barnacle is a sessile filter-feeding invertebrate in the rocky intertidal.

Hutchinson (1957) extended this concept in a more quantitative form, today known as the **ecological niche**: the set of biological and physical resources that determine growth, survival, and reproduction. For any species, we can identify a set of resources, each of which is important to these vital processes. We represent each resource on a separate axis. We can envision the niche as the volume defined by the tolerance limits along each axis (Figure 11.9). This figure shows a niche in three resource dimensions. However, in reality, each important resource is included as an axis in multidimensional space. The niche is an object known as a *hypervolume* bounded by the use of those *n*-dimensions. The axes that make up the ecological niche are both physical factors and biological resources. For example, for a trout living in a mountain stream, the physical factors that might determine its success include the oxygen concentration of the water, temperature, and pH. Prey size might be an important biological factor.

Obviously, it is nearly impossible to identify and quantify each of the resource axes important to a species. Fortunately, not every resource axis that constitutes a niche is critically limiting. For the study of competition, we can often reduce the ecological niche to the single axis of the resource that is most limiting. Figure 11.10

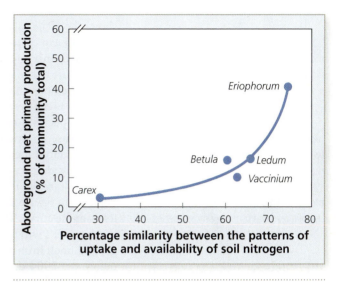

Figure 11.8 Competition for nitrogen. The production of tundra plants as a function of their similarity in the form of nitrogen utilized (from McKane et al., 2002). **Analyze: Why is the form of nitrogen used by these species important to interspecific competition?**

■ **functional niche** A descriptive definition of the niche as the ecological role of the species in the community.

■ **ecological (Hutchinsonian) niche** The set of biological and physical resources that determine the growth, survival, and reproduction of a species.

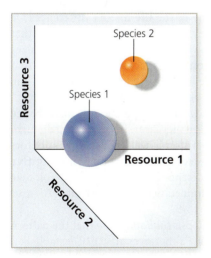

Figure 11.9 Ecological niche. An ecological niche based on three resource dimensions. Species are represented by spheres defined by the limits of their tolerance and use of each resource. **Analyze: What quantitative measures could you use to characterize the niches of these species and the relationships of their niches?**

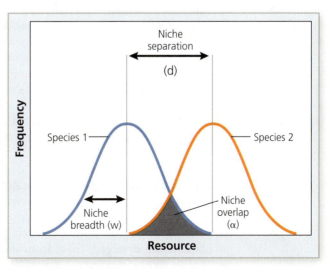

Figure 11.10 Niche parameters. The niche parameters of two species using a single resource. **Analyze: Where is competition between these two species most intense?**

shows the niches of two species along a single limiting-resource axis. From such a graph, we can quantify key aspects of the niches and their relationship. Not all individuals in a population use precisely the same resources. **Niche breadth** (w) is a measure of the variation among individuals in their use of the resource. Note that in previous chapters we used the same symbol, w, for relative fitness; here it measures niche breadth, an entirely new quantity. The difference in the niches of the two species is measured by the difference between the mean values of their resource use, the **niche separation** (d). The extent of common use of the resource by two or more species is quantified by the amount of **niche overlap** (a).

Finally, density factors into the intensity of competition. If a potential competitor is rare, its impact will be much less than if it is abundant. In the niches depicted in Figure 11.10, the size of each population is depicted by the area under its resource curve. Thus, species 2 is less abundant than species 1. We may expect that species 2 will have less impact on species 1 than the reverse. When the large ground finch immigrated to Daphne Island, its population grew for many years before the 2004 drought brought it into competition with the medium ground finch. If its numbers had been still small when the drought occurred, its competitive impact on the medium ground finch would have been much smaller. In other words, competition is a density-dependent phenomenon.

What Are the Effects of Disturbance on the Intensity of Competition?

Many species experience physical disturbance of one form or another. The immediate effect of physical disturbance such as fire, volcanic eruption, or windstorm is to reduce the population density. We now understand that in some habitats, disturbance is relatively infrequent; in others, it is common. For those species whose habitat is regularly disturbed, competition is a much less potent selective force. Colonists of disturbed systems inhabit a world of high resource abundance and low population density of their own and other species. They are adapted to exploit these conditions with rapid growth and reproduction. As this occurs, the habitat becomes more saturated and competition increases. Disturbance-adapted species do not fare well as the intensity of competition increases. Eventually their offspring must disperse to a newly disturbed habitat.

These differences between mature, undisturbed habitats and habitats that experience recurring disturbance underlie some of the life history differences we explored in Chapter 10. Specifically, the theory of r- and K-selection suggests that for species normally found at high density, the reproductive strategy emphasizes the competitive ability of the offspring. In contrast, species or populations that are frequently disturbed, and thus experience low density and little competition, invest their reproductive energy in large numbers of offspring.

In general, competition and disturbance work in opposition: if disturbance is infrequent, densities are high and competition increases. In contrast, if disturbance is frequent and densities are low, competition is weaker. Poor competitors that exploit disturbances where they face fewer competitors are known as **fugitive species**. Their preferred habitat eventually disappears and they must colonize newly disturbed sites. Dispersal ability is an important life history trait of fugitive species.

Experiments with the annual plant *Arabidopsis thaliana* show that the pattern of disturbance determines competitive ability. Fakheran et al. (2010) established populations of *Arabidopsis* that experience either a high or low disturbance regime. Static populations experienced no disturbance. The habitat of the dynamic populations was destroyed each generation. After five generations the populations had diverged: those from static populations tolerated competition better than those from dynamic populations. Specifically, static conditions selected for more rapid growth and thus preemption of resources. Dynamic populations were more sensitive to crowding. However, plants from the dynamic regime

Distinguishing Resource and Nonresource Competition

Plants face competition from nearby plants above ground as well as below ground. Below-ground competition is particularly hard to study. Plant roots face two forms of competition. One, known as resource competition, is for nutrients and water. The other, nonresource competition, is over space. Obviously the two are intimately connected.

Messier et al. (2009) examined these forms of competition in four tree species: a hybrid of cottonwood and balsam poplar (*Populus deltoides* x *P. balsamifera*), paper birch (*Betula papyrifera*), sugar maple (*Acer saccharum*), and white ash (*Fraxinus americana*). They addressed the question of the relationship between resource and nonresource competition among these species. These four species differ in the kind of forests they inhabit. The hybrid and birch are found in disturbed forests; sugar maple and ash are found in mature forests that have not been disturbed for some time. These differences led the researchers to analyze the effect of disturbance on competition.

HYPOTHESIS: Species adapted to disturbance are less competitive for soil resources than species that inhabit undisturbed sites.

The researchers devised a clever system for testing the effect of below-ground competition. They planted the trees in large containers divided into two halves separated by a solid wall. Roots of the trees grew into each half of the container. In this way they could manipulate the conditions the roots encountered in the two compartments. Four variables were used in all possible combinations: no competition, competition from the roots of grasses, no fertilization, and fertilization. At the end of the experiment, they carefully removed the roots from the containers and measured their length, biomass, and branching. Asymmetry between the two halves of the container in root production or architecture was attributed to the different environments the roots encountered. Thus, asymmetry between fertilized/nonfertilized containers was a measure of nonresource competition. Asymmetry between grass/nongrass roots measured the effect of competition.

PREDICTION: The disturbance species should have asymmetric root production in the competition and noncompetition halves of the containers. Nondisturbance species should have less root asymmetry between competitive and noncompetitive treatments.

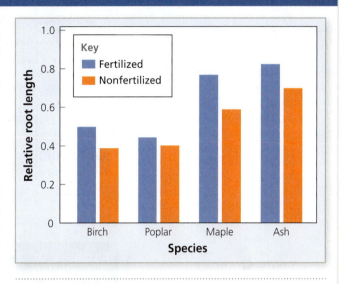

Figure 1 Root production. The root length (y-axis) for four species expressed as the ratio of vegetated half:nonvegetated half (that is, competition:no competition). Orange bars are nonfertilized experiments; blue bars are fertilized experiments (from Messier et al. 2009). **Analyze: What do we mean by "asymmetrical root production" in this experiment?**

The accompanying figure shows the results of the experiment. The prediction was confirmed: there was asymmetrical root production in the disturbance species exposed to grass and nongrass. This means that disturbance species compete poorly with other species. Their roots avoided those of other species and grew less in the presence of competitors. Maple and ash, species found in mature forests, had much less asymmetry in competition.

These results make sense in light of the ecological niches of the two groups of trees. In species whose niche centers on disturbed sites, fine root growth that avoids competing roots is an adaptation to maximize nutrient uptake for rapid growth. In essence these are fugitive species adapted to exploit noncompetitive environments. The root morphology of sugar maple and white ash responds to competitors, an important adaptation if they are to establish roots in the face of competition.

eventually grew taller, which enabled their seeds to disperse farther from the adult plant (Figure 11.11).

KEY CONCEPTS 11.2

- The ecological niche is central to competition because it is based on the use of critical resources.
- Resources exploited by a particular niche may be biological, such as prey type or size, or physical, such as a specific temperature regime.

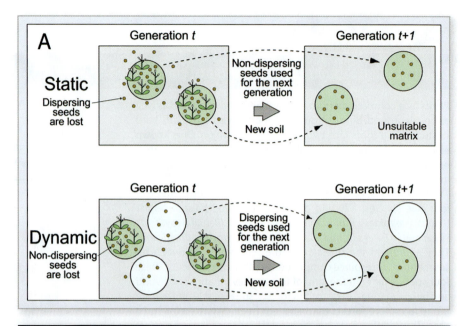

Figure 11.11 Diagrammatic (A) and photographic (B) depictions of *Arabidopsis* populations. Photo shows static (left) and dynamic (right) populations (from Fakheran et al., 2010). **Analyze: How do the differences between these treatments affect seed dispersal?**

■ Disturbance leads to higher resource abundance and lower competition. In undisturbed sites, the number of species and their densities increase, leading to competition for scarce resources. Species tend to be adapted to one or the other situation.

QUESTION:

In what way is the use of a single-resource axis appropriate for the analysis of competition between two or more species? In what way might it be inappropriate?

THE HUMAN IMPACT

Competition and Invasive Species

Human activity moves many species around the planet, introducing new species to well-established communities of native organisms. Plants and animals move in a variety of ways—seeds contaminate agricultural seed, marine organisms are released with ballast water in ships, and some are deliberately introduced. Nonnative species often proliferate rapidly in their new communities. Many have significant economic impact because they interfere with or outcompete important native plants and animals. For example, invasive marine species introduced to the Great Lakes by oceangoing vessels probably cost the commercial and sportfishing industries more than $200 million a year. The total losses in the United States from invasive species may be as high as $100 billion per year.

Why are these species so devastating to the local biota? Each introduction is in a sense ecologically unique. However, a few general principles have emerged. Because introduction by humans is rapid relative to natural invasions over longer evolutionary or geological time spans, the native species do not have time to adapt to the new competitor. If the invasive is a superior competitor, it may rapidly outcompete the native species. Research shows that the native species with a narrow ecological niche are particularly vulnerable (Bohn et al., 2008).

Many of the most successful invaders are species adapted to disturbance. This is particularly important for agricultural weeds. Invading plant species adapted for disturbance in their native range find an ideal habitat in cultivated fields—a form of artificial disturbance that opens the habitat and reduces competition. Velvetleaf (*Abutilon theophrasti*), imported from China, now causes millions of dollars of damage to corn and soybean crops. Grazing land disturbed by cattle is also ideal habitat for alien fugitive species that thrive in the absence of competition. Cheat grass (*Bromus tectorum*) and leafy spurge (*Euphorbia esula*) are both exotic species that colonize sites disturbed by heavy grazing.

There is also evidence that invasive species evolve rapidly in the new community. The evolution of increased competitive ability (EICA) hypothesis suggests that because invasive plants are released from their natural herbivores, they rapidly evolve increased competitive ability, perhaps because they do not need

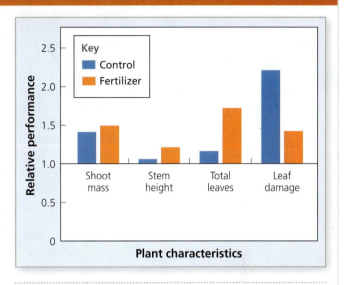

Figure 1 Invasive species. The performance of *Sapium sebiferum* from an invasive population relative to the native population in competition with one another. The performance of the native population was standardized as 1.0. Bars extending above the value 1.0 indicate that the invasive population outperformed the native population (from Zou et al., 2007). **Analyze: What do these results show about the adaptations of invasive species compared to natives?**

to invest as much energy in predator defense. Zou et al. (2007) compared invasive populations of the plant Chinese tallow (*Sapium sebiferum*) with populations from the native range. Rapid evolutionary shifts have clearly occurred. The invasive populations grow more rapidly and outcompete noninvasive populations. They also have lost the mechanisms that deter herbivores, but they tolerate herbivory better than the plants from the native range of the species.

QUESTION:

What biological interactions besides competition might determine the success of an introduced species?

11.3

What Are the Ecological and Evolutionary Consequences of Competition?

We consider the consequences of competition on two different time scales. The immediate effects occur in ecological time—periods of one or a few generations. They represent the immediate ecological impact of competition for resources. There are also long-term consequences that occur as a result of natural selection. These are multigenerational effects and thus occur on an evolutionary time scale.

What Is the Immediate Effect of Competition?

The niche reflects the pattern of resource use by the species. Figure 11.10 shows the relationship between two species that share a single limiting resource. Competition occurs where the two bell curves overlap. Both exploitation and interference competition can cause a change in the niches of competing species. For example, imagine a species that has the niche shown in Figure 11.12a. We refer to its niche in the absence of competition as its **fundamental niche**. If a competitor arrives in the habitat, the species may alter its resource use to avoid competition (Figure 11.12b). The resulting change in its niche is known as the **realized niche**.

In the scenario depicted in Figure 11.12, the two species both occupy the same habitat but their niche dimensions change to accommodate competition. However, there are limits to this process. If the niche shrinks too much, the population may not survive. This principle was first elucidated in a set of classic experiments with *Paramecium*. Gause (1934) grew *Paramecium aurelia* and *P. caudatum* in separate cultures. Each culture had the typical sigmoid growth pattern, described in Chapter 8, of a population in new and resource-rich habitat. However, when the two species were grown in the same culture, one increased to an asymptote whereas the other declined to extinction (Figure 11.13). The results were consistent: *P. aurelia* persisted and *P. caudatum* always went extinct.

Gause termed this phenomenon **competitive exclusion**: the extinction of one species as a result of competition with another. This process is intimately tied to the overlap in resource use. In the *Paramecium* cultures, both species were consuming bacteria as a food resource. With just that single resource and the two species' complete reliance on it, coexistence was not possible. The better competitor acquired sufficient bacteria to prosper; the other declined to extinction.

This led to the **competitive exclusion principle**, which states that no two species can long coexist on a single limiting resource. If the niche overlap of the two species is significant, one will outcompete the other. In other words, two species cannot occupy the same ecological niche. In nature we rarely find two species exploiting precisely the same limiting resource. When we look closely, we usually

fundamental niche The niche in the absence of competition.

realized niche The subset of the fundamental niche that results from competition.

competitive exclusion The extinction of one species due to competition from another.

competitive exclusion principle The concept that two species cannot coexist on a single limiting resource.

character displacement A shift in the niches of competing species that reduces competition among them. Natural selection favors those individuals in each species whose resource use overlaps less with that of the other species.

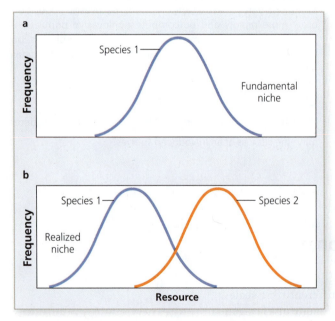

Figure 11.12 Fundamental niche. (a) A fundamental niche in the absence of competition. (b) In the presence of a competitor, the realized niche is a subset of the fundamental niche such where competition is reduced. **Analyze:** Are the two realized niches necessarily narrower than the fundamental niche?

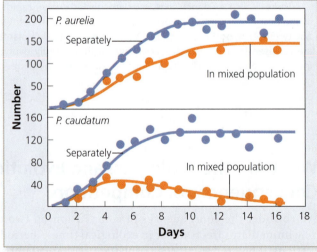

Figure 11.13 Limits of niche competition. Growth curves of *Paramecium aurelia* and *P. caudatum* separately (top) and together (bottom). (From Gause, 1934.) **Analyze:** What would you conclude if in some trials *P. caudatum* won and in other trials *P. aurelia* won?

find that there are subtle differences in their niches. Where more than one species seems to be using a resource, one of two mitigating factors generally occurs. Either the resource is abundant enough to support more than one species, or the resource is not critically limiting.

What Is the Evolutionary Effect of Competition?

The long-term effects of competition are due to selective forces acting on the competing species. One obvious result of adaptive change is an increase in competitive ability. We have alluded to this in the discussion of the difference between populations that inhabit stable high-density habitats compared to those that experience disturbance.

There is ample evidence for the evolution of competitive ability. Mueller (1988) showed that populations of *Drosophila*—the common fruit fly—could be selected for competitive ability. Mueller established two groups of populations: one set was maintained at high density, one at much lower levels. The high-density populations experienced food shortage for both larvae and adults. After 128 generations, Mueller tested the two groups for their competitive ability. When placed in direct competition for limited food resources, the flies that had experienced high density were significantly more competitive—they acquired the limited food supplies more efficiently than the low-density flies. This phenomenon is certainly important in nature. Many adaptations improve the efficiency of resource exploitation or ensure exclusive access to resources by actively excluding other species.

Competition also causes adaptive shifts in the niche. Specifically, competition may cause the niches of two competing species to diverge, a phenomenon known as **character displacement**. Consider two competing species as shown in Figure 11.14. Those individuals whose resource use overlaps with the other species will be at a selective disadvantage. If resource use has a genetic basis, directional selection will drive the niches of the two species in opposite directions, causing the niches to diverge.

Once again Darwin's finches provide a classic case of this important process. The medium ground finch (*Geospiza fortis*) and the small ground finch (*Geospiza fulginosa*) occur separately on Daphne Major and Los Hermanos Islands. Populations separated in this way are called **allopatric**. On Santa Cruz the two species occur together, or **sympatrically**. This provides a natural experiment in which we can examine the niches of the two species where they do and do not compete. Figure 11.15 shows their niches under these two conditions. In allopatry, the two species use similar seed resources; in sympatry, their seed-size niches have diverged by the process of character displacement.

Character displacement has been described for a number of competing organisms and for a range of critical resources. For example, Tyerman et al. (2008) observed character displacement in strains of the bacterium *E. coli* competing for different carbon sources. When strains experienced competition, they diverged. When competition was relaxed, their niches converged again. Any critically limiting resource can result in adaptive shifts in the niche. Song is an important means of communication in birds. For forest-dwelling birds, species with similar songs are at a disadvantage—their songs are less acoustically distinct, making them less effective. Kirschel et al. (2009) show that where more than one species of African tinkerbirds (*Pogoniulus sp.*) occur, their songs diverge (Figure 11.16).

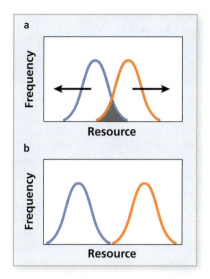

Figure 11.14 Character displacement. The individuals of each species that overlap with the other in resource use are at a selective disadvantage (top). Over time, directional selection will shift the niche of each such that competition is reduced (bottom). **Analyze: What factors will determine how fast this shift can occur?**

allopatric populations Populations occurring in different places.

sympatric Describes populations that occur in the same place.

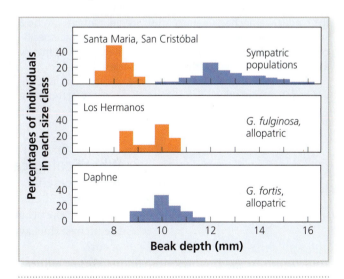

Figure 11.15 Allopatry and sympatry. Character displacement in the finches *Geospiza fulginosa* and *G. fortis*. Where the two species are allopatric, their niches are similar. In sympatry, their niches have diverged (from Schluter et al., 1985). **Analyze: What is the control for this experiment?**

THE EVOLUTION CONNECTION

The Effect of Competition on Evolutionary History

Competition is also important in the phylogeny of a species, the broad patterns and history of its evolution. There are two important mechanisms by which competition affects phylogenies. First, the mass movement of species into new habitats may put them in direct competition with other species. The formation of land bridges such as the one that connected North and South America 3.5 million years ago in the Pliocene permitted mixing of previously separate biotas. In this case, placental mammals invaded South America, where many niches were occupied by marsupials. Many of the North American placental mammals, especially the carnivores, were superior competitors and drove their South American counterparts to extinction. In effect, this history reflects a series of competitive exclusion events.

Second, invasions can lead to the origin of new species. If a species or taxonomic group colonizes a new region with little competition, the group may diversify into an array of new species, each of which occupies a new ecological niche, a process known as adaptive radiation. Many examples of this phenomenon are known from both the fossil record and extant species. When placental mammals invaded South America, one of the marsupial groups driven to extinction was a set of small granivorous and insectivorous species. The placental rodent family *Cricetidae*—including rats and mice—radiated into some 60 genera and 300 species in South America.

The ecology of islands is conducive to adaptive radiation. Because islands are typically colonized by just a few species who manage to disperse there, they face little competition. As a result, there are many examples of adaptive radiation on islands. The finches of the Galápagos are just one of many such cases. Figure 1 shows the diversity of vangids on Madagascar. One of the most spectacular adaptive radiations of birds is the honeycreepers of the Hawaiian Islands. The diversity of niches occupied by honeycreepers in Hawaii far exceeds that of their ancestors the cardueline finches. Character displacement appears to be a key component of this process. As species arise, interspecific competition leads to character displacement and ultimately adoption of new niches by the nascent species.

The fossil record documents the role of character displacement and competitive exclusion in the process of adaptive

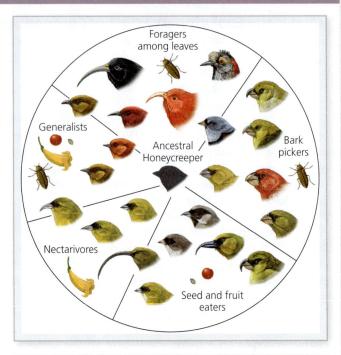

Figure 1 Adaptive radiation. Adaptive radiation of the honeycreepers in Hawaii.

radiation. Although there are spectacular examples of adaptive radiation in the Galápagos and Hawaii, there are taxa that colonized the islands that did not radiate as the finches and honeycreepers did. For example, mockingbirds reached the Galápagos but did not diversify even though they inhabit islands with what seem to be empty niches (Arbogast et al., 2006).

QUESTION:

What is the relationship between isolation of populations and adaptive radiation?

What Ecological Conditions Permit Coexistence?

We have described two important consequences of competition: competitive exclusion and character displacement. If niches are too similar, one species may drive the other to extinction. Or selection may result in the divergence of the niches, thus avoiding competitive exclusion. How do we know which will occur?

We can address this question using a set of graphical models developed independently by Lotka (1925) and Volterra (1926). These models are based on the fundamental growth equations for populations with overlapping generations (Chapter 8). However, we will be following the course of two competing populations, so we add a set of subscripts to denote each population. The growth rate of species 1 in the absence of competition is given by the equation

competition coefficient A measure of the competitive effect of one species on another, determined by the overlap in their resource use.

$$\frac{dN_1}{dt} = r_1 N_1 \frac{K_1 - N_1}{K_1}.$$

Similarly, the growth of species 2 without competition is

$$\frac{dN_2}{dt} = r_2 N_2 \frac{K_2 - N_2}{K_2}.$$

Now, if species 1 and 2 compete, we modify their growth equations as follows:

$$\frac{dN_1}{dt} = r_1 N_1 \left(\frac{K_1 - \alpha_{12} N_2 - N_1}{K_1} \right)$$

and

$$\frac{dN_2}{dt} = r_2 N_2 \left(\frac{K_2 - \alpha_{21} N_1 - N_2}{K_2} \right).$$

We have added the variable α and the population size of the competitor to each equation. The variable α_{21} is the **competition coefficient**, a measure of the competitive effect of species 1 on species 2 (note the order of the subscripts). The carrying capacity, K_1, is decreased by two factors: the population size of species 1 (N_1) and the competitive effect of species 2 on species 1 times the population size of species 2 ($\alpha_{12} N_2$). The larger the value of α and the larger the competitor population, the more the competitor decreases the carrying capacity.

In the simplest case of coexistence, the populations of the two species are stable (and nonzero). In other words, for each species, $dN/dt = 0$. The easiest way

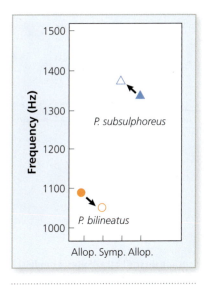

Figure 11.16 Song competition. Character displacement in the songs of African tinkerbirds. The arrows show the shift in song frequency when populations are sympatric (from Kirschel et al., 2009). **Analyze: What factors other than competition might explain these results?**

DO THE MATH

How to Quantify Niche Overlap

The value α is the competition coefficient in the Lotka-Volterra analysis. How do we calculate this variable such that it measures the competitive effect of one species on another?

If two species use parts of the same resource distribution and their resource curves are normal (bell shaped), their niche parameters determine the value of α. Specifically,

$$\alpha = e^{\frac{-d^2}{4w^2}}.$$

Qualitatively, we see that if the separation of the curves (d) is large and the variation (niche breadth (w) is small, the curves will have little overlap—the value of α is small. This means that exploitation of the resource by species 1 will have little impact on species 2.

Another way to think of the value of α is with direct measurement of the common uses of a resource. Imagine, for example, a food-resource axis based on n different prey size categories. We identify the proportional use of prey items of different sizes as shown in Figure 1. We calculate α, the measure of the overlap in prey-size categories, with the equation

$$\alpha_{12} = \frac{\sum_{i=1}^{n} p_{i1} p_{i2}}{\sum_{i=1}^{n} p_{i1}^2}.$$

The values p_{1x} and p_{2x} refer to the use of the series of prey size categories by species 1 and species 2, respectively. If the values p_{i1} and p_{i2} are similar, the value of α_{12} approaches 1.0, because the numerator and denominator become more similar. The more

different they are, the smaller the value of α_{12}. We calculate α_{21} in similar fashion. Note it is not necessarily the case that $\alpha_{12} = \alpha_{21}$. In other words, competitive impacts between two species are not necessarily symmetrical. This occurs if the two resource curves deviate from the assumption of normality.

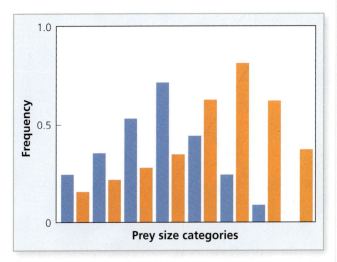

Figure 1 Prey size categories. The use of prey size categories by two species. **Analyze: Why do we depict these data as bar graphs rather than continuous curves?**

to describe that condition using these equations is to set the numerator term in the growth equation equal to zero. For species 1,

$$(K_1 - \alpha_{12}N_2 - N_1) = 0.$$

If this term equals zero, the entire right side of the equation is zero and growth rate is zero ($dN/dt = 0$).

For species 2,

$$(K_2 - \alpha_{21}N_1 - N_2) = 0.$$

Each of these equations represents a straight line in terms of the population sizes of species 1 and species 2. The first equation describes a straight line on which at all points, $dN_1/dt = 0$; the second represents a straight line on which all points represent $dN_2/dt = 0$. These lines that depict zero population growth for the species are known as **isoclines**.

We can represent an isocline on a graph whose axes are the numbers of individuals of species 1 and species 2. An example of such a graph is shown in Figure 11.17. For species 1, we obtain the intercepts by setting N_1 and then $N_2 = 0$. Thus, for species 1, if $N_1 = 0$,

$$(K_1 - \alpha_{12}N_2) = 0$$

or

$$K_1 = \alpha_{12}N_2$$

and

$$N_2 = \frac{K_1}{\alpha_{12}}.$$

If $N_2 = 0$,

$$N_1 = K_1.$$

We thus have the two intercepts for the species 1 isocline. A line joining these points represents the combinations of numbers of species 1 and species 2 at which species 1 is not changing. We do the same analysis for species 2 and determine that the intercepts of its isoclines are K_2 and K_2/α_{21}. We plot that isocline on the same graph.

With this graphical representation of the isoclines, we can determine the population trajectory for each species at any population size. If the population of species 1 is beyond its isoclines—that is, if the value on the x-axis is farther to the right than the corresponding point on its isocline—species 1 is beyond its carrying capacity and declines ($dN_1/dt < 0$). If the population of species 1 is below the corresponding point on its isoclines, it is below its carrying capacity and increases ($dN_1/dt > 0$). We analyze species 2 the same way using the y-axis. If the population of species 2 is larger than the corresponding point on its isocline—that is, higher on the y-axis—species 2 declines. If the population is below the corresponding point on the isoclines, species 2 increases.

isocline (in the Lotka-Volterra models) The line on a graph of the population size of species 1 and 2 that represents zero population growth of one of the species.

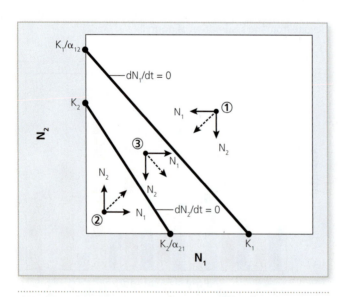

Figure 11.17 Isoclines. The isoclines of two competing species. At each point (1, 2, and 3), the trajectory of each population and their combined trajectories are shown. At all points on the graph, the trajectory of the populations is toward $N_1 = K_1$ and $N_2 = 0$. **Analyze: If the species here were *P. caudatum* and *P. aurelia*, which would be species 1 and which would be species 2?**

Now consider point 1 on Figure 11.17. Both species 1 and 2 are beyond their isoclines; the numbers of both species decline. We calculate the combined trajectory of the two species as the sum of two vectors—one depicts the decline of species 1, the other depicts the decline of species 2. The trajectory of the two populations is the sum of these vectors—a line diagonally toward the origin. At point 2, both species are below their respective isoclines and so both populations increase. The sum of the two trajectories is a line diagonally away from the origin. Now look at what happens at point 3. Here species 1 is below its isocline, and so its population increases. But species 2 is above its isocline, and its population decreases. The net movement of the two populations is diagonally down and right, toward K_1. If the trajectory hits $N_2 = 0$, only species 1 remains, and it grows until it reaches its carrying capacity, K_1. In fact, at any point between the two isoclines, the trajectory of the two populations inevitably leads to $N_1 = K_1$ and $N_2 = 0$. Species 2 has gone extinct and species 1 is at its maximum density.

There are four possible arrangements of the isoclines (Figure 11.18) depending on the relative positions of the intercepts. Figures 11.18a and b depict conditions that result in the extinction of one of the species. In fact, these arrangements of the isoclines represent competitive exclusion. In Figure 11.18a, species 1 always outcompetes species 2. The reverse occurs in Figure 11.18b.

Figure 11.18c represents an interesting case. Where the two isoclines intersect, neither population is changing; this is an equilibrium point. Note, however, that in one of the regions created by the intersection, species 1 declines to 0 and species 2 increases to K_2. In other words, species 2 competitively excludes species 1. In the other region, the reverse occurs. When the two species are either above or below both isoclines, they move toward the isoclines. If they happened to hit the equilibrium point, the two species would coexist. However, any perturbation away from that equilibrium point results in competitive exclusion of one of the species. Thus, Figure 11.18c represents an unstable equilibrium.

Now examine the trajectories of the two populations that occur in Figure 11.18d. No matter where we start on the graph, the trajectories of the two populations lead to the equilibrium point at the intersection of the isoclines. Any perturbation away from that equilibrium point leads to the immediate return to that point. Thus, Figure 11.18d represents the arrangement of the isoclines that always leads to stable coexistence of the two species.

This graphical analysis allows us to identify the conditions that permit coexistence. We obtain stable coexistence when the isoclines are arranged as in Figure 11.18d. The algebraic relationship that leads to that arrangement is that two conditions hold:

$$K_1 < \frac{K_2}{\alpha_{21}} \text{ and } K_2 < \frac{K_1}{\alpha_{12}}.$$

What does this mean biologically? For both inequalities to hold, the values of α, the competitive effect of each species on the other, must be small. Also, the values of K_1 and K_2 must not be too different. The more different they are, the less likely it is that both inequalities can hold. The similarity of the carrying capacities makes sense as well. If the carrying capacity of one species were radically smaller than that of the other, it would be much easier for competition to negatively impact the species with the smaller value of K, which in turn would make coexistence less likely.

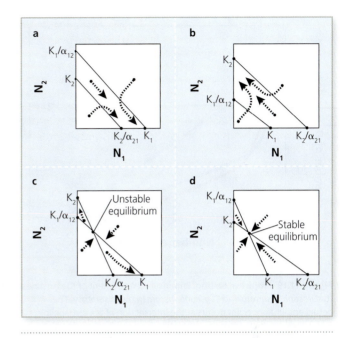

Figure 11.18 Isocline arrangement. The four possible arrangements of the isoclines. **Analyze:** How do the variables K and α determine the arrangement of the axes?

THINKING ABOUT ECOLOGY:

What fundamental assumption are we making about the nature of the carrying capacity (K) in this analysis? What would a graph of K over time look like in this analysis?

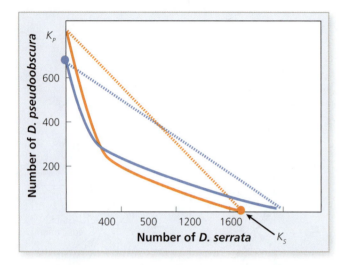

Figure 11.19 Empirical testing. Empirical measurement of the isoclines for *Drosophila serrata* and *D. pseudoobscura* in the laboratory. The actual equilibrium point occurs at the intersection of bowed isoclines (solid lines) below that for the theoretical isoclines (dashed lines). (From Ayala, 1969.) **Analyze: Why do bowed lines like this suggest that α is density-dependent?**

It has been difficult to test these predictions empirically with natural populations. This is largely because detailed quantitative information on growth rates, carrying capacity, and the values of α are extremely difficult to obtain. In addition, populations are affected by many interactions besides interspecific competition from just one other species. Consequently, it is difficult to devise experiments that isolate this one factor. Where it has been possible to experimentally assess this graphical analysis in the laboratory, populations seem to operate as expected, with some modification. For example, Ayala (1969) showed that for two laboratory populations of *Drosophila*, coexistence occurs at a point that requires both isoclines to bend downward relative to the predictions of the Lotka-Volterra equations (Figure 11.19). This kind of shift would occur if the values of α are not fixed but change with density. This is certainly biologically possible. This suggests that the Lotka-Volterra models are correct in broad outline but may need modification to include other factors.

KEY CONCEPTS 11.3

- Competition results in immediate and long-term effects on the species.
- Competitive exclusion is the most important immediate effect of competition. If the niches of two species are too close together, one drives the other to extinction.
- There are two important long-term (evolutionary) effects of competition: (a) increased competitive ability and (b) character displacement.
- The Lotka-Volterra graphical analyses show that coexistence is possible if the niche overlap of the two species is small and the carrying capacities for the two species are similar. If these conditions do not hold, the result is competitive exclusion or an unstable equilibrium condition.

QUESTION:

One adaptive response to competition is niche specialization, a narrower niche breadth.

Explain why there is a limit to this response. Why is there a limit to specialization?

Putting It All Together

Darwin's finches in the Galápagos inhabit an ecologically dynamic system. The abiotic conditions, especially rainfall, change markedly and unpredictably over time. Rainfall determines plant production and thus the seed resources available to the finches. However, it is not just weather and the food resources that change; species that co-occur also change. Although the islands are separated by some distance, the finches occasionally colonize a new island, suddenly throwing species into competition.

This is precisely what happened to the medium ground finch on Daphne. Not only did it have to cope with the vagaries of rainfall and seed production, in the 2004 drought it suddenly had to deal with competition from the large ground finch. This was particularly significant because the large ground finch has a more massive bill and is better able to open the large seeds that are left after drought has depleted the total seed supply. Here is a prime example of the ongoing nature of adaptive evolution. As the ecology of the Galápagos continues to change, so do the adaptive responses of its inhabitants.

In this way, the answer to the central question of the chapter—what are the effects of competition for resources?—is remarkably complex. Interacting forms of competition such as interspecific and intraspecific competition and interference and exploitation competition may be superimposed on one another. These interacting forces change the niche—sometimes temporarily, as in the shift from the fundamental to the realized niche, and sometimes permanently, as when character displacement occurs. If the niches are too similar, competitive exclusion removes one species. Thus, the niche is central to the interacting effects of competition. If the niche of the large ground finch centered on the use of insects as food, its competitive effect on the medium ground finch would be negligible. Had it been even more aggressive or efficient in acquiring its seed resources, it might have competitively excluded the medium ground finch from Daphne.

Moreover, the nonequilibrium nature of many ecological systems, driven by the prevalence of disturbance, opens an important set of niches—those of fugitive species. There is a whole set of species that, rather than compete for scarce resources in relative stable systems, is adapted to recurring disturbance. These species invest energy differently—in rapid reproduction to take advantage of the abundant resources for which there is little competition, and in dispersal systems to ensure that their offspring locate the next disturbed site. This, too, is an adaptive response to competition—their ecology, population dynamics, and life history strategies are geared toward noncompetitive, nonequilibrium conditions.

■ Summary

11.1 What Is Competition?

- Competition occurs within and among species when one or more crucial resources is limiting.
- Competition is empirically demonstrated by showing that (a) the resource is limiting and (b) the interaction negatively affects one or both species.
- Competition can be direct (interference competition) or indirect (exploitation competition).
- The negative impact of competition may arise from an interaction with a single competitor or from the summed effects of many competitors (diffuse competition).

11.2 What Determines the Intensity of Competition?

- The ecological niche can be defined in several ways:
  - The ecological role of the species.
  - The pattern of use of a critically limiting resource.
  - The summed use of biological and physical resources (**Hutchinsonian niche**).
- Niches are characterized by niche breadth and their separation from and overlap with the niches of other species.

- Species that inhabit systems with frequent disturbance often experience high resource levels and few competitors. Where disturbance is less frequent, competition is often more intense.

11.3 What Are the Ecological and Evolutionary Consequences of Competition?

- Competition alters the niche in ecological time (short-term effects) and over evolutionary time (long-term effects).
- In ecological time, competition may reduce niche breadth. This smaller niche, the realized niche, is a subset of the niche in the absence of competition (fundamental niche).
- Two species cannot long coexist on the same limiting resource. According to the competitive exclusion principle, one will drive the other to extinction.
- Character displacement is the shift in one or more species' niches when natural selection shifts the niches such that competition among them is reduced.
- Species are more likely to coexist if the values of α (overlap) are small and the carrying capacities (K) are similar.

Key Terms

Review Questions

1. Explain the potential changes in niche dimensions that result from interspecific competition.

2. Using Lotka-Volterra graphs, explain the relationship between niche overlap and carrying capacity that leads to (a) competitive exclusion and (b) coexistence.

3. What is the relationship between disturbance and competition?

4. Why do our analyses of competition tend to focus on (a) pairs of species such as Connell's barnacles and (b) a single-resource axis?

5. Why does intraspecific competition tend to increase niche breadth whereas interspecific competition tends to narrow it?

6. What is the relationship between fugitive species and the reproductive life history strategy (Chapter 9)?

7. What is the relationship between the competition coefficient and niche breadth?

8. What is the null hypothesis in a DeWitt replacement series? Why is it important to the analysis?

9. Is the shift from the fundamental to the realized niche an evolutionary or an ecological effect?

10. In Gause's experiments with competition in *Parame-cium*, the same species always drove the other to extinction. Why is this consistent result expected from competition theory?

Further Reading

Goldberg, D.E., and A.M. Barton. 1992. Patterns and consequences of interspecific competition in natural communities: a review of field experiments with plants. *American Naturalist* 139:771–801.

This paper provides an extensive review of the empirical work on interspecific competition in plants. The authors quantitatively analyze the literature for experiments that demonstrate competitive effects and organize those results into the forms and types of effects.

Weiner, J. 1995. *The beak of the finch*. New York:

This book is an excellent and thorough explanation of the long-term studies of Darwin's finches in the Galápagos by Peter and Rosemary Grant. Although it was written for the general public, it presents their scientific work in detail. It explains the range of work on the adaptive evolution and ecology of this group. It also provides insight into the nature of fieldwork in ecology.

Bolnick, D.I., et al. 2010. Ecological release from interspecific competition leads to decoupled changes in population and individual niche width. *Proceedings Biological Society B. London.* 277:1789–1797.

This elegant study demonstrates the relationship between intraspecific and interspecific competition on the niche of sticklebacks in relation to trout and sculpin competitors. They also address the time scale and mechanisms over which niche shifts occur.

MacArthur, R.H. 1958. Population ecology of some warblers of northeastern coniferous forests. *Ecology* 39:599–619.

This is a classic paper on niche differentiation due to competition. MacArthur studied the details of the microhabitat and foraging niches of a group of warblers that can be found foraging in the same tree. His observations detail how the species differ slightly in ways that, he argues, allow them to coexist.

Chapter 12

Coevolution

On a wet, humid night in the rainforest of Barro Colorado Island in Panama, mating frogs call from the margins of a small pond. The pond holds a mixture of frog species. Some are toxic—just a small amount of the neurotoxin they produce is lethal to almost any animal. Other species are palatable and contain no toxins. Bats dip and dive over the pond, moving through the vegetation structure by echolocation. One species, the fringe-lipped bat (*Trachops cirrhosus*), homes in on the call of one of the palatable frogs. In the pitch dark, the frog is unaware of the danger until the bat grabs the frog from its singing perch.

There are two remarkable facets of this predation event. First, the bat used its sophisticated hearing and echolocation system to locate a frog by its mating call. Second, the air was filled with the calls of both deadly and palatable frogs, yet the bat chose the palatable species and avoided the toxic ones. How does the bat's echolocation system find calling frogs? And how do the bats discriminate between toxic and palatable frogs? Why are some frogs toxic but others palatable? The ecological interaction, the predation event itself, is brief, lasting but a few seconds. But the roots of this event, and the answers to these questions, extend deep into the evolutionary history of bats and frogs.

Predation is just one of many ecological interactions among species. It exemplifies the fundamental issue of this chapter: the evolutionary events that underlie important ecological interactions between species. Virtually every species interacts in some way with others. These interactions can be exploitative, such as in predator-prey systems or parasites and their hosts. Others, like some flowers and their insect pollinators, are mutually beneficial. If we are to fully understand them, we must also understand the adaptive changes that underlie them. This leads us to the central question of this chapter: *How do interacting species adapt to one another?*

- **exploitative interaction** One species benefits by exploiting another as a food source.

- **predator** A species that obtains energy by capturing, killing, and consuming other organisms.

- **parasite** A species that lives in or on another species and obtains energy from the other species' living tissue.

Figure 12.1 Selective pressure. Cheetahs exert selective pressure on their prey.

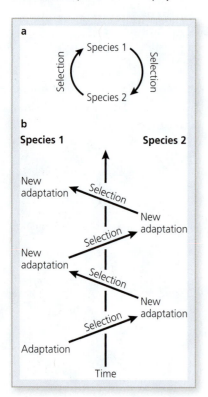

Figure 12.2 Coevolution. In coevolution, each species acts as a selective force on the other (a). Over time, each new adaptation by one species exerts selection pressure on the other species (b). The result is a series of adaptations and counteradaptations. **Analyze: How might a third species accelerate or retard this process?**

12.1 What Are the Major Forms of Exploitative Interaction?

We classify the interactions among species on the basis of their ecological and evolutionary impacts on the participating species, specifically their positive or negative effect on each species (Table 12.1).

We have already devoted an entire chapter to competition, in which the effect on both species is negative. In Table 12.1 we see that while predators, herbivores, and parasites benefit, the species they exploit are harmed. In contrast, both species in a mutualistic interaction benefit. In commensalism, one species benefits but there is no effect on the other.

In **exploitative interactions**, one species benefits from exploiting another as a food source. The negative impact on the exploited species ranges from negligible to lethal. Exploitative interactions are so varied that it can be difficult to define each form precisely. **Predators** capture, kill, and eat other organisms. **Parasites** are species that live in or on another species, the host, and obtain energy from its living tissue. Although parasitic infection causes harm to the host, it does not necessarily cause death. The term is sometimes applied more widely to include other forms of exploitation, such as nest parasitism in birds. Nest parasites lay their eggs in the nest of their host, where they receive food and care, often at the expense of the host's own young. We use the term **parasitoid** for a specialized group of insects whose offspring develop in the body of another species of arthropod. As they consume the tissue of the host from within, they eventually kill or sterilize the host species. Another major group of consumers is the **herbivores**, animals that consume plants. The action of some is like predation—they consume the entire plant or seeds (which contain embryonic plants). Others are more like parasites in that they consume only a portion of a living plant. Each of these forms of **heterotrophy**, the consumption of other organisms or organic forms of energy, is exploitative.

How Do Predators and Prey Adapt to One Another?

Predation represents one of the most important evolutionary interactions in ecology (Figure 12.1). The cheetah faces selective pressure to successfully locate, capture, and consume its prey. In the process it exerts a selective force on the prey; adaptations that reduce the probability of predation directly increase the antelope's fitness. The result is an array of adaptations on the part of both predator and prey to increase their success in this interaction. These affect virtually every aspect of both species, including their morphology, physiology, behavior, and social interactions.

Predator-prey interactions constitute an example of **coevolution**, the process in which each species acts as a selective force on the other (Figure 12.2). In predator-prey coevolution, each change in the predator results in new selective pressure exerted on the prey. As new adaptations arise in the prey, they in turn impose new selection pressure on the predator. As this process continues, we sometimes see a

TABLE 12.1 The effects on each species in coevolutionary interactions

	SPECIES A	SPECIES B
Parasitism	+	−
Commensalism	+	0
Mutualism	+	+
Predation/herbivory	+	−
Competition	−	−

phenomenon known as a **coevolutionary arms race**, in which a series of escalating adaptations and counteradaptations arises over time. There are two important aspects to coevolutionary arms races. First, the intensity of selection is not symmetrical. That is, the selective pressure on the prey is stronger than on the predator. This occurs because the consequences of failure are greater for the prey than the predator. If the prey's antipredator adaptations fail, it dies. If the predator fails to capture a prey item, it has the chance to try for another. This asymmetry is known as the **life-dinner principle**. Obviously, the predator cannot fail too often or it, too, will die. However, many predators can prey on a range of different prey items. If they fail with one individual or species, they generally have other options. Second, predator-prey coevolution may not always be at equilibrium. At any point in Figure 12.2, either the predator or the prey may have superior adaptations. If so, the other species is at a disadvantage.

How Do Predators Maximize Success?

Predation takes many forms (Figure 12.3). Nevertheless, there are a number of aspects of predation that are common to the range of types of predation and the many taxa that include predatory species. The act of predation contains two key events: detection of a prey item and the capture, killing, and consumption of the prey.

The predator must first detect a prey item in its environment. Different predators use different sensory modes to accomplish this. Generally, the mode is determined by the characteristics of the prey and the environment in which the predator hunts. For many diurnal predators, vision is an important sensory mode. Some species, such as the raptors, have extraordinarily acute vision. Hawks and eagles have two foveas, or light-gathering areas, on their retinas. Their eyes also contain large numbers of light-sensitive rods that provide images in fine detail as well as motion detection.

For nocturnal hunters, vision may be much less useful. Nocturnal predators are more likely to use sound or olfaction to detect prey. Owls have very acute hearing. By turning their heads, they gather sound from different angles, allowing them to determine the source location precisely. Bats have elaborated the auditory system such that they use an entirely different sensory mode: echolocation. Bats emit high-frequency sounds that are reflected by objects in the environment. By comparing the pulse of sound with the pattern of its echo, bats can maneuver through complex habitat structure in the dark. Moreover, they can locate, identify, and capture moving or stationary prey items.

Many predators, such as wolves and coyotes, hunt both diurnally and nocturnally. They have both acute vision and olfactory ability. Olfactory ability, like vision and hearing, is enhanced by morphological adaptations. The long snouts of coyotes and wolves provide a large surface for olfactory receptors, thus greatly increasing their ability to detect small amounts of scent.

As humans, our own experience is greatest with vision, hearing, and olfaction, but other important sensory modes are used by other species. For example, rattlesnakes and other pit vipers use thermal information to detect prey. The pits on their heads sense temperature differences (Figure 12.4). Their main prey, small nocturnal rodents, are homeotherms. As the desert cools at night, a small animal with high body temperature stands out against the cooler background. A few even use electric fields to locate prey. Some benthic fish that live at depths where light does not penetrate emit electric fields that extend outward from their bodies. Objects, including prey items, perturb this field and provide the fish with information on the size, speed, and direction of the object.

Once a prey item has been detected, the next step in the process is to capture and subdue the prey item. Predators adopt specific hunting strategies appropriate to the type of prey and the environment. **Ambush predators** (sometimes called sit-and-wait predators) remain stationary until prey is detected. Great blue herons exemplify this strategy: they remain motionless near the water until a small fish

Figure 12.3 Predation. Predation takes many forms, but in each case, the predator must detect the prey item, then capture and kill it.

Figure 12.4 Thermal senses. The pits on the head of this viper sense heat in its environment.

Figure 12.5 Hunting strategies. Great blue herons and anhingas (above) prey on the same fish species but by different methods. Herons wait patiently until a fish swims by, then grab it with their bill; anhingas swim actively after their fish prey.

■ **active predator** A predator that actively moves through the habitat searching for prey.

■ **solitary predator** A predator that hunts alone.

■ **group hunting** Predation by a group of conspecifics.

■ **cooperative hunting** A form of group hunting in which the group develops tactics that may include anticipation of prey behavior and division of labor.

Figure 12.6 Group hunting. Wolves hunt large formidable prey in groups.

swims near. Then a quick jab of the bill captures the fish. **Active predators** move across the landscape or through the environment searching for prey. The anhinga preys on fish in many of the same habitats as the great blue heron (Figure 12.5). It hunts actively, moving through the water, chasing down its fish prey. Wolves travel miles searching for their large ungulate prey. Although you may be more familiar with the large vertebrate predators that exemplify these strategies, many less-conspicuous, less-familiar species operate in the same fashion. Thus, for example, hydra are in fact ambush predators that await accidental contact with their prey. Rotifers are active predators of their protozoan prey.

The second key aspect of the predator strategy is the size of the hunting unit. **Solitary predators** hunt alone. In fact, many defend intraspecific hunting territories. This strategy is found in predators that can individually capture and subdue their prey. It is also common where the prey base is not large enough to support large groups of predators. A cheetah hunting antelope on an African savanna exemplifies this strategy. The cheetah can easily kill an antelope by itself. And this prey would be insufficient to sustain a large group of predators.

Group hunting typically occurs when the prey is large and perhaps dangerous. An adult moose is a formidable adversary for a predator. Wolves typically hunt moose and other large mammals in groups (Figure 12.6). In some predators, the group size and tactics are variable. Small packs of African wild dogs are more selective in their prey choices than large packs (Creel and Creel, 2002), presumably because small packs cannot bring down every potential prey species. Group hunting may also have advantages beyond prey capture. Hyenas are notorious for stealing prey from other predators. The larger the wild dog pack, the more able they are to fend off hyenas.

Cooperative hunting is a specialized form of group hunting in which the group employs tactics including anticipation of prey behavior and division of labor. For example, African wild dogs prey cooperatively on large mammals. In a coordinated hunt, wild dogs anticipate their prey's behavior and deploy individuals to counteract it. One group may chase the prey toward a cliff or toward the boundary of its territory, where other dogs are waiting in ambush. Cooperation has been noted in some predatory marine mammals as well. Orcas that feed on seals in the Arctic and Antarctic work together to isolate and capture their prey. They have even been observed tipping one edge of an ice flow so that a seal falls off to others waiting on the other side.

The predator's strategy includes a number of components: its sensory mode, its hunting style, and perhaps group or cooperative hunting. Other factors contribute to the predator's success. For example, if the predator is hunting in one area, how long should it continue to search before moving on? Should the predator specialize on a small number of prey types, or should it prey on a wide range of prey species? The evolution of the predator's full strategy requires choices such as these. How are these choices formed?

A body of theory known as **optimal foraging theory** addresses these questions. Much of optimal foraging theory is applied to carnivorous predators. However, because there are ecological similarities between predation, herbivory, and parasitism, this type of analysis can also be applied to these other exploitative interactions. Optimal foraging theory is based on the relative costs and benefits of the organism's actions. For a predator, fitness is maximized by making choices about what to eat and how to capture the prey so that the benefit exceeds the cost. The currency that underlies this cost-benefit analysis is energy—the net gain or loss of energy for different predation strategies. Optimal foraging theory makes an important assumption: that predation strategies are genetically controlled and thus subject to natural selection. The "decisions" the predator makes have fitness consequences. Note that these "decisions" are not conscious; they represent inherent foraging strategies that are either successful or not.

Optimal foraging theory recognizes two kinds of energetic constraints on predators. Some predators are **energy maximizers**. Their fitness is highest when they obtain the highest total energy return from predation. Others are **time maximizers**. For them, fitness is correlated with the highest rate of return of energy, the most calories per unit time. Predators that are calorie limited tend to be energy maximizers. Even if predation cuts into the time and energy available for other activities, their physiological energy requirements demand maximum return. For time maximizers, other tasks like courtship, nesting, or migration require that energy be obtained as efficiently as possible so there is time (and energy) for these other tasks.

We can illustrate the optimal foraging approach with the example of the breadth of the predator diet—whether to adopt a specialized or generalized diet. We break down the predation event into two components we have already described: searching for prey and capturing, subduing, and consuming prey. The former is the **search time** (s); the latter is the **handling time** (h).

For any diet, the average energy return for each item is represented by

$$\bar{E}/\bar{h}.$$

where $\bar{E}$ is the average energy gain from the item and $\bar{h}$ is the average handling time for the item.

Now imagine that the predator encounters another, less energetically rich prey item. Should it include it in the diet (broaden the diet) or ignore it (remain specialized)? The trade-off in this choice is that the average search time decreases as more items are included in the diet. However, some will have a lower energetic value. The additional energy input from the new item is

$$E_i/h_i.$$

If the predator chooses *not* to eat this item, it would need to continue searching for prey for the average search time. Thus, the energetic value of the specialized diet is

$$\bar{E}/(\bar{s} + \bar{h}).$$

This allows us to compute the conditions under which the new prey item should be added to the diet. It should be added if the total energy gain is larger with the new item than without it. Algebraically, we represent this inequality as follows:

$$E_i/h_i \geq \bar{E}/(\bar{s} + \bar{h}).$$

The diet will remain specialized unless the left side of this equation is larger than the right side. Each potential prey item is subject to this analysis. In general, the smaller the handling time of the new item, the more likely it is to be included in the diet. As handling times of potential items increase, the diet remains specialized. When the average search time for all items is very small, the right side of the equation approaches

$$\bar{E}/(\bar{h}).$$

New items will be added only if their handling time is smaller than the average. This means that predators with long search times relative to their handling times

---

optimal foraging theory A body of theory that tries to identify the best foraging strategy by which an organism can maximize total energy or energy per unit time.

energy maximizer A species for which selection favors those individuals that obtain the maximum total amount of energy.

time maximizer A species for which selection favors those individuals that obtain the most energy per unit time.

search time The time required for a predator to locate and identify an item as food.

handling time The time required for a predator to capture, subdue, and consume a prey item.

DO THE MATH

How Long Should a Predator Remain in a Patch?

Optimal foraging theory predicts how predators can maximize their energy intake. The body of theory attempts to explain the ecological basis for predator behavior, especially the "decisions" predators make about how specialized their diet should be, or which prey items to avoid. Many food items have a clumped distribution—the food resources occur in patches scattered across the landscape. When an animal forages in a patch, it gradually depletes the food resource. It must determine how long to stay in a patch and when to give up on a depleted patch to seek another. Part of the calculation includes the distance to another patch and the energy cost of moving relative to the diminishing gain where they are.

Charnov (1976) applied an important principle from economics to this problem: the marginal value theorem. We begin with the rate of energy gain as a function of time in a patch (Figure 1). We see that, over time, the extracted energy reaches an asymptote and the rate of energy gain declines. When should the predator move to a new patch?

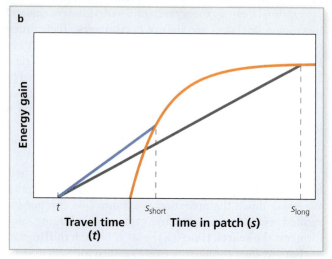

Figure 2 Asymptotic curve of energy gain. The rate of energy for short stays and long stays in a patch (from Charnoy, 1976). **Analyze:** Explain why s_{short} and s_{long} are not optimal staying times.

Shorter and longer stays result in lower energy gain (Figure 2). If the value of s is small (short stay), the line hits the curve at S_{short}. The rate of energy extraction (the slope of the line) is relatively low. If the value of s is very large (long stay), the line hits the curve at S_{long}. Its slope is also relatively low. Thus, the optimal staying time is shown in Figure 1. More or less time in the patch results in a lower rate of energy intake.

This theory makes predictions about the optimal strategy in different kinds of patches. If we hold travel time constant, patches with low energy availability (Figure 3) should be abandoned more rapidly than productive patches. The tangent line to the two energy-gain curves results in a shorter optimal stay in poor patches.

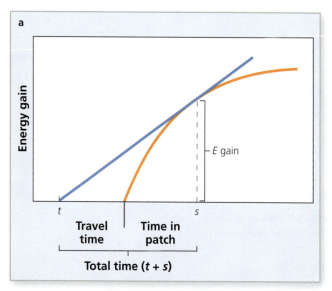

Figure 1 Rate of energy gain as a function of time in a patch. The rate of energy gain in a patch changes over time. This is determined by the relationship between travel time to a new patch (t) and the time the predator spends in a patch (s) (from Charnoy, 1976). **Analyze:** Why is the optimal staying time (s) determined by the tangent line to the curve?

We can answer this question with a detailed analysis of the asymptotic curve of energy gain (Figure 2). The variable t represents the average time of travel to a new patch; s is a measure of how long the predator stays in the patch. To forage optimally, the predator should maximize the rate of energy gain. This value is determined by the energy gain (E) divided by the total time:

$$E/(t + s).$$

If the value t is constant for the mosaic of patches, the optimal staying time is determined by the line tangent to the energy-gain curve because the rate of energy gain (slope) is greatest there.

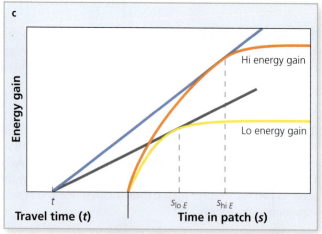

Figure 3 Energy availability. The optimal staying time also depends on the energy availability of the patch (from Charnoy, 1976). **Analyze:** Explain why this graph shows that staying time should be shorter in low-energy patches.

should be generalists. This makes intuitive sense. If it takes a long time to find a prey item but it can be eaten quickly, the predator should consume the item whenever it encounters it. Predators with long handling times relative to their search times should be specialists. For them, searching for an easier prey item is advantageous; they should specialize on these easier prey.

These predictions make intuitive sense. However, other predictions arise from this analysis that we might not expect without the mathematical logic. If, for example, prey density increases, search time decreases. Because search time is not included on the left side of the equation, the abundance of a new item is irrelevant. But the overall prey density affects the average search time, which is a variable on the right side of the inequality. This means that if all prey are abundant (small average search time), the diet should be specialized regardless of how abundant a new prey item might be.

Selection for optimal foraging interacts with selection pressure on predator body size and niche breadth. For example, in some birds, body size and niche breadth are positively correlated (Brandle et al., 2002). Larger birds eat a wider range of prey than smaller birds (Figure 12.7a). This occurs because larger birds include larger prey in the diet but continue to eat smaller prey because they can capture and subdue both large and small prey. On the other hand, small birds are limited to small prey. However, this correlation is not universal. Large lizards do not generally have larger niche breadths. As lizard size increases, the maximum and minimum prey size also increase (Figure 12.7b). This is thought to occur because larger lizards have difficulty handling small prey (Costa et al., 2008).

How Do Predators Respond to Changes in Prey Numbers?

The predator adaptations outlined above represent the means by which predators detect, capture, and consume their prey. Each is a response to the prey species' antipredator adaptations. We know from Chapter 9 that the populations of all species, including prey species, are dynamic. Thus, predators must respond to changes in the density of their prey populations. When prey populations decline significantly, they impose density-dependent limits on the predator population. Our discussion of population regulation in Chapter 9 also included the important role of predatory species as regulatory factors in prey population dynamics. However, in that discussion we did not address the mechanisms of the predator

THINKING ABOUT ECOLOGY:

Bluegills readily prey on *Daphnia*. Some *Daphnia* species are large; others are very small. How could you design an experiment to test the hypothesis that the abundance of a new prey item is irrelevant to the degree of diet specialization?

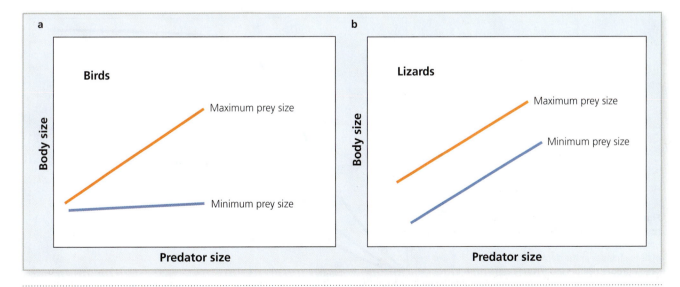

Figure 12.7 Maximum and minimum prey size. The change in maximum and minimum prey size for birds (a) and lizards (b) as a function of the predator's size (adapted from Costa et al., 2008). **Analyze: Why are the lines for maximum and minimum prey size parallel for lizards?**

functional response A predator response to increased prey numbers in which each predator consumes more prey.

numerical response An increase in predator number by reproduction or immigration in response to an increase in prey numbers.

response to changes in prey density. When the prey population increases, the predator may respond in either (or both) of two ways. First, each individual predator can consume more prey. We refer to this as the predator **functional response**. Or, the predator might exhibit a **numerical response**—an increase in the predator population through reproduction or immigration.

A classic study of predation by small mammals on pine sawfly pupae demonstrates these two predator responses (Holling, 1959). The pupae of the pine sawfly are distributed in the soil. Three species of mammals, the white-footed mouse (*Peromuscus leucopus*), the short-tailed shrew (*Blarina brevicauda*), and the masked shrew (*Sorex cinereus*), prey on the pupae. Each species opens the pupa in a distinctive manner, which allowed Holling to identify the source of predation. He tracked the population sizes of the predators (numerical response) and the numbers of prey consumed (functional response) at different prey population densities. The predator response to increasing prey numbers differed (Figure 12.8). The masked shrew had a large numerical response but a small functional response. In contrast, the

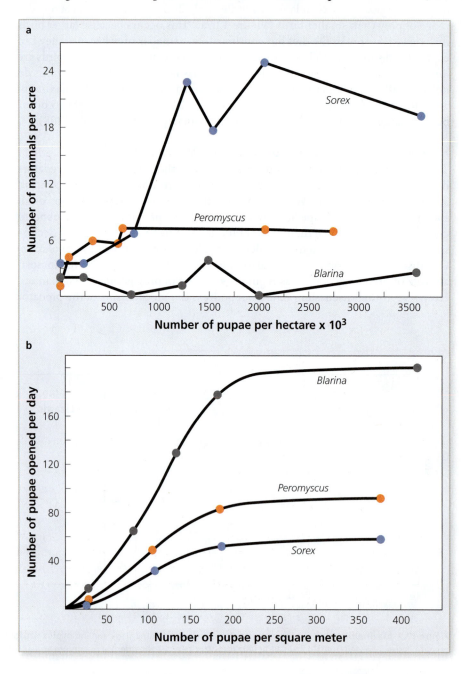

Figure 12.8 Numerical and functional response. The numerical (top) and functional response (bottom) of small mammals to changes in the density of the pine sawfly (from Holling, 1959). **Analyze: Is it possible for a species to have a large numerical response and a large functional response?**

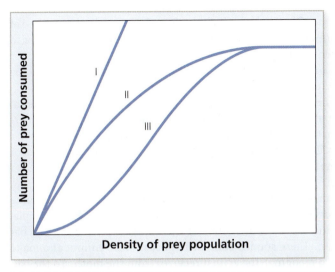

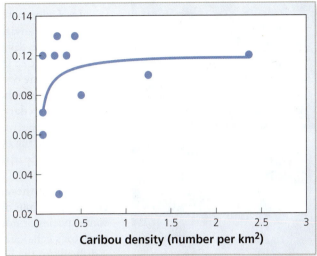

Figure 12.9 Types of functional response. The three types of functional response. **Analyze:** Why would you expect a Type I response in species with small handling time?

Figure 12.10 Type II functional response. The functional response of wolves preying on moose in Alaska (from Dale et al., 1994). **Analyze:** Why does the rate of consumption level off?

white-footed mouse exhibited a small numerical response but large functional response. The short-tailed shrew had intermediate responses of both types.

We recognize three types of functional response (Figure 12.9). In a Type I functional response, predator consumption increases linearly with prey density. This type of functional response is found primarily in species in which handling time is minimal. The Type II functional response is characterized by a low rate of consumption at low prey density. The rate of consumption is maximal at intermediate prey density, then levels off when prey are very abundant. The relative importance of search and handling time changes with prey density. At low prey density, consumption is low because search time is significant. As prey become more abundant, search time decreases and the rate of consumption increases. At still higher prey density, search time is very small but now the rate of consumption is limited by handling time. This is the most common form of functional response. It is exemplified by wolf predation on large ungulates (Figure 12.10). The Type III functional response is similar to the Type II response except that the curve is S-shaped. The same constraints affect consumption, but more stringently.

How Do Prey Species Avoid Predation?

Prey species, of course, have their own sets of adaptations to reduce the risk of predation. Most center on the predator's search and handling. We will consider prey adaptations that affect each.

Predator search time is affected by the ease by which the predator can detect its prey. **Crypticity** is the set of characteristics of a prey item that decrease the probability that a predator will detect it. Regardless of the sensory mode used by the predator, it is searching for a pattern in its environment that represents a prey item. For a bat foraging on insects, this is the nature of the echo of the sound pulse it emitted. For a hawk high above a meadow, it is the pattern, color, and movement of a mouse in the grass below. Predators learn to associate certain sensory patterns with prey items. Thus, the key to prey crypticity is to make it difficult for the predator to separate the pattern shown by the prey item from the general background information. In other words, cryptic prey represent a random sample of the background. How is this achieved?

It is perhaps easiest to understand this for visually hunting predators. Several general adaptations make it difficult for the predator to distinguish the prey item from a random sample of the visual input it receives (Figure 12.11).

crypticity A mechanism employed by prey species to avoid detection by predators. Cryptic species appear to the predator as a random sample of the background.

Figure 12.11 Crypticity. The principles of crypticity include background matching (a) of the largescale flounder, hiding the eye (b) of the emperor red snapper, and countershading (c) in the catfish.

▪ **predator swamping** A prey strategy in which the per capita predation rate is reduced at high prey density.

▪ **aposematic coloration** Distinctive, conspicuous color patterns that signal that the individual is toxic or unpalatable.

Figure 12.12 Aposematic coloration. Many toxic species, such as this poison dart frog, have conspicuous color patterns known as aposematic coloration.

1. Color matching. Obviously, many prey adopt the general color scheme of their environment. Some, such as anoles and some fish, can shift color to match different backgrounds.
2. Break up the outline. Patterns of color or light and dark help destroy the outline of the prey item.
3. Countershading. Since most animals are lit from above, their pigmentation is darker above and lighter below. This reduces the brightness difference between the top of the animal and its underside.
4. Hide the eye. The eyes of animals are surprisingly conspicuous. Lines or other patterns that obscure the eye increase crypticity.

Although these adaptations help prey hide from visual predators, the same general principle of crypticity applies to other sensory systems. For example, some species of noctuid moths emit sounds of the same frequency as their bat predator's echolocation. These sounds confuse the bat's sensory input and make it difficult for the bat to recognize the echo it receives from the moth. To the bat, the moth echo sounds like a random sample of background noise.

Another important means by which an individual can avoid detection is to reduce the probability that it is the specific individual the predator locates. Thus, herd and school behavior help individuals hide among many conspecifics. Detailed studies of predator attacks on schools of fish have shown that individuals at the periphery are more vulnerable than those in the center.

There are other intriguing ways that prey can use numbers to reduce the individual probability of predation. **Predator swamping** occurs when the numbers of prey are so large that although many are eaten, the per capita predation rate is quite low. The synchronous emergence of insects in large numbers is an example. For example, some species of periodic locusts emerge synchronously in phenomenal numbers. The sudden appearance of so many prey items overwhelms the numerical and functional response of their predators. Some waterfowl also adopt the predator-swamping strategy by nesting synchronously. Swamping tactics such as these exemplify strong stabilizing selection (Chapter 2). Individuals whose laying or emergence does not match the other members of the population are at greater risk of predation.

The second aspect of prey adaptations affects the predator's handling time. Once the prey item is detected, its only recourse is escape or defense. Speed is an important adaptation for many prey species. But there are many other mechanisms of escape. False targets, structures such as long or brightly colored tails, may attract the predators' attention, rendering their attack less lethal or successful. Lizard and amphibian tails can break off, leaving the predator with a portion of the tail while the animal escapes. Many fish and butterflies have large circular patches that resemble eyes. A butterfly resting on a tree hides these spots with the folded wings. When a predator appears, the butterfly quickly reveals them, which startles the predator and perhaps gives the butterfly a chance to escape.

Some prey species are certainly formidable opponents for predators. Hooves, horns, teeth, tough shells, and toxins all constitute defensive adaptations. Toxins are especially interesting antipredator adaptations. Some of these toxins are remarkably potent. Poison dart frogs in the genus *Dendrobates* are known for their extreme toxicity. The newt, *Taricha torosa*, produces tetrodotoxin, a neurotoxin that is among the most toxic compounds in nature. A small piece of a newt's tail contains enough toxin to kill a snake or raccoon. Many toxic species have striking and conspicuous color patterns known as **aposematic coloration** that serve as warning signals to predators (Figure 12.12). In most cases, predators are genetically programmed to avoid these color patterns. There is little opportunity for learned avoidance—the toxins are lethal.

The presence of toxic prey items with aposematic coloration opens the door to still more complex predator-prey coevolution. For example, the bulbar grasshopper

Eyespots as Deterrents to Predation

One of the most interesting antipredator adaptations is the presence of eyespots, paired circular spots that seem to represent eyes. This adaptation is especially common in butterflies and moths (Figure 1). The traditional view of these patterns is that they represent the eyes of the predator's own enemies. The sudden appearance of these patterns as a butterfly spreads its wings might startle the predator.

Figure 1 Eyespots. Many species of butterflies have eyespots on the wings.

This traditional explanation exemplifies an important question in evolutionary ecology: How can we experimentally verify that the explanation we devise for an adaptation like eyespots is correct? The eyespots do indeed resemble eyes. And they are often paired just like real eyes. It seems logical to assume that they mimic eyes. But of course this is actually a hypothesis that should be subject to the same kind of experimental verification as other hypotheses. Moreover, there are other potential explanations, such as the conspicuous signal hypothesis, which posits that eyespots function by strongly stimulating the predator's visual system. It is the visual impact, not the mimicry of another animal, that startles the predator and causes the hesitation that might allow the prey to escape.

Stevens et al. (2008) set out to test these hypotheses by analyzing the effect of the size and shape of eyespots on the predation rate of birds. Here is another way of stating the conspicuous signal hypothesis:

HYPOTHESIS: The size and conspicuousness of shapes is more important than their similarity to eyes.

In their experimental system, they pinned a series of different shapes and patterns to trees and measured the rate of predation on mealworms attached to them. If their hypothesis is correct, three predictions should be true:

PREDICTION 1: Two spots are not as effective as multiple spots.

PREDICTION 2: Circular spots should not be more effective than other shapes.

PREDICTION 3: There should be no effect of symmetry in the eyespots on predation rate.

In other words, eyespots composed of precisely concentric circles should be no more effective than those in which the interior circle is off center.

Each of these predictions relates to the importance of shapes that resemble eyes. The field experiments consisted of placing mealworms on trees against a triangular background on which eyespots of various shape and size were placed (Figure 2). The shapes were randomly assigned to trees. Predation on the mealworms was measured at regular intervals for 48 hours.

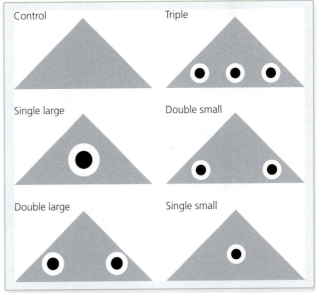

Figure 2 Eyespot patterns. The patterns of eyespots used in the experimental treatments (from Stevens et al., 2008). **Analyze:** Why were these patterns chosen as experimental treatments?

The researchers found that large spots led to better survival than small spots. Also, survival was related to the number of spots: multiple spots were more effective than pairs of spots. There was no difference in survival among different shapes. Bars and squares were as effective as circles. Finally, it did not matter if the rings of circles were centered or on the periphery (Figure 3). All these results are more consistent with the conspicuous signal hypothesis: survival was higher when the visual signal was strong and obvious, but it was not higher if the signal more closely resembled eyes.

continued

ON THE FRONTLINE   *continued*

Eyespots as Deterrents to Predation

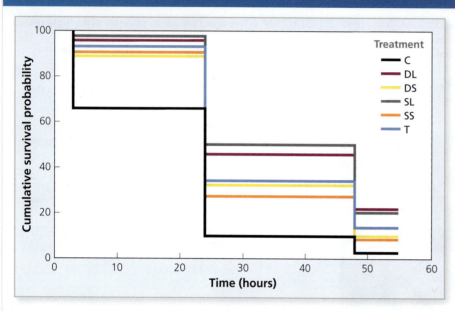

Figure 3 Eyespots and predation. Survival of mealworms as a function of the type of eyespot (from Stevens et al., 2008). **Analyze:** Which pattern of eyespots was most effective at deterring predation?

A question remains: Why are the patterns common in nature round shapes that resemble eyes? This study did not address this question experimentally. However, the authors point out that a circular spot can probably be produced rather simply by the outward diffusion of pigment, a much simpler developmental process than the production of a square.

This study reminds us to be cautious about interpreting the function of adaptations. Moreover, it reiterates that explanations for adaptations are in fact hypotheses that must be experimentally verified no matter how logical and intuitive our explanations might be.

■ **Mullerian mimic** A species that resembles other noxious species.

■ **Batesian mimic** A benign species that resembles a noxious species.

THINKING ABOUT ECOLOGY:

How would you expect a Batesian mimicry system to be affected if the mimetic species is more common than the species it mimics?

produces strong toxins that deter predation by most predators. One avian predator, the shrike, captures the grasshopper but does not immediately consume it. Instead, it impales the animal on a thorn and leaves it there for several days, during which the toxin degrades. Some garter snake populations have developed resistance to tetrodotoxin. These systems represent ongoing predator-prey arms races.

Mimicry represents another set of complex coevolutionary interactions that arise when prey are toxic or otherwise dangerous. There are two basic forms of mimicry. **Mullerian mimics** are groups of toxic or unpalatable species that resemble one another. The yellow and black bands of many bees and wasps exemplify this type of mimicry. Many species avoid any yellow-and-black-banded insect, and this benefits the entire group of mimics. The more coexisting noxious species with the same color pattern, the more quickly avoidance can either be learned or genetically ingrained.

In **Batesian mimicry**, a benign species resembles a toxic or unpalatable one. Coral snakes are among the most toxic venomous snakes in the New World. Their striking color pattern is an aposematic warning system (Figure 12.13). A number of species of nontoxic king and milk snakes closely resemble coral snakes. Each obtains some measure of protection because predators are genetically programmed to avoid the red, black, and yellow pattern of coral snakes. Some of the geographic patterns of mimicry in this system are quite precise: the mimics resemble the local subspecies of coral snake.

How Do Plants Respond to Herbivory?

Herbivores are also important exploitative consumers. Like carnivorous predators, they exert selective pressure on their "prey." Because plants are sessile, search times for herbivores tend to be low. And, of course, escape is not an option for plants. Thus, their defenses center on two features: deterrence (adaptations that prevent grazing) and tolerance (adaptations that help the plant withstand the effects of grazing).

Deterrence falls into two broad categories: structural and chemical. Thorns and spines are structural deterrents to grazing. Other structural features, such as lignin, cellulose, and silica, also offer protection. As the content of these compounds increases, the palatability of the plant and its nutritional value both decline, especially the ratio of nitrogen to carbon. Nitrogen is much more important to grazers than carbon, in most plant food, and a shift to a lower N:C ratio renders the plant less nutritious.

Chemical deterrence depends on the production of chemicals known as **plant secondary compounds** that render the plant toxic or undigestible. Originally, secondary compounds were thought to be the toxic by-products of normal cellular biochemistry that were used by the plant to thwart herbivores (thus the name *secondary* compounds). However, we now understand that many are produced by novel biochemical pathways specifically as deterrents.

Many thousands of secondary compounds are known. The alkaloids include substances such as morphine, caffeine, and nicotine, which have deleterious physiological effects on grazers. Others, such as phenols, reduce the digestability of the plant material, rendering it less nutritious. Secondary compounds are especially prevalent in the flora of tropical rainforests and tropical marine systems. Because of their physiological effect on vertebrates, a large number of pharmaceuticals are derived from plant secondary compounds, including many cancer-fighting drugs.

Tolerance of grazing is based on the plant's **compensatory ability**, its ability to change the amount and pattern of tissue production to reduce the impact of grazing. Some plants produce more leaf or stem tissue when grazed than ungrazed. For example, Stevens et al. (2008) showed that defoliation of aspens (*Populus tremuloides*) results in increased photosynthesis and allocation of energy to leaves (Figure 12.14). The mechanisms behind compensatory changes are not fully understood and probably vary greatly with the plant's physiology and ecology. Still, some mechanisms have been identified. Removal of leaves at the top of the plant increase light penetration to lower leaves, leading to a net increase in photosynthesis despite the loss of some photosynthetic tissue. If apical buds are nipped by grazers, branching increases due to the removal of the plant hormone auxin.

Plant responses to herbivores represent another example of evolutionary trade-offs among different strategies. Finite energetic and physiological resources dictate that adoption of one strategy precludes the simultaneous adoption of another. Thus, plants that invest heavily in secondary compounds to deter grazing are likely to have a smaller compensatory ability, and vice versa. In the mustard, *Brassica rapa*, lines artificially selected for increased glycoside production show much greater loss of fitness if they are grazed (Stowe, 1998).

Of course, herbivores adaptively respond to these challenges as part of their coevolutionary relationship with plants. Two main theories have been developed to explain the patterns of plant-herbivore interactions. The first is the **phytochemical coevolution theory**. This idea was developed to explain specialization in

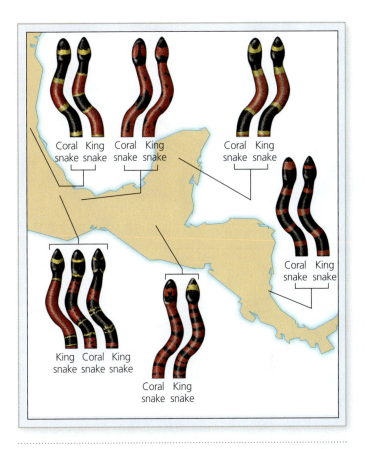

Figure 12.13 Batesian mimicry. Geographic variation in the color pattern of king snakes and coral snakes. The local coral snake pattern is on the left in each set; the mimic is on the right. In the set with three species, two king snakes (left and right) mimic one coral snake (center). (From Green and McDiarmid, 1981.) **Analyze: Why are the two king snake mimics in E different from one another?**

■ **plant secondary compounds** Compounds produced by plants that deter herbivores.

■ **compensatory ability** The ability of a plant to alter tissue production to counteract herbivory.

■ **phytochemical coevolution theory** A coevolutionary interaction between plants and their herbivores in which plant secondary compounds select for specialized herbivores that can counteract these deterrents.

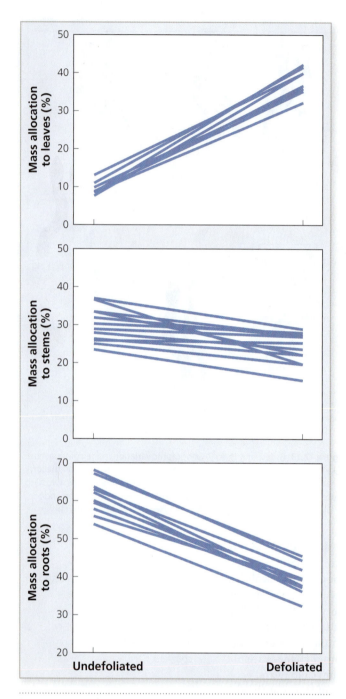

Figure 12.14 Compensatory ability in aspens. The allocation of energy to leaves, stems, and roots in defoliated and undefoliated aspens (from Stevens et al., 2008). **Analyze: Is decreased allocation to the roots of defoliated plants an example of compensation?**

apparency theory A theory for the evolution of specialized herbivores. Conspicuous plants are protected by compounds that reduce herbivore growth; inconspicuous plants are protected by acutely toxic compounds. The latter selects for specialized herbivores.

herbivores in response to plant secondary compounds. Many herbivores are able to detoxify plant secondary compounds or sequester them in their bodies in ways that minimize their harmful effects. Insects have co-evolved with their plant hosts for at least 400 million years. Detailed study of these systems has demonstrated two key facts. First, most herbivorous insects are specialists that feed on just a few host species. Second, plant secondary compounds are toxic to some species and harmless to others. These two properties are the direct result of plant-insect coevolution. Selection favors plants that produce secondary compounds, and insects adapt with mechanisms that detoxify the compounds. The result is specialization in the insects—specific bio-chemical detoxification pathways reduce their ability to feed on plants with other compounds.

One effect of this process is the diversification of both plants and insects. If a plant develops a novel secondary compound that protects it from herbivory, it may be able to diversify rapidly and form many new species. Eventually the insects adapt to this new chemical protection, and this opens up a new group of plant species on which to specialize (Ehrlich and Raven, 1964).

The other main theory, known as the **apparency theory** (Feeny, 1976; Cornell and Hawkins, 2003), helps explain the evolution of generalist insect herbivores. According to this theory, there is a correlation between the plant's conspicuousness, its "apparency," and its means of chemical protection. Inconspicuous plants tend to be protected by qualitative chemical defenses, specifically by acutely toxic secondary compounds. Most of their herbivores are specialists that can detoxify these chemicals. In contrast, conspicuous plants tend to be protected quantitatively—that is, by compounds that reduce digestibility or the growth rate of the herbivore. Herbivores that attack these plants are less able to circumvent these compounds. Consequently, they tend not to be specialists; they feed on many different plant species.

How Do Parasites and Their Hosts Coevolve?

Parasite-host coevolution shares an important characteristic with predator-prey and plant-herbivore systems: the host is under strong selection to thwart the attack of its parasite, and the parasite faces selection pressure to circumvent the host's resistance.

The taxonomic and ecological diversity of parasites is enormous. Parasitic forms are known from virtually every animal phylum as well as many plant families. Parasitism takes many forms. Some, like trematode worms (Figure 12.15), have complex life cycles and multiple host species. They directly exploit the energy stores of their hosts, often with lethal effect. In contrast, avian brood parasites engage in a much simpler form of energy acquisition from their hosts—they lay their eggs in their hosts' nest, where their offspring compete with the host offspring for food. Their effects can be just as lethal. Some juvenile nest parasites actively push the host offspring from the nest. Still other parasites, such as ticks,

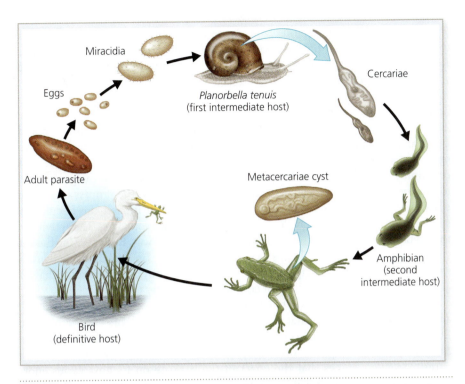

Figure 12.15 Trematode life cycle. The life cycle of a trematode parasite includes a number of intermediate hosts and infectious forms. **Analyze: What are the fitness costs and benefits of a complex life cycle?**

extract such small amounts of blood from their hosts that, except in unusual circumstances, their direct impact is relatively small.

There are three fundamental aspects of parasite-host coevolution:

1. Mechanisms by which parasites locate and infect the host
2. Adaptive changes in parasite virulence and host resistance
3. Selection for general or specialized parasites

The parasite must be able to locate and enter the host. In essence, host species are "islands" of suitable habitat that must be discovered and colonized by the parasite. You might wonder why the life cycle of the trematode parasite is so complex. This parasite illustrates an important set of selective forces that shape the parasite's ability to find and infect its host. In this example, the definitive host, the species it must infect if it is to reproduce sexually, is a bird. The bird occurs in low numbers relative to the many abundant species in that environment. Moreover, it is not possible for an adult trematode to locate and enter a bird. The parasite infects an **intermediate host**, a secondary species that prepares the parasite to locate and infect the bird. Specialized stages of the parasite are adapted to enter each of the intermediate hosts. In addition, they utilize the energy reserves of these intermediate hosts to produce and release huge numbers of infective forms, helping to ensure that at least some locate and infect the next host. In this trematode, the motile miracidia locate and infect the first intermediate host, a snail. That infection leads to the production of large numbers of cercariae that infect the second intermediate host, a frog. Those metacercariae remain in the frog in an encysted (inactive) form until the frog is eaten by a bird, such as a great blue heron, in which the adult worm emerges. The adult obtains energy from the tissue and food consumed by the bird. Large infections may be severely debilitating. Much of the energy acquired by the parasite is devoted to the production of large numbers of eggs, which, when released, begin a new cycle.

intermediate host A host of a parasite in which the parasite produces forms that can locate and infect a different host species.

pseudoflowers Structures produced in a plant infected by a rust parasite that superficially resemble the plant's flowers but contain the spores of the rust.

Figure 12.16 Pseudoflowers. The rust *Puccinia* produces pseudoflowers, structures that resemble the flowers of angiosperms.

evolutionary theory of virulence The theory that selection favors decreased virulence in the parasite because this increases the probability that the parasite can exploit the host and reproduce.

Parasites use some remarkable means to increase the probability of transfer to a new host. The rust fungus *Puccinia monoica* infects buttercups in the genus *Ranunculus*. The rust induces the host to produce **pseudoflowers**, flower-like structures that produce both nectar and scent and contain the reproductive spores of the rust (Figure 12.16). Insect visitors to these "flowers" spread the spores to new plants (Roy, 1994).

Another trematode fluke (*Dicrocoelium dentriticum*), which infects sheep and cattle, also exemplifies these elaborate mechanisms. The first intermediate host is a terrestrial snail. It is infected by consuming feces of the vertebrate definitive host containing eggs of the fluke. Encysted cercariae are excreted into the grass. Ants follow the moisture in the slime trails of the snails and in the process ingest the cysts. These develop into metacercariae, most of which inhabit the circulatory system. One, however, travels to a set of nerves near the esophagus. The activity of this metacercaria alters the ant's behavior in a remarkable way. Each evening the ant climbs a blade of grass and clamps its mandibles onto the tip. It remains there until morning, then releases its grip and returns to the ant colony. It repeats this behavior and is eventually eaten by a sheep or cow, which completes the life cycle.

The adaptations of the host and those of the parasite constitute a coevolutionary arms race, much like that of predators and prey or plants and herbivores. Each new host adaptation constitutes a new selective force on the parasite. In turn, the parasite's response forces adaptive changes in the host. A survey of host-parasite systems shows that the effect of the parasite on its host varies enormously. Why are some parasites so virulent whereas others have only mild effects on their hosts?

The **evolutionary theory of virulence** suggests that a highly virulent parasite would be less successful than one with less devastating effect. Severe virulence might kill the host before the parasite can exploit opportunities for transmission to new hosts, in which case it would die with the host. Or if the parasite debilitates the host, its restricted movements might decrease the probability of transmission to new hosts. Thus, for some time it was thought that parasite virulence should decrease toward benign coexistence with their hosts.

We now understand, however, that the evolutionary theory of virulence has relatively limited application. One reason is that changes in virulence must be evolutionarily stable strategies (Chapter 2). That is, mutations that increase or decrease virulence must be advantageous when rare. A genetic change that decreases virulence might be advantageous once established in the species. But these individuals will lose out to other, more virulent types in the short run—decreased virulence is not an evolutionarily stable strategy. Second, parasites vary enormously in the nature and ecology of their life cycles and host characteristics. Thus, parasites that utilize transmission vectors that infect new hosts even if the current host is immobilized and severely affected can maintain virulence with relatively little cost. For example, the malaria parasite (*Plasmodium*) is transmitted by mosquitoes, and host condition has little impact on the probability of transmission. Finally, the evolutionary theory of virulence does not consider the impact of multiple infections. Multiple infections by the same species may lead to competition among the populations of the parasite. If so, there is no selective advantage to decreased virulence; strains of the parasite that can rapidly and extensively exploit the host resources are favored.

Thus, host resistance and parasite virulence are part of a complex and dynamic coevolutionary interaction. In this context, the wide range of parasites' impacts on their hosts is not surprising. First, parasite-host systems are so diverse that the selective forces for or against virulence vary from system to system. In addition, coevolution is an ongoing process. It may be that the parasite or host is in the process of responding to new challenges from the other. Thus, we might expect that few parasite-host systems are at equilibrium.

Finally, parasites vary in the diversity of hosts they can infect: some are generalists, and others are specialized. What ecological factors determine the optimum strategy? Specialized parasites benefit in two important ways. First, they face a smaller range of host defense systems. Their adaptations to infect the host

THE EVOLUTION CONNECTION

The Phylogenetic Relationship of Parasites and Hosts

This chapter explores the intimate evolutionary connections between species, some of which result in coevolutionary arms races. Over a few hundred generations, each species adapts and counteradapts. However, the evolutionary interactions between these species may extend far into the past and reveal a long evolutionary history. Modern molecular techniques allow us to trace the history of these interactions.

It is common for the phylogenetic tree of a group of parasites to match closely the phylogenetic tree of their hosts. This provides information about the time scale of the parasite-host relationship. It also reveals that the species diversity of the parasite is dependent on the species diversity of the host.

This kind of evolutionary analysis also reveals times in the past when parasites have switched to new hosts. This can occur within the broad confines of a host lineage, or it may represent a significant shift in host taxa. The first possibility is shown in Figure 1. A phylogenetic tree of the primates, based on morphological and molecular data, is shown on the right side of the figure. The left side of the figure depicts a phylogenetic tree for the primates based entirely on the parasites they host. In other words, the data for this tree is not morphological and molecular information; instead, it is the parasitic species that infect each primate. Note that the two trees differ. The difference between the two trees is due to host switching by the parasites. The more congruent the two phylogenies, the less switching has occurred.

Sometimes parasites make major jumps to entirely new taxa of hosts. An example is shown in Figure 2. This phylogenetic tree shows the relationships among a diverse group of fleas. Note that most are parasites of mammals. In fact, other molecular information indicates that the fleas first parasitized mammals, especially rodents. However, it is also clear that clustered within fleas that infect mammals are a few groups that switched to birds. Note, too, that the bird fleas are not all derived from a single ancestral group. Bird fleas arose at four different places in the tree. In other words, bird fleas did not arise from a single switch followed by diversification to new bird hosts. Rather, several lineages of mammalian fleas independently switched to bird hosts. We hypothesize that these switches occurred where birds and mammals are found in close proximity so that mammalian fleas occasionally found themselves on birds. An obvious possibility would be birds that inhabit burrows, where they might contact mammalian fleas. We know, for example, that burrowing owls inhabit abandoned rodent burrows. The comparative phylogenetic histories of hosts and parasites allow us to develop hypotheses regarding the history and ecology of their coevolution.

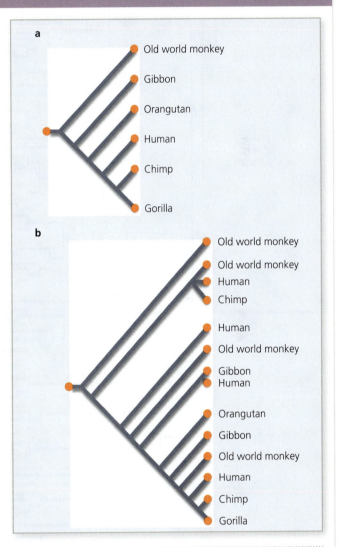

Figure 1 Phylogeny of primates and parasites. Phylogenies of primates based on morphological and molecular data (a) and their parasites (b). (From Whiting et al., 2008.) **Analyze:** How does this figure demonstrate host switching by the parasites?

QUESTION:

How do generalist and specialist parasites affect phylogenetic analysis of parasites and their hosts?

continued

The Phylogenetic Relationship of Parasites and Hosts

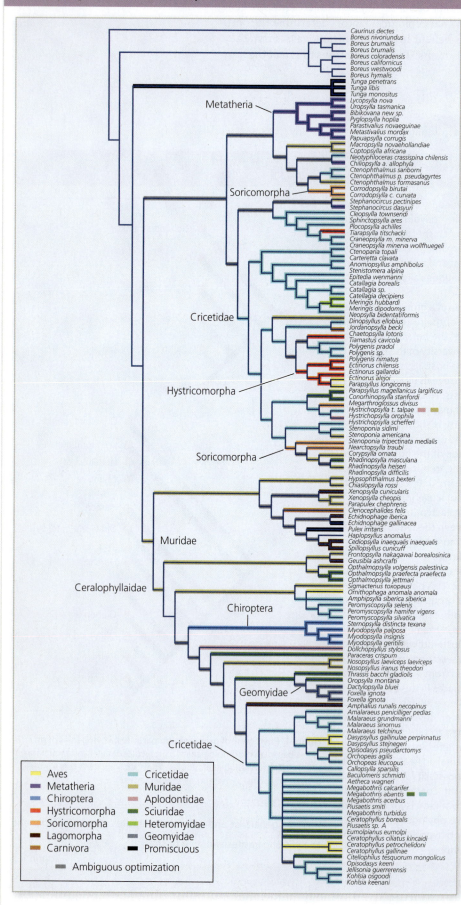

Figure 2 Phylogeny of fleas and hosts. The phylogeny of fleas and their bird and mammal hosts (from Whiting et al., 2008). **Analyze:** What is the significance of the multiple origins of bird fleas?

and thwart its defenses can be very specific. Second, they face a single set of selection pressures. Generalists may face conflicting selection pressure from their array of hosts. Finally, competition theory (Chapter 11) suggests that a narrower niche is one means by which the effects of competition may be reduced. Thus, specialized parasites may benefit from reduced competition from other species (Poulin, 2007).

However, there are costs associated with specialization as well. Specialization may put the parasite at ecological risk if the population of the host species declines or goes extinct. There is some evidence that at least for some groups, specialization is the ancestral condition and over time the parasite has become more generalized. For example, phylogenetic analysis of the lice that infect birds shows that host specificity is the ancestral condition (Johnson et al., 2009). Generalist species that can infect many bird species have evolved multiple times from specialized ancestors. And in this case, competition seems to drive the trend away from specialization. The origin of generalists is correlated with the presence of competing species. Apparently, generalists are able to avoid intense competition on one species by utilizing another.

KEY CONCEPTS 12.1

- The major exploitative interactions include predation, herbivory, and parasitism.
- Exploitative coevolutionary relationships result in adaptations and counteradaptations known as coevolutionary arms races.
- Key predator adaptations affect the detection, capture, and consumption of prey. Antipredator adaptations can interfere with any or all of these processes.
- Plants deter herbivores by means of morphological and chemical adaptations.
- The key parasite adaptations enable them to locate and infect the host.
- There are costs and benefits to a generalized versus specialized parasitic strategy.

QUESTION:

In what ways are predation, herbivory, and parasitism ecologically and evolutionarily similar? In what ways are they distinct processes?

12.2 What Species Interactions Are Mutually Beneficial?

Mutualistic interactions benefit both participants. In **facultative mutualisms**, the interaction is not necessary for either species' success. For example, cowbirds often eat insects from the backs of large ungulates such as bison. The birds find a ready source of insects attracted to the bison, and the cowbirds provide the bison with at least some relief from their insect pests. Still, neither species absolutely requires the presence of the other.

In **obligate mutualisms**, neither species can survive or reproduce without the actions of the other. The yucca (*Yucca spp.*) and yucca moth (*Tegeticula yuccasella*) exemplify an obligate interaction between a plant and its pollinator (Figure 12.17). The moth gathers pollen from the stamens of the yucca and deposits it directly into the stigma, effecting pollination. As the yucca's only pollinator, its action is essential for seed production. After pollinating the plant, the yucca moth deposits its eggs in the yucca's ovary, where its larvae consume all but a few of the yucca seeds before emerging as adults. It, too, requires the yucca—this plant is the only one on which the moth can successfully reproduce.

In some mutualisms, both species benefit equally. For example, one well-known mutualistic interaction is the ant-acacia mutualism in the neotropics

facultative mutualism A mutualistic interaction that is not absolutely required for either species.

THINKING ABOUT ECOLOGY:

Cowbirds are nest parasites—they lay their eggs in the nests of other species that unwittingly raise the young as their own. Bison herds were nomadic, wandering over vast reaches of grassland. What advantage would nest parasitism provide to cowbirds?

obligate mutualism Describes interaction that is necessary for the survival and reproduction of at least one of the species involved.

Figure 12.17 Obligate mutualists. The yucca and yucca moth are obligate mutualists.

Figure 12.18 Ant-acacia mutualism. The ant-acacia mutualism in the neotropics has been studied extensively. The acacia provides protection and nutrition for the ants; the ants protect the tree from herbivores.

■ **protection mutualism** The protection of one species by another.

Figure 12.19 Protection mutualism. The effect of ant removal on acacia growth (a) and survival (b). (From Janzen, 1966.)
Analyze: How could you measure the benefit to ants of living on an acacia?

(Figure 12.18). Some species of bullthorn acacia produce thorns with enlarged bases that house colonies of ants. Foliar nectaries produce nectar, and specialized structures on the leaf tips, known as Beltian bodies, produce lipids. The ants use both as food sources. In return, the ants defend the tree vigorously and aggressively against both herbivores and other plants whose branches touch the acacia. Of course, not all mutualisms benefit both species equally as in this example. There is a spectrum of symmetry/asymmetry in the relative benefits such that true mutualism grades into commensalism.

What Forms Do Mutualisms Take?

The adaptive needs of species vary widely depending on the other interactions in which they participate—that is, competition, predation, or parasitism. Most fall into three broad categories. In **protection mutualisms**, one species provides a benefit or reward to another that provides protection from predators, herbivores, or parasites. The ant-acacia mutualism described earlier exemplifies a protection mutualism. In this case, both species derive some protection—active defense of the plant by the ants and thorny nest sites for the ants. Detailed experiments demonstrate that the ants do indeed increase the fitness of the acacia (Janzen, 1966). If ants are experimentally removed, the survival of new plants and their size is compromised (Figure 12.19).

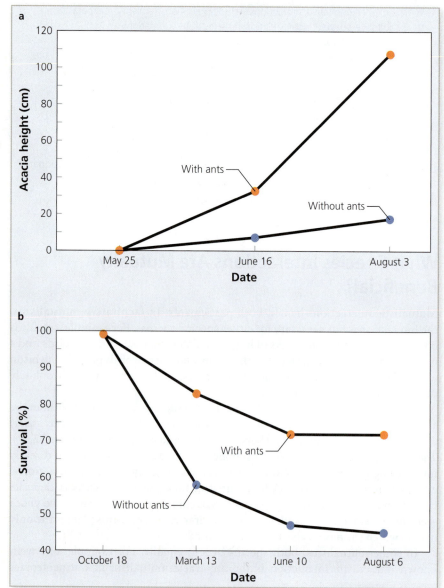

Transportation mutualisms are based on a mobile species that is rewarded for moving gametes or individuals of a sessile species. Pollination and seed dispersal are the two most common forms of this mutualism. These interactions range from obligate and highly specific, like the yucca–yucca moth interaction, to generalized and facultative. Many pollinators are generalists that visit many flower types. Both flowers and insects benefit, but the coevolutionary relationship among them is much less specific.

The flower and pollinator have different ecological goals in these systems. The flower requires transport of pollen; the pollinator is foraging for its food. Thus, the pollinator follows the precepts of optimal foraging theory: it is under selection to maximize its total energy intake or its energy intake per unit time. The components of a successful strategy are much like those of a predator and include factors such as the breadth of the diet (specialist/generalist) and how long to stay in a patch of flowers. Also, like predators, the pollinator must be able to detect the flowers of species that provide important rewards.

For its part, the flower must attract species that potentially effect pollination. Adaptations to accomplish this include the size, shape, and color of the flower, as well as the rewards offered to visiting pollinators. These features determine the type of pollinator the flower attracts (Figure 12.20). The categories of flower characters that attract certain pollinator types are consistent enough that we refer to them as **pollination syndromes**. Certain floral traits make visitation by some pollinators easier or more rewarding. For example, many bat-pollinated flowers open only at night when bats are active. They also tend to be fragrant and located near the top of the plant, where the bat can access it easily. Many butterfly-pollinated flowers have large landing platforms on which the butterfly can alight while it gathers nectar. Some bee-pollinated flowers have landing guides that direct the bee's attention into the flower and the nectarines.

For some time it was thought that pollinators have an innate attraction to certain flower types, resulting in the set of pollination syndromes. We now understand that learning is important to many pollinators. The floral characteristics to which the insects respond are **floral advertisements**—flower characteristics that indicate to the pollinator that the plant provides a nectar reward it can exploit (Melendez-Ackerman et al., 1997). This does not negate the concept of suites of characters associated with certain pollinators. Rather, it changes our explanation of their evolutionary origin. Specifically, we now understand that many pollinators sample flowers in their environment and center their foraging activity on those that provide useful rewards. For example, hummingbird-pollinated flowers tend to be red and produce much nectar in long floral tubes. Hummingbirds learn from sampling many flower types that red tubular flowers advertise the fact that they contain a significant nectar reward. And red is a good advertisement color because the birds see this portion of the spectrum well. Similarly, bees see particularly well in the blue and violet regions of the spectrum, hence the blue color of many bee-pollinated flowers.

Some pollination mutualisms are among the most intimate, mutually beneficial interactions in nature. However, both plants and pollinators sometimes exploit one another without mutual benefit. In **deceptive pollination**, the flower tricks a pollinator into visiting it, gaining the benefit of pollen transfer without the cost of offering a reward. Some orchids look like wasps (Figure 12.21) and produce fragrances that mimic the sex pheromones produced by female wasps. Male wasps are attracted to these flowers and even attempt copulation. In the process, they effect pollination. Some insects manage to steal nectar without accomplishing pollen transfer. Ants may reach a flower's nectaries without touching either the stamen or pistil. Other insects even chew through the base of the flower to reach the nectar without contacting the reproductive parts of the flower. These examples reinforce the evolutionary concept that the plants and the animals that visit them act in their own evolutionary self-interest.

transportation mutualism A mutualistic interaction in which one species provides transport of gametes or individuals for another.

Figure 12.20 Pollination syndromes. Flowers exhibit pollination syndromes—morphological adaptations for specific pollinators.

pollination syndrome A set of flower characteristics associated with a particular type of pollinator.

THINKING ABOUT ECOLOGY:

What would you predict to be the relative costs and benefits of pollinator specialization (a) for the flower and (b) for the pollinator?

floral advertisement A set of floral characteristics that advertise that the flower provides a reward to pollinator visitors.

deceptive pollination A set of floral characteristics that falsely advertise a reward to pollinator visitors.

Figure 12.21 Deceptive pollination. This orchid tricks a pollinator into visiting it by mimicking a female wasp.

nutritional mutualism Describes an interaction in which one or both species provide nutrition for the other.

mycorrhizae The mutualistic interaction between plants and fungi.

Like pollination mutualisms, seed-dispersal interactions range from highly specific to more general connections. The latter even merge into commensalism, because the plant benefits but the animal that transports its seed is neither benefited nor harmed. For example, the seeds of many species have hairs or barbs that cling to fur. The animal disperser is essentially unaffected, but the plant's seed may be moved long distances. Some seed-dispersal mutualisms are as specialized as some of the pollination mutualisms. The mutualism between mistletoe and the tanagers that disperse its seeds exemplifies this kind of highly coevolved interaction. Mistletoe, a parasitic plant species, produces conspicuous fruits high in the forest. The outer pulp layers of the fruit have high sugar content. Inside the pulp layer is a coat of viscin, a substance that protects the seed from the bird's digestive tract. It also causes the seed to cling to the bird's cloaca, which causes the bird to rub the seed off on a tree branch, where a new mistletoe plant can germinate. The tanager also has adaptations that facilitate the interaction. Its gizzard is positioned as a side pouch of the gut, and sphincter muscles regulate whether material enters the gizzard. If the tanager eats an insect, the sphincter opens to allow the insect to enter the gizzard, where its powerful muscles and grit grind it up. But if the tanager eats a mistletoe fruit, the sphincter closes and the seed passes through the gut without being destroyed in the gizzard.

Nutritional mutualists exchange nutrients, or one species is rewarded in some way for improving the nutritional status of the other. These mutualisms occur across many taxonomic groups. The nutritional mutualism between plants and bacteria or fungi is ecologically very important. Among the most widespread and significant of these are the mycorrhizal fungi. The hyphae of ectomycorrhizal fungi penetrate the intercellular space of the plant's roots (Figure 12.22). The fungus digests dead plant material in the soil and provides nitrogen to the plant in exchange for carbon compounds from the plant. Arbuscular **mycorrhizae** produce hyphae that enter the cells of the root. Most of these species provide the plant with phosphorus from the soil and receive carbon compounds in return. The more we study these mutualistic fungi, the more common they appear to be—at present, some 80 percent of plant species have mycorrhizal mutualists of some type. The legumes are known for their mutualistic interaction with *Rhizobium* bacteria. The bacteria live in nodules that form on the root of the plant, where they receive protection and carbohydrates. The bacteria have enzymes that can convert atmospheric nitrogen, which the plant cannot use, into ammonia, which it can.

Animal-plant nutritional mutualisms are crucial to the function of coral reef systems. Most of the reef-building corals, the large species that develop the basic structure of the reef, have mutualistic algae called zooxanthellae in their cells. The zooxanthellae receive protection and carbon compounds necessary for photosynthesis. In return, the zooxanthellae supply the coral with glucose, glycerol, and amino acids, the precursors of carbohydrates, lipids, and proteins, respectively.

How Does Mutualism Evolve?

The evolution of mutualism is complicated by the fact that each interacting species is offering something of value to another species. In effect, a mutualist reduces its fitness by the cost of the service or material it provides the other species. There are two key requirements for mutualism to be a successful strategy. First, the benefit the species receives must outweigh the cost of the service or material it provides in return. Second, the relationship must be an evolutionarily stable strategy for both species.

Zooxanthella

Endoderm
Mesogloea
Ectoderm
Organic layer
Lime skeleton

Figure 12.22 Zooxanthellae. Zooxanthellae are mutualistic algae that inhabit corals.

THE HUMAN IMPACT

Mutualism and Species Conservation

The intimate connections among coevolved species affect the human impact on communities and our efforts to conserve them. Specifically, a functional community requires its full complement of species. The absence of mutualists may have a significant effect on the function of the system.

The problem of species loss is compounded by the webs of coevolved interactions among species. The complexity is illustrated by the Brazil nut tree, an economically and ecologically important tree of the New World tropics. Its only pollinator is a euglossine bee. Its seeds rely on agoutis (rodents) for dispersal and germination. When the agoutis eat a pod, they break up the hard seed coat, facilitating germination. Seeds are dispersed by defecation, sometimes great distances from the parental tree. This web of mutualistic interactions, even though comprising only three species, illustrates the potential risk. The loss of any of these species ripples throughout the web of interaction. The fact that the euglossine bee is also an important pollinator of orchids and bromeliads extends the web to these species and their own webs of mutualistic interactions. Thus, the bees are a *mobile link species*, a species whose mobility across the landscape is ecologically essential. Because the Brazil nut tree is so important to the bee, we refer to it as a *keystone mutualist*, a mutualist essential in that community.

The tallgrass prairies of the Midwest are highly fragmented—just 0.1 percent of the original prairie remains in a matrix of agriculture. The eastern prairie fringed orchid is an endangered species characteristic of the tallgrass. This species is pollinated only by night-flying hawk moths. Remnant prairies containing the orchid are tiny and widely scattered. Most are not large enough to offer the hawk moths many orchids. In effect, optimal foraging theory suggests that it is disadvantageous for the hawk moth to seek small prairie fragments. The result is that the orchid often does not set seed, further endangering it. The recovery plan for this species requires enormous volunteer effort: plants

Figure 1 Prairie fringed orchid. The prairie fringed orchid is an endangered species of the tallgrass prairie that relies on hawk moths for pollination.

are cross-pollinated by hand to ensure seed set and the future of the species.

QUESTION:

What characteristics of mutualists would you expect to determine their importance to the rest of the species in the community?

A simple set of equations describes these two principles. The fitness of individuals in a nonmutualistic population is w_{nm}. Now imagine that mutants for mutualism arise. Successful variants offer a benefit to the other species and receive benefit in return. Some variants may be unsuccessful: they provide a benefit but do not receive benefit in return. The average fitness of a population that contains mutualists is $w_m = pw_{sm} + qw_{um}$ where p is the proportion of successful mutualists (w_{ms}) and q is the proportion of unsuccessful mutualists (w_{um}). Mutualism will arise only if

$$pw_{sm} + qw_{um} > w_{nm}.$$

The fitness of the unsuccessful mutualists is an important component of this equation because they have lower fitness than successful mutualists. This equation contains both principles: mutualism arises if its benefits outweigh its costs, and when those benefits are realized, even if rare. Under what conditions does this inequality hold?

mobile link species A species whose mobility across the landscape to a series of dependent mutualists is important to the normal function of the community.

keystone mutualist A species whose participation in webs of mutualistic interactions is vital to the function of the community.

The phylogenies of mutualists suggest two patterns in the history of these interactions. First, mutualisms can arise from exploitative interactions such as herbivory or parasitism. If an insect forages on and consumes plant pollen, it decreases the plant's fitness even if some pollen is occasionally transferred to another flower. If the plant produces nectaries and as a result the insect forages on nectar rather than pollen, the interaction shifts from exploitation to mutualism. Clearly, the key component of this evolution is whether the fitness cost of nectar production is lower than the benefit of pollen transfer. In the equation above, w_{sm} must be large.

The second pattern that emerges is that many mutualisms involve the interaction of three species, one of which is exploitative. Thus, the ant-acacia mutualism is adaptive only if the acacia faces exploitative interactions with herbivores. The cost of providing Beltian bodies and large hollow thorns is significant. But if the loss of tissue to herbivores is larger, this cost may be less than the benefit, so that the inequality $pw_{sm} + qw_{um} > w_{nm}$ is satisfied. Thus, a third exploitive species contributes to the fitness values of this equation.

Recent studies confirm the importance of this cost-benefit approach to the evolution of mutualism. The experimental removal of large herbivores from an African savanna community significantly altered the ant-acacia mutualism (Palmer et al., 2008). After 10 years without mammalian herbivores, the acacia reduced both nectar production and housing for the ants. At the same time, the ant shifted from mutualism to exploitation. Clearly the third species group of exploitative mammalian herbivores was central to the ant-acacia mutualism.

KEY CONCEPTS 12.2

- Mutualisms generally evolve to facilitate reproduction, energy acquisition, or protection from exploitative interactions.
- Mutualisms can be obligate or facultative. They may also be specialized or general.
- Mutualisms evolve when the cost of the benefit provided is less than the benefit received by each species.

QUESTION:

What is the connection between the evolution of mutualism and exploitative interactions?

Putting It All Together

Each organism faces a range of selective pressures imposed by its environment. Some of the most intense selection faced by an organism is due to other species with which its ecology is intimately connected. This is the case for the bat, *Trachops*, and the frogs it preys on. These species are enmeshed in a complex web of interaction. Each member of this web faces its own selection pressure to find and capture prey or avoid detection and capture. We can imagine a series of adaptations and counteradaptations. An ancestral bat switched from insect prey to a new prey item—frogs. The frogs, already vulnerable to predation while mating, call near running water that masks their location to predators. The bat's auditory ability was already so sophisticated that it was a short step to homing in on the mating calls of these frogs even when near water. One means by which some tropical frogs thwarted predation is the development of potent toxins. Eventually, *Trachops* was able to distinguish between the calls of toxic and nontoxic frogs.

This predator-prey interaction exemplifies the process of coevolution. Each species constitutes a selective force on the other, and as each adapts, the other must respond. For exploitative interactions such as predation, parasitism, and herbivory, the antagonistic species engage in an escalating series of adaptations and counteradaptations known as a coevolutionary arms race. Other coevolutionary interactions, such as flower-pollinator mutualisms, benefit both species.

Coevolutionary interactions are important in ecology for two reasons. First, they involve processes crucially important to the fitness of the participating species: most involve reproduction, energy acquisition, or predator/parasite/ herbivore avoidance. Second, virtually every organism on earth participates in coevolutionary interactions with other species. Probably only a handful of species, such as those in extreme environments, truly live in isolation from all other species. The rest are deeply embedded in the full range of coevolutionary interactions. Together with another coevolutionary interaction, competition, these interactions affect nearly every ecological process described in this text.

It is easier to describe and analyze these interactions in isolation. However, it is also rare that a species participates in just one such relationship. The vast majority are connected to many other species as competitors, predator or prey, host or parasite, and mutualists. And any one interaction, such as a bird eating the fruits of a plant, is likely to be connected to many other interacting species (Figure 12.23). There are both theoretical and practical reasons to study and understand this complexity. If we are to fully understand the nature of ecological communities, we must understand them as networks of interactions. And as human activities affect more species and more communities, it is imperative to understand the ways these impacts ripple through the webs of interaction.

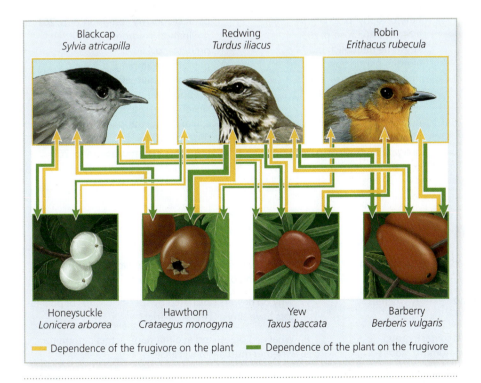

Figure 12.23 Interaction web. A web of interacting fruit-eating birds in montane forest in Spain (from Thompson, 2006). **Analyze: What are the consequences of webs of interaction for conservation?**

 Summary

12.1 What Are the Major Forms of Exploitative Interaction?

- Herbivores, predators, parasites, and parasitoids exploit other species for their food resources.
- Exploitative species and their prey or hosts constitute selective forces on one another, leading to coevolutionary arms races.
- Predators must detect their prey (search time), then capture and subdue it (handling time).
- Prey avoid predation by crypticity (appearing as a random sample of background) and with adaptations that thwart capture. Plants employ structural and chemical deterrents to herbivores.
- Parasites must locate and infect host species. Complex life cycles facilitate these tasks by (a) including specialized forms that can infect intermediate and definitive hosts and (b) producing large numbers of infectious forms.

- Predators respond to increased prey density by increasing the rate of consumption (functional response) and by increasing their numbers (numerical response).

12.2 What Species Interactions Are Mutually Beneficial?

- Most mutualisms fall into three main categories: protection mutualisms, transportation mutualisms, and nutritional mutualisms.
- Flowers and their pollinators represent specialized transportation mutualisms. Flowers have pollination syndromes—characteristics that advertise a reward to their preferred pollinators.
- The evolution of mutualism depends on the relative costs and benefits of participation in the interaction for both species.

Key Terms

active predator p. 252
ambush predator p. 251
aposematic coloration p. 258
apparency theory p. 262
Batesian mimicry p. 260
coevolution p. 251
coevolutionary arms race p. 251
compensatory ability p. 261
cooperative hunting p. 252
crypticity p. 257
deceptive pollination p. 269
energy maximizer p. 253
evolutionary theory of virulence p. 264
exploitative interaction p. 250
facultative mutualism p. 267

floral advertisement p. 269
functional response p. 256
group hunting p. 252
handling time p. 253
herbivore p. 251
heterotrophy p. 251
intermediate host p. 263
keystone mutualist p. 271
life-dinner principle p. 251
mobile link species p. 271
Mullerian mimic p. 260
mycorrhizae p. 270
numerical response p. 256
nutritional mutualism p. 270
obligate mutualism p. 267

optimal foraging theory p. 253
parasite p. 250
parasitoid p. 251
phytochemical coevolution theory
    p. 261
plant secondary compounds p. 261
pollination syndrome p. 269
predator p. 250
predator swamping p. 258
protection mutualism p. 268
pseudoflowers p. 264
search time p. 253
solitary predator p. 252
time maximizer p. 253
transportation mutualism p. 269

Review Questions

1. What general principle underlies adaptations that make prey items cryptic?

2. Why do some predators maximize energy whereas others maximize time?

3. What prey characteristics lead to solitary, group, cooperative, ambush, and active hunting modes?

4. What factors determine the virulence of a parasite?

5. What are the major plant adaptations that thwart herbivores?

6. What advantage do Mullerian mimics receive from resembling other noxious species?

7. Why do flowers pollinated by specific groups of pollinators have similar characteristics?

8. What ecological and evolutionary factors determine whether mutualism will arise?

9. Why is the selection pressure on predators and their prey asymmetric?

10. Why do many parasites include intermediate hosts in their life cycles?

Further Reading

Bronstein, J.L. 2009. The evolution of facilitation and mutualism. *Journal of Ecology* 97:1160–1170.

This paper reviews current hypotheses for the evolution of mutalistic interactions. It also discusses unanswered questions that are the focus of current research in the field.

Creel, S., and N.M. Creel. 2006. *The African wild dog: behavior, ecology, and conservation*. Princeton, NJ: Princeton University Press.

This book compiles the Creels' extensive research on African wild dogs. It includes discussion and examples of many of the important predator-prey interactions discussed in this chapter.

Esch, G.W. 2004. *Parasites, people, and places*. Cambridge: Cambridge University Press.

This book summarizes the key ecological and natural historical aspects of parasites, with emphasis on their relationship to humans. It also provides insight into the nature of fieldwork in parasitology.

Real, L. 1983. *Pollination biology*. Waltham, MA: Academic Press.

This classic book summarizes many important aspects of plant-pollinator interactions.

Chapter 13

The Structure of Communities

The Everglades are a vast wetland ecosystem. Water flows south from Lake Okeechobee across southern Florida, eventually to Florida Bay and the Atlantic Ocean. But the flow is not confined to rivers and streams. Rather, it is known as sheet flow—the gradual movement of water across the breadth of the landscape. The land slopes so gradually that the flow is almost imperceptible. Here precipitation is seasonal. Summer rains wash through the Everglades, raising water levels all across the ecosystem. But in the winter dry season, water remains only in the deepest pockets and sloughs.

Visit one of these sloughs in late winter, the driest time, and you will find a rich community of wading birds. Egrets, herons, rails, bitterns, limpkins, roseate spoonbills, wood storks—all congregate in the shallow waters of the Everglades to feed on the small fish and invertebrates concentrated by the disappearing waters. If you carefully watch these birds forage, you will be impressed by two contradictory ideas. First, you will note how similar they are in where they hunt, how they perceive their prey, and how they capture that prey. But you will also detect subtle differences among them. Some stay close to dense vegetation; some move slowly, hoping to flush prey from the vegetation; some probe into the mud for burrowing invertebrates. They are similar, yet different.

At this time of year their food resources are both dwindling and concentrated. The stage is set for intense competition. We know from Chapter 11 that for two species with similar niches, a limiting resource imposes strict limits on both. We also know that if that competition is too intense and their niches too similar, competitive exclusion may eliminate one. Or competition might push their niches further apart by the process of character displacement. How do these processes work in complex communities composed not of pairs of species but of many species living together? If competition is so important, how can all these species coexist in these small isolated patches of food? What other ecological processes determine which species are present? In this chapter we expand the discussion of competition in this way and ask: *How are ecological communities organized?*

13.1 What Processes Determine Community Composition?

A community is a group of species that co-occur in a particular place. If we are to understand how communities are organized, we must address two components of the community: (1) its composition—that is, which species are present—and (2) its structure—that is, the relationships of the co-occurring species.

In Chapters 4 and 5 we described the basic terrestrial and aquatic community types. Much of that discussion centered on the role of the physical environment as a determinant of the species that comprise a particular community. We now expand on that previous discussion.

What Is the Role of the Physical Environment in Community Composition?

The biological interactions among species are irrelevant unless they are physiologically and morphologically suited to that environment. The Gleasonian approach to communities emphasizes the role of the physical environment. Today we refer to this process as **environmental filtering**—the sorting of species' presence or absence according to their physiological tolerance of the abiotic conditions. All of the mechanisms of physiological tolerance that determine species' broad geographic ranges apply on the local scale as well. Fine-scale matches between genotypes and the physical environment are well known. For example, in many ocean ecosystems, the alga *Prochlorococcus* is the dominant phytoplankton species. There are six known ecotypes of this species, each adapted to a slightly different temperature regime. In the open ocean, phytoplankton drift over great distances, but local temperature acts as a filter determining which ecotype predominates in a particular area (Johnson et al., 2006).

Because the species constituting the local community potentially interact, the effect of physical factors on one species may ripple throughout the community. Local ecotypes of a plant, the evening primrose (*Oenothera biennis*), correspond to local variation in the abiotic environment. Johnson and Agrawal (2005) showed that the members of the arthropod community that feed on this plant are determined by the genotype of the primrose. Thus, the physical environment filters the arthropod community as well—but the filtering is indirect, mediated by the plant host.

Other kinds of biotic interactions that affect the pool of potential colonists also affect the filtering process. For example, predation can reduce the size of the species pool available for colonization. Recent studies demonstrate that predation limits the potential constitution of sessile marine invertebrate communities, particularly in the tropics, where predation is intense (Freestone and Osman, 2010). Mathematical models show the role that mutualism plays in the structure of communities in harsh environments. Specifically, when mutualism increases the probability that a species can persist in a harsh environment, the resulting community contains more species and other interactions than would otherwise occur (Filotas et al., 2009).

environmental filtering The presence or absence of species according to their physiological tolerance of abiotic conditions.

What Is the Relative Importance of Random and Deterministic Processes?

Neutral models of community composition propose that communities are random combinations of species physiologically able to live in the same place. We refer to these models as neutral because they invoke no deterministic mechanisms such as competition or predation to explain which species make up a community. The alternative, **niche-based models** of community structure, emphasize the role of deterministic biological processes. Competition is central to niche-based models because it shapes the niches of coexisting species. Imagine a community of species exploiting two key resources such as prey size and preferred ambient temperature. We create a graph with one resource as the x-axis and the other as the y-axis. For each species, we calculate the mean value of its use of each of the two resources. We plot these mean values on this two-dimensional graph. The position of each species on this graph is its position in two-dimensional **niche space**.

This graph allows us to determine if the niches of these species conform to a niche-based model or a neutral model. Under the neutral model, niches should be arranged at random in niche space. On the other hand, if competition structures the community, the niches should be widely separated: competitive exclusion removes similar species and character displacement separates niches. The result is that the distance between species is maximal; that is, the niches are overdispersed.

Dispersal plays a central role in neutral models: the species that constitute a community must first have the means of arriving there. Then environmental filtering determines which can stay. For example, assume that the rocky intertidal community of sessile invertebrates is determined entirely by where each species colonizes the substrate. If the drift of the pelagic larvae of these species is random, each locality receives a random assortment of species. Some are physiologically able to persist there; others die out. The result is a neutral community.

Some communities conform to a neutral model. Lamb and Cahill (2008) showed that the root systems in grassland communities show no evidence of competitive effects on local growth or community composition. Also, diverse communities of herbivorous insects coexist on single leaf blades in some tropical forest species. It is hard to imagine that competition is structuring such a community. However, this does not mean that competition has played no role in these communities. If competition organized the set of potentially coexisting species in the past, extant communities might appear random even though we do not see evidence of competition today. This occurs because some species were eliminated from the community by competition or their niches shifted by character displacement. The resulting species *appear* random even though competition actually structured the community. This phenomenon is known as the **ghost of competition past**.

Under neutral models, community structure is a random process influenced by dispersal and colonization. In that case, neutral theory also predicts that if communities are altered and then reorganize, they should not necessarily return to the same community structure as before. Stochastic processes may lead to a different pattern each time the community forms. In contrast, if niche-based processes underlie community composition, we should expect communities to converge on a similar composition unique to the local area. This is what actually occurs in the rocky intertidal. Across large areas of coastline, intertidal communities converge on a small number of community structures (Caro et al., 2010). These results are not consistent with the neutral model; they support the role of deterministic processes such as competition.

We now understand that both niched-based communities, such as the rocky intertidal, and neutral communities, such as grassland root systems, exist. However, the preponderance of evidence from a wide array of studies argues for the importance of competition as a key structuring force. Two examples will illustrate the kind of analysis and data that demonstrate the role of competition and niche-based structure.

neutral model The concept that communities are random assemblages of species physiologically tolerant of the abiotic conditions.

niche-based model A model of community structure in which biological processes, especially competition, act on species' niches to permit coexistence.

niche space A multidimensional representation of a species' use of a set of resources.

ghost of competition past An explanation for communities with niches that appear to be randomly associated because competition eliminated some species in the past.

DO THE MATH

The Null Hypothesis in Community Structure

As we have seen, communities may be organized by deterministic processes such as competition, or they may simply be random assemblages of species that happen to be able to live in a particular region. The difference between these two views illustrates an important feature of scientific and statistical analysis: the role of the null hypothesis. The null hypothesis states that the pattern we observe in a data set is random, the result of chance effects. Scientists are obliged to reject the null hypothesis before accepting alternative, deterministic explanations for their data. The analysis of community structure illustrates this principle and the conceptual basis of testing the null hypothesis.

Imagine an archipelago of small islands and a set of granivorous bird species that inhabit them. Each island has the same total range of microhabitats and seed-producing plant species. And each has a community of birds. Islands differ in both the number of species they contain and the particular species present. We might hypothesize that the bird fauna of each island is organized by competition, a deterministic process. Species whose niches are too similar cannot coexist; competition weeds out those species too similar to others on an island. The null hypothesis in this scenario is that the community of birds on each island is simply random. How can we distinguish among these alternatives?

There are many approaches to this problem. However, all are based on some means of calculating what a random assemblage of species would look like, then comparing the actual assemblage to it. For example, imagine that we quantify the niche of each bird species by its position on two niche axes: microhabitat use and the seed sizes it consumes. Each species occupies a position in this two-dimensional niche space. What predictions do the competition and null hypotheses make about the arrangement of these niches?

If niches are arranged at random on each island, we expect that some niches will be very similar, others very different. That is, niches should be arrayed randomly in niche space as in Figure 1a. On the other hand, we expect that competition will not allow species with similar niches to coexist on the same island. Under that hypothesis, niches maintain a certain minimum distance from one another in niche space (Figure 1b); we say they are *hyperdispersed in niche space*. Now we have a graphical and mathematical mechanism for distinguishing between these two hypotheses. We calculate the mean distance between niches for all species inhabiting the same island and repeat the process on all islands. This represents the observed niche separation on the actual islands. To generate a null community for comparison, we take the pool of all species in the archipelago and randomly select a number from this group equal to the actual number of species on an island. We calculate the mean niche separation for this random subset. We repeat this 1,000 times and calculate the mean niche separation for random assemblages of species. We do this for each island. We can now compare the observed niche separation on each island to the average niche separation of the corresponding random community.

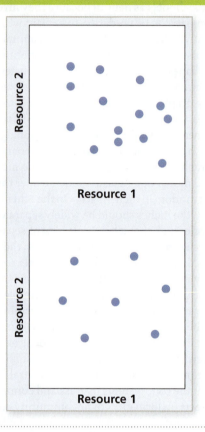

Figure 1 Neutral model and competition. The arrangement of niches in niche space comprising two different resources important to the organism, according to (top) the neutral model and (bottom) competition. Under the neutral model, niches (and species) are arranged randomly in niche space. If competition structures the community, niches should have a minimum separation; they should be overdispersed. **Analyze:** Explain the role of dispersal in the neutral model.

If niches are random, we expect that on some islands the observed niche separation will be greater than the random group and on others it will be smaller. We can develop a table of the relative sizes of random and observed distances (Table 1).

Each island is represented by a plus if its observed distances are greater than the random distances; it is represented by a minus if the observed distance is smaller. If the null hypothesis is correct, the number of pluses and minuses should be approximately equal: by chance, some niches are far apart, others are close. We can test this prediction with the chi square, which detects deviations from a 50:50 ratio of plus and minus values.

If there is no difference between the random communities and the actual island communities, we accept the null hypothesis. If so, there is no reason to infer a deterministic process such as competition. On the other hand, if we reject the null hypothesis, we conclude that some biological process is operating. For

TABLE DTM-1 Mean niche separations on a series of hypothetical islands. Islands on which the actual mean separation is greater than that of a randomly generated community are marked with a $+$; those on which the actual mean separation was smaller than expected receive a $-$. Under the null hypothesis, the number of pluses and minuses should be equal. The chi square calculation shows the following:

ISLAND NUMBER	OBSERVED NICHE DISTANCE	EXPECTED NICHE DISTANCE	DIFFERENCE
1	8.4	7.2	+
2	11.16	10.0	−
3	9.9	9.6	+
4	6.4	6.3	+
5	7.8	7.6	+
6	6.4	6.2	+
7	8.0	7.4	+
8	6.4	6.5	−
9	6.4	5.4	+
10	5.8	5.0	+
11	6.2	6.6	−
12	11.6	11.2	+
13	6.2	5.1	+
14	12.1	11.3	+

$$x^2 = \sum \frac{(O-E)^2}{E}$$
$$= (11-7)^2/7 + (3-7)^2/7 =$$
$$= 2.29 + 2.29 = 4.58$$

For this chi square, there is 1 degree of freedom. This is calculated as the number of classes −1. There are two classes (+ or −), so there are 2 − 1 = 1 df. For p = 0.05, the critical value of the chi square is 3.84. If the calculated chi square is larger than 3.84, we conclude that the differences are not due to chance. Since 4.58 > 3.84, we conclude that the number of + values is larger than expected by chance. The actual distance between niches is significantly larger than the distances in random assemblages. We reject the null hypothesis that species are found on islands at random.

example, if the observed niche separation is consistently greater than in random communities, the results are consistent with the competition hypothesis. This in itself does not prove that competition is the driving force; additional experiments would be required to show that competition has driven the niches farther apart.

First, studies of coexisting granivorous rodents reveal size relationships that support the role of competition in community structure. Granivorous desert rodents exhibit remarkably similar size relationships across the deserts of the southwestern United States (Figure 13.1). Could this be due to competition separating the niches of coexisting species? If so, it would be an example of competition structuring the composition of a community.

Bowers and Brown (1982) tested three predictions based on the hypothesis that this pattern is due to competition:

1. Rodents of similar size in the same **guild**—that is, coexisting species with similar niches—should coexist less frequently than expected by chance.
2. Prediction 1 should hold more strongly for species of similar body size than for species of very different size. This prediction is based on the expectation that the more similar the species, the stronger the competition.
3. If species from other guilds, which would not be expected to compete directly, are included in the analysis, co-occurrence should be random.

guild A group of ecologically similar, coexisting species.

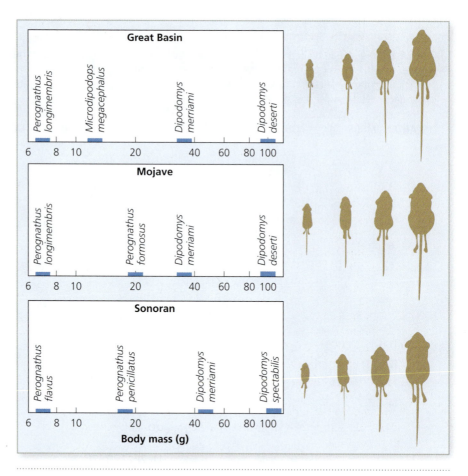

Figure 13.1 Size relationships. The size relationships of coexisting granivorous rodents in deserts in the southwestern United States. Each community has a set of rodents with similar size relationships even though the actual species differ (from Bowers and Brown, 1982). **Analyze: What would these size relationships look like under a neutral model?**

The researchers surveyed 95 desert sites for the presence or absence of rodents. These data allowed them to analyze whether species coexist at random or less frequently than expected by chance. They accomplished this by first calculating the proportion of the 95 sites in which each species was found. From this information, they developed a table of pairwise species co-occurrences. In other words, they calculated the proportion of the 95 sites in which each pair of species was found together. This constituted the observed pattern of co-occurrence. Then they calculated the expected number of sites where two species should co-occur based on chance. If, for example, one species is found in 70 percent of the sites and another is found in 50 percent of the sites, we expect them to occur together by chance in $0.7 \times 0.5 = 0.35$, or 35 percent, of the sites. This calculation is based on the product rule of probability: the probability of two independent events occurring together is the product of their independent probabilities. Next they compared the observed co-occurrence for each pair of species with the expected value if they occur together by chance. Each pair was scored as a $+$ if it occurred more frequently than expected by chance or a $-$ if it occurred less frequently than expected by chance.

To test Prediction 2, they performed separate analyses of pairs of species whose ratios of body mass were greater than 1.5 and less than 1.5. For Prediction 3, they included more species, those from other guilds, in the analysis.

The results are shown in Table 13.1.

For similar-sized granivores, all pairwise associations were negative; none were positive. In other words, pairs of species this similar did not co-occur (Prediction 1). There were also significantly more negative associations for granivores whose

TABLE 13.1 Positive and negative associations of rodents based on presence or absence at 95 sites (from Bowers and Brown, 1982)

BODY MASS RATIO	GRANIVORES		ALL GUILDS	
	−	+	−	+
<1.5	27	0	93	15
>1.5	65	28	274	98

ratio of body size was greater than 1.5, although the pattern was not as strong (Prediction 2). When rodents from other guilds were included, the co-occurrence of pairs of species was not significantly different from chance (Prediction 3). These results are consistent with the idea that competition structures the size relationships in desert rodent communities.

A second example of the importance of competition comes from stunningly diverse tropical rainforest communities. In this case we find evidence of the importance of competition even in systems where it is difficult to conceive how niches might be separated. Tropical rainforests are among the most species-rich forests on earth: communities are known in the Amazon that contain 1,100 tree species in just 25 hectares! Recent studies of these diverse forests show that the niches of locally co-occurring trees differ importantly from one another on the basis of light regime, nutrient requirements, and response to soil moisture. Similar species simply do not occur together as expected under neutral theory (Kraft et al., 2008). Even in these diverse communities of tropical rainforest trees, we find evidence that competition determines which species can coexist.

competitive hierarchy Linear competitive dominance relationships among species; species A dominates species B, which dominates species C, etc.

....

KEY CONCEPTS 13.1

- The first determinant of species presence or absence in a community is whether or not it can survive the physical conditions. This is the process of environmental filtering.
- Neutral models of community structure posit that communities are random assemblages of species that can tolerate the abiotic conditions.
- Dispersal is a key component of neutral models. In neutral models, the composition of each community is determined by environmental filtering of the species that disperse there.
- Although neutral communities are known, the evidence suggests that competition structures many communities.

QUESTION:

How do we define the null hypothesis in community structure?

....

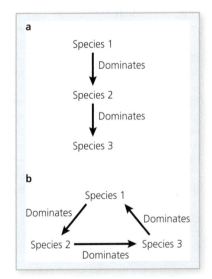

Figure 13.2 Competitive hierarchy and network. (a) A competitive hierarchy is a linear pattern of competitive dominance. (b) In a competitive network, the competitive interactions are not linear. **Analyze: In a very species-rich community, would you expect competition to be hierarchical or a network?**

13.2 How Does Competition Structure the Community?

If many communities are structured by competition, we must then address the mechanisms by which this occurs. In Chapter 11 we discussed how the process of competition operates between pairs of species. If we are to understand its role in community structure, we must expand the analysis to groups of species. The potential interactions increase rapidly as we consider competition among more than two species. In simple pairwise competition, one species is typically competitively superior to another. If this same pattern holds among more than two species, the result is a **competitive hierarchy** (Figure 13.2a), in which there

THINKING ABOUT ECOLOGY:
....

How would you expect character displacement to affect niches in competitive networks?

is a linear pattern of competitive dominance—one species dominates another which dominates a third, and so on. Alternatively, multispecies competitive interactions can take the form of **competitive networks** (Figure 13.2b), in which Species A dominates Species B, and Species B dominates Species C, but Species C dominates Species A. As the number of species increases, the potential interactive networks may become extremely complex.

How Are Niches Arranged in Communities?

Regardless of the competitive network that arises in multispecies communities, the law of limiting similarity states that no two species can long coexist on the same limiting resource. Competitive networks may increase the potential for coexistence, but they cannot prevent competitive exclusion when niches are too similar.

Ecologists have documented many instances of subtle niche differences among coexisting species, such as those described above for tropical forest trees. These studies show that virtually any critically limiting resource can be subdivided.

THE HUMAN IMPACT

Competition from Invasive Species

Humans have a long history of accidentally or intentionally moving species from one part of the world to another. How do these invasions change community structure? And what role does competition play in invasions?

First it is important to understand the magnitude of the problem. Thousands of species have been introduced to the United States from other places. Of these, approximately 15 percent have established populations that cause ecological or economic damage. These impacts are not always predictable. For example, highly flammable plants, such as Australian melaleuca (*Melaleuca quinquenervia*) and Australian pine (*Casuarina equisetifolia*), increase the risk of catastrophic fire in native communities.

Too often, introduced plant species overrun the native community, forming dense stands that exclude other species. The flammable species mentioned above also draw tremendous amounts of groundwater and in the process outcompete native species. Dense stands of others cast shade that inhibits the growth of native seedlings. In the deciduous forests of the Midwest, buckthorn (*Rhamnus cathartica*) and Asian honeysuckle (*Lonicera maackii*) form thickets so dense and block so much light that the ground beneath them is devoid of vegetation.

Often the effect is peculiar to the physiological ecology of the invader. For example, in California, the introduced African crystalline ice plant (*Othonna capensis*) grows in coastal dune communities. This species outcompetes native plants in this community because the salt it accumulates leaches from the leaves at the end of the growing season, suppressing the growth of competitors.

The competitive impact of introduced species is not limited to plants. Brown trout (*Salmo trutta*), introduced from Europe throughout North America, compete with native species such as cutthroats (*Oncorhynchus clarkii*). The California golden trout (*Oncorhynchus mykiss*) is already threatened with extinction due to declining habitat and overharvest. These problems are compounded by competition with brown trout.

Apparently, competitive ability per se is not always characteristic of invasive species. The initial invasion of a community often conforms to a neutral model (Davis, 2005). Most successful invaders have two key characteristics: they are adept at dispersal and they can exploit disturbed habitat. These *invasive species* thrive in disturbed sites where resources such as soil moisture, nutrients, and light are abundant. Many are able to tolerate high levels of pollution (Karatayev et al., 2009). Their competitive impact on the native communities occurs secondarily. That is, they exploit disturbed conditions and increase their numbers rapidly, thereby increasing their competitive impact on native species. Melaleuca exemplifies this process. These trees can survive in both terrestrial and shallow-water aquatic systems. Thus, they invade the marshes of the Everglades and disturbed forest sites. They produce large numbers of tiny seeds that disperse widely. Once established, the population increases exponentially and forms a monoculture that competitively excludes most all other plant species (Figure 1). In the Everglades ecosystem, melaleuca is increasing at the rate of 35 acres per day!

Figure 1 Melaleuca. Melaleuca is an introduced species that outcompetes other plants in the Everglades.

QUESTION:

Would you expect most invasive species to have broad or specialized niches?

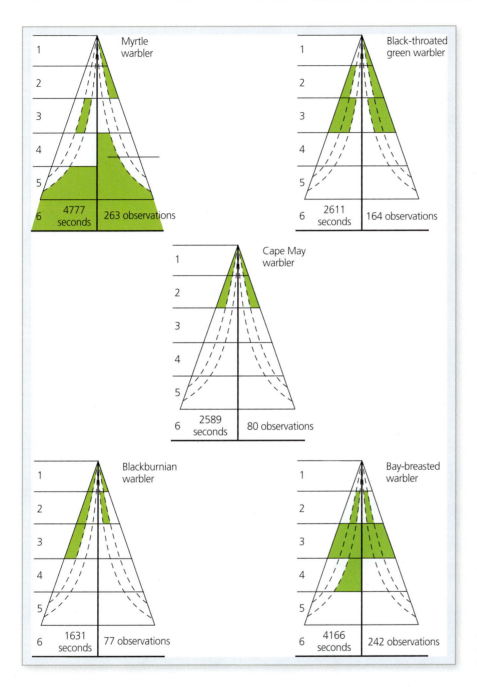

Figure 13.3 Warbler foraging. Feeding niches of coexisting warblers in a single conifer tree. Each species forages in a different microhabitat in the tree (from MacArthur, 1958). **Analyze:** What factors could explain why there is more overlap among some of these species than others?

In some cases, the crucial component of the niche is indirectly related to the limiting resource. For example, warblers, small insect-gleaning birds of conifer forests, utilize unique and highly specific microhabitats (Figure 13.3). Detailed observations of their foraging patterns show that even when they forage in the same tree, they are using specific and different portions of the tree (MacArthur, 1958).

The niches of species in diverse communities change in one of three ways (Figure 13.4). First, niches may have high overlap (within the constraints imposed by competitive exclusion). Recall from Chapter 11 that the Lotka-Volterra models show that coexistence is possible if resources are abundant and niche overlap does not exceed a maximum value. For a long time, ecologists searched for general rules for the limit to overlap. G. E. Hutchinson (1957) proposed, based on a number of empirical studies, that species cannot be more similar than a ratio of 1.4 in body size. He reasoned that species more similar than that share too much of the prey resource and cannot coexist. However, Hutchinson's "rule" has

Figure 13.4 Niche changes. Three effects of competition on the niches of species in a community: (top) increased niche overlap, (middle) decreased niche breadth, and (bottom) increase in the range of resources used by some species. **Analyze:** What limits each of these mechanisms' ability to allow coexistence?

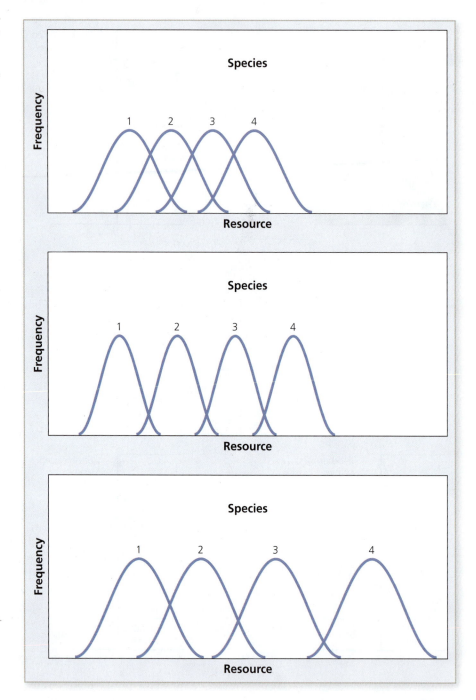

THINKING ABOUT ECOLOGY:

The analysis of the role of competition on niches depends on which species are included in the analysis—that is, the species included in the same guild. What would you predict would be the effect of including species too different in the guild? What would be the effect if not all the actual members of the guild were included in the analysis?

phylogenetic overdispersion The relationship among coexisting species when competition is the dominant structuring mechanism.

phylogenetic clustering The relationship among coexisting species under environmental filtering.

been abandoned. Too many exceptions have been found. And it is unlikely that a relationship that simple holds among the great diversity of communities and webs of interacting species. For example, consider the situation if more than one niche axis is limiting. Figure 13.5 shows that overlap in one dimension is significantly reduced when we consider overlap in a second dimension. Thus, we know that there is a limit to overlap, but the factors that determine that limit are complex and vary from one community to another.

Second, overlap may be possible in diverse communities if niche breadths are small. In other words, increased specialization among species permits more species to coexist without significant niche overlap (Figure 13.4 [middle]). Of course, there is a limit to specialization. If organisms use a portion of the resources dimension that is too narrow, the resource may not be sufficient to support the species.

Third, diverse communities may include species whose niches extend farther out on the resource axis, where they avoid competition by utilizing resources other species cannot (Figure 13.4 [bottom]). Examples of each of these three types of niche shift are known. Together they constitute the primary means by which competition structures communities. They provide boundaries by which competition limits the species that coexist. They also provide examples of how niches are arranged in different communities to permit species to coexist.

How Do Community Structure and Evolutionary History Interact?

We have seen thus far that community composition is the product of two ecological forces: environmental filtering and competitive interactions. A new synthesis of the study of phylogenies and community organization shows just how tightly linked evolution and community structure are (Johnson and Stinchcombe, 2007).

Environmental filtering and competition have opposing effects on the species present in a community (Figure 13.6). If environmental filtering dominates, coexisting species should be similar. That is, there should be clusters of species with similar traits that permit them to tolerate the abiotic conditions. This phenomenon is known as **phylogenetic clustering**: in two-dimensional niche space, species are found clustered in groups of species using similar resources. If, on the other hand, competition dominates, coexisting species should be less similar, a condition we call **phylogenetic overdispersion** (Figure 13.7). In this case, niches are widely separated. Both phylogenetic clustering and phylogenetic hyperdispersion are a result of natural selection operating on the species in a community.

Communities of New World monkeys exhibit phylogenetic hyperdispersion (Cooper et al., 2008). Phylogenetic distance, a measure of the evolutionary distance

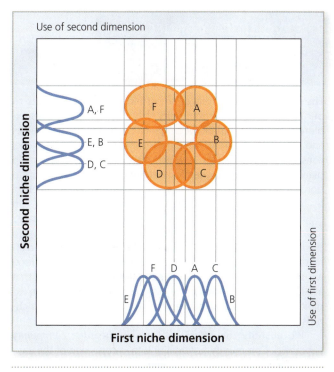

Figure 13.5 Niche overlap reduction. Significant niche overlap in any one dimension may be reduced when two-dimensional niches are considered. The graphical projections of niches that overlap in one dimension are less in two dimensions. **Analyze: Does this mean that neither resource is limiting?**

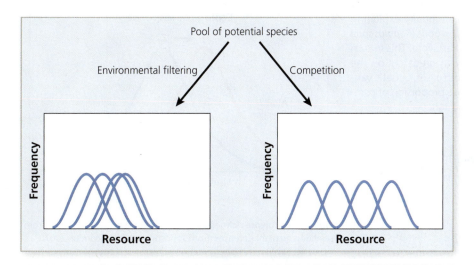

Figure 13.6 Niche arrangements. If a community is formed from a pool of source species, the niche arrangements will differ depending on whether environmental filtering or competition is the more important process (Lovette, 2006). **Analyze: Would you expect environmental filtering to be more important at high latitude or in the tropics?**

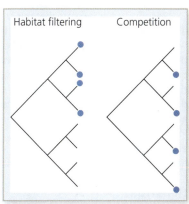

Figure 13.7 Environmental filtering vs. competition. The difference between environmental filtering and competition on the patterns of species across phylogenies. When environmental filtering is important, species with similar phylogenetic backgrounds have traits that enable them to persist in the community (left). When competition is important, phylogenetically and ecologically different species form the community (right). (From Cooper et al., 2008.)

THE EVOLUTION CONNECTION

Community Heritability and Structure

The phylogenetic analysis of coexisting species contributes to our understanding of community structure. These analyses explore the relationship between genetic variation within species and the organization of ecological communities. For example, consider a plant species in which there is genetic variation in plant defenses such as chemical deterrents or structural adaptations like spines (Figure 1). We expect that directional selection by herbivores should shift the bell curve of plant defenses. In Figure 1.Aa, the left-hand tail of the defense curve would have lower fitness, and over time, selection should shift the curve to the right. This single-species adaptive shift can have profound effects on the rest of the community. If the new defenses are effective, their impact will extend to the community of herbivores that attack the plant. Specifically, the number of herbivores able to exploit the plants will decline (Figure 1.Ab: graph of herbivore richness vs. plant defense).

We quantify this effect with a new measure: *community heritability*. Community heritability is the proportion of the variation in a characteristic of the community (such as herbivore diversity) that is explained by genetic variation in the population of a particular species (Johnson and Stinchcombe, 2007). In our example, the herbivore component of the community is determined in part by the genetic variation for herbivore deterrence in a plant.

Using the same conceptual basis as standard heritability of traits, community heritability is quantified by the equation

$$H^2_C = V_G/V_t$$

where V_G is the genetic variation for a trait in a population and V_t is the variation in some community-level trait.

Recent studies indicate that this kind of analysis helps explain differences in community structure. For example, Johnson and Agrawal (2005) showed that identification of the genotypes of the evening primrose (*Oenothera biennis*) explains 41 percent of the variation in the arthropod herbivore community exploiting that species. Similarly, the genetics of cottonwoods (*Populus spp.*) determines the density of its main insect herbivore. This in turn affects the density and community of avian predators on the insect community (Bailey et al., 2006). In both cases, genetic variation in a single species accounts for a large proportion of community variation: community heritability is high.

QUESTION:

How does community heritability differ from the concept of population heritability? What are the essential differences?

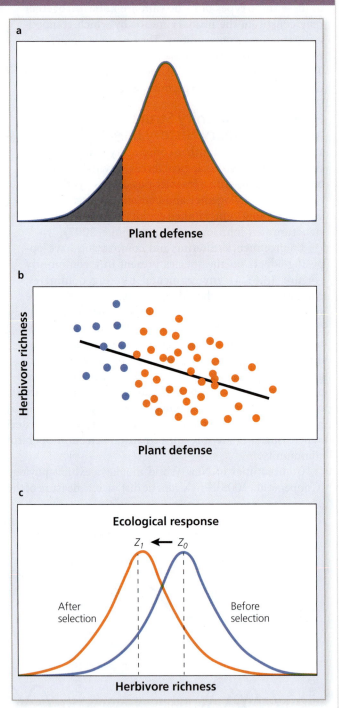

Figure 1 Herbivore richness. Shift in herbivore richness resulting from selection on plant defenses (from Johnson and Stinchcombe, 2007). **Analyze:** Explain this phenomenon in terms of a coevolutionary arms race.

between groups based on the branch points and branch lengths in a phylogeny, is greater in coexisting monkey communities than expected by chance. Only distantly related—and hence very different—monkeys occur together. Similarly, coexisting species of oaks in Florida are phylogenetically hyperdispersed—close relatives are less likely to occur in the same community (Figure 13.8). The niche separation of warblers described earlier (Figure 13.3) allows them to coexist. Analyses of the genetic relationships among these species show that the divergence in their feeding niches is also seen in their phylogeny (Lovette, 2006). That is, those species that coexist locally are only distantly related (Figure 13.9). The phylogeny of host species can also play a role in the structure of an herbivore community. Figure 13.10 shows that community similarity in herbivores decreases as a function of the phylogenetic distance among the species on which they graze (Weiblen et al., 2006).

What Is the Effect of Spatial Scale?

We define the community as a group of coexisting species inhabiting a particular area. However, that definition leaves a crucial component poorly defined—the size of the area and thus the spatial extent of the community. Just as it is difficult to precisely define the boundaries of a population, it is difficult to determine where one community ends and another begins. Movement of individuals and changes in the habitat make community boundaries imprecise and fluid. Still, we can address the spatial component of the community by analyzing how community processes change as a function of the physical distance between communities.

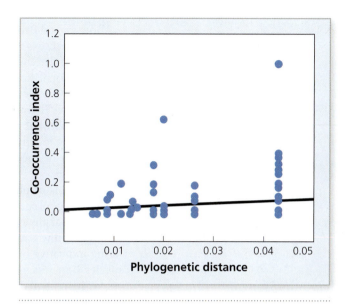

Figure 13.8 Phylogenetic hyperdispersal. The more phylogenetically distinct the oak species in a community are, the more likely they are to coexist (from Cavender-Bares et al., 2004). **Analyze:** Why does environmental filtering cluster species together?

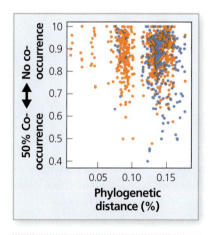

Figure 13.9 Co-occurrence of warbler species. The co-occurrence of warblers as a function of phylogenetic (genetic) distance between species. Each dot is the pairwise association between phylogenic distance and co-occurrence of two species. Blue dots represent species that forage in different microhabitats; orange dots are species that forage in similar microhabitats (from Lovette, 2006).
Analyze: Which species of warblers studied by MacArthur (Fig. 13.4) do you think are most closely related?

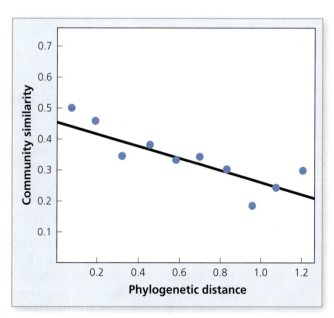

Figure 13.10 Phylogeny of host species. The similarity of herbivores in a community as a function of the phylogenetic distances among their host plants (from Weiblen et al., 2006). **Analyze:** Are the kinds of plant defenses likely to increase or decrease with plant phylogenetic distance?

Bowers and Brown's analysis of competition in desert rodents was based on many sites across a large geographic region. This allowed them to address the matter of scale. They did the same analysis described for local coexistence on a larger regional scale. They predicted that species co-occurrence patterns should show less evidence of competition when analyzed at the larger regional spatial scale. Their reasoning was that across this larger landscape, small differences in microhabitat or resource abundance might permit coexistence that is not possible locally. This is indeed what they found at the larger spatial scale: the pattern of species co-occurrence was random rather than structured by competition.

The kinds of phylogenetic analysis described earlier can also be performed across a range of spatial scales. For example, Gomez et al. (2010) examined changes in an index of phylogenetic community structure as a function of the spatial scale over which it was measured. On the local scale, they observed phylogenetic clustering indicative of environmental filtering. At the largest spatial scale, related species were less likely to co-occur. At the intermediate scale, there was neither convergence nor dispersion. This suggests that in this system, competition is more important at the largest spatial scale; environmental filtering is more important locally.

KEY CONCEPTS 13.2

- Communities structured by competition are affected by both competitive exclusion and character displacement.
- Competition in a community leads to increased niche overlap, decreased niche breadth, or an increase in the range of resources used. There are limits to all three possibilities.
- The phylogenetic history of the members of a community is consistent with the paradigm that competition structures the community. Coexisting species in communities structured by competition show phylogenetic hyperdispersion.

QUESTION:

What criteria might you use to define the spatial boundary of a community?

Figure 13.11 Giant kelp forests. Giant kelp forests affect competition among other species by limiting the cover of potentially competing algae.

 ## 13.3 What Other Ecological Factors Structure the Community?

A wealth of empirical studies and examples shows that competition is an important component of community organization. However, we know that other kinds of interactions also play a role in community structure. Because of the central role played by competition, many of these other interactions have indirect effects; that is, they alter the force of competition.

How Do Coevolutionary Interactions Affect Community Structure?

Facilitation is a process in which the presence of one species increases the probability that another occurs in the community. This phenomenon is exemplified in giant kelp forests and the rich, diverse communities associated with them (Figure 13.11). Competition for space on the substrate between algae and sessile invertebrates is intense. However, the density of kelp mediates this competition. The kelp forest casts shade on the substrate, which reduces the cover of the algae. This in turn releases the invertebrates from competition and increases the number of species present (Figure 13.12).

facilitation A mechanism in which the presence of one species increases the probability that another species is present.

keystone predator A predator whose presence is central to the structure of the community.

apparent competition When two similar species are both negatively affected by a shared predator or herbivore. The negative impact on both species mimics the effect of competition.

Mutualisms are known to affect community structure as well. For example, some ant species are facultative mutualists with wild cotton (*Gossypium thurberi*). In this mutualistic relationship, the ants consume nectar produced by the plant and in turn protect the plant from herbivores. However, the mutualism varies geographically with the number and abundance of herbivores: where herbivores are abundant, the ant provides protection and nectar production increases. Nectar production and ant protection decline where herbivores are rare. Recent studies show that this relationship determines the composition and diversity of the arthropod community associated with the plant (Rogers et al., 2010). The number and diversity of arthropods increase where nectar and ant protection are high.

Mullerian mimicry systems depend on the shared benefit of aposematic coloration. If these benefits are significant, mimicry systems permit species to coexist that competition might otherwise eliminate from the community. For example, in the neotropics, butterflies in the family Nymphalidae constitute a system of some 350 mimetic butterflies that contain potent alkaloids that provide protection from predators. Communities contain as many as 58 species, with considerable niche overlap among them. The summed mimicry benefit to each of these coexisting species outweighs the competitive disadvantages each faces from so many similar species (Elias et al., 2008).

Predation can also shift the competitive balance, facilitating the co-occurrence of species otherwise limited by competition. Classic experiments by Paine (1966) show the importance of predation by the starfish *Pisaster* on the rocky intertidal invertebrate community. When *Pisaster* was experimentally removed, barnacles rapidly increased. These were then replaced by mussels that gradually dominated the substrate, competitively excluding other invertebrates and algae. The result was an overall decrease in the number of species in the community. In this system, *Pisaster* is a **keystone predator**, a species whose impact is central to the structure of the community (Figure 13.13). When present, the starfish reduces the densities of other invertebrates such that none is able to outcompete the others, thus maintaining the diversity of the community.

Apparent competition is another indirect interaction that determines the structure of communities. Apparent competition occurs when two similar species are negatively impacted by the action of a shared predator or herbivore. We refer to this as "apparent competition" because, like true competition, an increase in one species leads to a decrease in another species. However, the mechanism does not involve competition. If one of the prey species increases in abundance, it causes an increase in the abundance of the shared predator. When the predator is more numerous, it drives down the abundance of the second prey species (Figure 13.14).

What Are the Roles of Resources and Disturbance?

The resource base for the community is central to the process of competition. Indeed, competition is the result of a *limiting* resource. Logically, any process that increases the resource base or decreases the degree to which a resource is limiting

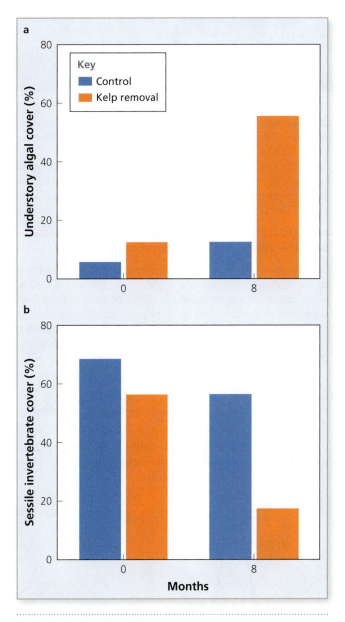

Figure 13.12 Kelp removal. The effect of kelp removal on (top) the understory alga cover and (bottom) sessile invertebrates (from Arkema et al., 2009). **Analyze: Why does invertebrate cover increase when kelp is present?**

Figure 13.13 Pisaster. *Pisaster* is a keystone predator in the rocky intertidal.

ON THE FRONTLINE

Niche Shifts After Competitive Release

We know that competition is an important structuring force in communities because it determines the niche relationships among coexisting species—the niche dimensions of each species reflect the pattern of competition it experiences.

Bolnick et al. (2010) examined this process in populations of the three-spined stickleback (*Gasterosteus aculeatus*) that compete with cutthroat trout (*Oncorhynchus clarkia*) and the prickly sculpin (*Cottus asper*).

HYPOTHESIS: Significant changes in stickleback niche occur when they are released from competition.

In testing this hypothesis, they made an important addition to the traditional treatment of niche parameters in the face of competition: they considered the niches of *individuals* as they contribute to the overall niche of the population. The *total niche width (TNW)* of the population was partitioned into the summed contributions of individuals. This allowed them to analyze the three ways that niches shift when released from competition.

Their hypothesis led to three testable possible consequences of competitive release:

PREDICTION 1: TNW increases due to increases in individual niche breadths.

PREDICTION 2: Individual niche breadths increase, but there is no change in TNW because of increased overlap.

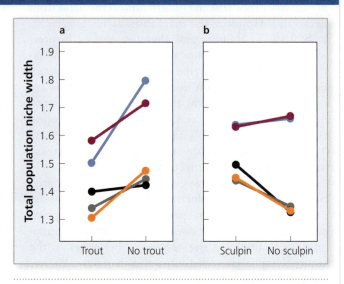

Figure 2 Release from competition. The effect of the release from competition (a) from trout and (b) from sculpins on the stickleback's total niche width (from Bolnick et al., 2010). **Analyze:** How do these results reject Predictions 1 and 2?

PREDICTION 3: TNW increases, but overlap among individual niches decreases.

(This requires a decrease in individual niche breadth—that is, less specialized niches.)

They tested these predictions in a set of enclosures constructed in Blackwater Lake, Vancouver Island, in which all three fish species occur together. Blocks of enclosures received four treatments on the sticklebacks: (1) competition from both trout and sculpins, (2) release from sculpins (trout present), (3) release from trout (sculpin present), and (4) total release (no competitors). Each experiment ran for 15 days, after which the sticklebacks were collected and their stomach contents analyzed for the identity and quantity of the prey items that constitute their dietary niche.

When released from trout competition, sticklebacks' TNW increased significantly. The researchers found that individuals were slightly more specialized; there was no expansion of the individual diet niches (Figure 3). These results are consistent with Prediction 3. In contrast, when released from sculpin competition, the sticklebacks' TNW decreased (Figure 2). In the process, individual specialization increased (Figure 3). Thus, release from sculpin competition matches Prediction 2.

It is not clear why the effect of ecological release differed for trout and sculpin. The researchers suspect that the difference is associated with the difference in competitive intensity between trout

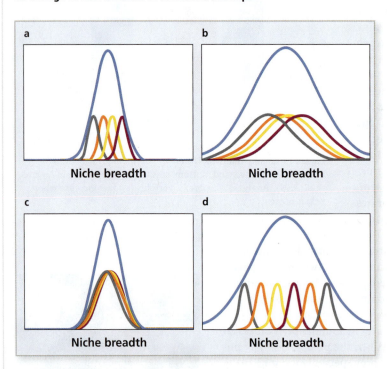

Figure 1 Niche shift. The ways niches might shift when released from competition (from Bolnick et al., 2010). **Analyze:** Explain how each of these niche shifts corresponds to the researchers' predictions.

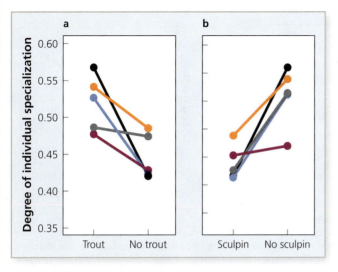

Figure 3 **Release from competition and niche specialization.** The effect of release from (a) trout and (b) sculpin competition on the niche specialization of sticklebacks (from Bolnick et al., 2010). **Analyze: Which prediction do these data support?**

and sculpins. Nevertheless, this study clearly demonstrates that there are multiple means by which a species' niche is affected by the competitors it faces in the community—and thus multiple effects of the composition of a community on the niche structures of the component species.

reduces the impact of competition. This is the mechanism by which the indirect interactions discussed in the previous section operate. The resource base may increase for other reasons as well. Thus far, we have focused on the competitive interactions in a community at equilibrium. As we know, however, many ecological systems are not in equilibrium. Change, due to disturbances like fire or storms, is common, and these may in turn affect limiting resources.

One potential effect of disturbance is to temporarily increase the resource base. For example, light is an important limiting resource in the understory of mature forest communities. If a fire or storm removes some or all of the shading canopy trees, light may no longer be limiting. The result is the development of a new community with many species despite significant overlap in their niches. In fact, such a community would appear to be a neutral community. Disturbance can also delay the process of competitive exclusion. The chance elimination of a dominant competitor can allow species to persist that otherwise would be competitively excluded.

In other cases, disturbance reduces the resource base, leading to increased competition and ultimately to a different community structure. The **core-satellite hypothesis** (Hanski, 1982) was devised to explain the structure of communities in which this occurs. Communities with a core-satellite structure have two key characteristics. First, the abundance of species is bimodal: most species are either rare or common (Figure 13.15). Second, the mean abundance of a species at a site and the number of sites it occupies are linearly related. Core species are defined as those found in many sites and at high abundance. Satellite species are found at low density in the few sites in which they are found. Moreover, satellite species periodically become locally extinct and then reappear.

The interactions of bison, fire, and tallgrass prairie plant communities illustrate the nature of nonequilibrium community dynamics and core-satellite structure. Tallgrass prairie is a community in which fire is a recurrent disturbance. The plants of the tallgrass prairie conform to a core-satellite structure (Collins and Glenn, 1991). The grasses tend to be core species; many of the forbs are satellite species. Fire stimulates the growth of the grasses. When fire is frequent, the grass biomass increases so much that grass leaves and stems reduce the light intensity

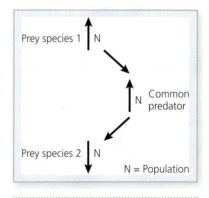

Figure 13.14 **Apparent competition.** Apparent competition between two prey species. If two prey items share a common predator, an increase in one prey item can cause an increase in the predator population. This may drive down the population of the second prey species just as competition among the two prey species might. **Analyze: Is it possible that apparent competition caused by predation can also be true competition between the prey species?**

core-satellite hypothesis Describes a community composed of (1) species that are either rare or common and (2) core species with high abundance and frequency, or satellite species found at low abundance and in few sites.

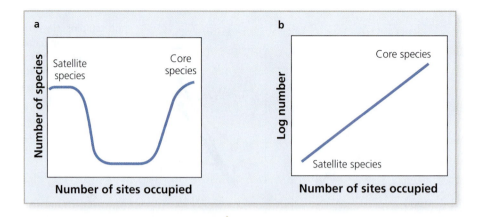

Figure 13.15 Core-satellite hypothesis. (a) The number of common and rare species in the community. (b) The log number of individuals as a function of the number of sites a species occupies (from Hanski, 1982). **Analyze: What characteristics would you predict for core species and for satellite species?**

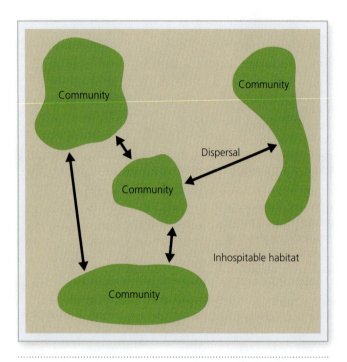

Figure 13.16 Metacommunity. A metacommunity: individual communities are found in a larger landscape. They are separated by inhospitable habitat and connected by dispersal. **Analyze: How does such a system include equilibrium and nonequilibrium processes?**

near the ground to levels that competitively exclude the forbs, thus reducing the number of species in the community. However, bison preferentially graze on the grasses. As they reduce the core grass species biomass, satellite forbs are competitively released.

These processes reiterate the importance of spatial scale. Like the concept of a population, the ecological community is a difficult entity to precisely delineate. Communities do not all have clear, discrete boundaries. Habitat and abiotic conditions shift, sometimes abruptly, sometimes gradually. Moreover, the movement of individuals connects distant communities. The result is a spatial structure known as a **metacommunity**: a set of similar communities that vary over space but are connected by dispersal (Figure 13.16). If there are discrete regions of habitat separated by unsuitable habitat, the communities in a metacommunity are more island-like. If the habitat is continuous but subtly shifts in quality, spatial differences in the metacommunity are much smaller. This concept of community spatial organization also incorporates equilibrium and nonequilibrium processes. Some of the components of the metacommunity may be at equilibrium, whereas others are in a state of flux following disturbance. Community structure shifts across this heterogeneous landscape.

THINKING ABOUT ECOLOGY:

How could you experimentally distinguish between delayed competitive exclusion and truly neutral community structure?

▪ **metacommunity** A set of communities with similar composition connected by dispersal.

KEY CONCEPTS 13.3

■ The effect of competition on community structure can be modified by other interactions among species, such as predation, facilitation, or apparent competition.

■ If resources are so abundant that they are not limiting, the role of competition structuring the community is reduced.

■ Many communities are not at equilibrium. In such systems, competition may be less likely to structure the community. Disturbance increases the levels of important resources, thus reducing the effect of competition. Disturbed communities often fit the neutral model of community structure.

■ Local communities occur in a broader landscape. The collection of communities connected by dispersal is known as a metacommunity. Not all member communities in a metacommunity are structured the same way.

QUESTION:

What is the relationship between competition, resources, and the equilibrium/non-equilibrium dichotomy in ecology?

Putting It All Together

The diverse community of wading birds in the Everglades is, at first glance, a paradox. If competition is so important, how can this guild contain so many species in such a small area? Careful observation of this community reveals at least two factors that permit them to coexist. First, there are subtle differences in their foraging patterns and behavior that reduce competition. These differences include niche dimensions such as microhabitat selection and prey size. Roseate spoonbills sift through the mud to filter out small invertebrates. The wood ibis forages in similar microhabitat, but it captures small fish that pass through its submersed bill. Great blue herons stand quietly watching water and occasionally grab larger fish prey. Little blue herons move quietly through the shallow water, one wing extended to cast a shadow and flush small fish from the vegetation that they grab with their bills. In other words, we can identify differences in the niches of these birds. Some niches are narrow and overlap little with other species. Other species shift their feeding or microhabitat niches into regions of the resource axis where they face less competition.

Second, the resource base is so large in some places and at some times that prey are not limiting. During the dry season, fish and invertebrates are highly concentrated in the few remaining pools of standing water. Although there is certainly overlap in the feeding niches of these birds, prey abundance is so great that competition is not the central structuring force in the community. The guild is a random sample of species for which environmental filtering is more important than competition. Disturbance, in the form of hurricanes and fire during the dry season, locally alters the prey base, the microhabitat types and availability, and the number of potential competitors. Periods of equilibrium, when competition is important, are interrupted by conditions when it is relaxed.

Summary

13.1 What Processes Determine Community Composition?
■ Environmental filtering determines which species can tolerate the local abiotic conditions.
■ In neutral models, communities are random assemblages of species that (a) can disperse to the region and (b) are tolerant of the physical conditions there.
■ Competition is an important deterministic structuring process.

13.2 How Does Competition Structure the Community?
■ Competition among groups of species (guilds) forms competitive hierarchies and networks.

■ Niches in competitively structured communities are altered in three ways:
  • Increased overlap
  • Decreased niche breadth (increased specialization)
  • Increased range of resources exploited
■ Communities in which environmental filtering is most important exhibit phylogenetic clustering. Communities in which competition is important exhibit phylogenetic hyperdispersion.
■ The structuring effect of competition decreases with increasing spatial scale.

13.3 What Other Ecological Factors Structure the Community?

- Coevolutionary interactions such as mutualism and mimicry may lead to facilitation, the presence of one species increasing the probability that another is present.
- Other coevolutionary interactions, especially predation, decrease the competitive dominance of the species' prey, thus reducing competitive exclusion.

- Apparent competition occurs when an increase in one species leads to an increase in a predator, which in turn drives down the population of another prey species.
- Resource levels are dynamic, and thus competitive structuring of communities varies over time and space. Disturbance is an important component of this variation.
- Spatial variation in resources and disturbance result in metacommunity structure.

Key Terms

Review Questions

1. Why is environmental filtering so crucial to community structure?

2. How are dispersal and environmental filtering related?

3. What is the relationship between nonequilibrium communities and neutral communities?

4. How does the resource base affect community structure?

5. Explain the relationship between phylogenetic clustering and environmental filtering.

6. Explain the relationship between phylogenetic overdispersion and competition.

7. How does competition modify the characteristics of niches in communities?

8. What is the effect of the spatial scale at which we study the community on its structure?

9. How do members of other guilds affect community structure?

10. What role does dispersal play in neutral communities?

Further Reading

Cody, M.L. 1974. *Competition and the structure of bird communities*. Princeton, NJ: Princeton University Press.

This classic monograph outlines the fundamental nature of communities structured by competition. It provides an interesting contrast with Ritchie (2010), showing the evolution of our thinking about community structure.

Emerson, B.C., and R.G. Gillespie. 2008. Phylogenetic analysis of community assembly and structure over space and time. *Trends in Ecology and Evolution* doi:10.1016/j .tree.2008.07.005.

This review paper introduces and summarizes the relationship of phylogeny and competition in community structure.

Ohgushi, T., et al. 2007. *Ecological communities: plant mediation in indirect interaction webs*. Cambridge: Cambridge University Press.

This book explores the noncompetitive interactions that determine plant community structure.

Ritchie, M.E. 2010. *Scale, heterogeneity, and the structure and diversity of ecological communities*. Princeton, NJ: Princeton University Press.

This book extends the discussion of competition as a structuring force in communities by exploring the importance of the spatial distribution of resources.

Ecological Succession

On Saturday, August 20, 1988, fires raged across Yellowstone National Park with a speed, intensity, and size unknown in modern times. The fires that had been burning for weeks reached a crescendo as the flames consumed 70,000 hectares (170,000 acres) *in that single day*. Fire suppression was hopeless. Driven by self-generating winds of 60 mph, the fire raced through the crowns of 300-year-old lodgepole pines. Embers carried miles beyond the fire front ignited new blazes. Only heroic efforts by the fire crews saved Old Faithful Lodge and other historic buildings. Weeks later, when the first snows of autumn finally extinguished the flames, 283,000 hectares (700,000 acres), or nearly one-third of the park, had burned. Visitors the following year drove past mile after mile of blackened forest.

Now, 26 years later, the park is green again. But the effects of the fires of 1988 are apparent everywhere. Yellowstone today is a complex mosaic of nearly impenetrable stands of young trees, open fire-cleared understory beneath mature lodgepoles, unburned forest, and forests of fire-killed snags. Elk and bison graze the lush grass in the burns. Mountain bluebirds nest in cavities in standing dead trees.

Yellowstone was set aside as the world's first national park in part because of the scenic beauty of mile after mile of green conifer forest. And for more than 100 years, visitors to Yellowstone could count on that stable landscape. Then, in one summer, that landscape was altered so radically that, decades later, the effects are still obvious. It often takes a major ecological event like this to focus our attention on the patterns of change in plant communities. Now, all these years later, we can appreciate how Yellowstone exemplifies the nature of change in natural habitats. And this seminal ecological event and the new vegetation that followed lead to the central question of this chapter: *How does the plant community respond to disturbance?*

14.1 # What Are the Effects of Disturbance?

The fires of 1988 resulted in catastrophic changes to the conifer forests of Yellowstone. They exemplify an **ecological disturbance**—a physical or biological factor that alters the structure and composition of the community. Although the Yellowstone fires clearly fit this definition, many other phenomena meet these criteria without causing the massive changes that occurred in Yellowstone. Disturbances range from large-scale cataclysmic events to minor, local incidents. Central to the definition is the impact on the community. In fact, one reason disturbance events are so important is that they initiate changes in the community. The sequence of changes in a community following a disturbance is known as **succession**. When disturbance of a plant community initiates succession, the new plant communities that arise are known as seral stages; the entire set of stages is referred to as a **sere**. Disturbance and succession are not restricted to plant communities. Other systems, such as coral reefs, experience disturbance and go through a successional sequence as a result.

Although some disturbances are large and dramatic, many smaller, less intense disturbance events occur, too. These also play a role in the dynamics of the community, especially if they occur frequently. In any forest, old trees eventually weaken and die. Wind throw of these individuals represents a local disturbance that creates a **light gap**, or **canopy gap** (Figure 14.1). Because susceptible old trees are distributed randomly and widely across a forest, gap formation is random within the community, leading to a heterogeneous mix of mature forest and gaps where young trees grow. The entire forest has not been disturbed by a regional event like fire. Nevertheless, the community is a mosaic of smaller disturbances.

One important aspect of disturbance is its effect on soil and soil nutrients. Some major disturbance events, such as volcanic eruptions or glaciers, not only destroy the vegetation but also remove or bury the soil. As a glacier advances, it scrapes the existing soil away, exposing the bedrock below (Figure 14.2). When the Mount St. Helens volcano erupted, it deposited many meters of pulverized rock and ash on top of the soil. After events like these, the new substrate lacks most of the life-sustaining characteristics of soil, especially water-holding capacity, organic material, and the organisms important in nutrient cycling. Some disturbances, such as lava flows, create particularly difficult conditions for plants. Hot, black lava absorbs and holds heat but does not hold water. In contrast, a crown fire may have devastating effects on the trees, but when the fire passes, the soil, including its key components and functions, remains.

Not surprisingly, the pattern and rate of succession depends on whether or not soil and nutrients remain in the system. **Primary succession** is a successional sequence following a disturbance so severe that the new conditions are nearly abiotic. Often the soil is lost or destroyed. The slow process of soil formation and accumulation of organic nutrients is an important part of primary succession. If the disturbance is less severe, such that functional soil remains, the pattern of vegetation change is known as **secondary succession**. Although there is considerable variation in the rate of change in secondary succession, it generally proceeds faster than primary succession.

Biological factors may contribute directly or indirectly to disturbance as well. In the montane conifer forests of the western United States, insects that attack conifers create a heterogeneous mix of healthy untouched trees and stands with significant mortality from insects. Among the most important such insects is the mountain pine bark beetle (*Dendroctonus ponderosae*; Figure 14.3).

ecological disturbance A physical or biological factor that alters the structure and species composition of the community.

succession The sequence of changes in a community following disturbance.

sere The sequence of stages or community types that occur during succession.

light gap (canopy gap) An opening in the tree canopy in a forest.

Figure 14.1 Light gap. When a tree dies and falls, it opens a light gap in the forest canopy.

primary succession The pattern of succession that occurs following a major disturbance that reverts the environment to nearly abiotic conditions.

Figure 14.2 Glacier. An advancing glacier is a significant disturbance that removes the vegetation and soil.

Trees killed by beetles are easily wind thrown or burned and thus contribute to the disturbance regime. Some biotic disturbance factors operate in conjunction with physical factors. For example, in northern and mountain conifer forest systems, beavers kill trees in the course of their foraging and dam building. When they dam a stream, the impounded water floods the forest above the dam, killing some tree species and encouraging the growth of other plant species, especially willows and alder.

How Frequent Is Disturbance?

One indication that disturbance is a relatively common event is that so many plants exhibit adaptations to the changes in the environment disturbances bring about. For example, the cones of some conifer species, such as lodgepole pine (*Pinus contorta*) and jack pine (*Pinus banksiana*), are serotinous—they do not release their seeds until they experience the high temperatures of a forest fire.

Moreover, the successional sequence following disturbance may determine the timing of the next disturbance. Figure 14.4 shows the area burned in subalpine forests in Wyoming as a function of fire interval. Note that the largest fires tend to occur when the interval between fires has been greatest. One reason for this phenomenon is the susceptibility of different successional stages to fire. The forests of Yellowstone illustrate this concept. The dominant tree in much of Yellowstone is the lodgepole pine (*Pinus contorta*). We recognize four successional stages in lodgepole pine forest following fire (Figure 14.5). The early stages (LP0 and LP1) are not very flammable, because there is little fuel on the forest floor. The large trunks and high branches characteristic of LP1 make it difficult for fire to reach the canopy. In stages LP2 and LP3, the lodgepoles are reaching an age at which they are vulnerable to wind throw. Moreover, they are vulnerable to pine bark beetles that kill trees, resulting in even more fuel. In these later stages, the probability that a fire can begin and carry into the crown increases significantly. In 1988, much of Yellowstone was in LP2 and LP3, the stages most conducive to fire. When a hot, dry summer arrived, a few lightning strikes ignited a major conflagration. (Some of the fires that summer were started by human activity as well. By late summer, these fires and lightning-caused fires had merged.)

Researchers can use the presence of fire scars in the growth rings of very old trees to determine the dates and sizes of previous fires. This work shows that a large proportion of what is now Yellowstone National Park burned in the late 1600s. In other words, a fire much like that in 1988 occurred approximately 300 years earlier. The return time of fire in this system closely matches the time required for the lodgepole pine successional sequence (Romme and Despain, 1989).

The frequency of disturbance events is important to the nature and pattern of the successional changes that occur in the vegetation. For example, the average return time for a hurricane in the Caribbean Basin is approximately 60 years. Even though some tropical forest systems there recover relatively quickly, this interval is in fact short relative to the time it takes the vegetation to fully return to the prehurricane state. Consequently, these communities are in the process of succession much of the time.

As long-term studies accumulate for more and more plant communities, we see that even systems that appear to be quite stable are in fact disturbed regularly. For example, the lowland tropical forests of the interior Amazon Basin are

■ **secondary succession** The pattern of succession that occurs when following a disturbance that leaves the soil and some parts of the community in place.

Figure 14.3 Pine bark beetles. Pine bark beetles (*Dendroctonus ponderosae*) cause significant mortality in conifer trees.

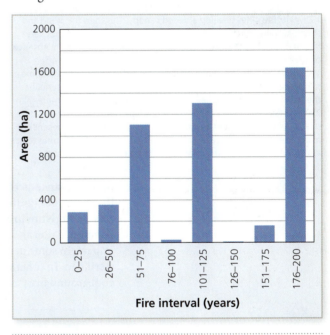

Figure 14.4 Fire interval. The effect of fire interval on the area burned in subalpine forests of southeastern Wyoming (from Kipfmueller and Baker, 2000). **Analyze:** What factors cause fires to be larger when the interval between fires is greater?

Figure 14.5 Successional stages. Lodgepole pine succession in Yellowstone National Park. (a) Stage LP0. In the first 50 years after a fire, extremely dense even-aged stands of pines develop. There is little fuel, and the forest is not very flammable. (b) Stage LP1. Between 50 and 150 years, the pines develop a closed canopy. This thins the stand and shade-kills the lower branches on the trees. This stage is still not very flammable due to the low fuel levels on the forest floor and the absence of lower branches to carry a fire into the crown. (c) Stage LP2. After about 150 years, dense shade inhibits new lodgepole seedlings. Some mature trees die, and canopy gaps begin to open. As fuel accumulates, the stand becomes more flammable. (d) Stage LP3. In stands more than 250 years since fire, the canopy has many light gaps. Young and mature subalpine fire, lodgepole pines, and Engelmann spruce are present in many age classes and sizes. Fuel is abundant, as well as small trees to act as conduits for fire into the crown. This stage is highly flammable.

considered to be among the most stable systems on earth. They are not vulnerable to many of the common disturbance forces such as fire, glaciation, landslides, or hurricanes. The many ancient trees in these communities are evidence of a stable forest. However, we now understand that flooding and wind are important disturbance factors. Severe wind and rain storms defoliate and topple canopy trees in Brazilian rainforest. Although large-scale blowdowns are relatively infrequent, over time most of the rainforest is eventually affected. Flooding is an important disturbance factor in lowland rainforest in Peru. More than 26 percent of the forest has experienced significant erosion or received newly deposited soil from flooding (Salo et al., 1986).

Coral reefs are also superficially stable communities. However, they, too, are subject to frequent disturbance. Figure 14.6 shows an 89-year history of cyclones in the region of the Great Barrier Reef in Australia. Many of these passed over and impacted the reef, often multiple times, during this 89-year period.

How Large Are Disturbances?

As with frequency, the size of disturbance varies enormously depending on the habitat and the nature of the disturbance. Table 14.1 shows the size distribution of the patches of succession initiated by disturbance in mixed hardwood/conifer forest in Minnesota. Note that within a particular forest type, fire appears to affect a larger area on average than wind.

One of the difficulties in assessing the size distribution of disturbance is that within a region affected by a specific disturbance event, there may be significant variation in its intensity. The Yellowstone fires illustrate this point. Note in Figure 14.7 that the effect of fire was not homogeneous. The complex patterns in these burns were the result of varying fire intensity and impact. In some parts of this photo, all the trees were killed and even the soil was sterilized. In others, the trees were damaged but recovered. In still others, only a cool ground fire passed through.

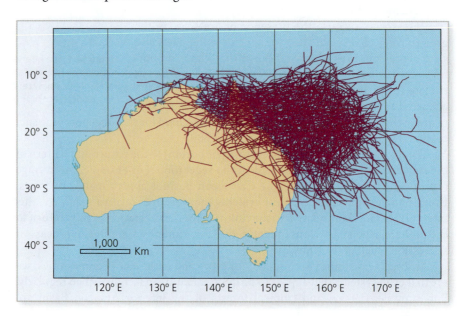

Figure 14.6 Tropical cyclones. The paths of tropical cyclones over the region of the Great Barrier Reef in Australia from 1910 to 1999 (from Poutinen, 2004). **Analyze:** How might a cyclone disturb a coral reef, most of which is underwater?

TABLE 14.1 The sizes of fire and wind disturbances in conifer in Minnesota

Subsection	FIRE		WIND	
	Mean	Maximum	Mean	Maximum
Border Lakes	235	8148	30	1165
Tamarack Lowlands	226	4004	22	398
Chippewa Plains	220	6306	119	2260
Nashwauk Uplands	415	2826	207	4151
Pine Moraines	89	4302	30	2395
North Shore Highlands	54	2671	13	88
Laurentian-Toimi Uplands	75	634	17	146
St. Louis Moraines	23	324	30	203

Source: (White and Host, 2008)

How Does the Intensity of Disturbance Vary?

The spatial variation in disturbance documented earlier is a result of local variation in intensity. This point is illustrated by a comparison of three spatially large disturbance events: the Yellowstone fires, the eruption of Mount St. Helens in 1980, and Hurricane Hugo in Puerto Rico in 1989 (Table 14.2). Mount St. Helens erupted explosively with a blast of superheated air and rock debris. The impact varied with proximity to the mountain. Immediately adjacent to the blast zone, winds and rock debris at temperatures as high as $450°C$ sterilized the ground. Further away, scorching winds and debris at temperatures between $40°C$ and $180°C$ blew down trees over many square kilometers. Elsewhere, mud flows buried vegetation under up to 150 m of rock and mud. Hurricane Hugo struck the montane tropical forest of the northeast corner of Puerto Rico as a Category 4 hurricane with winds of 166 kph. The storm dropped some 339 mm of rain over a few hours. Canopy trees were stripped of leaves. In the hours and days after the storm passed, landslides further impacted the vegetation on the steep mountain slopes.

Each of these events had a significant effect on the vegetation. However, note in Table 14.2 how differently each affected the local community. Recovery varied significantly among the three systems. More importantly, *within* each system the impact on the vegetation varied significantly and led to a mosaic pattern of recovery.

Figure 14.7 Yellowstone fires. The 1988 Yellowstone fires did not affect all areas equally. The intensity of fire was heterogeneous.

TABLE 14.2 Effects of three major disturbance types on the vegetation and succession

DISTURBANCE	CROWN FIRE	HURRICANE	VOLCANO
Variation in impact	High: Intense crown fire to cool ground fire	Low (within boundaries of storm; higher at distance from eye)	Highly variable as function of distance from blast zone
Delayed mortality	No	Yes	No
Mode of recovery	Dispersal	Resprouting + seeds	Dispersal
Recovery rate	Rapid	Very rapid	Slow but variable
Plant species dominance	Changes in herbaceous species; not trees	Shifts due to pulse of early successional species	Shifts with succession
Regional diversity	Increases	Increases	Increases
Local diversity	Initially decreases	Little change	Initially decreases

Source: (Turner et al., 1997)

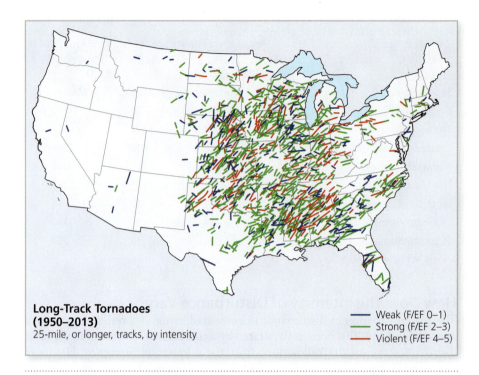

Long-Track Tornadoes (1950–2013)
25-mile, or longer, tracks, by intensity

— Weak (F/EF 0–1)
— Strong (F/EF 2–3)
— Violent (F/EF 4–5)

Figure 14.8 Tornadoes. The paths of tornadoes in the United States. Although each tornado affects a relatively small area, tornadoes are sufficiently frequent that many areas are affected (data from http://www.ustornadoes.com). **Analyze: What patterns can you identify in these tracks?**

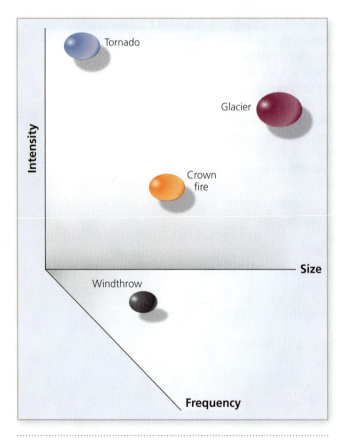

Figure 14.9 Characteristics of disturbance. Disturbances are characterized along three key axes: size, intensity, and frequency. **Analyze: Where would disturbances that result in primary succession fall in this graph?**

For most all of the important disturbance factors, there is a spectrum of intensity that varies regionally. Figure 14.8 shows the frequency of violent wind events in the United States. Clearly, most are concentrated in the Great Plains. The most violent of these wind events, tornadoes, are extremely intense but local in their impact.

We have seen that disturbance events vary in frequency, size, and intensity. Figure 14.9 summarizes the key characteristics of disturbance and their potential effects on the vegetation along three axes: frequency, spatial scale, and intensity. For example, a hurricane is very intense and may be spatially very large. Still, compared to other, more minor events, like thunderstorms, they are relatively infrequent. In contrast, lightning strikes and wind from a thunderstorm affect any given place more frequently. The impact is generally less severe, but the vegetation will have less time to recover before the next event.

Not surprisingly, the response of the community depends on the nature of the disturbance. The response varies in two key ways: (1) the degree to which the community withstands disturbance—that is, its resistance—and (2) the speed with which it can recover from disturbance, or its resilience. For example, the Everglades are adapted to hurricane disturbance. The marsh vegetation tolerates the flooding, the storm surge, and the high rainfall of a hurricane. Sawgrass (*Cladium jamaicense*), the dominant plant in this system, is herbaceous, and its flexibility decreases the damage from hurricane-force winds. Consequently, the Everglades have both high resistance and resilience to hurricanes. In contrast, the pine forests of this region have low tolerance and resilience, because the trees cannot tolerate flooding and their woody trunks and branches are more vulnerable to breakage in high winds.

THINKING ABOUT ECOLOGY:

What relationship would you expect between resistance and the frequency of disturbance?

KEY CONCEPTS 14.1

- Primary succession occurs following disturbances that revert the ecosystem to abiotic conditions; secondary succession follows less severe disturbances.
- Disturbances vary according to intensity, spatial scale, and frequency.
- Although many intense, widespread disturbances are relatively rare, most communities are affected by frequent disturbance of some kind.
- Communities are typically a mosaic of recently disturbed and undisturbed sites.

QUESTION:

What factors determine the relative abundance of disturbed and undisturbed sites in a community?

14.2 What Are the Patterns of Vegetation Change Following Disturbance?

Disturbance often reduces, sometimes drastically, the number of species in the community. As succession proceeds, plants colonize the site according to their ability to tolerate the physical and biological conditions there and as a function of their dispersal ability. Over time, early colonists die out and are replaced by other species. Conditions immediately after disturbance may be tolerable for relatively few species compared to later in succession. As a result, species diversity is often low immediately after disturbance, even in secondary succession, where the conditions are not so harsh. There is a general trend toward increasing species diversity as succession proceeds, at least until late in the process, when the number of species either levels off or declines somewhat.

THINKING ABOUT ECOLOGY:

Are introduced species more likely to be found early or late in succession?

How Does the Vegetation Change During Primary Succession?

Primary succession is typically a slow process because the disturbance is so severe. Few plant species can survive in the difficult new conditions. Two classic systems, the Lake Michigan dunes and Glacier Bay, Alaska, illustrate the point. Each system represents a **chronosequence**, a series of communities arrayed along a linear sequence of increasing age (time since disturbance).

The basin that now contains modern Lake Michigan was formed by the advance of the Wisconsin glacier, which dug out the lake basin and whose meltwater initially filled the lake. When the glacier retreated, an enormous lake, Lake Chicago, extended considerably farther south than the current Lake Michigan. The river draining this lake flowed to the south and cut into soft glacial deposits. As the waters drained and the lake dropped, the river cut deeper. Occasionally, it would strike harder material that took longer to cut through. When this happened, the lake drained more slowly and the shoreline remained stable for some years (Figure 14.10). Later, the river again cut through into softer material and the lake level dropped rapidly. Eventually, the drainage shifted to the eastern Great Lakes and the Saint Lawrence River—the current topography. The ancient drainage pattern left a series of sand benches at the south end of Lake Michigan, each of which represents a period when the shoreline was stable for some time.

chronosequence A series of communities arrayed linearly from young to old and increasing distance from recent or current disturbance.

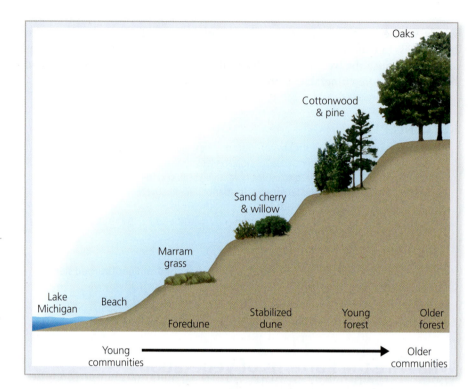

Figure 14.10 Lake Michigan dunes. The chronosequence at the Lake Michigan dunes. As the lake level dropped, new sand substrate was exposed on which primary succession occurred. Benches at increasing distance from the lakeshore are older (Cowles, 1901). **Analyze: What fundamental assumption underlies the concept of a chronosequence showing the stages of dune succession?**

■ **pioneer species** The initial colonists following a disturbance.

Figure 14.11 Glacier Bay. The glaciers in Glacier Bay, Alaska, receded rapidly from 1978 (a) to 1997 (b), exposing new substrate on which primary succession occurs.

The communities on these benches represent a chronosequence: the set of communities of increasing age at increasing distance from the lakeshore.

The part of the current beach that was most recently exposed represents the abiotic conditions immediately after the disturbance—that is, the youngest plant community. This is a difficult environment for plants. The sand contains little organic matter; there is no true soil. In addition, sand does not hold water well, and plants struggle to obtain enough moisture. Finally, the exposed dunes are unstable. Windblown sand buries newly colonizing plants. Consequently, the early stages of succession develop slowly. The set of plants able to tolerate the harsh conditions early in primary succession are known as **pioneer species**. For example, marram grass (*Amophila breviligulata*) pioneers in the dry, poor, shifting substrate early in dune succession. This species produces adventitious roots, roots that develop from the stem, when buried by shifting sand. As marram grass colonizes, lives, and dies, it gradually stabilizes the sand and adds organic matter to the developing soil.

Eventually, the harsh conditions of early primary succession are improved to the point that other species can colonize. The result is a successional sequence of plants in the dune complex, each suite of plants at a different distance from the beach. H. C. Cowles (1901) first envisioned the plant communities at greater distance from the shore as a chronosequence. His great insight was that each community had passed through the same stages as the younger communities closer to the shore. As we shall see later in this chapter, that interpretation is not quite correct. Nevertheless, the dune system does illustrate a key feature of primary succession—the difficult abiotic conditions immediately after disturbance.

Glacier Bay, Alaska, also represents a chronosequence following the rapid retreat of the glaciers in this portion of Southeast Alaska. As the glaciers retreated, they exposed new substrate for plants to colonize (Figure 14.11). As in the case of the dunes, there is a sequence of communities found along the length of Glacier Bay in regions that were exposed at different times. Near the mouth of Glacier Bay, where the glaciers have been absent for the longest period, we find a spruce-hemlock forest (Figure 14.12).

The substrate there is granite, which, like the sand in the dune system, is not particularly conducive to plant growth. There is no soil or organic matter in this hard, bare rock to nourish plants, let alone allow them to root. And water simply runs off, creating xeric conditions. Here, too, early colonists must be tolerant of these difficult conditions and modify the substrate in ways that permit other species to colonize. Among the earliest colonists are lichens that can cling to the hard, smooth granite. Lichens begin the slow process of breaking down the rock and forming a thin soil. As other plants colonize, soil formation accelerates. As in many other primary successional series, the developing soil in the Glacier Bay sere is low in nitrogen. Pioneer species there include plants with symbiotic nitrogen-fixing bacteria that add organic nitrogen to the soil.

How Does the Vegetation Change During Secondary Succession?

Less intense disturbances such as fire leave soil relatively intact, which greatly accelerates the rate of succession. Many plants can colonize such a site, although some are better adapted to recent disturbance and thus are more successful. We have already examined one secondary successional sequence—the development of a lodgepole pine forest following fire.

Secondary successional sequences known as **old field succession** occur following the abandonment of agricultural fields. In the eastern United States, where deciduous forest is the natural community type, old field succession eventually leads to the development of a forest. There are many variants on the sequence, but the general pattern is depicted in Figure 14.13.

Figure 14.12 Spruce-hemlock forest. Spruce-hemlock forest is the climax vegetation in the oldest sites in Glacier Bay.

old field succession The sere following abandonment of an agricultural field.

0–5 years	5–15 years	15–30 years	30–60 years	>60 years
Herbaceous annuals	Herbaceous perennials	Shrubs	Softwood trees	Hardwood trees
Horseweed ragweed	Asters Goldenrod	Dogwood Blackberry	Red maple Sassafrass Tulip	Sugar maple Beech Oaks

Figure 14.13 Old field succession. Old field succession follows a set pattern of change from herbaceous annuals to hardwood forest. In each locale, the species comprising the stages and the relative length of the stages vary.

The successional sequences in regions where forest occurs share a number of characteristics. The conditions immediately following disturbance differ markedly from those in a mature forest. Far more light is available early in secondary succession than in a mature forest, where light is limiting. Consequently, the characteristics of species that occur early and late in the sequence differ

Figure 14.14 *Cecropia*. *Cecropia* is an important early successional species in rainforest. It can exploit high light conditions and grows extremely rapidly.

■ **climax community** The final stage of succession that can replace itself if no new disturbance occurs.

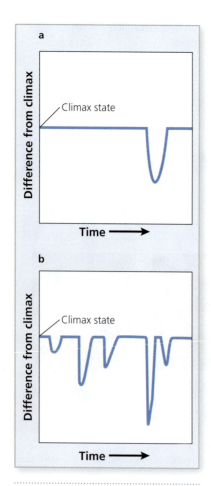

Figure 14.15 Clementsian paradigm. (a) The view of disturbance in the Clementsian paradigm: disturbance is rare; equilibrium is the norm. (b) The current view of disturbance: disturbance is frequent and many communities are undergoing succession much of the time. **Analyze:** How do the disturbance parameters, size, and intensity relate to these patterns?

markedly. Recall Grime's plant classification from Chapter 10. Competitive plants occur in stable communities. Their life history traits, such as iteroparity and large seeds with high energy stores for the embryo, are adaptations to a competitive environment. The life history strategy of ruderals, species that occupy disturbed sites, includes features such as rapid growth, short life span, semelparity, and long dormant stages.

Thus, in a recently disturbed community, selection favors those species with the ability to colonize quickly and exploit the high light environment. We have seen the advantage that serotinous cones play in lodgepole pine colonization following fire—they ensure that this species colonizes before its potential competitors are able to. Fireweed also exemplifies this ability. Fireweed is a poor competitor; it is restricted to the burned site, where there is no competition. It cannot invade the unburned meadow. Fireweed produces large numbers of wind-borne seeds that colonize recent burns. Each plant community appears to have its own set of species adept at colonizing following disturbance. *Cecropia spp.* is an important early successional species in tropical rainforest (Figure 14.14). This species colonizes rapidly and grows phenomenally fast in the open sun, up to several meters per year. Recent light gaps are often pure stands of *Cecropia*.

Species that occur later in succession are adapted to a competitive environment. They tend not to disperse their seeds widely. Rather, they produce seeds or propagules that can survive and reproduce in competition with other individuals. Growth may be slower because light becomes limiting as the forest matures. In deciduous forest, sugar maple (*Acer saccharum*) and American beech (*Fagus grandifolia*) produce relatively small numbers of large, well-provisioned seeds. Beech also reproduces vegetatively from the roots of mature trees. These young individuals have access to resources from the established tree that improve their competitive ability.

What Is the Nature of the Final Stage of Succession?

In each of the examples of primary and secondary succession we have discussed, the sequence ends with a specific community. The dune sequence leads to an oak-hickory forest. At Glacier Bay, the oldest community is the spruce-hemlock conifer forest. Old field succession leads to hardwood deciduous forest. Until that final plant community, each successional stage is gradually replaced with another. The final stage of succession, the community that replaces itself, is referred to as the **climax community**. This community persists until the next disturbance restarts the successional process.

Our understanding of the successional process and particularly this end point has changed significantly in the past few decades. For many years, plant succession was dominated by the **Clementsian paradigm**. Recall that Clements described the major plant communities, or biomes, of North America. He believed that each biome is the climax plant community adapted to the particular soil and climate in its region. According to the Clementsian paradigm, disturbance is infrequent; stability of the climax is the norm. In fact, Clements believed that mature biomes are resistant to change, but if disturbance does occur, succession leads to a series of plant communities that gradually converge on the biome—that is, the climatically determined climax. In fact, the root of the term *climax* in this context is *climate*. Clements's paradigm emphasized the relative rarity of disturbance and a predictable, repeatable chronosequence converging on a single climax.

Two important sets of observations have gradually refuted the Clementsian paradigm in its purest form. First, as documented earlier, disturbance is by no means a rare event (Figure 14.15). The frequency of disturbance events, including those that scale high on the intensity and size axes in Figure 14.9, changed our view of succession. Moreover, most plant communities are undergoing successional change much of the time; few are at equilibrium for long.

Second, the classic chronosequences, such as the Lake Michigan dunes and Glacier Bay, Alaska, have been found to be far less straightforward than originally

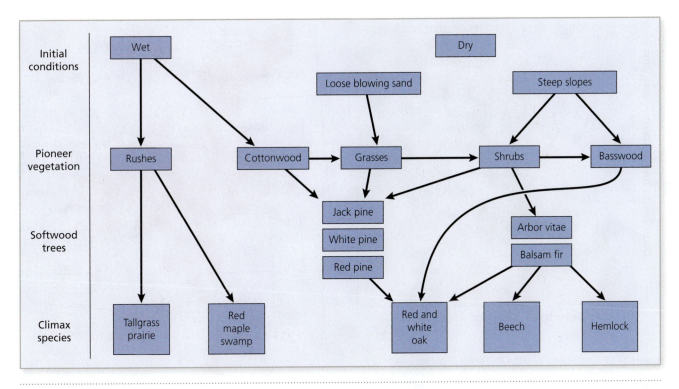

Figure 14.16 Chronosequences. Variation in the pattern of primary succession at the Lake Michigan dunes. Rather than a single, linear sequence that converges on a single climax vegetation, there are many pathways. Chance and the local starting conditions determine which sequence occurs (from Olson, 1958). **Analyze:** Do you think chance is more important in the initial conditions or later in succession?

understood (Johnson and Miyanishi, 2008). Specifically, the relationship between space (distance along a transect) and time (since disturbance) is not as clear as originally imagined. We now understand that the Glacier Bay chronosequence is not a simple, linear change represented by distance from the receding glacier. For example, studies have shown that some sites do not pass through an alder stage (Fastie, 1995). The cottonwood stage is not actually a consistent part of the sequence, either. Rather, it is peculiar to dry sites close to seed sources on nearby unglaciated slopes. Central to the dunes' chronosequence is the importance of the grass and shrub stage stabilizing the dune and developing soil so that tree species can eventually colonize. The common occurrence of cottonwoods on unstable dunes refutes this concept. More detailed studies of this system also reveal not just a single chronosequence but many, each of which is dependent on the particular starting conditions (Figure 14.16).

Many factors, including human impacts on the environment, change the nature and pattern of succession. For example, on the eastern shore of Hudson Bay, rapid warming and post–ice age glacial retreat have led to a rapid shift in tree line closer to shore. White spruce (*Picea glauca*) has moved shoreward in the last 50 years. Soil formation has increased underneath the advancing trees as well (Payette and Laliberté, 2008). While this advance is clearly successional following glacial retreat, it has been accelerated by recent climate change.

These observations imply that the process of succession is much more complex, unpredictable, and site-specific than we once believed. Community stability varies with the size of the disturbance event relative to the size of the community (disturbance size/community size) and with the frequency of the disturbance relative to the time required for recovery (disturbance interval/recovery interval; Figure 14.17). Note that when the disturbance is relatively large and the recovery interval is small, the community is unstable. In contrast, for relatively small disturbances from which the community recovers rapidly, the community has high stability, perhaps even a long period of equilibrium.

Clementsian paradigm The concept that mature, climate-determined plant communities resist change. Most are at equilibrium most of the time; disturbance is rare.

THINKING ABOUT ECOLOGY:

What characteristics of early successional species would you expect to contribute to the role of chance in chronosequences?

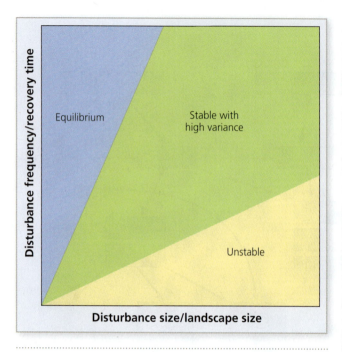

Figure 14.17 Community stability. The effect of disturbance frequency and size relative to recovery time and landscape size affects the stability of a plant community (adapted from Perry, 2002). **Analyze: Why does equilibrium occur in the triangle on the left of this graph?**

Figure 14.18 Prairie peninsula. The prairie peninsula is an extension of prairie east into the deciduous forest biome in Illinois and Indiana caused by recurring fires. This region has sufficient rainfall to support trees. **Analyze: Is it likely or unlikely that the eastern boundary of the prairie was stable?**

We see communities within the rather narrow time window of our own experience or recorded history. We may thus be led to believe a community is at equilibrium when in fact, in the context of a longer time frame, it is not. Two examples illustrate this important point. The very existence of tallgrass prairie in the eastern Great Plains and the Midwest is dependent on historical disturbance patterns. Figure 14.18 shows the distribution of prairie in the eastern portion of this biome. Note the *prairie peninsula*, an extension of prairie into northeastern Illinois and northwestern Indiana. This region lies in a climatic regime with sufficient rainfall to support deciduous forest. However, prairie penetrated so far east due to fires that pushed east into the forest. In most prairie plants, the growing point, or meristem, is located beneath the soil surface, where it is protected from fire. This is an adaptation to the fires common in grasslands. Trees do not have this adaptation and so succumb easily to fire. Much of the eastern portion of the North American prairie would have been forested had it not been for the recurring disturbance of fire. A community that persists due to recurrent disturbance is known as a **disclimax**.

In other communities, the return time for disturbance is so long that other ecological processes confound the analysis of successional dynamics. For example, in the northern hardwood forests south of Lake Superior, American beech

disclimax A community that persists and appears to be a climax because it replaces itself due to recurring disturbance.

Equilibrium and Climax in Spruce-Fir Forests

Clements's view of the climax was based on the idea that communities are at equilibrium most of the time. According to this view, the observed community composition should represent a climax state. This idea was tested in spruce-fir forests in Colorado by Applet et al. (1988). In this community, fir is usually numerically superior to spruce. The researchers addressed the fundamental question: Does the present composition of this community type represent an equilibrium (climax) condition?

The researchers recognized that there are at least two possible scenarios that could lead to the present community structure (Figure 1).

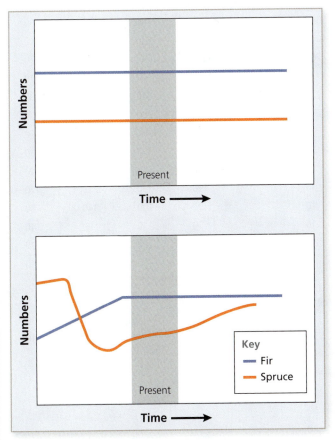

Figure 1 Spruce and fir. Two hypotheses for the relationship between spruce (solid line) and fir (dashed line) in Colorado: (top) the current forests represent an equilibrium (climax) system, and (bottom) the current relationship is a result of ongoing succession following major fire many years ago (from Applet et al., 1988). **Analyze: If this were an equilibrium system, what factors might cause fir to be numerically dominant over spruce?**

HYPOTHESIS 1: The present numbers of spruce and fir represent a long-term equilibrium condition.

Hypothesis 1 suggests that the relative numbers of spruce and fir do not change over time once equilibrium has been

reached. When a fire occurs, both spruce and fir colonize, grow, and eventually achieve stability.

PREDICTION: Both spruce and fir colonize after fire in the same relative numbers as occur at climax. Their relative numbers are constant thereafter.

However, it is also possible that these forests are not in equilibrium.

HYPOTHESIS 2: The relative numbers of spruce and fir change over the long-term successional process.

PREDICTION: The populations of spruce and fir change over time. Their numbers today represent a specific stage in a long-term sequence.

The researchers examined the age structure of spruce-fir forests that differed in the time since the last fire. They aged living trees with an increment borer, a device that removes a small-diameter core from the tree. The growth rings of the tree are visible in the core, which allows the investigator to assign a precise age to each tree. Applet and his coworkers cored all the trees in stands of spruce and fir that had developed for 175 years, 275 years, 375 years, and 575 years. These data are presented in Table 1.

The numbers of spruce and fir change over the 575 years following fire. Thus, Prediction 1 is false; Prediction 2 is supported. Three key observations from these data reject Hypothesis 1 and support Hypothesis 2. First, note in the data outlined in red in the table that in young stands (175 years), *both* spruce and fir have many individuals that are just a little younger than 175 years. In other words, both species colonized soon after fire. Second, note that after this initial colonization, spruce recruitment, the addition of new individuals to the population, stops. This is illustrated in the series of blue boxes that show the absence of young spruce trees in stands of increasing age. For example, in 175-year-old stands, there are no spruces between 0 and 50 years of age. This means that once the stand reaches that age, spruce recruitment stops. In the 275-year-old stand, there are no spruces of 100–125 years of age, again indicating lack of spruce recruitment at about 175 years. The absence of young spruce in stands of 175 years of age is shown as a diagonal of missing young spruce trees. In other words, after the initial colonization, new individuals do not enter the population. In contrast, there are young firs in stands of all ages; fir does not have the diagonal of missing young trees. Third, note that once a stand reaches 275 years of age, young spruces begin to appear again. This is indicated by the green boxes showing many young spruces in 275- to 575-year-old stands.

Thus, the results are consistent with the second hypothesis. Both species colonize immediately after fire. Spruce recruitment then stops until the stand is at least 275 years old. Spruce-fir forests are not at long-term equilibrium. Rather, they are dynamic, with much change over a period of several hundred years. The researchers explain the pattern of change as follows.

continued

ON THE FRONTLINE   *continued*

They suggest that spruce requires relatively high light to colonize and reproduce. Thus, immediately after fire, there are many spruce trees (and firs as well). As the canopy closes, shade increases and spruce recruitment is suppressed. In contrast, fir is able to reproduce in the shade of a maturing forest. After nearly 300 years, some of the original spruces and firs begin to die and fall, opening up light gaps in the canopy. These light gaps allow spruce to reproduce and colonize again.

TABLE 1   The numbers of individuals of spruce and fir in stands of different ages

	SPRUCE					FIR				
Stand	1	2	3	4	5	1	2	3	4	5
Stand age (yr)	175	275	375	575	575	175	275	375	575	575
AGE CLASS					STEMS IN AGE CLASS (NO/HA)					
0–25		240	240	240	5040	1120	720	5790	4370	9600
26–50		480	240	240	4080	2440	5280	11820	7440	9360
51–75	20		490		2200	480	4560	1690	4100	4370
76–100	160				1720	1280	1680	720	2170	2040
101–125	340				770	2220	480	740	1000	680
126–150	1160	260			260	860	50	280	20	1370
151–175	280	10				140	520	360	490	370
176–200	20	30	10	10	20		390	320	540	430
201–225		160	10				630	780	630	340
226–250		610	10	10	10		190	130	130	170
251–275	20	370	30		10		190	100	340	80
276–300		20	50		10		10	70	80	40
301–325			120	30	10			100	120	30
326–350			140					50	40	10
351–375			160	30					70	10
376–400			20	30	10				60	
401–425				30	10					10
426–450				30	10				20	10
451–475				40	40				20	
476–500				60						
501–525				40	30					
526–550				20	50					
551–575				90	10					
576–600										
601–625					10					

(*Fagus grandifolia*) is invading stands that have not been disturbed for more than 400 years. Its increasing dominance resembles changes late in succession. However, in this case, beech invasion is actually a result of a gradual range expansion over the past several hundred years, rather than successional change (Woods, 2000).

Markov Chains

A key feature of the Gleasonian approach to plant community dynamics is the random nature of successional changes. According to this view, the transition from one seral stage to another occurs not with certainty but with a particular probability. This kind of stochastic (random) process can be modeled using a mathematical concept known as a Markov chain. A Markov chain is a series of states (s_1, s_2, s_3, etc.) that occur in discrete time periods. Successional stages might constitute a set of states. The transition from one state to another has a certain probability, p_{ij}, in which j and i denote the beginning and ending state, respectively. In other words, the transition from State 2 to State 3 occurs with the probability p_{23}. A key feature of a Markov process is that the probability of any transition from one state to another depends *only* on the current state, not any previous state. If all the transition probabilities are known, we can generate a transition matrix for all the possible changes.

Ecologists recognized some time ago that this process has direct application to ecological succession. This can best be demonstrated with a specific example. Valverde and Silvertown (1997) used Markov chains to analyze the formation and closure of canopy gaps in a deciduous forest in Britain. In their system, large gaps are formed primarily from disturbance events in closed-canopy forest. Succession then begins to close the gap via growth from bordering plants and growth of new trees in the gap. Figure 1 shows the essential features of their Markov analysis: (1) a set of states and transition probabilities among them, and (2) the resulting matrix of transition probabilities. The transition probabilities represent the probability of one state changing to another. For example, the probability that a small gap begins to

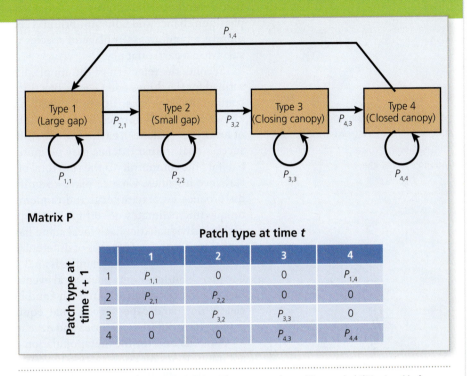

Matrix P

Patch type at time *t*

		1	2	3	4
Patch type at time *t* + 1	1	$P_{1,1}$	0	0	$P_{1,4}$
	2	$P_{2,1}$	$P_{2,2}$	0	0
	3	0	$P_{3,2}$	$P_{3,3}$	0
	4	0	0	$P_{4,3}$	$P_{4,4}$

Figure 1 States and transition probabilities. The states and transition probabilities in a Markov chain (a). Each type of state has a probability of transition to another or to itself. (b) The matrix of transition probabilities among the states (Valverde and Silvertown, 1997). **Analyze:** Why are the zero values in the matrix characteristic of a Markov chain?

close is p_{23}. The probability that a gap opens in closed forest (p_{14}) is actually the probability of disturbance. From these transition probabilities we can create a matrix that shows the chance of any set of changes from one time to the next.

Markov chain analysis applies to more complex successional pathways as well. Consider the set of dune chronosequences shown in Figure 1. Each vegetation type represents a state with a certain probability of transition to other states. The fit to a Markov process does not mean that there are no biological causes for succession. Each value of p_{ij} is based on chance events *and* the biological features of the system. Together, the matrix of transition probabilities conforms to the Markov system—that is, a random process.

The shift from the Clementsian equilibrium paradigm represented a fundamental change in our understanding of plant community dynamics and succession. Two key conceptual changes result. First, we understand that chance plays a more significant role in plant community organization than under the Clementsian paradigm. Instead of a highly regulated, predictable pattern of vegetation, community composition is dependent on chance: the random occurrence of disturbance events and the chance events associated with colonization and individual successional sequences. We have seen this in several ways so far. Canopy gaps form and fill in at random points in a forest, creating a mosaic of sites of different age and successional status; many systems have multiple pathways of succession that depend on chance events of colonization or site characteristics. This emphasis on chance and nonequilibrium is characteristic of a Gleasonian plant community, a random assemblage of species that happen to be able to colonize and survive in a particular place.

Second, the rejection of the simple, fixed chronosequences as originally described for the dunes and Glacier Bay shifts our focus from *time* to *space* in plant community dynamics. This is manifest in the growing emphasis on an approach known as **patch dynamics**, in which plant communities are described as mosaics of local sites (patches), each of which differs in disturbance history and current successional status. Patches may be quite small relative to the total size of the habitat. The community is the sum total of these patches and their individual disturbance histories. Chance plays a significant role in patch dynamics, because disturbance events are local and random. Moreover, because patches occur in unique microclimates or soil types, the plant communities they constitute are highly individualistic. Each local stand has a different fire history. The habitat as a whole is a mosaic of patches of different age.

The concept of patch dynamics is analogous to the metapopulation concept in Chapter 7. Both systems place local events into the larger scale of the landscape. Just as a metapopulation can be in equilibrium even when local populations are clearly not, landscape successional equilibrium may occur even when local patches have different histories and developmental statuses (Gravel et al., 2010).

As is the case for the metapopulation concept, patch dynamics have important implications for conservation biology. For example, Kirtland's warbler (*Dendroica kirtlandii*) nests in pure, even-aged stands of jack pines (*Pinus banksiana*) that develop after fire. Succession eventually renders the habitat marginal or inappropriate for the warblers. Recent studies show that patch dynamics predict the presence of nesting warblers (Donner et al., 2010). Specifically, larger patches were colonized earlier than small patches. Thus, birds exploiting large patches had a longer window of appropriate habitat available to them. Isolated patches were colonized later and were abandoned earlier. Like many other neotropical migrant birds, Kirtland's warblers are in decline. One contributing factor is the decrease of recently burned stands due to fire suppression in the northern boreal forest. Conservation biologists can use this patch dynamics data to design management plans for this species.

patch dynamics A concept of community structure in which a region is a mosaic of sites, each with its own disturbance history and successional status.

KEY CONCEPTS 14.2

- When the time since disturbance changes linearly over space, a chronosequence develops: a set of communities in different stages of succession as a function of distance from the disturbance.
- In primary succession, pioneer species must be able to tolerate harsh abiotic conditions.
- Secondary succession is generally more rapid, because the soil is relatively intact and the physical conditions are more conducive to plant growth.
- The final, self-perpetuating stage of succession is the climax community.
- The theory of patch dynamics characterizes communities on the basis of the size, isolation, and successional status of patches within the habitat.

QUESTION:

How are patch dynamics and metapopulation theory related?

THE HUMAN IMPACT

Succession and Conservation Biology

Habitat loss is among the most important reasons that so many species face extinction today. And because succession has such a profound effect on the habitat, we should expect that the successional status of the habitat is important in conservation biology.

Some of the most well-known examples of threatened or endangered species are those that require large tracts of a mature or climax community. The spotted owl (*Strix occidentalis*) of the conifer forests of the Pacific Northwest is a prime example. Each nesting pair of spotted owls utilizes a home range of 550 hectares of old-growth forest. As these mature forests are cut, the number of potential home ranges accommodated in the fragmented landscape declines. Climax forest in this region is crucial to many other threatened or endangered species. Some, like the marbled murrelet (*Brachyramphus marmoratus*), spend most of their time on the nearby ocean, where they feed on invertebrates and small fish. However, they nest far inland and only in large trees in old-growth forest.

Figure 1 Forked blue curl and puccoon. The forked blue curl (a) requires disturbance, whereas the puccoon (b) occupies stable sites.

Although human disturbance of climax habitat is perhaps the most obvious threat, we know that many species are adapted to disturbed sites and a particular stage of succession. If those successional stages disappear, these species, too, are at risk. We have seen in this chapter that the endangered Kirtland's warbler nests only in the large tracts of young jack pine forest that arise following fire. The prime cause of the warbler's decline is fire suppression and the elimination of this seral stage.

In the nineteenth and early twentieth centuries, a large proportion of the boreal forests of the northern United States and southern Canada were logged. The result was extensive tracts of successional forest. This constituted prime habitat for two important game birds, the American woodcock (*Scolopax minor*) and the ruffed grouse (*Bonasa umbellus*), and their populations increased significantly. Today, as these forests mature, both species are in decline. Their rise was due to human activity; their current decline is due at least in part to our efforts to protect these forests so that they can mature.

Species that occur in the same general habitat but require different seral stages can present a dilemma to conservation biologists. For example, in parts of the eastern prairie, sand barrens habitat constitutes an unusual variant of this biome. Sand barrens occur where sand, carried by melting glaciers, was deposited by wind or water. They are characterized by a mix of open sand and stands of sparse, xeric-adapted grasses. In parts of Indiana, two species of plants endemic to sand barrens are threatened, primarily because so much of this habitat has been destroyed. Populations of the forked blue curl (*Trichostema dichotomum*) and the fringed puccoon (*Lithospermum incisum*) are in decline. In the remaining sand barrens, managers face a difficult problem: the forked blue curl requires areas of disturbed, bare sand; the fringed puccoon requires habitat in which the sand has been colonized and stabilized by grasses and other species.

The best management practice for the forked blue curl would be to create more disturbed, open-sand habitat. For the puccoons, these disturbed areas should be allowed to reach a later successional stage. The dilemma is particularly acute because the total area of sand barrens habitat is so small that managing intensely for one species would negatively impact the other.

QUESTION:

How could patch dynamics analysis contribute to conservation of the forked blue curl and the fringed puccoon?

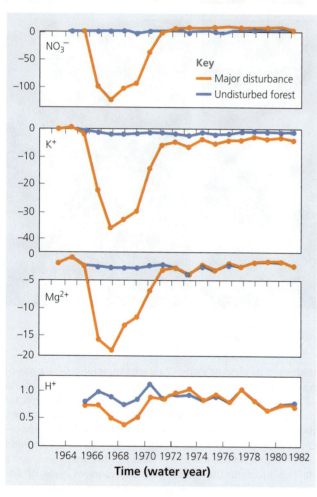

14.3 What Causes Successional Change?

We have seen that in many systems there are specific sequences of successional change. Moreover, plants have adaptations to the specific stage of succession at which they typically occur. Disturbance, in concert with the activity of successional plant species, drives the pattern of succession. What are the specific mechanisms that underlie the changes in the community?

How Does Disturbance Change the Abiotic and Biotic Environment?

Although there is variation among disturbances, some consistent effects occur. Three aspects of the postdisturbance environment are particularly important to the mechanisms of succession:

1. Increased light availability
2. Loss of nutrients or interruption of nutrient cycles
3. Succession itself: the biotic community in flux as succession proceeds

In many mature plant communities, light is a limiting factor. When disturbance removes the climax vegetation, light availability increases significantly. In forest systems, light may have been limiting on the forest floor for centuries. A light gap or a more widespread disturbance opens the canopy, and suddenly light is abundant. Moreover, the immediate loss of species by the disturbance itself means that there may be relatively little competition for the newly available light. The result is a very different competitive environment than in the climax.

There is often a net loss of nutrients from the soil after disturbance, even in secondary succession. Some nutrients leach from the soil because plant diversity or density is so low that there are few roots to absorb them. Nutrients may also be released by decomposing plants killed by the disturbance. These processes have been well documented at the Hubbard Brook Experimental Forest in New Hampshire by experimental and natural disturbance of mature forest (Figure 14.19). Over time, nutrient levels generally rise again, but initially some important plant growth factors may be in short supply.

Thus, after a disturbance, light availability tends to be high whereas nutrient levels tend to be low; these resources tend to be inversely related throughout succession (Figure 14.20). Each new community that arises during succession changes both the light and nutrient budget. Thus, as succession proceeds, plants face new sets of competitive interactions. The preferred conditions for one species are, by definition, changing. The preferred habitat of each successional species is gradually disappearing out from under it.

Figure 14.19 Nutrient gain and loss. The pattern of nutrient gain and loss following a major disturbance in 1966–1967 (solid line) compared to undisturbed forest (dotted line) at the Hubbard Brook Experimental Forest. Following disturbance, there is a net loss of nutrients before the system recovers (from Likens, 1985). **Analyze: What do you think are the mechanisms of nutrient loss after a disturbance?**

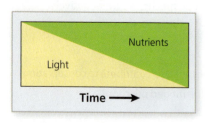

Figure 14.20 Light and nutrient levels following disturbance. Immediately after disturbance, light is often abundant whereas nutrients are low. The reverse is true in older communities. **Analyze: How do the plants change the light and nutrient conditions as succession proceeds?**

What Plant Characteristics Are Adaptive as Conditions Change?

For plants that colonize after disturbance, the three changes outlined earlier constitute a strong selective force. Early successional plants must be able to exploit a high light environment, but often with low nutrient levels. Late-successional and climax species are adapted to the reverse. Table 14.3 illustrates the relative photosynthetic rates of early- and late-successional species. Early plants can exploit

TABLE 14.3

SPECIES	RATE
Early-succession species:	
Common ragweed (*Ambrosia artemisiifolia*)	35
Giant ragweed (*A. trifida*)	28
Foxtail (*Setaria faberii*)	38
Daisy fleabane (*Erigeron annuus*)	22
Aster (*Aster pilosus*)	20
Mid-succession species (softwood trees):	
Eastern red cedar (*Juniperus virginiana*)	10
Cottonwood (*Populus deltoides*)	26
Sassafras (*Sassafras albidum*)	11
Tulip poplar (*Liriodendron tulipifera*)	18
Late-succession species (hardwood trees):	
Red oak (*Quercus rubra*)	7
American beech (*Fagus grandifolia*)	7
Sugar maple (*Acer saccharum*)	6
White oak (*Quercus alba*)	4

Source: (Bazzaz, 1979)

a high light environment and can photosynthesize at a prodigious rate. However, as shown in Figure 14.21, as the community develops and light becomes limiting, the photosynthetic rates of early species decline. At low light intensity, they may even enter a negative energy balance in which their tissues' respiration exceeds the input of energy via photosynthesis. Late-successional species cannot match the photosynthetic rate of the early species when light is abundant. But when light intensity declines, they can maintain a positive energy balance.

What accounts for these differences? A number of physiological and anatomical differences between early- and late-successional species result in their differential responses to light intensity. Tree species characteristic of the later stages of succession tend to have very efficient photosynthetic machinery. Some of the red light important to photosynthesis is absorbed as light passes through layers of leaves in a mature forest, leading to both quantitative and qualitative changes in light availability. Late-successional species can gather and utilize this light because of adaptations that include the number and arrangement of chloroplasts and accessory photosynthetic pigments that gather a wider array of wavelengths.

There are also anatomical differences among these species. For example, the branching structure of early- and late-successional species differs. We quantify this with the concept of the *effective layers of leaves* on a tree (Figure 14.22). We measure the amount of light absorbed by the leaves on a single branch. Then we measure the amount of light absorbed by the entire tree. The number of layers of leaves required to absorb that total amount of light is the number of effective layers of leaves. For example, if one layer absorbs 5 percent of the light it receives (it passes 95 percent), and the entire tree absorbs 40 percent of the light (60 percent passes through), we calculate the number of effective layers:

$$(0.95)^X = (0.60)$$

Solving for X:

$$\log (0.95)\, X = \log (0.60)$$

$$X = \log 0.60 / \log 0.95 = 9.95 \text{ effective layers of leaves.}$$

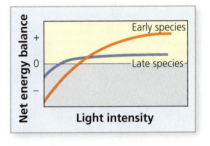

Figure 14.21 Photosynthetic rates. The rate of photosynthesis of early and late successional tree species as a function of light intensity. Where photosynthesis exceeds respiration, the plant has a net positive energy balance. Where respiration exceeds photosynthesis, the plant is in negative energy balance and may not survive. The late successional species cannot photosynthesize as rapidly in high light as the early species, but it is more efficient in low-light conditions (after Bazzaz, 1996). **Analyze:** What are the competitive relationships between early and late species at the point the two lines cross?

Figure 14.22 Layers of leaves. The concept of effective layers of leaves. A single layer of leaves absorbs some fraction of the light it receives (point 1 to point 2). Of the total light reaching the tree, some fraction is absorbed between points 3 and 4. The number of effective layers of leaves is the number of layers that reduce the light by the amount absorbed from 3 to 4. **Analyze:** Why do we refer to this concept as the "effective" layers of leaves?

Early-successional species tend to have more effective layers of leaves than late-succesional species (Table 14.4). This accounts for both their ability to photosynthesize rapidly in high light and their inability to compete in low light conditions. Each leaf on a tree photosynthesizes *and* respires. If the amount of photosynthesis exceeds respiration, the leaf provides a net energy benefit to the tree. If, on the other hand, the amount of photosynthesis is lower than the amount of respiration, that leaf is a net energy drain on the tree.

Now consider an early- and a late-successional species in a high light environment. Under these conditions, there is so much light that even the lower,

TABLE 14.4 The number of effective layers of leaves in early-, mid-, and late-successional tree species

SPECIES	NUMBER OF EFFECTIVE LAYERS
Early-succession species:	
Gray birch (*Betula populifolia*)	4.3
Aspen (*Populus tremuloides*)	3.8
White pine (*Pinus strobus*)	3.8
Mid-succession species:	
White ash (Fraxinus americana)	2.7
Black gum (Nyssa sylvatica)	2.6
Red maple (Acer rubrum)	2.6
Late-succession species:	
Sugar maple (*Acer saccharum*)	1.9
Hemlock (*Tsuga canadensis*)	1.6
American beech (*Fagus grandifolia*)	1.5

Source: (Bazzaz, 1979)

shaded leaves on the early species receive enough sunlight that photosynthesis exceeds respiration. This species will grow rapidly, especially compared to the late-successional species with just a few effective layers of leaves to utilize the abundant sunlight. Now consider the situation late in succession, when the forest canopy is closed and light becomes limiting. The lower leaves in the early species are no longer receiving enough light that their photosynthesis exceeds their respiration; they become a net energy drain on the tree. The late species will not grow fast under these conditions, but it will still have a positive energy balance. When light is abundant, an early species will outcompete a late species; when light is limiting, the reverse occurs.

Many plants found early in succession are adapted to absorb and retain nutrients at low soil concentrations thanks to a number of anatomical and physiological adaptations. In some, deeply penetrating or highly branched (high-surface-area) root systems prevent nutrients from leaching away. One important set of adaptations that increase absorption of nutrients are the mutualistic interactions between plants and fungi associated with the root system. The fungi receive carbon from the plant in return for nutrients such as nitrogen. Other plants, such as legumes, fix nitrogen due to mutualistic interactions with nitrogen-fixing bacteria associated with their roots.

Nitrogen fixation due to mutualistic bacteria and legumes was significant early in primary succession after the Mount St. Helens eruption in 1980 (Brotney and Bliss, 1999). However, as has been observed in other, similar, successional sequences, the rate of nitrogen fixation eventually slowed. Recent work (Menge and Hedin, 2009) shows that the decrease in nitrogen fixation in the community is not the result of a negative feedback system that reduces the rate of fixation in the species already present. Rather, it is the result of colonization by new species that do not fix nitrogen. Nitrogen fixation is energetically expensive. Once soil nitrogen has accumulated, species that do not fix nitrogen can outcompete those that do.

The other suite of adaptations important in the successional sequence centers on the fact that by definition, successional communities are changing. The offspring of climax species are adapted to the conditions of a mature system. Following disturbance, the environment favors a new set of species. Early-successional species face the problem that succession gradually eliminates their preferred environment. We refer to species dependent on colonizing a new disturbance as fugitive species.

Fugitive species must have adaptations that allow their offspring to "find" the next disturbance. This occurs in one of two ways. The species can either colonize another disturbed site, or it may already be present as seeds when the disturbance occurs. Fireweed exemplifies the first pattern by producing huge numbers of wind-dispersed seeds. The life history strategy of this species emphasizes the production of many seeds with little energetic investment in each. No one seed has a high probability of landing in a disturbed site, but their sheer numbers ensure that a few do. The chance nature of seed dispersal and site colonization is an important feature of successional dynamics, especially the differences among sites recovering from a common disturbance. The local succession is highly dependent on the proximity of seed sources.

An alternative adaptation seen in fugitive species is the production of seeds that can persist in the soil for long periods of time. They may be produced soon after a disturbance, yet remain in place for decades until another disturbance creates favorable conditions again. In some cases, germination is inhibited by the conditions in a mature community and promoted by the conditions immediately following disturbance (Figure 14.23).

A number of interesting variations on this concept are known. For example, in the cloud forests of Costa Rica, some successional tree species develop on "nurse

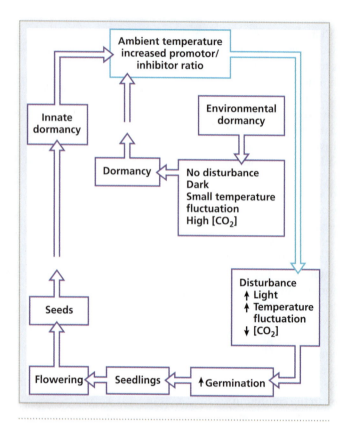

Figure 14.23   Ragweed seeds. Ragweed seeds persist in the soil until disturbance creates favorable conditions. Germination is inhibited by the abiotic conditions in a mature forest; it is promoted by the conditions that occur immediately after disturbance (from Bazzaz, 1996). **Analyze:** Why is temperature variation different in disturbed and undisturbed sites?

logs," the trunks of mature trees felled by disturbance (Figure 14.24). Remarkably, some of these species germinate among epiphytes in the crowns of trees in mature forest. When these old trees eventually fall and create a gap, they serve as nurse logs for the saplings (Lawton and Putz, 1998).

How Do Changing Resources and Physical Conditions Drive Succession?

We have seen that light and nutrient resources tend to be inversely related during succession. Tilman (1985) developed a graphical model of species replacement during succession based on changes in two key resources, such as light and nutrients, whose abundance limit plant growth (Figure 14.25). If both resources are below a critical level, the population declines; if they are above those levels, the population increases. If the resources lie precisely on the lines, known as isoclines, the population is stable.

Figure 14.25a shows this model for a single species. However, to explain succession, we must explain the replacement of one species by another over time. A mechanism for this is modeled in Figure 14.25b (Tilman's resource model of succession for two species), in which isoclines for four species are shown. Because the regions of positive growth and decline differ among species, there are resource levels at which each species either increases or decreases. The result is a sequential rise and fall of each of the four species as resources change over time (Figure 14.25c: replacement resulting from b). This model explains the success of some species early in succession and others later.

What Mechanisms Effect Change?

Connell and Slatyer (1977) organized the mechanisms of succession into three main categories depending on how the plants respond to and change the physical and biological environment. Each model begins with a disturbance event that

Figure 14.24   Nurse log. A nurse log in tropical rainforest. Seedlings germinate on mature trees before they die. When they fall to the forest floor, the nurse logs provide nutrients for the seedlings.

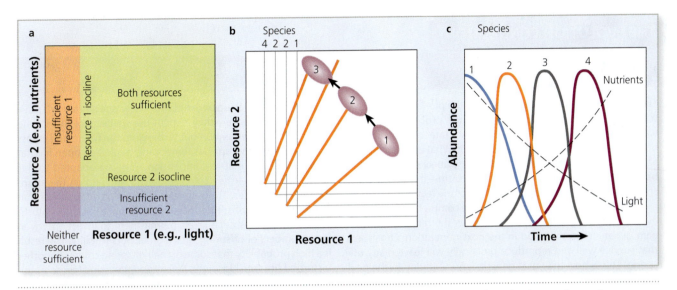

Figure 14.25 Limiting resources. (a) The effect of two limiting resources on plant growth. If both are abundant, the plant survives and grows. If one or both is scarce, the plants decline. Labeled lines indicate the rate of use of the resources. The heavy lines are isoclines, abundances of the nutrients at which the plant population is stable. (b) Tilman's resource model of succession for two species. The different resource requirements of the two species means that at different resource levels, one may outcompete the other, leading to replacement of one species by the other during succession. (c) The pattern of replacement of the four species in (b) (from Tilman, 1985). **Analyze: How does this model relate to the simple light budget model in Figure 14.21?**

eliminates competition and significantly alters the physical environment. They differ, however, in the subsequent mechanisms by which species either persist or are replaced as succession proceeds.

The **facilitation model** applies if the disturbance produces a physical environment in which colonization and survival are especially difficult. Species typically seen later in succession cannot colonize until earlier species modify the harsh initial conditions. We have already alluded to this phenomenon in our discussion of the early stages of primary succession. In dune succession, for example, lack of soil and shifting sand make it difficult for many species to colonize soon after disturbance. Pioneer species such as marram grass not only survive in shifting sand, they stabilize the substrate and gradually add organic matter to the developing soil as individuals die and decay. In the process, they facilitate colonization by other species. Nitrogen fixers play the same role: they add sufficient organic nitrogen to developing soil that other nitrogen-limited species can colonize.

In the forest-grassland ecotone in Montana, limber pine (*Pinus flexilis*) facilitates growth of other species, especially Douglas fir (*Pseudotsuga menziesii*). The physical environment in this ecotone can be harsh. High winds, high light, and heavy snowpack make it difficult for young Douglas fir to establish. However, in the protection of limber pine, it can flourish. Experimental analysis showed that shade was a crucially important component of facilitation. Protection from snow load and high light significantly enhanced survival of young Douglas fir, but only when the plants were shaded.

In the **tolerance model**, species replacement occurs when new species more tolerant of the newly changed conditions replace earlier species. We alluded to this process in the discussion of the light requirements of early- and late-successional species. Recall that early species are adapted to grow rapidly when light is abundant, but they lose in competition to later species when light becomes limiting. In this model, succession is a series of replacements by increasingly shade-tolerant species. The early species that effectively exploit a high light environment decrease the available light and gradually give way to species adapted to lower light.

facilitation model A model of successional change in which early colonists improve conditions for other species that subsequently colonize.

tolerance model A model of successional change in which early species change the abiotic conditions and are replaced by new species able to tolerate those conditions.

THINKING ABOUT ECOLOGY:

What is the relationship between competitive exclusion (Chapter 11) and succession?

THE EVOLUTION CONNECTION

Pioneer Species

The facilitation model of succession, and the pioneer species central to it, presents an interesting evolutionary problem. You might wonder why a plant species would facilitate colonization by other species that will ultimately replace it. What is the evolutionary advantage of such behavior? At first glance, this appears to be an extreme form of evolutionary altruism—it benefits other species and assures the local extinction of the facilitators.

The difficulty is resolved with a more comprehensive view of evolutionary fitness. We define fitness as the relative ability to gain genetic representation in the next generation. Organisms that survive well and reproduce prolifically will leave many offspring. This is how they gain genetic representation in the next generation. We have used variables such as the intrinsic rate of growth (r) that incorporates both reproduction and survival as a measure of fitness. However, fitness is actually more complex than this—and more difficult to measure. Fitness extends beyond just the next generation: it is the total genetic contribution to the future by an individual. This is obviously a difficult variable to measure. The intrinsic rate of growth or the number of offspring produced in the next generation is correlated with the fitness in this broader sense, but these short-term measures may be misleading, as in the case of pioneer species.

The keys to the full concept of fitness and the solution to the paradox of pioneer species are a longer time scale and a broader spatial scale. The characteristics of pioneer species are adaptive in this broader sense—they benefit the species not just in one locale over one generation but farther into the future and across a wide area. Pioneer species temporarily exploit an open habitat in which they face little competition. They are adapted to this specialized niche—they colonize quickly and exploit the abundant light resources. These species also exhibit adaptations that ensure that their offspring locate and colonize another disturbance. The high resource levels permit them to invest energy in large numbers of offspring with high dispersal ability. The result is a high probability that some of their offspring colonize another disturbed site with high light, harsh conditions, and few competitors. Their success is limited in any one site, but over the mosaic of disturbances across the landscape, they are as successful as a species whose offspring persist in their parents' habitat.

Thus, there is no evolutionary paradox here: the pioneer species are evolutionarily successful in their particular niche. We simply must consider their niche, and their fitness, more broadly in space and time.

QUESTION:

What other seemingly altruistic actions have we studied that are explained by a broader concept of fitness?

---

inhibition model   A model of successional change in which early colonists inhibit colonization by new species.

The **inhibition model** is driven by the ability of species to colonize the disturbed site and then resist invasion by other species. A classic example of this process occurs in the rocky intertidal zone, where wave and storm action disturb the substrate and initiate a successional sequence of algae (Sousa, 1979). In this environment, space is the key limiting factor. The green alga, *Ulva spp.*, is an early colonist of recently decimated rocky substrate. Its presence inhibits invasion by the red algae common later in succession. So long as *Ulva* remains vigorous, it dominates the site. If *Ulva* is experimentally removed, red algae quickly colonize (Figure 14.26).

KEY CONCEPTS 14.3

- Resource conditions change as a result of disturbance.
- Plants that colonize early in succession are adapted to exploit the high light conditions. They may have to be able to thrive in low-nutrient or other difficult conditions.
- Plants that occur later in succession are more efficient in their use of light. They can compete with early species under low light conditions but cannot grow as fast when light is abundant.
- Some early-successional species are fugitive species: their habitat disappears as succession proceeds.
- Three key models of successional change are the tolerance, facilitation, and inhibition models.

QUESTION:

Why are light and nutrients so important to the mechanisms of succession?

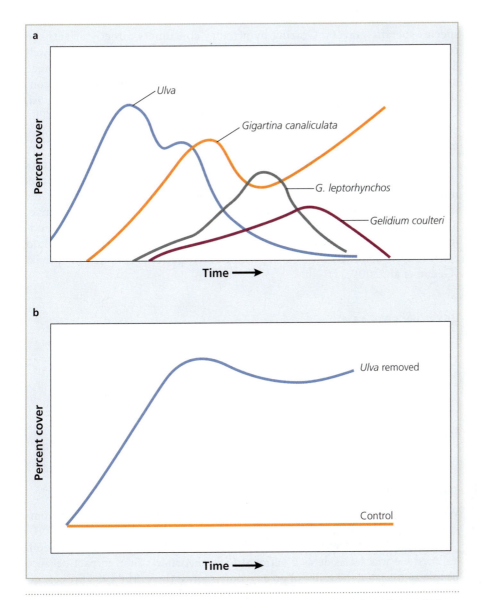

Figure 14.26 Intertidal algal succession. The pattern of algal succession in the rocky intertidal zone. (top) *Ulva* is an early colonist that can inhibit colonization by the red algae species that occur later in succession. (bottom) If *Ulva* is experimentally removed, red algae colonize (from Sousa, 1979). **Analyze:** Many successional sequences are altered by consumers that affect the early colonists. What ecological features of the early species contribute to this phenomenon?

14.4 What Role Do Consumers Play in Succession?

The mechanisms of succession outlined earlier are based on the complex interactions among plant species competing for light, space, nutrients, and other limiting resources. However, these competitive interactions do not occur in a vacuum. They are embedded in many other ecological interactions, especially grazing by primary consumers. For example, we have seen that insects such as pine bark beetles constitute a disturbance: they cause tree mortality, which in turn leads to wind throw and light gaps. Conifers killed by beetles dry out and provide a conduit for a small ground fire to climb into the canopy and initiate a major crown fire.

What Are the Direct and Indirect Effects of Consumers?

Early-successional plant communities are often characterized by high densities of a few species—those few species adapted to either survive the harsh conditions or exploit newly abundant light resources. Dense populations are attractive to species-specific grazers. Grazing by primary consumers directly affects the course and speed of successional change. This effect has been particularly well documented in marine systems, especially the rocky intertidal. The dominance of *Ulva* discussed earlier, and its ability to inhibit later red alga species, is constrained by grazing. The crab *Pachygrapsus crassipes* selectively grazes on *Ulva*, ultimately releasing the red algae from *Ulva's* dominance (Sousa, 1979). The course and time scale of succession depend on the activity of primary consumers of the algal species.

As is common in many ecological processes, biotic and abiotic factors interact. In New England salt marshes, disturbance by wave action and storms initiates succession. Lower in the intertidal zone, where flooding by seawater is frequent and prolonged, salinity determines which species can colonize. Small mammal herbivores play a central role higher in the zone, where flooding is less common. Preferential grazing on species that otherwise would competitively dominate the system permits other species to colonize and persist (Gedan et al., 2009).

KEY CONCEPTS 14.4

- Consumers alter the rate and pattern of successional change. Grazers increase plant diversity if they specialize on competitively dominant plant species.
- Dense populations of early-successional species attract consumers, which in turn affect the persistence of early species.

QUESTION:

Why are early-successional species able to reach such high densities?

Putting It All Together

We can now return to the basic question of this chapter: How does the plant community respond to disturbance?

The cataclysmic fires in Yellowstone in 1988 led us to address this question. Yellowstone was set aside as a national park, the world's first, in part because people appreciated the miles of green, conifer forest there. The 1988 fires exemplify an important aspect of plant community ecology: disturbance events are inevitable. Although fires of that scale had not been witnessed in the history of the park, we now understand that the history of fire in that system matches the time scale of lodgepole pine succession. The many plant adaptations to fire—whether the serotinous cones of lodgepole pines or the wind-dispersed seeds of fireweed—testify to the fact that disturbance is an important selective force acting on plants.

The Clementsian paradigm emphasized an equilibrium view of ecology rooted in the notion that disturbance is the exception rather than the norm. Today we understand that the Yellowstone situation is typical—most plant communities, even those like tropical rainforests that appear to be very stable, are not in equilibrium for long. Rather, they are regularly affected by disturbances large and small, mild and intense. Thus, the successional changes that follow disturbance take on special importance: most communities are in a state of change much of the time.

Each community, each disturbance, and each successional pathway is in some sense unique. Nevertheless, there are mechanisms common to both primary and secondary successional sequences even when they occur in vastly different communities. These follow from the adaptive responses of individual plant species to the suite of conditions, biotic and abiotic, they experience following disturbance.

THINKING ABOUT ECOLOGY:

Explain how animals might contribute to each of the Connell-Slatyer mechanisms of succession.

We have emphasized that three key selective forces drive the process of succession. These conditions select for three fundamental traits:

1. The ability to disperse widely to colonize a new disturbance or to persist in the community until a disturbance occurs
2. The ability to tolerate the abiotic conditions following a disturbance event
3. The ability to compete for limiting resources such as light and nutrients

The differences among disturbances among communities and the ensuing successional sequences may be large, but these three basic sets of plant adaptations drive the process of succession.

Summary

14.1   What Are the Effects of Disturbance?
- The intensity and spatial extent of disturbance determine the type of succession that follows.
- We categorize disturbance along three axes: intensity, spatial scale, and frequency.
- Most communities experience some form of disturbance and thus succession.

14. 2   What Are the Patterns of Vegetation Change Following Disturbance?
- Pioneer species in primary succession are adapted to (a) survive the harsh conditions that often occur there and (b) can rapidly exploit the abundant light resources in the absence of competitors.
- If the disturbance leaves the soil intact and functional, succession proceeds more rapidly.
- The end point of succession is the climax, a community that replaces itself until another disturbance alters it.
- Although there are general patterns of successional sequences, each site is unique and there are many variants in the specific species that occur in these sequences.

14.3   What Causes Successional Change?
- Resources change over the course of succession. In general, light and nutrients often vary inversely during succession.
- Many early-successional species participate in mutualistic interactions with bacteria or fungi to acquire limiting nutrients.
- Early-successional species can exploit high light conditions but do not use light as efficiently as late-successional species. Thus, as light becomes limiting, early species are replaced.
- Early-successional species are fugitives: they are poor competitors whose habitat is disappearing.
- The tolerance, facilitation, and inhibition models comprise the major mechanisms of successional change.

14.4   What Role Do Consumers Play in Succession?
- Consumers may be attracted to large populations of plant species exploiting abundant resources.
- Grazing can alter succession by reducing populations of competitively dominant species.

Key Terms

canopy gap p. 300
chronosequence p. 305
Clementsian paradigm p. 309
climax community p. 308
disclimax p. 310
ecological disturbance p. 300

facilitation model p. 321
inhibition model p. 322
light gap p. 300
old field succession p. 307
patch dynamics p. 314
pioneer species p. 306

primary succession p. 300
secondary succession p. 301
sere p. 300
succession p. 300
tolerance model p. 321

Review Questions

1. Explain the Clemenstian paradigm in the context of disturbance and succession. Why have we rejected this idea? What has replaced it?

2. Why are some disturbance events random whereas others are predictable?

3. Why are the patterns of succession observed at the Lake Michigan dunes and in Glacier Bay, Alaska, imperfect chronosequences?

4. What is the relationship between the concepts of the niche, interspecific competition, and competitive exclusion (Chapter 10) to succession?

5. How do changes in resource abundance affect the process of succession?

6. How do plants and animals interact in the course of disturbance and succession?

7. Are the Tilman resource model, the facilitation model, the tolerance model, and the inhibition model mutually exclusive? Can more than one operate during succession?

8. In what ways is succession a random process? How is it deterministic—that is, with predictable cause and effect?

9. How is it adaptive for a species to change the abiotic conditions so that a new species replaces it?

10. Compare and contrast the adaptations important for early- versus late-successional species.

Further Reading

Moore, S.A., et al. 2009. Diversity in current ecological thinking: implications for environmental management. *Environmental Management* 43:17–27.

This important paper reviews the shift from equilibrium ecology to a nonequilibrium paradigm. It explains the ways in which this change will require a fundamental shift in how we think about conservation and environmental issues.

Tilman, D. 1988. *Plant strategies and the dynamics and structure of plant communities*. Princeton, NJ: Princeton University Press.

This book is the classic description of Tilman's enormously influential and important resource model.

Walker, L.R. et al. 2007. *Linking restoration and ecological succession*. New York: Springer.

As in the paper by Moore et al., this book explores the environmental implications of systems in which disturbance is the norm rather than the exception. Restoration ecology, the science of bringing back threatened or absent communities, is complicated by the frequency of disturbance in natural systems.

Wallace, L.A. 2004. *After the fires: the ecology of change in Yellowstone National Park*. New Haven, CT: Yale University Press.

This book collects a set of major research papers on the impact of the Yellowstone fires of 1988. It also surveys the patterns of forest succession following the fires. Although the book deals with the specific successional process in Yellowstone, a number of generally important principles emerge in the discussion.

Communities and Ecosystems

Sunlight streams into the foliage of a tropical forest, one of the most species-rich ecosystems on earth. In these three chapters, we examine the physical and biological processes that determine how species are distributed across the globe. These patterns play a crucial role in how energy is captured in ecosystems and how materials cycle within them.

Chapter 15

Species Diversity

If the traveler notices a particular species and wishes to find more like it, he may often turn his eyes in vain in every direction. Trees of varied forms, dimensions and colour are around him, but he rarely sees any one of them repeated. Time after time he goes towards a tree which looks like the one he seeks, but a closer examination proves it to be distinct.

Alfred Russel Wallace

Alfred Russel Wallace, who developed the same concept of natural selection as Darwin, traveled widely in the New World and Old World tropics. He describes perhaps the most striking feature of tropical forests—the astounding diversity of species. Biologists have long known that the tropics hold far more species than temperate and Arctic regions. But we are still learning just how stunningly diverse these regions are. Consider the Yasuni National Park in Ecuador. Recent surveys identify this site as perhaps the most species-rich forest in the world. Two survey plots totaling 50 hectares contain some 1,300 tree species (Bass et al., 2010). A single hectare of this forest—approximately the area of two football fields—contains at least 655 tree species and more than 900 vascular plants. This represents more tree species than are found in all of the United States and Canada!

This striking diversity is not confined to plants. The 150 species of amphibians and the 121 species of reptiles both represent world diversity records for a landscape of this size. To put this in perspective, Yasuni constitutes only 0.15 percent of the area of the Amazon, yet it harbors about one-third of all the amphibians found in the Amazon Basin. The bird species richness is similar—some 596 species that represent a third of all the birds of the Amazon. No other site holds anywhere near the 200 species of mammals found there or the 12 species of coexisting primates.

The Yasuni forest represents one extreme of a well-documented ecological pattern: the latitudinal trend in species diversity. Moving north or south from the equator, the number of species in virtually all taxonomic groups declines rapidly. In the Arctic, and especially in the Antarctic, species diversity is a tiny fraction of what we find in the tropics. This is perhaps not too surprising given the extraordinarily harsh climates of the high Arctic and Antarctic. But the trend also progresses through the relatively mild latitudes of the temperate zone.

What ecological processes underlie this striking pattern? In Chapters 13 and 14 we examined the processes that determine the structure of ecological communities. We now expand that analysis to a larger spatial scale. When we do so, we encounter a number of new questions. How do we measure species diversity? What patterns emerge from these measurements? Ultimately, this analysis will lead us to an explanation of the latitudinal trend in species diversity and, most important, allow us to answer the general question: *What factors determine species diversity?*

15.1   How Do We Measure Species Diversity?

We define **species diversity** as the number of species (**species richness**) present and their relative abundance. The latter term, relative abundance, is also called **species evenness**. A community with high evenness contains similar numbers of each species. Low evenness means that one or a few species are numerically dominant. Species diversity reflects the current ecological conditions as well as evolutionary history. The latter is reflected by the **species endemism**, the number of species endemic to a region—that is, found there and nowhere else. Endemism is the product of the isolation and evolutionary history of the local flora and fauna. Another related term, **biodiversity**, is more inclusive—it represents the total biological diversity in a community. It includes the species diversity in all taxonomic groups, the amount of genetic variation in their populations, and the ecotypes that have arisen.

Species diversity can be measured on a range of spatial scales. **Alpha diversity** is the species richness of a single community in one locale. The tree species richness in a 2 ha plot in Yasuni constitutes its alpha diversity. The species richness across several adjacent communities is termed the **beta diversity**. The diversity among a group of different communities on a landscape scale is the **gamma diversity**.

At first glance it seems a simple matter to measure species richness—we simply count up the number of species we encounter. However, just as was the case in measuring population density, this is a more difficult task than it might seem. Many species are cryptic or difficult to find. Others are present only seasonally, because of either migration or hidden resting stages. In some cases, we are limited by the extent to which the taxonomy of the groups is known and understood. In tropical regions, the sheer magnitude of the richness makes simple counting almost impossible; some type of sampling scheme must be employed to find species and measure their numbers.

The difficulty of this problem can be seen in the species-sample curve (Figure 15.1). If we plot the number of species encountered as a function of the number of samples taken, we find that species richness increases toward an asymptote. The asymptote usually indicates that all species in the community have been discovered and counted. We can use the species-sampling curve to determine whether or not enough samples have been taken. If the number of species

species diversity The combination of the number of species (species richness) and their relative abundance (species evenness).

species richness The total number of species in a community.

species evenness The relative abundance of each species in a community; high evenness means most species occur in similar numbers; low evenness means that a few species are numerically dominant.

Species endemism The number of species found in one place and nowhere else.

biodiversity The total biological diversity of a community.

alpha diversity Local species richness.

beta diversity Regional species richness.

gamma diversity The landscape-level species richness.

discovered is still rising, more samples are needed. If the asymptote has been reached, additional samples will yield few new species.

The sampling problem is further complicated in communities with low evenness. Any sampling system will emphasize the abundant species and may miss rare ones unless a large sampling effort is made. We see the difficulty in a graph of the number of species as a function of their abundance. Traditionally, the x-axis is the geometric abundance classes of the species in the community in increasing powers of 2 (Figure 15.2). In other words, the x-axis is a plot of abundance classes on a scale of $2^1, 2^2, 2^3....2^n$, (where n is the largest abundance class). Each abundance class on the axis represents twice the abundance of the preceding one. The y-axis measures the number of species in each of these abundance classes. In most communities, this results in a bell-shaped curve. This means that some species are abundant; others are rare. But most, the peak of the bell curve, occur at moderate abundance. However, the most abundant species, those on the far-right tail of the curve, will be represented most in any sampling scheme. The rare species, those on the left tail, will only be discovered with significant sampling effort.

As noted earlier, the other difficulty inherent in measuring species richness is the quality of our taxonomic understanding of the group. For poorly studied regions like the tropics, many species are not well defined and described. Prokaryote diversity is especially difficult to measure. Several features of this group complicate measuring species richness (Fraser, et al., 2009). First, the number of species is huge. Second, and perhaps more importantly, bacteria and other prokaryotes do not fit the traditional concept of a species. For plants and animals, species represent reproductively isolated groups that share recognizable morphological traits. However, bacteria reproduce by fission, eliminating the concept of reproductive isolation. Moreover, they are known to share genes horizontally—that is, among unrelated species—thus complicating classification based on genetic information.

The log-normal species abundance curve illustrates the connection between species richness and evenness. If species differ in abundance—that is, if evenness is low—we may presume that they differ functionally in the community as well. One index of species diversity that incorporates both richness and relative abundance is the **Shannon/Weaver index of species diversity**. This index was originally devised in the field of information theory as a measure of "uncertainty" in the transfer of information by messages. Its application in community ecology equates the concept of uncertainty with the functional diversity of a community. Communities with many species of approximately equal abundance are likely to be more functionally diverse than communities in which a single abundant species dominates. Imagine that you take a sample from two communities, one in which a single species is very abundant, and a second in which all the species are equally abundant. In the sample of the first community, there would be a high probability that you would capture the abundant species in the sample. But in the second, you would be less certain which species your sample might yield; the "uncertainty" of this community is high. The Shannon/Weaver index quantifies that type of uncertainty.

To illustrate, consider the two communities in Table 15.1.

The index is calculated

$$H = -\sum_{i=1}^{n} p_i \ln p_i$$

where n is the number of species in the community and p_i is the frequency of the *i*th species. For Community A,

$$H = -\sum_{1}^{10} [(0.01)(-4.6)] + [(0.01)(-.4.6)....[(0.91)(-0.09)]$$

$$= 0.50$$

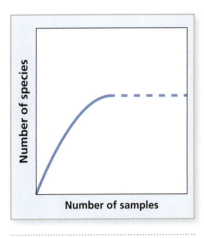

Figure 15.1 Species-sample curve. In a species-sample curve, the number of species discovered is plotted as a function of the number of samples that have been taken. When all species have been discovered in the samples, the curve reaches an asymptote. **Analyze:** What is the optimal level of sampling in this curve?

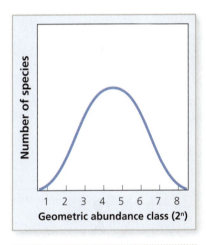

Figure 15.2 Log-normal species curve. The number of species in each abundance class is plotted against the geometric abundance classes, usually in the form of factors of 2^n. Each point on the x-axis is an abundance class (n) in which the number of individuals is 2^n. **Analyze:** What is the total range of abundances on this x-axis?

■ **Shannon/Weaver index of species diversity** A measure of diversity that incorporates both the number of species and their evenness.

TABLE 15.1 Two hypothetical communities. Each column represents the numbers of individuals of each species.

	NUMBER OF INDIVIDUALS IN COMMUNITY A	NUMBER OF INDIVIDUALS IN COMMUNITY B
Species 1	1	10
Species 2	1	10
Species 3	1	10
Species 4	1	10
Species 5	1	10
Species 6	1	10
Species 7	1	10
Species 8	1	10
Species 9	1	10
Species 10	91	10
Total individuals	100	100

THINKING ABOUT ECOLOGY:

Given what you know of the tree species richness of the Yasuni forest, would you expect this system to have high or low evenness? What would you predict for a high-latitude forest?

rank-abundance curve A graph of the log proportional abundance of each species as a function of the rank of its abundance.

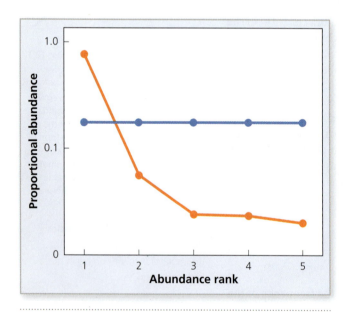

Figure 15.3 Rank-abundance curves. Two hypothetical rank-abundance curves. The log proportional abundance of each species is plotted against the species' abundance rank. The greater the slope, the more the community is dominated by one or a few species. **Analyze: Explain why a higher slope indicates a community dominated by a few species.**

For Community B,

$$H = -\sum_1^{10} < (0.1)(-2.31) + < (0.1)(-2.31)\ldots < (0.1)(-2.31)$$
$$= 2.30.$$

The communities have equal species richness (10 species), but the dominance by a single species in Community A is reflected in smaller value of the Shannon/Weaver index. Community B has a higher value; it has a higher level of uncertainty because of the even abundance of its species.

Another graphical procedure for assessing functional diversity is the **rank-abundance curve** (Figure 15.3). The rank-abundance curve is a plot of the log proportional abundance of each species as a function of its abundance. The slope of this curve is a measure of community evenness: the lower the slope, the more even the distribution; the steeper the slope, the more the community is dominated by a few species.

This kind of analysis allows us to address questions of community function. For example, Raybaud et al. (2009) developed rank-abundance curves for the phytoplankton and their zooplankton consumers in the Mediterranean Sea. Both groups have low evenness: the five most abundant species in each account for 80 percent of all the individuals. This is reflected in the shape of the rank-abundance curves, which are similar for the two functional groups (Figure 15.4). In this system, the community organization of the phytoplankton is the same as the organization of the herbivores (zooplankton) that consume them.

KEY CONCEPTS 15.1

■ The number of species and their relative abundance comprise species diversity.

- Species diversity is measured on several spatial scales (alpha, beta, and gamma diversity).
- Differences in species abundance, taxonomic confusion, and cryptic habits complicate sampling for species diversity.
- Species diversity and evenness reflect the functional ecology of the community.

QUESTION:

Explain the relationship between uncertainty, evenness, and community function.

15.2 How Do Communities Vary in Species Diversity?

If we hope to explain the processes underlying species diversity, we first need to understand how it varies. As in many areas of ecology, it is variation over space or time that helps us to understand ecological processes. The patterns of variation hint at their underlying causes and allow us to test hypotheses experimentally.

What Are the Spatial Patterns of Species Diversity?

The latitudinal trend in species diversity is among the most striking and thoroughly documented patterns, in part because of the extreme differences between the tropics and the Arctic. Although tropical communities on the whole contain more species than those at higher latitudes, not all tropical communities are equally diverse. Figure 15.5 shows some of the variation in species richness across

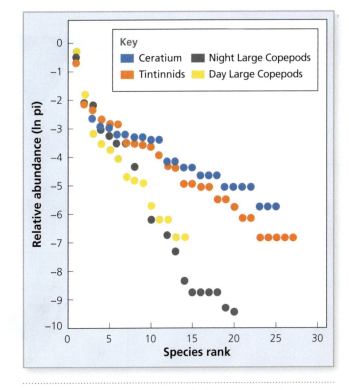

Figure 15.4 Phytoplankton and zooplankton rank-abundance curves. Rank-abundance curves for phytoplankton (*Ceratium* and tintinnids) and their zooplankton predators (copepods) in the Mediterranean Sea (from Rayaud et al., 2009). **Analyze:** What does this curve imply about the nature of this community?

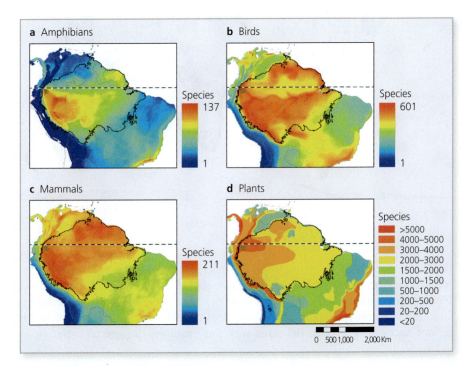

Figure 15.5 Species diversity. Variation in species diversity across the Amazon Basin for four taxonomic groups (from Bass et al., 2010). **Analyze:** What factors might account for the fact that each group has a slightly different pattern?

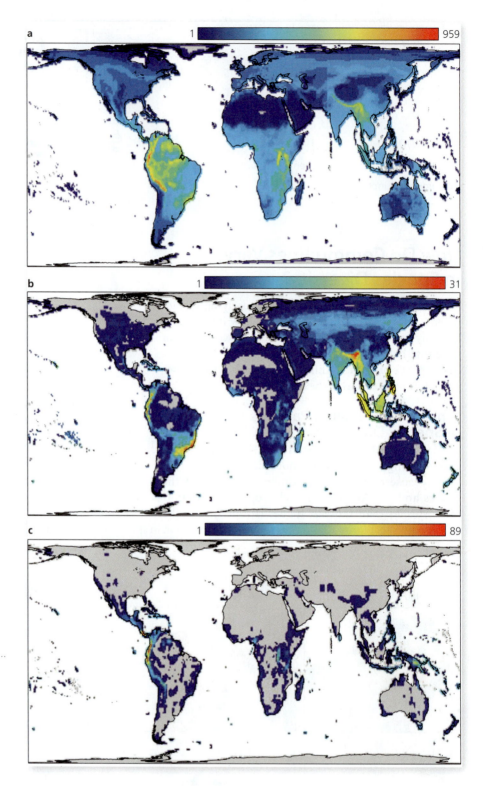

Figure 15.6 Measures of species diversity. Three different measures of species diversity: (a) total species richness, (b) threatened species richness, and (c) endemic species richness (from Orme et al., 2005). **Analyze:** Why are endemic species important to conserving threatened species?

the Amazon region. The Yasuni forest lies in a region of extreme species diversity. Nevertheless, even within Yasuni, there is variation associated with topography. Upland, drier sites contain more species of most taxa than lowland, flooded sites. In addition, across the tropics (and elsewhere), the different measures of diversity do not necessarily vary in the same way. For example, the spatial variation in

Figure 15.7 Species diversity of coral reefs. The high species diversity characteristic of coral reefs is shown in this sample of a single cubic foot from a reef.

species richness does not coincide with the spatial pattern of species endemism (Figure 15.6).

Ecotones, the boundaries between terrestrial habitats, usually have high species richness. We identify ecotones on many spatial scales, from the boundaries between two habitat types to broad regions on the borders of biomes. Regardless, ecotones typically contain more species than simply the sum of the diversities of the two habitats. Ecotones support plant and animal species specifically adapted to the habitat gradient between the two systems.

In marine ecosystems, two community types are known for their high species diversity. Coral reefs are among the most species-rich communities in either terrestrial or aquatic systems. Reef systems vary enormously in spatial scale—from small atolls to the extensive Great Barrier Reef of Australia. All house many species in layer upon layer of interactions, ultimately determined by the structure and production of the coral (Figure 15.7). The other species-rich marine system is, surprisingly, the deep-ocean floor. Ecologists long assumed this habitat, far below the depth to which light penetrates, to be a biological desert. However, more systematic sampling of benthic habitats and construction of species-sample curves indicate that these communities are remarkably diverse.

There are also communities that are consistently lower in species diversity. On average, peninsulas tend to hold fewer species than other comparable areas. Thus, for example, the Sonoran Desert of the Baja Peninsula is less diverse than the same desert across the Sea of Cortez in the large land mass of Mexico.

Islands also hold fewer species than mainland communities. Early in the study of this phenomenon, a pattern emerged that, as we will see, has great explanatory importance in species diversity: the **species-area effect**, in which larger islands contain more species than small islands. Moreover, the mathematical relationship between area and species diversity is remarkably constant across many different island systems. Species diversity and area are related according to the equation

$$S = cA^z$$

where S is the number of species, A is the area, z is the slope of the line, and c is a constant (Figure 15.8). The value of z varies only slightly, from 0.20 to 0.35, across a wide range of island systems. Another important factor in the number of species on an island is its distance from the mainland. More distant islands hold fewer species than nearby islands.

THINKING ABOUT ECOLOGY:

What is the significance of the slope of the diversity-sample curve?

■ **species-area effect** The increase in species richness as a function of area.

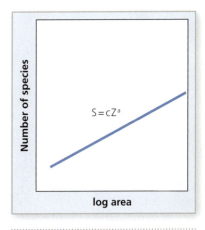

Figure 15.8 Species-area curve. The number of species increases linearly with the log area. **Analyze:** What is the significance of the y-intercept in this graph?

ON THE FRONTLINE

What Determines the Length of a Food Chain?

The relationship between species diversity and stability has been a matter of interest and debate among ecologists for more than 50 years. However, relatively few manipulative field experiments have been employed to address this relationship, and the few that have were short-term studies. Tilman et al. (2006) set out to test the key hypothesis in this relationship.

HYPOTHESIS: High species diversity increases the stability of the community.

The researchers addressed the resistance and resilience of communities that differ in diversity. One difficulty in field tests of this hypothesis is the design of a measure of stability that incorporates both components. Tilman et al. measured the temporal stability of aboveground biomass. Their measure of stability, S, quantified the constancy of the biomass relative to its mean. Specifically, they defined $S = \mu/\sigma$ where μ is the mean biomass and σ is its standard deviation. This variable was the basis of the prediction that follows from their hypothesis.

PREDICTION: S increases as a function of the number of species in the community.

Their long-term study was carried out at the Cedar Creek Ecosystem Science Reserve in Minnesota, one of many National Science Foundation Long-Term Ecological Research sites established for precisely this kind of long-term ecological fieldwork. They established 168 9 x 9 meter plots and seeded them with 1, 2, 4, 8, or 16 randomly chosen perennial grassland species. Each species diversity treatment was replicated at least 29 times. Each plot was carefully hand weeded to maintain the proper number of species. Sampling occurred in mid-August, when biomass reached its annual maximum. All aboveground plant biomass was clipped, dried, and weighed. Their experiment ran for 10 years.

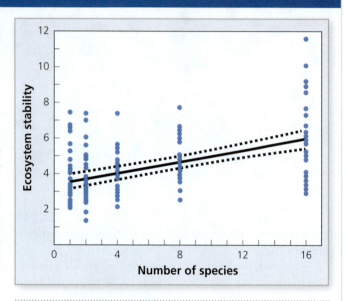

Figure 1 Ecosystem stability. Ecosystem stability as a function of the number of species in tallgrass prairie (from Tilman et al., 2006). **Analyze: Do you think that some species are more important to ecosystem stability than others?**

The key result is shown in Figure 1. The value of S increased significantly as a function of the number of species in the plot. High-diversity plots were 70 percent more stable than low-diversity plots. S also increased slightly as each plot matured. This is an important result for two reasons: First, it suggests that developmental processes in communities interact with factors such as species diversity. Second, it emphasizes the importance of long-term studies: the change in stability they observed might have been missed or deemphasized in a shorter-term study.

What Are the Ecological Consequences of Differences in Species Diversity?

We might expect that communities of different diversity should be functionally different. Charles Elton (1927) was the first to address this relationship. Elton reasoned that diverse systems have more complex webs of interaction among species. He proposed that this should make diverse systems more stable, because any single species in a complex network is less important than it would be in a simpler system. In effect, more species and interactions contribute functional redundancy. Perturbations, even the elimination of species, should have a smaller impact in diverse systems.

Elton's reasoning set the stage for a long and contentious debate among ecologists. Some theoretical work suggested that Elton's logical argument is not supported mathematically. Part of the difficulty lies in precisely defining **ecological stability**. Two general concepts of stability have emerged. One centers on a system's pattern of change, its dynamic stability. According to this concept, stable communities are those in which the variation in community function is small

■ **ecological stability** The tendency of community composition and function to remain constant.

■ **resistance** The tendency of a community to remain constant when disturbed.

and centered on an equilibrium value. The other concept is based on the community reaction to perturbation. In this view, stability has two important components: **resistance**, the tendency of a system not to change when disturbed, and **resilience**, the ability of the system to recover from disturbance.

■ **resilience** The ability of a community to return to equilibrium following disturbance.

THE HUMAN IMPACT

The Sixth Mass Extinction

The history of life on Earth is a long series of changes in the planet's species diversity. A major increase in biological diversity, known as the Cambrian explosion, occurred some 530 million years ago (Figure 1). Many new groups, some radically different from anything known today, arose in a relatively short geological period. This was a time of great evolutionary innovation: most extant animal phyla arose during the Cambrian explosion.

Once these new forms arose, they were sorted by the ecological and selective forces that have operated over the millennia. The fossil record documents the emergence of new species (and taxonomic groups) as well as the extinction of many. The fossil record also indicates that the origin and extinction of species did not proceed at a steady rate. Rather, there were periods of sharp declines in diversity, sometimes followed by bursts of new species filling the empty niches. Five mass-extinction events have been documented in Earth's history (Figure 2). The causes of these changes are varied—they range from the tectonic movement of

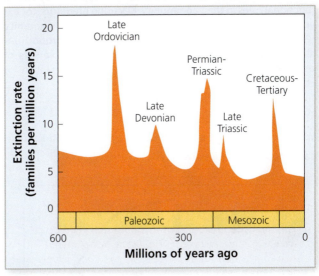

Figure 2 Five mass extinctions. The five mass extinctions in geological history. **Analyze: What is the background level of extinction in this figure?**

continents that radically reduced the extent of shallow seas to the impact of an asteroid that caused the extinction of the dinosaurs.

Today we seem to be in the middle of a sixth mass-extinction event. This one, however, is anthropogenic—the direct result of human activity. This extinction event began some 100,000 years ago, when *Homo sapiens* first migrated from Africa to all parts of the Earth. The second component of it began about 10,000 years ago and coincided with the increase in the human population and the development of agriculture. Currently, we are in a third phase as a result of the Industrial Revolution and its impact on environmental quality and climate. Different taxonomic groups and the inhabitants of different ecosystems are differentially impacted by human activity. Approximately 70 percent of coral reefs, among the most species-diverse systems, have been significantly degraded, and at least one-fifth of our reefs are in imminent danger of collapse (Avise et al., 2008). Amphibians are particularly vulnerable to anthropogenic effects on habitat and climate: more than one-third of the 6,300 species of amphibians are at risk of extinction.

The result is a significant loss of global species diversity. We can now document the changes in species diversity among ecosystems and taxonomic groups well enough to identify the trends (Figure 3). These data suggest that we are indeed in the midst of a sixth mass-extinction event. For example, Barnosky et al. (2011) compared the current extinction rate in mammals

continued

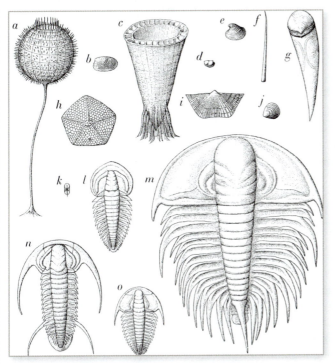

Figure 1 Cambrian explosion. The Cambrian explosion resulted in many novel forms of life. **Analyze: Is it possible that the extinction rate is not a linear function of the number of species?**

THE HUMAN IMPACT   *continued*

The Sixth Mass Extinction

with that in the fossil record. Over the last 500 years, the extinction rate for mammals increased by more than an order of magnitude compared to any previous time in the fossil record.

We know that species diversity plays a role in the function of ecological communities, specifically that diverse communities are often more stable. In addition, some species play crucial functional roles in their community. Keystone predators, for example, determine the competitive interactions in a community (Chapter 13). Their loss by extinction would have a proportionally large impact on the community, affecting whole sets of other species. Thus, the loss of species at this rate is reason for concern.

QUESTION:

Why do you think the origin of human agriculture had a significant impact on species diversity?

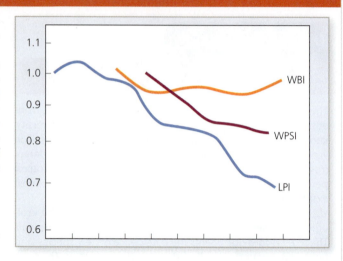

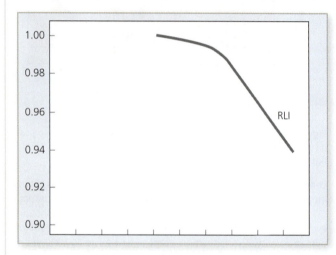

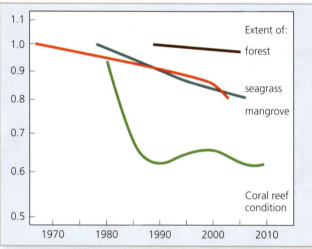

Figure 3 Loss of species diversity. Changes in biodiversity over time for several groups of organisms and habitats. WBI = wild bird species index; WPSI = waterbird population status Index; LPI = living planet index, a measure of vertebrate diversity; RLI = red list index, a measure of extinctions (from Butchart et al., 2010). **Analyze: What information would you need to compare the magnitude of these losses to the historical mass extinctions?**

Careful empirical work supports the general principle that high species richness makes communities more stable. A number of studies show that resource use is more efficient in more diverse communities. For example, microbial diversity increases the efficiency of resource acquisition, which in turn stabilizes the composition of the community (Ptacnik et al., 2008). The stability of European grassland communities is related to the diversity of mutualistic mycorrhizal fungi (van der Heijden et al., 1998): variation in plant diversity and stability are related to fungal diversity (Figure 15.9). A review of the many studies of this hypothesis shows that a large proportion of published studies (69 percent) demonstrate a positive relationship between diversity and stability (Ives and Carpenter, 2007).

KEY CONCEPTS 15.2

- Species diversity varies widely across community types and with geography.
- Low-latitude ecosystems, ecotones, coral reefs, and marine benthic communities are known to have high species richness.
- Islands and peninsulas tend to have fewer species than other comparable communities.
- Species diversity is related directly to area in a similar pattern across a wide array of island systems.
- Species diversity increases the stability of the system.

QUESTION:

How do resistance and resilience each relate to the logic of the diversity-stability paradigm?

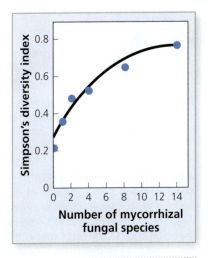

Figure 15.9 Plant and fungal diversity. The diversity of grasslands plant community increases as a function of fungal diversity (from van der Heijden et al., 1998). **Analyze:** How could you tell if plant diversity affects fungal diversity or vice versa?

15.3

Why Do Communities Differ in Species Diversity?

The species diversity in any region is explained by just three processes: colonization, extinction, and the origin of new species. Species richness increases by colonization and the formation of new species; it decreases by the process of extinction. As you know, an important dichotomy pervades ecology: equilibrium and nonequilibrium systems. Although the three fundamental processes apply to all communities, their operation and relative importance differ in equilibrium and nonequilibrium communities.

What Determines Species Diversity in Equilibrium Communities?

The phenomenon of low species richness on islands has been an important area of inquiry in ecology. Originally it was thought that islands hold few species simply because they have not had time to accumulate since the island's origin. In effect, this is a nonequilibrium explanation: given sufficient time, islands would eventually achieve the species richness of the source pool of species on the mainland.

However, seminal theoretical work by MacArthur and Wilson (1963) demonstrated that island species diversity is best explained as an equilibrium process. This theory, now known as the **equilibrium theory of island biogeography**, holds that the low species richness of islands is not simply a matter of insufficient time; it is an inherent property of islands. MacArthur and Wilson's theory is based on the relationship between the two forces that determine the number of species on an island—the immigration rate and the extinction rate. Colonization is a random process. Dispersing individuals traverse open water to arrive at an island. As species gradually colonize, the rate of immigration declines because there are fewer species that have not yet arrived (Figure 15.10). As the number of species on an island increases, so too should competitive interactions and the extinction rate. An equilibrium occurs where

- **equilibrium theory of island biogeography** An explanation of species richness on islands based on a balance between colonization and extinction.

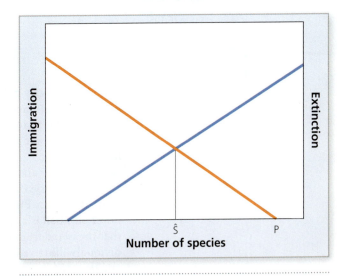

Figure 15.10 Theory of island biogeography. In the theory of island biogeography, immigration decreases as a function of the number of species on the island; extinction increases as the number of species increases. The equilibrium number of species on the island Ŝ occurs where immigration exactly balances extinction. P is the number of species in the mainland source pool (MacArthur and Wilson, 1963). **Analyze:** Is it possible that the extinction rate is not a linear function of the number of species?

DO THE MATH

The Equilibrium Number of Species on an Island

The theory of island biogeography is based on the species area. As we have seen, the species-area relationship is described by the power function

$$S = cA^z$$

where S is the number of species on the island, A is the island's area, and C and Z are constants. MacArthur and Wilson's theory explains island species diversity as a result of an equilibrium between colonization (λ) and extinction (μ). The pool of species on the mainland (P) is the set of species that can potentially colonize the island. We can depict the rate of change of species number on the island as dS/dt, which is the difference between the colonization and extinction rate:

$$dS/dt = \lambda - \mu.$$

Figure 15.10 shows the change in immigration as a function of the number of species on the island. When there are no species on the island, the immigration rate is maximal (I). When all P mainland species have arrived, the immigration rate must fall to 0. These two points define a straight line in which the intercept is I and the slope is $-I/P$:

$$\lambda = I - (I/P)S.$$

The extinction rate is an increasing function of the number of species. In a manner analogous to the immigration rate, the maximum extinction rate occurs when the island contains as many species as possible (P). It is zero when there are no species on the island. These two points also define a straight line:

$$\mu = (E/P)S.$$

We can now use these equations to measure the rate of change of species on the island:

$$dS/dt = \lambda - \mu$$

$$= I - (I/P)S - (E/P)S.$$

The equilibrium number of species on the island occurs when immigration and extinction are zero. In other words, we can determine the equilibrium number of species by setting $dS/dt = 0$:

$$0 = I - (I/P)S - (E/P)S.$$

We obtain the equilibrium number of species by solving this equation for S:

$$I = (IS/P) - (ES/P)$$

$$I = S(I - E)/P$$

$$S = IP/(I - E).$$

What does this tell us biologically? The equilibrium number of species on an island is determined by three variables. The larger the mainland pool (P) and the immigration rate (I), and the smaller the extinction rate (E), the more species the island will hold at equilibrium. This makes intuitive sense. The mathematics allows us to quantify the relationship between these three variables. And it allows us to ask relevant research questions such as, what biological factors determine the immigration rate? And what processes increase the extinction rate? Or what is the size of the mainland pool of species contributing to the island?

extinction exactly balances immigration. The number of species at equilibrium will be smaller than the mainland pool. No matter how long we wait, the species richness of the island no longer increases. The actual species present may change over time as some go extinct and others colonize. But the *number* of species no longer changes.

MacArthur and Wilson's theory also explains two key observations of island species diversity: that small and distant islands have fewer species than large and near islands. In Figure 15.11 we see that the immigration rate is consistently lower on small islands and on distant islands. This makes sense because the farther the island from the mainland, the lower the probability that a dispersing individual will arrive. Similarly, small islands make a smaller "target" for immigrants, and they, too, have a lower immigration rate. There is no effect of distance on the extinction rate. However, small islands have a higher extinction rate, because the limited resources on a small island cannot support as many species. The theory predicts smaller equilibrium species richness for small and distant islands.

Because of extensive empirical support, the equilibrium theory of island biogeography is widely accepted. One of the key elements of this theory is that it makes testable predictions. For example, according to the model, the equilibrium is dynamic—species continue to colonize and go extinct. Consequently, although the model predicts an equilibrium number, the particular species comprising

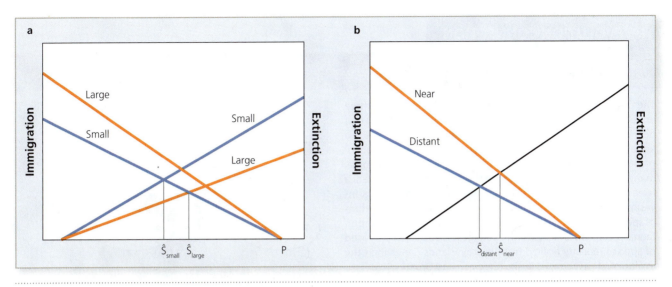

Figure 15.11 Equilibrium diversity. The effects of size (a) and distance (b) on equilibrium diversity on islands. (a) The immigration rate is lower on small islands than large; the extinction rate is higher on small islands than large. The result is a lower equilibrium diversity on small islands. (b) The immigration rate is lower on distant islands. The result is a lower equilibrium value on distant islands (MacArthur and Wilson, 1963). **Analyze: Why is there no effect of distance on extinction?**

that equilibrium are expected to change. This contrasts with the expectation of a nonequilibrium explanation, in which species gradually accumulate on islands. Diamond (1969) used an early twentieth-century survey of the bird species diversity to test this prediction of the model. In 1917, Howell surveyed the Channel Islands off the coast of California and determined the bird species diversity of each. Diamond repeated this work some 50 years later. He found that, just as predicted by the MacArthur-Wilson model, some species had gone extinct and some new species had colonized, but the total number of species on each island was essentially unchanged (Figure 15.12).

Others used manipulative experiments to test the theory. Simberloff and Wilson (1969) surveyed the arthropod fauna on a series of small mangrove islands in Florida Bay. As expected, the islands differed in species richness according to size and distance from the mainland. They exterminated all species from the islands and monitored their recolonization. The islands returned to their preextermination diversities (Figure 15.13). However, the actual species on each island differed from the composition before the original species were removed.

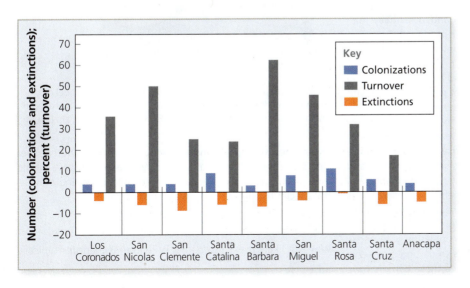

Figure 15.12 Channel Islands bird species diversity. Colonizations (blue columns), extinctions (red columns), and the percent turnover in the number of species present from 1917 to 1967 for the Channel Islands of California (data from Diamond, 1969). **Analyze: What should be the values of turnover if this were a nonequilibrium rather than an equilibrium system?**

THE EVOLUTION CONNECTION

How Species Arise

The diversity of species in a community is determined by three important processes—immigration, extinction, and the origin of new species. Biogeographic analyses indicate that species are added to communities first by immigration from a source pool of preadapted species (Emerson and Gillespie, 2008). For example, many of the plant species found in the Andes of South America have close relatives in high-altitude regions of North America. Invasion of South America by these species determined the pattern of species diversity in the new habitat. Similarly, in the arthropod fauna of the Canary Islands, the youngest islands are composed of species recently arrived from nearby islands rather than the evolution of new species in situ (Emerson and Oromi, 2005).

As species accumulate in a region, the initial determinants of diversity are immigration and extinction. Over time, however, new species that evolved in situ may also contribute to species diversity. In some groups, such as the genus *Drosophila* on the Hawaiian Islands, new species arise very rapidly, and this process contributes importantly to species diversity.

How do new species arise? We begin with the standard definition of a *species*: a group of interbreeding individuals that is reproductively isolated from other such groups. In other words, a species is an isolated gene pool; gene flow from other groups does not occur. Thus, we can address the question of how *speciation*, the processes by which new species arise, occurs by asking how a gene pool becomes isolated.

Although there are a number of mechanisms of speciation, evolutionary biologists believe the *allopatric speciation model* is particularly important. In allopatric speciation, a geographic barrier separates two populations of a species. This may arise in a number ways, such as a shift in a river's course, the uplift of a mountain range, or the advance of a glacier. The effect is to separate the populations and restrict gene flow between them. If the environmental conditions are different on the two sides of the barrier, each population will undergo adaptive evolution to the specific conditions it faces. When this occurs, a new selective force emerges: selection against hybrids formed by mating of an individual with one from the other environment. Hybrid matings reduce the fitness of the parents because their offspring carry genes adapted for another region and potentially maladaptive for the habitat where they are produced. Selection favors individuals that mate only with individuals from their same region.

This new pattern of selection produces *reproductive isolating mechanisms*, features of the organism that prevent mating with individuals from another region. These occur in two basic forms, *pre-mating isolating mechanisms* and *post-mating isolating mechanisms*. Post-mating mechanisms often arise first because of genetic or developmental anomalies that arise in hybrid matings. Selection then favors pre-mating barriers so that gametes and mating activity are not wasted on failed hybrids. Pre-mating isolating mechanisms include the complex behavioral interactions between males and females known as courtship rituals. Each member of the pair must perform the correct behavior in sequence and in response to the other if copulation is to occur. Morphological and physiological barriers to mating also serve as pre-mating isolating mechanisms. Once these barriers have

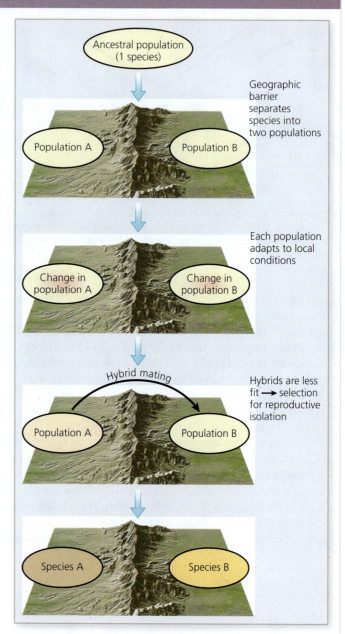

Figure 1 Allopatric model of speciation. The allopatric model of speciation depends on physical separation of two populations and adaptation to the local conditions. **Analyze: Explain why hybrids are less fit.**

arisen, the two populations are, by our definition, different species. Their gene pools are now entirely independent; even if the barrier changes to allow more frequent contact, the single ancestral species has given rise to two new species.

QUESTION:

Why is it likely that post-mating isolating mechanisms typically arise first?

We know that competition plays a key role in sorting the species that can persist in a community and how the niches of coexisting species are structured. Species-rich communities are organized according to three aspects of their species' niches: decreased niche breadth, increased overlap, and an increase in the range of resources utilized. We know that there is a limit to niche overlap; competitive exclusion prevents this mode from progressing past some limit. The other two are, in fact, aspects of niche specialization. Thus, environmental factors that promote specialization lead to increased species diversity.

Among the most important of these factors is **habitat complexity**. Habitat complexity refers to heterogeneity in the physical, topographic, or functional components of the habitat. As complexity increases, opportunities for unique and specialized niches increase. So, for example, forest communities that contain complex layers of vegetation—herbs, shrubs, saplings, and large trees—have higher habitat complexity compared to a salt marsh with a single, uniform plant structure. Bird species diversity increases with measures of the structural diversity of the vegetation (Recher, 1969; Figure 15.14). The physical structure of the environment also plays a role in this process. A **keystone structure** is a spatial structure, physical or biological, that provides resources, habitat, or shelter (Tews et al., 2004). Savanna trees in the arid grasslands of southern Africa serve as keystone structures in that community (Figure 15.15). They provide focal structural resources for many species as nest sites, shade, food, and lookouts. Feces or carcass remains under these trees constitute important resources for another whole set of species. Thus, this structural element, large isolated trees in a matrix of grass, constitutes the resource base for many niches. Large complex corals play the same role in coral reefs. Their structure provides habitat for the rich reef community. The decline of coral reef communities is associated with the death of the corals that constitute keystone structures (Alvarez-Philip et al., 2009).

Biotic interactions are another key factor promoting niche specialization and thus species diversity. As species diversity increases, so too do the opportunities for specialized biological interactions. The species diversity of insects in the tropics is staggering—estimates of the number of herbivorous insects range from 3 to 30 million species. There is evidence of extreme specialization among these insects. For example, a single population of plants can have associated with it several insects specific to just the female flowers and others that specialize on male flowers (Condon et al., 2008). This is just one example of the myriad insect-plant interactions that suggest that tropical insect diversity is a response to plant phylogenetic and taxonomic diversity in the tropics (Novotny et al., 2006). The biotic diversity of plants provides opportunities for a diverse insect fauna. And, of course, these species and their interactions generate new potential niches for predators, parasites, and a host of other species.

What Determines Species Diversity in Nonequilibrium Communities?

Of course, many ecological systems are not at equilibrium, or if they are, that equilibrium is short-lived. A different set of processes is important to species diversity in disturbed communities. These processes vary according to the time scale on which we focus. We distinguish between the Red Queen model and the

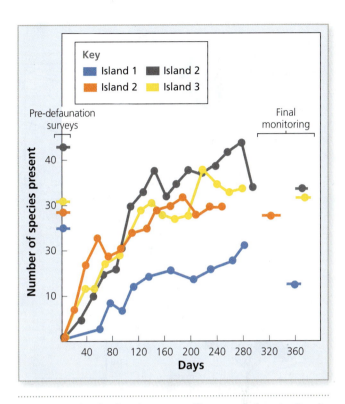

Figure 15.13 **Species diversity patterns.** The patterns of species diversity accumulation on defaunated mangrove islands. Values on the y-axis represent the diversity before defaunation. Lines represent the accumulation of species after defaunation (from Simberloff and Wilson, 1969). **Analyze: Would you expect all the islands to recolonize at the same rate?**

habitat complexity Variation in the abiotic, topographic, or functional components of the habitat.

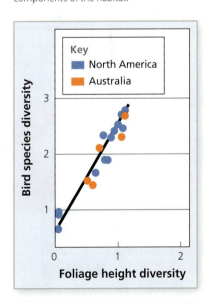

Figure 15.14 **Habitat complexity.** Bird species diversity increases as a function of the diversity in foliage height, the physical structure of the vegetation (from Recher, 1969). **Analyze: What relationship would you expect between bird niches and vegetation structure?**

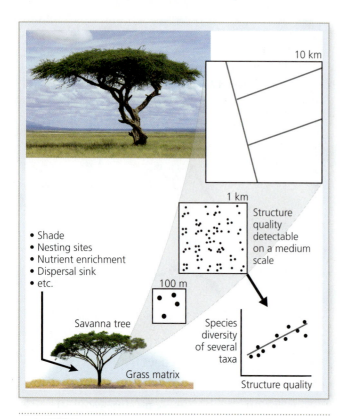

Figure 15.15 Savanna trees. In African savanna, large isolated trees constitute keystone structures (Tews, 2004).

keystone structure Biological or physical habitat structure that provides important resources to a species.

Court Jester model The idea that species diversity is a product of stochastic disturbance events.

Court Jester model of diversity. We have previously encountered the Red Queen model in the context of the evolution of sexual reproduction. Biotic interactions of all kinds are central features of ecological communities. Each species faces increasing selective pressure from the other members of the community. The Court Jester model, named for the unpredictable comedians of medieval times, emphasizes random changes in the physical environment, disturbances that impact the number of species in the community. The two models differ somewhat in the spatial and temporal scales over which they operate (Figure 15.16). The Court Jester model is applicable to events on longer time scales and in larger geographic areas.

The disturbance events central to the Court Jester model affect the key determinants of species diversity: extinction, colonization, and speciation. The main difference from equilibrium communities is the greater impact of chance in nonequilibrium communities. Extinction due to disturbance is stochastic—it is the result of both the random nature of the disturbance itself and chance factors that permit some species to survive while others go extinct. When a disturbance reduces species diversity, immigration is the short-term means by which diversity is restored. As in the case of immigration to islands, this process has an important stochastic component. Speciation plays a role in species diversity only over long time scales.

There have been five mass-extinction events in geological history (see "The Human Impact"). Each drastically reduced species diversity on a global scale. In some cases, the radical loss of certain habitats, such as shallow seas, led to a collapse of the diversity of species dependent on that habitat. In other cases, a cataclysmic event caused environmental changes whole taxonomic groups could not tolerate. The asteroid impact 65 million years ago apparently resulted in climatic changes that the dinosaurs could not survive; this entire taxonomic group abruptly disappeared.

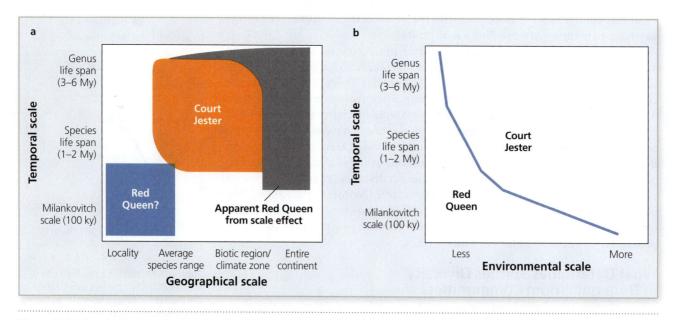

Figure 15.16 Models of species diversity. The Court Jester model and the Red Queen model operate at different geographic and temporal scales (from Benton, 2009). **Analyze: Why should these models differ in this way?**

The Court Jester model applies to the patterns of species diversity across this history of cataclysmic disturbance. The severity of the disturbances and the subsequent recovery of the flora and fauna differ markedly in the fossil record. For example, following the asteroid impact 65 million years ago, the recovery of insect herbivore species richness was relatively slow (Wilf et al., 2006). Thus, cataclysmic disturbances can have long-lasting effects on species diversity.

Speciation has played a key role in generating species diversity following major disturbance events. The extinction of the dinosaurs following the asteroid impact vacated a number of ecological niches. These were filled by a series of speciation events in the surviving mammal fauna. This group was present prior to the mass extinction but at low diversity. The diversification of a taxonomic group to fill a set of newly available niches is known as an **adaptive radiation**. Some adaptive radiations occur remarkably quickly, such as when islands are colonized for the first time. The adaptive radiations of Darwin's finches on the Galápagos and the honeycreepers on the Hawaiian Islands (see Chapter 11) exemplify this kind of rapid speciation.

Disturbance events also occur locally and across short time scales. The immediate effect of disturbance is often to reduce species diversity. Consequently, many of the biotic interactions among species, such as competition, predation, and parasitism, are reduced or eliminated. For example, a disturbance such as fire often leads to a temporary reduction in species diversity, because many species are driven locally extinct by the fire itself and because a small number of species dominates in the immediate post-fire ecological conditions. Species adapted to exploit the abundant resources available when most competitors have been removed can be extraordinarily successful. If they are able to exploit the resources in the absence of competition, they can dominate the community, depressing its species diversity for the period of their dominance.

In high-diversity communities such as coral reefs, local disturbance followed by colonization is an important process that promotes regional diversity. The **lottery model** was developed to explain such a pattern of species diversity (Sale, 1979). On the Great Barrier Reef in Australia, many species of fish have such high niche overlap that local coexistence is not possible. However, local reef patches experience regular disturbance. They are recolonized at random from a large pool of potential species. A subset of this pool successfully colonizes—that is, "wins the lottery." The result is a mosaic of patches, each with its own complement of species, and thus high beta diversity for the reef system.

Not surprisingly, the magnitude and frequency of disturbance contribute to the pattern of species diversity. According to the **intermediate disturbance hypothesis** (Connell, 1978), species diversity is highest at moderate levels of disturbance. If disturbance is rare, species diversity declines, because a few species competitively dominate the community, excluding others. When the frequency of disturbance is so high that few species can persist, the community has lower diversity (Figure 15.17). This phenomenon has been well documented in a number of communities. For example, the forest-grassland ecotone in Minnesota (Peterson and Reich, 2008) conforms to the intermediate disturbance hypothesis. Fire frequency determines the species diversity of the understory vegetation at the margins of the forest (Figure 15.18). The canopy tree species competitively suppress the understory vegetation. Fire reduces their dominance, thus increasing understory diversity. The intermediate disturbance hypothesis has also been documented in stream communities where drought and flood constitute important disturbances (Lake, 2000).

adaptive radiation The evolutionary diversification of a group of species, often in newly colonized regions, into a number of new species in new niches.

lottery model A model in which species diversity is the result of disturbance followed by colonization.

intermediate disturbance hypothesis The concept that species diversity is low where disturbance is either frequent or infrequent and maximal at intermediate levels.

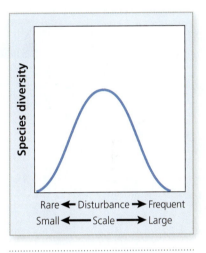

Figure 15.17 Intermediate disturbance hypothesis. The intermediate disturbance hypothesis proposes that species diversity is maximal at intermediate levels of disturbance (Connell, 1978). **Analyze: Why should rare disturbance lead to fewer species?**

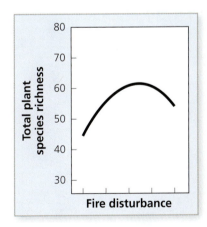

Figure 15.18 Forest-grassland ecotone. The understory species richness in a grassland-forest ecotone is highest at intermediate levels of fire disturbance (from Peterson and Reich, 2008). **Analyze: How does fire affect grassland and forest systems?**

KEY CONCEPTS 15.3

- The theory of island biogeography explains the pattern of species diversity on islands in terms of the relationship between colonization and extinction.
- Species diversity increases with the complexity of the habitat and potential biotic interactions.

THINKING ABOUT ECOLOGY:

You are studying the diversity of the Great Barrier Reef. You find that each patch has a constant number of species even though the actual species present change over time. This is expected under both the lottery model and the theory of island biogeography. How would you test whether the pattern is due to the lottery model or the hypothesis that each patch acts like an island?

- Many ecological communities are not at equilibrium. They are dominated by disturbance events. These occur on a range of spatial and temporal scales.
- Chance events are important in nonequilibrium communities. Disturbance events are stochastic in terms of occurrence and severity. Colonization following disturbance is also a random process.
- The same key processes determine species diversity in equilibrium and nonequilibrium communities: speciation, extinction, and colonization. Equilibrium and non-equilibrium communities differ in the relative importance of these processes and the time scale over which they are important.

QUESTION:

How do the processes underlying species diversity differ with respect to the role of chance?

Putting It All Together

The latitudinal trend in species diversity is among the most thoroughly documented patterns in ecology. And the difference between the tropics and higher latitudes is so striking that we might expect its explanation to provide insight into fundamental ecological processes. As is the case for many complex ecological phenomena, no single explanation is sufficient. More important for our purposes here, the explanation for tropical species diversity summarizes many of the key elements of this chapter.

The latitudinal trend in species diversity ultimately results from spatial variation in the key determinants of diversity: colonization, extinction, and speciation. All three, especially the extinction and colonization rates, are affected by the fact that the tropics comprise a larger land area than found in higher latitudes (Rosenzweig, 1992). This is especially true in the Southern Hemisphere, where the continents taper to the south.

As we know from the species-area curve, larger areas hold more species. The larger area provides more resources and supports larger populations. The larger land mass also contains more habitat types and greater topographic diversity, both of which support more species. Consequently, the extinction rate is lower in the tropics.

Unlike the Arctic and much of the temperate zone, the tropics have not had a major geological disturbance for millions of years. Not only has this also reduced the extinction rate, the region has had ample time for the accumulation of species via new speciation events. Once this speciation-extinction differential led to higher diversity, biotic interactions could further increase diversity as new species arose to fill niches created by others.

Although the tropics have been geologically and climatically stable for a long time, disturbance does occur on short time scales and over small spatial scales. Major river systems flood regularly, disturbing lowland tropical forest. Montane tropical forests experience repeated landslides that disturb the vegetation. And although the tropics do not have winter, many parts of the tropics have seasonal variation in rainfall in which relatively dry periods are interspersed with wet seasons. As a result, fire periodically disturbs even some tropical forest and shrublands. As we have seen, the lottery model can generate high beta diversity in this kind of scenario. Moreover, in flood- and landslide-prone habitats, the frequency and intensity of local disturbance appear to result in high alpha diversity, as predicted by the intermediate disturbance hypothesis.

As with many ecological phenomena, the answer to the question "What ecological factors determine species diversity?" is both simple and complex. Just three key processes—colonization, extinction, and speciation—explain any pattern of diversity. The complexity arises from the variety of factors and mechanisms that determine the relative importance of these three processes in any particular place.

Summary

15.1 How Do We Measure Species Diversity?
- Species diversity is composed of the number of species and their relative abundance.
- Species diversity changes with the spatial scale over which it is measured.

15.2 How Do Communities Vary in Species Diversity?
- Species diversity varies geographically and among community types.
- The tropics, ecotones, coral reefs, and marine benthic communities typically have high diversity; high latitudes, islands, and peninsulas tend to have fewer species. Species richness is related to area on islands (and many mainland systems)

by the relationship $S = cA^z$. The exponent (z) is similar across a wide array of island systems.
- There is a complex, positive relationship between species diversity and community stability.

15.3 Why Do Communities Differ in Species Diversity?
- Species diversity is ultimately determined by the processes of colonization, extinction, and species formation.
- Island species richness is explained by the equilibrium theory of island biogeography, specifically the balance between immigration and extinction.
- Habitat diversity and webs of biological interactions increase species diversity.
- Stochastic processes are particularly important in nonequilibrium communities.

Key Terms

adaptive radiation p. 345
alpha diversity p. 330
beta diversity p. 330
biodiversity p. 330
Court Jester model p. 344
ecological stability p. 336
equilibrium theory of island
  biogeography p. 339

gamma diversity p. 330
habitat complexity p. 343
intermediate disturbance
  hypothesis p. 345
keystone structure p. 344
lottery model p. 345
rank-abundance curve p. 332
resilience p. 337

resistance p. 336
Shannon/Weaver index of species
  diversity p. 331
species-area effect p. 335
species diversity p. 330
species endemism p. 330
species evenness p. 330
species richness p. 330

Review Questions

1. Why do we need several measures of species diversity?

2. How can we determine when we have taken enough samples to determine the species diversity of a community?

3. Why do the extinction and immigration rates change as a function of the number of species on an island?

4. Why do the actual species present on an island change when the island is at equilibrium?

5. What are the key elements of community stability?

6. Which of the spatial patterns of species diversity are related to the species-area curve?

7. What are the long-term effects of disturbance on species diversity?

8. What are the short-term effects of disturbance on species diversity?

9. Why do intermediate levels of disturbance lead to higher species diversity?

10. Which explanations of species diversity rely in chance events; which emphasize deterministic processes?

Further Reading

Adams, J. 2008. *Species richness: patterns in the diversity of life.* New York: Springer.

  This book is an excellent summary of the patterns and explanations of species diversity across an array of ecosystems.

Carson, W.P., and S.A. Schnitzer. 2008. *Tropical forest community ecology.* Panama City: Smithsonian Tropical Research Institute.

  This book explores tropical community ecology. There are important sections on large-scale and local processes that promote species diversity. There is also an excellent section on conserving the spectacular diversity of tropical systems.

Kolbert, E. 2014. *The sixth extinction: an unnatural history.* New York: Henry Holt & Co.

  This popular book is an account of the current extinction crisis. The author travels with a number of researchers studying the loss of species due to human activity.

Markussen, M., et al. 2005. *Valuation and conservation of biodiversity: interdisciplinary perspective on the Convention on Biological Diversity.* New York: Springer.

  This book looks at biodiversity from disciplinary perspectives beyond biology. The International Convention on Biological Diversity was designed to conserve diversity. This book addresses the challenges to implementing this agreement.

Chapter 16

Energy Flow and Trophic Structure

The blue whale, *Balaenoptera musculus*, is the largest animal ever known to live on Earth. Adults reach lengths of more than 33 meters and attain weights in excess of 200 tons. Before their numbers were decimated by whaling, blue whales reached their highest density in the Antarctic Ocean—estimates place their numbers at nearly 300,000. Imagine that many huge mammals in the coldest ocean on the planet. To maintain a constant body temperature, feed, and reproduce, a blue whale requires an astounding 1.5 million kilocalories, some 3,600 kilograms of food, every day. How is this energetically possible? How is that much food available, and how can the whale satisfy its caloric needs?

The answer lies in the food base to which the blue whale is specially adapted. Blue whales are one of many species of baleen whales that feed by filtering small invertebrates from the water. Their mouths are lined with rows of baleen (Figure 16.1) that act as strainers. The whale ingests a large volume of water and the animals it contains. Pressure from the tongue pushes the water past the baleen and out the mouth, leaving the food trapped on the baleen. In the Antarctic, the main food source for blue whales is krill (*Euphausia spp.*), small shrimp-like crustaceans (Figure 16.2). In Antarctic waters, krill are phenomenally abundant—aggregations of krill are sometimes visible from space. Some scientists believe that krill are the most abundant macroscopically visible animal on Earth. They constitute the energy source that powers the blue whales of the Antarctic. An adult blue whale eats up to 40 million krill per day.

Figure 16.1 Baleen whales. Baleen whales contain plates of baleen composed of keratin. These plates act as strainers that trap small invertebrates.

■ **heterotroph** An organism that obtains energy from organic compounds.

■ **autotroph** An organism that obtains energy from inorganic sources.

Why are krill so abundant? Krill consume phytoplankton (and some zooplankton), which are also extraordinarily abundant. The immensity of the Southern Ocean, its abundant nutrients, and the constant daylight of the austral summer make this one of the most productive oceans in the world. As a result, the phytoplankton amass a great quantity of energy via photosynthesis. This energy flows to the krill and ultimately to the whales.

The mass and density of blue whales depend on this chain of photosynthesis by phytoplankton and consumption by krill (Figure 16.3). Although this Antarctic system is unusual in the numbers of organisms and the amount of energy flowing through the system, the general pattern is characteristic of many ecosystems. The unique biology of the photosynthesizing organisms, the krill that graze on them, and the whales that consume krill by the ton lead to this particular pattern of energy flow—and to the more general question that we will address in this chapter: *How does energy flow through ecosystems?*

16.1   What Are the Energy Sources in Ecosystems?

■ **primary production** Energy acquisition by autotrophs.

The organisms in ecosystems fall into two broad categories. **Heterotrophs** are organisms that obtain energy from organic compounds, generally from other organisms, living or dead. Organisms that obtain energy from inorganic sources, such as by photosynthesis, are called **autotrophs**. When life first evolved on Earth, all organisms were heterotrophs. They obtained energy from organic molecules in the primordial soup or by consuming other organisms. As the organic sources of energy eventually became scarce, biochemical processes emerged that permitted autotrophic energy acquisition.

Autotrophs are the ultimate biological source of energy for ecosystems. The energy they acquire from the inorganic environment is referred to as **primary production**. The total rate of energy capture by autotrophs is the **gross primary production (GPP)**. Some of this energy is used up by the autotrophs' own respiration; that which remains is the **net primary production (NPP)**. It is the NPP that is available to heterotrophs. The acquisition of energy by heterotrophs is **secondary production**.

What Are the Inorganic Sources of Energy?

The most ancient autotrophs were **chemoautotrophs** and are known from both the eubacteria and the archaea. These organisms derive energy from the oxidation of electron donors in the environment. Energy-rich inorganic compounds such as H_2, hydrogen sulfide (H_2S), or methane (CH_4) serve as an electron donor; the ultimate electron acceptor may be one of many compounds, such as oxygen, NO_3^{2-}, SO_4^{2-}, or CO_2.

Some ecosystems, such as the strange hydrothermal vent communities deep in the ocean, are powered almost entirely by chemoautotrophs (Chapter 5) that use the hydrogen sulfide pouring from the vents as an energy

Figure 16.2 Krill. Krill (*Euphausia spp.*) are an important food source for blue whales as well as for many other organisms in the Antarctic Ocean.

source (Corliss et al., 1979; Karl et al., 1980). The Galápagos Rift ecosystem contains "forests" of tube worms and giant clams, as well as shrimp, crabs, and other invertebrates. Sunlight does not penetrate this deep, so the discovery of a diverse community there was a surprise. Many of the invertebrates in these systems obtain energy by filtering chemoautotrophic bacteria from the seawater or from chemoautotrophic symbiotic bacteria.

As noted earlier, early life on Earth was powered by heterotrophs that obtained energy from energy-rich organic molecules, and later by chemoautotrophs. As organic energy sources were depleted, intense competition for energy led to selection for the ability to use other sources of energy. Approximately 3.8 billion years ago, a new set of biochemical processes—photosynthesis—arose that drastically altered the Earth's ecology. We can date this biochemical innovation in a number of ways. Photosynthesis produces molecular oxygen. We find the first evidence of oxidized iron compounds (rust) in rocks of about this age. Cyanobacteria have enzymes that convert atmospheric nitrogen (N_2) into organic forms (nitrogen fixation). Because these enzymes are poisoned by oxygen, they are protected in thick-walled heterocysts, which first appear in microfossils of cyanobacteria at about 3.8 billion years ago (Figure 16.4).

Cyanobacteria were also among the first photosynthetic organisms. Remnants of vast mats of cyanobacteria are preserved today as **stromatolites**, fossilized layers of thin biofilms containing cyanobacteria. Over thousands of years, layer upon layer of these ancient biofilms built up to form rock-like structures that persist today in several locations around the world (Figure 16.5).

The evolution of photosynthesis revolutionized life on Earth. Photosynthetic organisms could flourish anywhere there was sufficient light, water, a carbon source, and nutrients. Most importantly, the amount of energy generated by this process exceeded that available from chemoautotrophs by at least two to three orders of magnitude! All that new energy made possible an array of complex, mobile, active life forms. It also provided sufficient energy for rich, diverse,

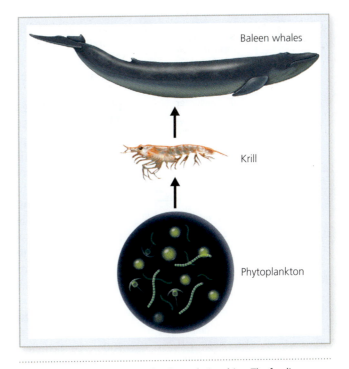

Figure 16.3 Antarctic Ocean feeding relationships. The feeding relationships in the Antarctic Ocean that support the blue whale depend ultimately on photosynthetic plants and the krill that graze on them. **Analyze:** How do you suppose this feeding system changes in the Antarctic winter?

Figure 16.4 Heterocysts. Heterocysts (the solid green cells at the top of the figure) first appear in fossil cyanobacteria at about the same time that we believe photosynthesis evolved and increased the atmospheric concentration of O_2.

gross primary production (GPP) The total rate of energy acquisition by autotrophs.

secondary production The acquisition of energy by heterotrophs.

net primary production (NPP) GPP minus the losses to respiration.

chemoautotrophs Organisms that derive energy from the oxidation of electron donors.

stromatolite Fossil biofilm of cyanobacteria.

Figure 16.5 Stromatolites. Stromatolites are the fossil remains of ancient cyanobacteria.

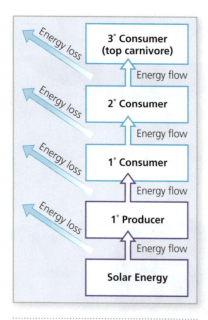

Figure 16.6 Food chain. A simple food chain showing the key components. Energy is lost at each transfer from one trophic level to the next. **Analyze: Can energy enter an ecosystem by any process other than photosynthesis?**

■ **primary consumer** An organism that consumes primary producers.

■ **secondary consumer** An organism that consumes primary consumers.

■ **tertiary consumer** An organism that consumes secondary consumers.

■ **top carnivore, or apex predator** The final trophic level in a food chain.

complex ecological systems in virtually every environment on the planet. It even contributes to the functional diversity of life at the hydrothermal vents in the deep, dark oceans, because some of the oxygen used in respiration in these systems is photosynthetically produced at the ocean surface.

How Does the Energy of Primary Production Flow Through Consumers?

The fundamental pattern of energy movement through an ecosystem is the **food chain**. A food chain is the set of hierarchical energy transfers from the primary producers to the consumers that feed on primary producers or on other consumers. Each step in a food chain is known as a **trophic level**, a feeding level. A simple food chain is shown in Figure 16.6. In this case, energy enters the ecosystem from the sun via photosynthesis. The NPP is available to the first consumers, the **primary consumers**, in the system. These are typically grazers, or herbivores that consume plants or phytoplankton. The primary consumers are in turn eaten by **secondary consumers**. The process continues, as **tertiary consumers** use the energy available in secondary consumers. The final trophic level is the **top carnivore**, or **apex predator**.

Note in the figure that energy is lost at each transfer between trophic levels. Ultimately this is a result of the laws of thermodynamics: no energy transfer is 100 percent efficient. Consequently, the total energy content of each trophic level is less than the preceding one (Figure 16.7). These losses are crucially important: they mean that energy is gradually dissipated as it moves up the food chain. Ultimately, so much energy is lost that there is not sufficient energy to support another, higher, trophic level. Note, too, that in general, the number of individuals in each trophic level tends to be less than in lower trophic levels. Thus, the structure of ecosystems is governed in the most fundamental way by energy transfer and

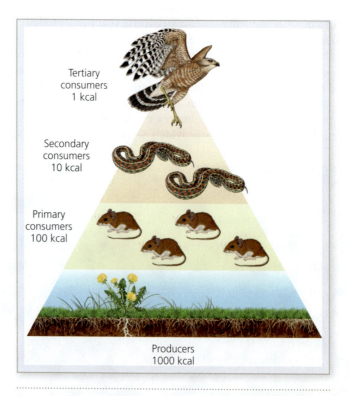

Figure 16.7 Pyramids of energy. The energy relationships in a food chain can be represented by pyramids of energy. The energy content of each trophic level is less than the one below. **Analyze: What hypothesis can you devise to explain why there are only three trophic levels in the blue whale food chain depicted in Figure 16.3?**

loss. We will explore the details of these processes later in this chapter.

The gradual loss of energy and the resulting organization of ecosystems are important to human environmental concerns in two ways. First, they affect which species we can sustainably harvest. Just as the amount of energy decreases as we progress up the food chain, the number of individuals in each trophic level, or their biomass, decreases as well. For example, consider a marine food chain like that in Figure 16.8. Each trophic level requires the energy input from many organisms in lower trophic positions. A tuna represents the accumulated energy high in the marine food chain. In this case, production of 1 pound of tuna requires 10 pounds of pollock, its prey, which is in turn dependent on 100 pounds of anchovies. The anchovies are supported by 1,000 pounds of phytoplankton. Clearly, it will be more difficult to sustain high catch rates of tuna compared to anchovies or pollack.

Second, although energy is lost as we move up the food chain, elements and compounds can become concentrated in the organisms of higher trophic levels, a process known as **trophic magnification** or **biomagnification**. This occurs when chemicals in lower trophic levels are funneled into smaller numbers of individuals higher in the food chain. As a result, toxins like heavy metals or DDT pose greater problems for animals in higher trophic levels, especially in apex predators. Before DDT was banned in the United States, many apex predator birds, such as bald eagles, osprey, and pelicans, were decimated by DDT as it was concentrated in the food chain.

We began the chapter with the remarkable food chain in the Antarctic Ocean in which tiny phytoplankton primary producers support the largest animal ever to live on Earth. However, ecosystems are usually far more complex than depicted in a simple food chain. Each trophic level may contain a number of different species. Moreover, some organisms feed on more than one trophic level, a process known as **omnivory**. For example, in a grassland ecosystem, mice may consume grass directly, as well as some of the insects that feed on the grasses. The mice are acting as both primary and secondary consumers. The result of all these interactions is the connection of multiple food chains into a **food web**. A food web is the set of pathways of energy transfer from primary producers through the apex predators. Figure 16.9 shows the simple food web of blue whales, krill, and phytoplankton embedded in the complexity of the Antarctic food web and the many other pathways of energy transfer.

Omnivory significantly complicates understanding and modeling food chains and webs. As food webs become more complicated, it is increasingly difficult to determine on which level any particular organism feeds. Consider the food web shown in Figure 16.9. How can we work out the myriad possible feeding interactions in a complex system like this? If an organism feeds at a single trophic level, the number of trophic levels from which it obtains energy is an integer. For example, if the organism feeds only as a secondary consumer, its energy comes from two transfers—from primary producer to primary consumer and from primary consumer to secondary consumer. Because omnivores obtain energy from more than one lower trophic level, their trophic position is not an integer value. An omnivorous secondary consumer that feeds on both primary producers and primary consumers would occupy a trophic level somewhere between 1 and 2, depending on how much it feeds on producers versus primary consumers.

Fortunately, an important physical/chemical technique, stable isotope analysis, is applicable to this ecological problem. Stable isotopes are nonradioactive forms of biologically important elements, especially carbon (^{12}C and ^{13}C) and nitrogen (^{14}N and ^{15}N). The stable isotopes of nitrogen have an important property that

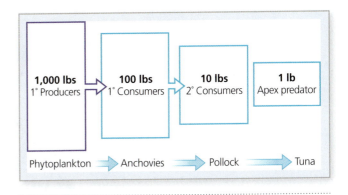

Figure 16.8 Marine food chain. A marine food chain indicating the numbers of fish required to support a large apex predator like tuna. The higher we feed in this food chain, the fewer individuals will be available for harvest. **Analyze: What health consequences might there be for humans that consume tuna compared to anchovies?**

trophic magnification, or biomagnification The increased concentration of chemical elements or compounds at higher levels in a food chain.

omnivory Feeding at more than one trophic level.

food web The connection of many food chains.

THINKING ABOUT ECOLOGY:

What are the energetic advantages and disadvantages to the organism of omnivory compared to specializing at a single trophic level?

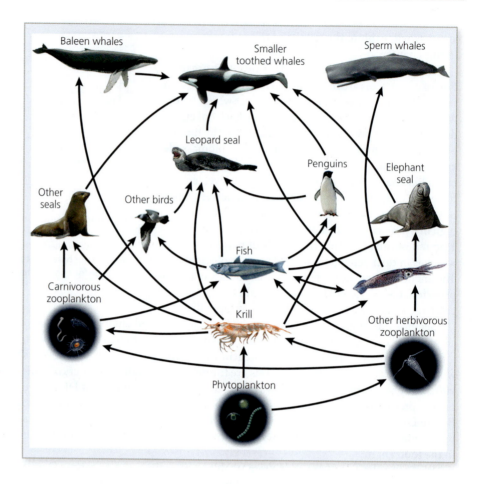

Figure 16.9 Antarctic Ocean food web. The diatom, krill, blue whale food chain is a small part of this complex web. **Analyze:** How might the interacting food chains in this system affect the number of blue whales?

allows ecologists to study food webs: during protein synthesis the lighter isotope (^{14}N) is incorporated into proteins at a higher rate than ^{15}N. Consequently, the ratio of ^{15}N/^{14}N gradually increases as nitrogen moves up the food chain. As it moves up the food chain, ^{15}N is enriched relative to ^{14}N by 3.2 percent at each transfer. This allows us to quantify the trophic position of an organism on the basis of its ^{15}N/^{14}N ratio.

If each integer transfer enriches ^{15}N by 3.2 percent, we can quantify fractional—that is, noninteger—energy transfers due to omnivory. Thanks to this kind of analysis, we now understand that omnivory is prevalent in food webs (Thompson et al., 2007). Some 36 percent of all organisms are omnivorous. Of the 64 percent that have integer trophic positions (nonomnivores), most are species that are either primary producers or primary consumers. At higher trophic levels, omnivory is the norm. According to Thompson, above the herbivore level, food webs are really just a tangled web of omnivores.

KEY CONCEPTS 16.1

- Organisms obtain energy in one of two fundamental ways: (a) from inorganic sources such as light or the energy-rich inorganic compounds, or (b) by consuming other organisms.
- Energy passes from producers to consumers in a hierarchical pattern known as a food chain.
- Ultimately, energy limits the addition of other trophic levels to the food chain.

QUESTION:

How is it possible for massive animals like whales to be apex predators?

Biomagnification

In the 1950s, people near Minamata Bay in Japan began experiencing significant health problems, including numbness and tremors in the extremities, tunnel vision and blindness, ataxia, and in some cases paralysis and death. Babies were particularly affected: many showed symptoms of cerebral palsy. Eventually it was discovered that these disorders were the direct result of biomagnifications of methyl mercury in the food chain of Minamata Bay. Some 27 tons of methyl mercury had been dumped into the bay by a petrochemical company. The mercury accumulated in the food chain and ultimately in the fish that were a staple of the local diet. Some 900 people died. Thousands of others were permanently affected.

The well-known impact of DDT on avian apex predators was also the result of biomagnification. Bald eagles, peregrine falcons, osprey, and brown pelicans became endangered because DDT accumulated in their tissues. DDT that entered aquatic systems accumulated through the food chain and eventually in the fish these birds consume. DDT interferes with calcium metabolism in birds and leads to abnormally thin eggshells, breakage, and ultimately reproductive failure. Only when the use of DDT was banned in the United States did these species recover.

Bioaccumulation is the absorption of chemical compounds by plants or animals from air, soil, or water. Once the material is in organisms low in the food chain, biomagnification through the food chain can increase its tissue concentration by several orders of magnitude. For example, DDT was applied to Clear Lake in California at a concentration of just 0.05 ppm. An apex predator in this system, the western grebe, eventually contained tissue concentrations 32,000 times that level. The rate of accumulation depends on a number of variables. Some compounds are volatile and can be transported great distances in the atmosphere, only to enter ecosystems far from their point of origin. Compounds that persist in the abiotic environment reach higher concentrations than those that readily break down. *Persistent organic pollutants (POPs)*, compounds that are particularly resistant

Figure 1 Biomagnification of DDT. Osprey (shown), bald eagles, peregrine falcons, and brown pelicans were all endangered by the biomagnification of DDT in their tissues.

to degradation in the environment, include many pesticides and industrial solvents. Because some important POP toxins, such as methyl mercury, PCBs (polychlorinated biphenyls), dioxin, and DDT, are so persistent, they can affect animals and humans long after their use or release into the environment has been curtailed. The rate at which organisms can metabolize the chemical also determines how much is passed on to the next trophic level. Fat-soluble compounds, such as DDT, build up in fat deposits over the life of the organism so that if an individual is consumed, significant amounts are passed on to the next trophic level. Long-lived species and older individuals in the population accumulate higher levels of toxins. For example, in the state of Minnesota, mercury advisories for human fish consumption vary with species and the size of the individual.

QUESTION:

Why do you suppose biomagnification tends to be a more significant problem in aquatic than terrestrial systems?

16.2

What Limits Primary Production?

If energy is gradually dissipated in the web of transfers, we might expect that the amount of energy entering the food chains via primary production is crucial to the organization of the ecosystem. Thus, the limits to primary production are an important aspect of any ecosystem.

How Is Primary Production Measured?

Primary production is the rate at which energy is captured by autotrophs. We express this rate in terms of the amount of energy captured per unit time over a specific spatial scale, such as kcal/area/time. Because much of the Earth's primary production occurs by photosynthesis, techniques for measuring primary production focus on quantifying the rate of photosynthesis. The technique employed to measure this rate depends on the system and especially the scale of the study.

biomass The total mass of organic material of biological origin.

EDDY analysis A technique to measure primary production by measuring the rate of uptake of CO_2 from the atmosphere.

At the level of an individual plant or a sample of phytoplankton, direct physiological measurements of photosynthesis may be possible. One way to accomplish this is to allow the plant to take up ^{14}C, the radioactive form of carbon. The rate at which ^{14}C is incorporated into plant tissue is a measure of photosynthesis and thus of gross primary production. The advantage of this technique is that it provides a direct and precise measure of the rate of photosynthesis. The difficulty is that it requires a way of enclosing the plant or phytoplankton sample so that the amount of ^{14}C available and incorporated is precisely controlled. This has been achieved in terrestrial systems by enclosing an entire tree in a large plastic bag—a procedure that is possible but cumbersome. In aquatic systems, a sample of water and phytoplankton can be inoculated with a known amount of ^{14}C. Isolating and containing the phytoplankton is much easier, but there are complications. Isolating the system may prevent circulation of nutrients or waste products, which may affect the results. If the goal is to measure production in a large ecosystem, the data from a small sample must be extrapolated, which may introduce error.

In ecosystems in which most of the primary production is concentrated in yearly plant growth aboveground, harvest techniques provide an estimate of production. For example, in an annual grassland, each year's production is harvested at the end of the growing season. The **biomass**, the total weight of this biological material, is an index of primary production. Because the plant biomass is decreased by the plants' respiration over the course of the growing season, this technique measures net primary production. The estimate is confounded by losses to herbivores and the amount of tissue growth that occurs belowground and remains unharvested. However, ecologists have devised methods for estimating herbivory and the growth of roots.

In recent years, improvements in the sensitivity of instruments that measure CO_2 concentrations and in computing power have led to new techniques for measuring production, especially at larger spatial scales. **EDDY analysis** is a technique that measures primary production by estimating the uptake of carbon dioxide from the atmosphere. Sensors are positioned in the vegetation and in the open atmosphere above the plants. These sensors provide a continuous record of the CO_2 concentration of the air in both places. During the day, CO_2 is used by photosynthesis. Although some CO_2 is also produced by plant respiration, the concentration measured among the plants drops compared to the concentration in the atmosphere above the plants, where photosynthesis is not occurring. At night, photosynthesis stops but respiration continues, causing the CO_2 concentration in the samples taken among the plants to increase compared to the atmosphere above. The greater the relative decrease in CO_2 during the day, the higher the primary production.

The second technique that depends on instrumentation and computing power uses remote-sensing measurements of chlorophyll as an index of production. Satellites can measure the reflected light from small samples of the Earth's surface. Chlorophyll absorbs most strongly in the blue and red portion of the visible spectrum. The rest of the visible spectrum is reflected, resulting in a distinctive pattern, or "signature," that can be measured by sensors on orbiting satellites. All other factors being equal, a high concentration of chlorophyll signals a high level of primary production. This technique is used to measure production on the largest geographic scales.

What Are the Global Patterns of Primary Production?

The geographic patterns of primary production on Earth are shown in Figure 16.10. Clearly, there is significant variation in the productivity of both terrestrial and marine ecosystems. Note in terrestrial systems that production decreases with latitude; the highest production is in tropical forest. Some regions are a complex mosaic of high and low production. For example, in western North America there are regions of relatively high production near to regions with very low values. This is due to the mountainous topography, which may put alpine forests relatively near low, unproductive desert.

Much of the open ocean has low primary production. The highest marine production is near shore. Note, too, that not all near-shore regions have equal production. For example, the west coast of South America has much higher production than the east coast. Moreover, some of the most productive waters are at high latitude, such as in the North Sea.

What Factors Determine the Global Patterns of Primary Production?

In essence, this question addresses the limits to photosynthesis. Indeed, we can explain most of the patterns in Figure 16.10 by determining what limits photosynthesis in that region.

Obviously, light is of paramount importance to photosynthesis. Geographic variation in light occurs due to both physical and biological factors. At high latitudes, summers are characterized by long days and short nights, winters by long nights and short days. Consequently, there is strong seasonal variation in production at high latitude. In contrast, the difference in day length over the course of the year in the tropics is minimal, and there, production is much less seasonally variable.

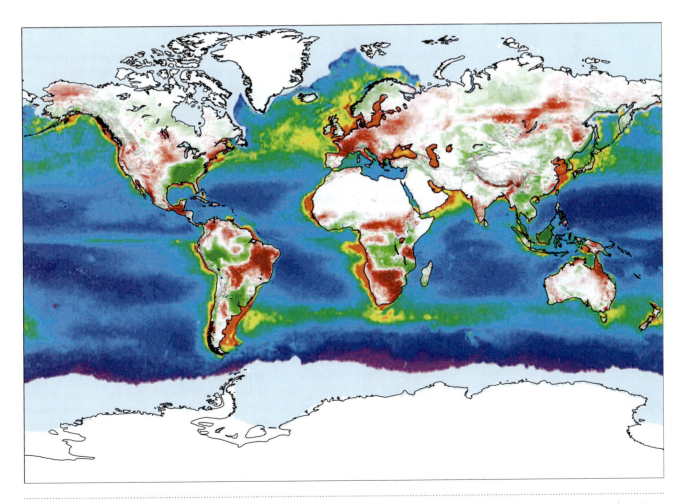

Figure 16.10 Global patterns of primary production. In terrestrial systems, regions of highest productivity are shown in green; low productivity is red. In marine systems, high productivity is red, moderate production is yellow, and low production is blue. **Analyze: What general patterns and trends can you see in these maps?**

In aquatic systems, water and suspended materials absorb light, which limits the depth at which photosynthesis can occur. Below the compensation point (Chapter 5) there is insufficient light to support production. There are qualitative changes in light with depth as well: blue light penetrates more deeply than red light.

It is also possible for there to be too much light. The photosynthetic apparatus in a plant leaf can be overwhelmed and damaged by high light intensities. Habitats like the white sands of New Mexico, which are composed of white gypsum sand, may have light levels so high that they limit which species can survive and photosynthesize.

Biological factors influence the quantity and quality of light as well. We saw the impact of this in Chapter 14, in which the light environment is a crucial factor in the development of a plant community. In mature communities, the canopy absorbs most of the photosynthetically useful light, limiting the production of subcanopy trees.

Water is a crucial limiting factor in photosynthesis and thus in primary production. You probably noticed the low production in desert systems and in Arctic systems in Figure 16.10. These patterns are the result of the interaction of temperature and precipitation. Net primary production increases with temperature. It also increases with precipitation. These correlations make sense because water is so important to plant growth. However, NPP declines at high levels of precipitation, probably because other limiting factors are affected by such high levels of rainfall. Specifically, the cloud cover associated with high precipitation may limit light availability. Nutrients may also be lost from systems continually washed by rains.

The effects of temperature and water also interact. **Actual evapotranspiration (AET)** is the total amount of water that evaporates and transpires from a region per unit time. This measure includes the physical process of evaporation from soil and standing water and the biological process of water loss through the stomata of plant leaves. Net primary production correlates with AET (Rosenzweig, 1968; Webb et al., 1978; Figure 16.11). For example, the high temperature and precipitation in tropical forest results in high AET, whereas Arctic tundra is characterized by low precipitation and temperature and thus low AET.

Nutrients also potentially limit primary production. On average, plants require a number of different nutrients, each in specific amounts. Macronutrients are those nutrients that plants must have in relatively large quantity. For example, plants require about 15,000 mg of nitrogen per kilogram of soil. Potassium, another macronutrient, is needed at about two-thirds that amount. Other elements that are necessary but only in trace amounts are micronutrients. Boron, zinc, and molybdenum are examples. These nutrients are required at one to two orders of magnitude lower concentration than macronutrients. The key point is that any of these nutrients can be limiting. Nitrogen, phosphorus, and potassium are among the most important limiting macronutrients in soil.

In aquatic systems, water is obviously not limiting. And in near-shore or shallow systems, light is abundant. Plant nutrients, especially nitrogen, are often the limiting factor for production in these ecosystems. Tropical marine systems have high light but are relatively nutrient poor, hence the low production shown for them. Productive regions like the North Sea and the west coast of South America are the result of currents and upwelling that bring nutrients from great distances and depths where they can support production by phytoplankton.

Nutrients limit some freshwater systems as well. In a classic experiment by Schindler (1974), the addition of phosphorus to part of a lake stimulated a significant increase in primary production (Figure 16.12). We know, too, that the nutrient content of some lake systems changes over time (Chapter 5). For example, lakes formed by glaciers gouging basins in granite substrate are initially very low in nutrients. Lakes with low nutrient concentrations and thus low primary production are known as **oligotrophic**. **Eutrophic** lakes are those with high nutrient

actual evapotranspiration (AET) The total amount of water that evaporates and transpires from a region per unit time.

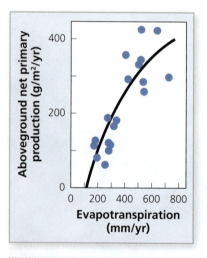

Figure 16.11 Net primary production and actual evapotranspiration. Ecosystems with higher AET tend to have higher NPP (from Webb et al., 1978). **Analyze:** How do temperature and precipitation affect AET?

oligotrophic Aquatic ecosystems with low nutrient content and low rates of photosynthetic production.

eutrophic Aquatic systems in which nutrients have accumulated and stimulated high rates of plant photosynthetic production.

concentration and production either because of the substrate on which they were formed or because they have gradually accumulated nutrients by inflow from terrestrial systems. Humans sometimes accelerate the rate of eutrophication when runoff from agricultural land or other sources contains phosphorus or nitrogen. This process is known as **cultural eutrophication**.

Production may also be limited by elements that are required only in very small amounts. There is growing interest in the role of these micronutrients, especially in marine production. Martin et al. (1994) showed that marine production can be stimulated in sample systems by the addition of iron. They suggest that the low concentration of iron is ultimately responsible for the low production of the open ocean.

The factors limiting production have human consequences as well, because our food supply is dependent on the net primary production of agricultural crops. Climate change in particular may lead to significant changes in crop production and thus food supplies. Because of the relationship between AET and NPP, we can predict how crop yields will be affected. Increased CO_2 may mitigate some of the effects of AET, because production may actually increase due to the higher CO_2 concentrations. Nevertheless, ecosystem ecologists are actively modeling these potential effects on world crop production (Schmitz, 2003).

Figure 16.12 Lake fertilized with phosphorous. In this whole-lake experiment, the portion of the lake fertilized with phosphorus has a large, visible algal bloom. In this aquatic system, production is nutrient limited.

cultural eutrophication Eutrophication caused by anthropogenic inputs of nutrients.

What Plant Adaptations Increase Production Efficiency?

Even under optimum light conditions, photosynthesis is a relatively inefficient process. Of all the light of the appropriate wavelengths reaching a plant, much is reflected. Some is absorbed by the leaf but not the photosynthetic apparatus. Of the light that reaches the chloroplast, much does not reach the photosynthetic pigments that actually capture the photons' energy. The maximum efficiency of plant photosynthesis is between 4.6 and 6.0 percent. Only a tiny fraction of incident energy is converted into biomass. Plants face strong selection pressure to increase their production and growth relative to other species. Their adaptive responses to this pressure play a significant role in the productivity of the ecosystems they inhabit. We discussed some of these adaptive responses in detail in Chapter 3.

The combined effect of limited water and high temperature is one of the most important selective forces on plants. Plants in hot, dry climates face a physiological dilemma. They obtain CO_2 for photosynthesis through the stomata. However, open stomata are also a major pathway of water loss. If the plant closes the stomata to avoid water loss, it cannot obtain CO_2 for photosynthesis. C4 photosynthesis is an important adaptation that helps solve this dilemma in hot, dry climates. The enzyme that picks up CO_2 in C4 plants (PEP carboxylase) has a high affinity for CO_2. Consequently, it can fix CO_2 at very low concentrations. The stomata can be closed for longer periods to limit water loss, but the plant can continue to fix CO_2 even when its concentration falls in the leaf.

Light is so important to primary production that plants have an array of adaptations that increase the efficiency with which they capture light. In Chapter 14 we discussed the adaptive geometry of early- and late-successional trees: early-successional trees have more effective layers of leaves that allow them to exploit the high light environment early in succession. As a result, early-successional forests can have high rates of primary production.

Chlorophyll a is the main form of chlorophyll found in plants, algae, and cyanobacteria. This molecule has two absorption peaks, 420 nm (blue) and 660 nm (red).

THINKING ABOUT ECOLOGY:

What are the potential consequences for animal populations if humans feed high or low in the food chain?

THE EVOLUTION CONNECTION

The Evolution of C4 Photosynthesis

C4 photosynthesis is an important adaptation to hot, dry climates. Although C4 plants represent only 3 percent of all vascular plant species, they are responsible for about 25 percent of primary production in terrestrial ecosystems. We can trace the origin of this pathway from C3 photosynthesis thanks to another application of stable isotope analysis. C4 plants incorporate ^{13}C, the stable isotope of carbon, at a higher rate than C3 plants. Ancient soils preserve the ratio of $^{12}C/^{13}C$ and thus the relative contributions of C3 and C4 photosynthesis. The information generated by these analyses, in conjunction with new data on historical changes in temperature, precipitation, and the atmospheric CO_2 concentration, allow us to reconstruct the evolution and spread of the C4 pathway (Edwards et al., 2010). This history illustrates two important aspects of adaptive evolution. The first, convergent evolution, is the phenomenon in which an adaptation arises more than once in unrelated groups. Second, the origin of C4 photosynthesis exemplifies preadaptation. Preadaptation is the process in which the initial adaptive value of a feature is different from its current function.

Convergent evolution occurred in the origin of C4 photosynthesis in the grasses. Some 60 percent of all C4 species are grasses. You might assume that this represents a single evolutionary event in an ancestral grass species that subsequently spread to other species in the group. In fact, the pathway evolved numerous times among different genera of grasses. Phylogenetic studies show that among the grasses, the C4 PEP carboxylase arose at least 8 times from C3 ancestors. Overall it has emerged 45 different times among all plant lineages. It thus represents a classic example of convergent evolution.

We see the process of preadaptation in the relative advantages of C3 and C4 photosynthesis. We know that C4 photosynthesis has an advantage over the C3 pathway in hot, dry climates. However, the abiotic conditions early in the evolution of C4 photosynthesis suggest that the current adaptive advantage is probably not the same as when it first appeared. C3 photosynthesis evolved at a time when the atmospheric CO_2 concentration was considerably higher than it is today. At the beginning of the Oligocene, 30 million years ago, tectonic activity led to a significant reduction in the CO_2 concentration (Figure 1). Rubisco, the carbon-fixing enzyme in C3 plants, has a relatively low affinity for CO_2. When CO_2 becomes scarce, rubisco picks up O_2 instead, leading to photorespiration (Chapter 3). The energy cost of photorespiration may reach 40 percent and presents a significant energy drain on the plant. The global drop in the CO_2

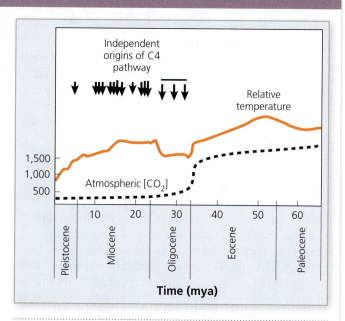

Figure 1 Origins of the C4 pathway. The patterns of change in temperature and atmospheric CO_2 concentration. Multiple origins of the C4 pathway (indicated by arrows) were associated not with a change in temperature but with a decrease in CO_2 availability (from Edwards et al., 2010). **Analyze: What would be the adaptive advantage of C4 photosynthesis when CO_2 is limiting?**

concentration favored plants that could pick up CO_2 even when its atmospheric concentration was low. Thus, the initial selective advantage of the C4 pathway was probably associated with low atmospheric CO_2. The pathway later turned out to be advantageous in hot, dry climates.

We now know that C4 grassland communities proliferated long after the initial origin of the C4 pathway. This expansion of C4-dominated biomes was associated with climatic changes that increased the proportion of the planet with hot, dry climates about 8 million years ago. The C4 grasses that constitute these grasslands were preadapted for this climatic shift.

QUESTION:

How would the anthropogenic increase in CO_2 that is occurring now affect the adaptive value of C4 photosynthesis?

accessory pigments Pigments other than chlorophyll that capture light of different wavelengths.

A number of other forms of chlorophyll (b, c, and d), as well as other pigments, such as phycobiliproteins and carotenoids, are found in different photosynthetic organisms. These molecules are collectively known as **accessory pigments**. For example, cyanobacteria contain chlorophyll a and phycobiliproteins. Each type of accessory pigment has a unique absorption spectrum. Accessory pigments capture light energy in wavelengths chlorophyll a cannot and send that energy to the photosynthetic machinery. Some accessory pigments, such as carotenoids, also function to dissipate light in high light environments.

Plants face strong selection pressure to acquire nutrients from the environment; their success determines primary production. One important adaptation is mycorrhizae, the mutualistic interaction between plants and fungi (Chapter 12). The fungal hyphae significantly increase the absorptive surface area of the plant roots, thus improving water and nutrient uptake. The fungi, in turn, receive carbohydrate from the plant. Extensive hyphae outside the plant transport materials to the root. We are only now beginning to understand the prevalence and importance of this mutualism. We do know, however, that the majority of plant species that have been studied participate in mycorrhizal associations.

KEY CONCEPTS 16.2

- Ecosystems vary widely in the rate of primary production.
- Any factor that limits photosynthesis or plant growth may limit primary production. The geographic differences in the key limiting factor determine the patterns of global production.
- Plants face strong selection to increase their acquisition of light, water, and nutrients.

QUESTION:

What factors limit production in terrestrial environments? In aquatic environments?

16.3 What Controls Energy Flow Through Consumers?

Energy is lost from each transfer in a food chain. Ultimately it is this loss of energy that limits the biomass or numbers of organisms high in the food chain. In addition to this physical constraint, there are important biological reasons for the loss of energy through the food chain.

How Is Energy Lost from Ecosystems?

The most important biological factor in the gradual dissipation of energy in a food chain is the simple fact that at each trophic level, organisms are using energy for their life processes. Energy is devoted to somatic growth, foraging, predator escape, reproduction, social interactions—in sum, all the biological activities in which an organism engages. Some of this energy, such as that invested in growth and reproduction, becomes available to the next trophic level; much is lost.

Figure 16.13 depicts the three efficiencies that determine the amount of energy that flows through one trophic level in a food chain. A certain amount of energy is available to a consumer from the trophic level below. Of that energy, some proportion is actually ingested by the organisms of the next level; some is not. The fraction of the energy available that is actually consumed is the **ingestion efficiency**. A number of ecological interactions determine the value of the ingestion efficiency. For instance, not all the biomass of primary producers is eaten by grazers. There may be more than the herbivores can consume, or perhaps some is unpalatable or has adaptations that deter grazing. Many of the coevolutionary interactions discussed in Chapter 12 determine the value of the ingestion efficiency.

Of the energy that is ingested, some proportion is assimilated by the consumer. This represents the **assimilation efficiency**. The nature of the material consumed plays a major role in determining this value. For example, grass is difficult to digest because much of the energy is locked up in the cellulose that comprises the primary cell wall of plants. Most animals cannot digest cellulose on their own. They rely on symbiotic gut bacteria to break down the cellulose so that

ingestion efficiency The amount of energy ingested relative to the amount available.

assimilation efficiency The amount of energy assimilated by the consumer relative to the amount ingested.

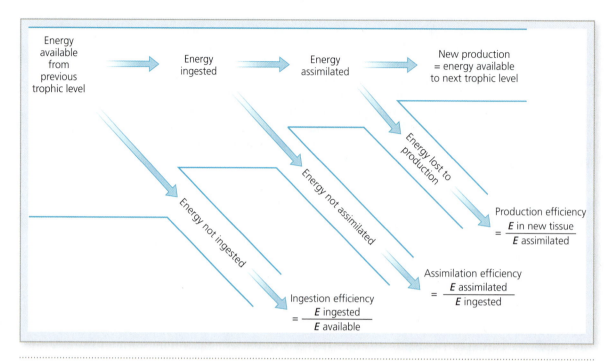

Figure 16.13 Efficiencies of energy transfer. The efficiencies of energy transfer determine the amount of energy that is ultimately transferred from one trophic level to the next. **Analyze: How is energy lost in the process of production (the production efficiency)?**

the energy and nutrients of the plant can be absorbed. Herbivores typically ingest a large amount of biomass (and excrete much) because the efficiency of assimilation is so low, often around 30 percent. In contrast, predators usually have higher assimilation efficiency because they can more readily and thoroughly digest the flesh of their prey. Still, there is variation among prey types due to characteristics such as exoskeletons that affect assimilation efficiency.

The energy actually assimilated by the organism is available for its own biological needs. This is the energy that fuels movement, behavior, growth, and reproduction. The proportion of the assimilated energy that is used to produce new tissue, either by growth of the individual or by reproduction, is the **production efficiency**. This is the energy available to the next trophic level for ingestion.

The fundamental physiology of the organism affects production efficiency as well. Poikilotherms do not invest energy in maintaining a constant body temperature. They generally have higher production efficiency than homeotherms, in which the energy devoted to maintaining the body temperature is not available for growth or reproduction.

Among homeotherms, production efficiency is also a function of the organism's size. The rate at which an animal loses heat is allometrically related to its size. Allometry, meaning different measures, is the relative scaling of different features of an organism. Figure 16.14 shows the relationship between surface area and heat loss. Heat production is proportional to the volume of the animal; it increases as the cube of the animal's linear dimension. Heat loss is proportional to the surface area of the animal; it increases as the square of the linear dimension. As a result, large organisms have a large volume relative to the surface area. For example, an animal as large as a blue whale has a small surface area compared to its volume. For an animal that inhabits the cold Antarctic Ocean, this is an important adaptive feature. A small animal faces a heat-loss problem: its surface area is relatively large for its volume. Consequently, a small homeotherm like a shrew or a hummingbird loses proportionally more heat than a larger animal. Its production efficiency is thus much lower—much heat is lost and unavailable to the next trophic level.

production efficiency The amount of energy devoted to new tissue (growth and reproduction) relative to the amount assimilated.

THINKING ABOUT ECOLOGY:

What is the relationship between production efficiency and the life history strategy?

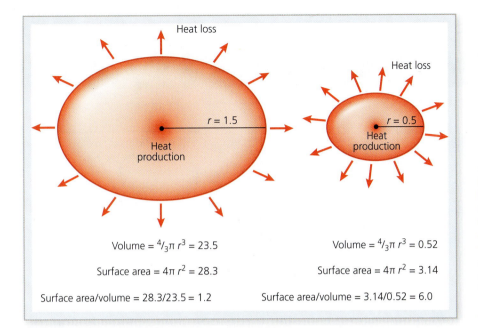

Figure 16.14 Surface area, volume, and heat loss. The allometric relationship between surface area, volume, and heat loss. Heat is produced as a function of the mass of the organism (which is related to its volume). Heat is lost as a function of surface area. The surface area of a small sphere is larger compared to its volume than that of a large sphere. **Analyze: Explain why the S/V ratios differ in large and small spheres.**

DO THE MATH

Allometry

Production efficiency is affected by allometric effects on body size. Biologists study the phenomenon of allometry in the contexts of both the phylogeny and evolution of taxa and the development of individuals. Let us explore how this phenomenon is measured and analyzed.

We begin with a general equation for the relationship between two components of an organism x and y.

$$y = bx^a$$

For example, let y be a measure of metabolic rate (a key component of production efficiency) and let x be a measure of body size. If two aspects of an organism change in the same way, their relationship is *isometric*. Figure 1a shows a hypothetical isometric relationship between basal metabolic rate and body size. In this example, each increase in body size leads to a twofold increase in metabolic rate. The equation

$$y = 2x^1$$

describes this isometric situation. The value of $a = 1$ defines an isometric relationship—the value of y is determined only by the multiplicative coefficient b. The key feature of isometry is that the *proportional* relationship of x and y does not change as x increases; in this example, y is always related to x by a factor of 2.

Allometric relationships occur when $a \neq 1$. When this occurs, y and x are no longer related in linear fashion (Figure 1b). When $a > 1$, the allometric relationship is called *positive allometry* or *hyperallometry*. The relationship is *negative allometry* or *hypoallometry* if $a < 1$.

Note that allometric relationships result in linear relationships when the two traits are plotted logarithmically (Figure 1c). They can thus be represented in the logarithmic equation below.

$$\log y = a \log x + \log b$$

Production efficiency is affected by allometry. For example, metabolic rate affects how much energy can be devoted to somatic growth, reproduction, or maintaining a constant body temperature. Metabolic rate (kcal hr^{-1}) for homeotherms shows positive allometry with body size (g):

$$y = .02x^{0.75}$$

The surface-to-volume ratio also plays a role in the production efficiency of homeotherms. Let us explore the mathematics of that relationship further. Surface area scales as the square of the linear dimension; mass (volume) scales as its cube. Thus, the surface/volume ratio of small animals is large compared to that of large animals. We can place this relationship in the mathematical context of allometry as follows. The volume of an organism increases as the cube of its linear dimension (l):

$$V = b\,l^3$$

(b is a constant). The surface increases as the square of l:

$$S = b\,l^2$$

In order to determine the allometric equation, we manipulate the equation for surface area in order to represent it with the same

continued

DO THE MATH   *continued*

Allometry

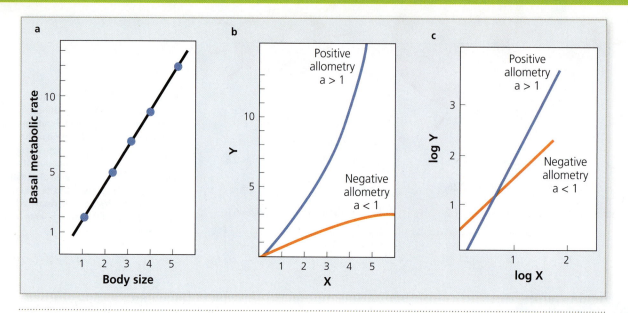

Figure 1 Possible relationships between body size and metabolic rate. (a) An isometric relationship. (b) Positive and negative allometry. (c) On a log-log plot, allometric relationships result in straight lines. **Analyze: Explain the difference between positive and negative allometry.**

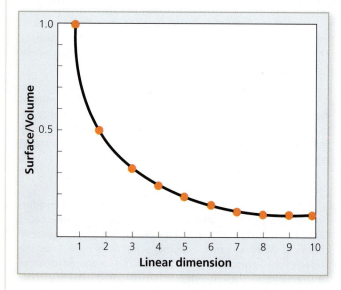

exponent as the volume equation. Because surface area grows only two-thirds as fast as volume, the surface area is also given by the equation

$$S = b\,(l^3)^{2/3}$$

Because $l^3 = V$, we can substitute

$$S = b\,V^{2/3} \text{ or } S = b\,V^{0.67}$$

to obtain an equation in the standard allometric form $y = b\,x^a$ where $a \neq 1$. The allometric change in surface/volume ratio as a function of linear dimension is depicted in Figure 2.

Figure 2 Allometric change in surface/volume ratio as a function of linear dimension. The change in the surface/volume ratio as the linear dimension increases. The surface area is largest relative to the volume in smaller organisms. The surface/volume area is allometrically related to the linear dimension. **Analyze: Is this a positive or negative allometry?**

KEY CONCEPTS 16.3

- Each transfer of energy among trophic levels results in energy lost to higher trophic levels.
- Physical and biological factors determine the efficiency of energy transfer among trophic levels. Three key efficiencies determine the overall efficiency of energy transfer:
  a. Ingestion efficiency
  b. Assimilation efficiency
  c. Production efficiency

QUESTION:

Explain the physical constraints and biological factors that determine energy flow in a food chain.

16.4 How Does Energy Flow Structure Ecosystems?

Thus far, we have examined the patterns and constraints on energy flow in ecosystems. In the process, we have discovered a key aspect of how energy flow structures ecosystems: the loss of energy through the food chain eventually limits the length of the food chain. We can now address the fundamental organization of ecosystems as controlled by energy.

What Determines Food Chain Length?

Ecosystems vary in the length of their food chains (Figure 16.15). The modal number of trophic levels is 3–4 (Pimm, 1982). Vander Zanden and Fetzer (2007) found that streams have shorter food chains (3.5 trophic levels) than marine and lake ecosystems, which average 4.0 levels. Although the range of food chain lengths among ecosystems is relatively small, approximately 2–6, the food chain is such a fundamental ecosystem characteristic that small differences reveal important ecological principles.

From your understanding of transfer efficiencies and the factors limiting primary production, you might generate some logical hypotheses regarding the control of food chain length. For example, you might hypothesize that the larger the primary production, the longer the food chain. The logic of this hypothesis is that the larger the initial input of energy to the food chain, the more will be available for higher trophic levels despite the inevitable losses along the way. Or you might hypothesize that ecosystems in which there are a large number of poikilotherms might have longer food chains. The low production efficiency of homeotherms might reduce the number of possible trophic levels. However, neither of these hypotheses is supported by comparison of ecosystems (Pimm, 1988; Pimm, 2002). Something else must determine how many links there are in a food chain.

One possibility is another manifestation of the nonequilibrium nature of ecological systems. Stuart Pimm (1988) has proposed that disturbance plays a central role in the determination of food chain length. He argues that populations higher in a food chain are more vulnerable to disturbance (Figure 16.16). Not only are their smaller populations more vulnerable, they do not have the capacity to recover quickly by reproduction or immigration. As a result, ecosystems in which disturbance is frequent should have shorter food chains. Some support for this hypothesis has emerged. Townsend et al. (1998) examined food chain length in streams that differ in the disturbance regime of the stream bed. Food web size and the number of trophic links among species were negatively correlated with disturbance intensity, as predicted by Pimm's hypothesis. Other studies have failed to support the disturbance hypothesis. Walters and Post (2008) experimentally disturbed stream systems in the Yale Experimental Forest. Their results show no significant effect of disturbance on the length of food chains.

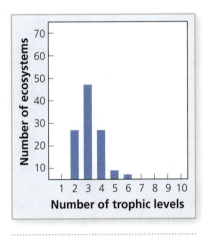

Figure 16.15 Food chain length. The numbers of trophic levels in various ecosystems (data from Pimm, 1982). **Analyze: Why do you think there is such a narrow range of numbers of trophic levels?**

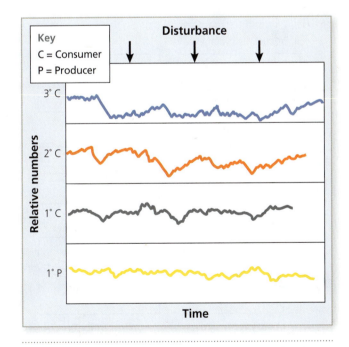

Figure 16.16 Pimm's hypothesis. Pimm's hypothesis to explain the number of trophic levels in an ecosystem is based on the effects of disturbance. In systems in which disturbance is prevalent, the smaller number of individuals in high trophic levels are most affected—they show the greatest variation and lowest relative population size. Food chains are shorter when the higher-level consumers are continually reduced by disturbance (Pimm, 1988). **Analyze: Why is the *variation* in population size important in this hypothesis?**

The Length of Food Chains

We have explored some factors that potentially affect the length of a food chain in an ecosystem. One potentially important factor that determines food chain length is primary production. The logic behind this idea is that if more energy enters the food chain through production, there will be more energy available to support higher trophic levels, despite the inevitable loss of energy at each transfer. Some researchers have also suggested that the physical size of the ecosystem is an important factor—size per se results in more energy available throughout the system and thus longer food chains. It is also possible that the interaction of production and ecosystem size is important.

Post et al. (2000) addressed these factors by analyzing a series of lakes in New York that differ primarily in size and productivity. The lakes were located in the same geographic region and thus were similar in chemistry, origin, and geology. They tested three hypotheses:

HYPOTHESIS 1: Ecosystems with higher primary production have longer food chains.

HYPOTHESIS 2: Larger ecosystems have longer food chains.

HYPOTHESIS 3: Production and ecosystem size interact.

Each of these hypotheses makes a specific prediction about the lengths of food chains in these lakes (Figure 1).

PREDICTION 1: Food chain length increases with primary production but not with ecosystem size.

PREDICTION 2: Food chain length increases with ecosystem size but not with primary production.

PREDICTION 3: Food chain length increases with both primary production and ecosystem size.

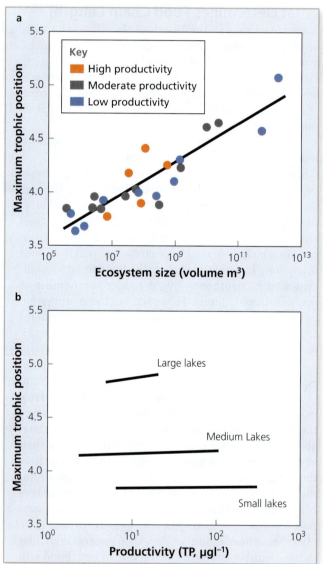

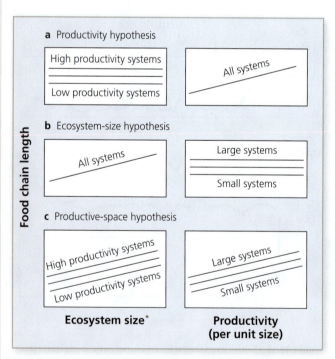

Figure 1 Hypotheses for food chain length. The predictions of the three hypotheses to explain food chain length in lake ecosystems (from Post et al., 2001). **Analyze: Explain why each hypothesis leads to different slopes and intercepts.**

Figure 2 Comparison of food chain length. The results from comparison of food chain length in lakes of different size and production. Food chain length was not affected by primary production but did increase with lake size (from Post et al., 2000). **Analyze: Which hypothesis do these data support?**

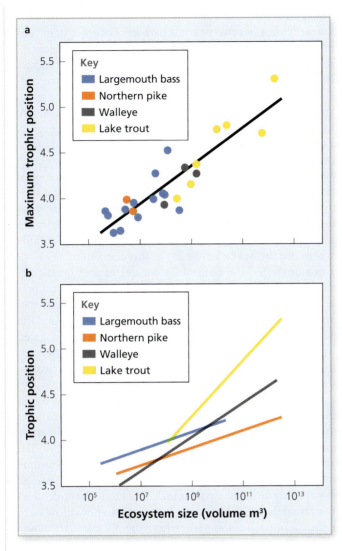

As in many ecosystems, omnivory complicates measurement of the number of food chains. In order to incorporate omnivory in the measure of food chain length, Post employed stable isotope analysis as described earlier. This analysis led to a new variable, maximum trophic position (MTP), the maximum number of energy transfers measured by the maximum enrichment of ^{15}N in the fish of the lake. This maximum number of transfers represents the length of the food chain.

The results are shown in Figure 2. There is no support for either the productivity hypothesis or the production-size hypothesis: there is no correlation between either production or production size and the maximum trophic position. However, there is strong support for the ecosystem size hypothesis. The largest lakes had the most trophic levels regardless of the lake's production.

How does the size of the lake affect MTP? Figure 3 shows that a new apex predator, lake trout, appeared in the largest lakes. This species did not occur in the smallest lakes. Something associated with lake size allowed this new predator to occur. Note, too, that in larger lakes, all species increased their trophic position. In other words, species that feed lower in the food chain in small lakes are feeding at a higher level in large lakes.

Figure 3 Changes in trophic position. Changes in the trophic position of consumers in lakes of different sizes (from Post et al., 2000). **Analyze: What is the significance of a change in trophic position?**

Some evidence suggests that size of the ecosystem determines food chain length. Takamoto et al. (2008) examined food chain lengths on barrier islands in the Bahamas. These islands vary both in size and disturbance by wind and wave action. Disturbance did not affect the number of trophic levels, only the type of apex predator present. In contrast, islands larger by an order of magnitude supported an extra trophic level.

This is clearly an area of research in which a consensus understanding has not yet emerged. The goal of this research is to discover general rules for the organization and structure of ecosystems. The controversy over the determinants of food chain length illustrates an important aspect of ecosystem ecology. Ecosystems are so complex and so variable even within biogeographic regions that it is extremely difficult to devise well-controlled experiments. Consider just the complexity of quantifying the trophic level of a particular species. Across ecosystems, the number of species that occupy one feeding level varies enormously. Superimpose on that the variation in the population density within and between species, and you begin to see the experimental difficulties.

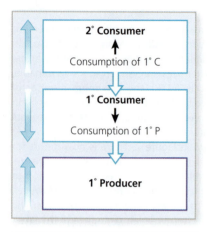

Figure 16.17 Trophic cascade. In a trophic cascade, changes in the top predator affect lower trophic levels. If the predator numbers increase, it depresses the numbers of its prey. If the prey species are primary consumers, this may release the producers, increasing their abundance. **Analyze: In longer food chains, would you expect these effects to alternate between even- and odd-numbered trophic levels?**

top-down control A process in which organisms higher in the food chain control the density and diversity of lower trophic levels.

bottom-up control A process in which organisms lower in the food chain control the density and diversity of higher trophic levels.

What Controls Energy Flow?

The organization of food webs suggests two possible pathways of ecosystem control. Top-down control occurs when organisms higher in the food chain control the density and diversity of organisms at lower trophic levels. If grazers impact the diversity of the plants on which they graze, they are exerting top-down control. Alternatively, if higher trophic levels are affected primarily by the food available to them (or to their prey), the system is under bottom-up control. We encountered these concepts previously (Chapter 9) in the context of the regulation of populations. In that case we were concerned with the population dynamics of individual species. Here we expand our analysis to the ecosystem level.

Instances of both **top-down** and **bottom-up control** abound in the ecological literature. The importance of keystone predators (Chapter 11) such as the starfish, *Pisaster*, is an example of a strong top-down ecosystem effect. *Pisaster* reduces the density of a number of other sessile invertebrate species in the rocky intertidal. Consequently, none of its prey species can competitively exclude others; top-down control maintains species diversity in this ecosystem.

Báez et al. (2006) demonstrated the bottom-up control of arid grassland/shrub ecosystems. In this ecosystem, primary production is dependent on precipitation, a factor that varies over space and time. The density and diversity of higher trophic levels were determined by the amount of plant production. The top-down effect of rodents on the producers was experimentally investigated with exclosures that reduced rodent density relative to control plots to which rodents had access. Rodents had virtually no impact on the abundance or structure of the vegetation. Control in this system is thus largely bottom-up.

An important area of current research focuses on effects that operate across several trophic levels. A **trophic cascade** occurs when a top-down process affects several lower levels in a food web. If a consumer reduces the density or diversity of its prey, the effect may continue on down the food chain (Figure 16.17). The reintroduction of wolves to Yellowstone National Park in 1995 provided a unique opportunity to examine a trophic cascade. Wolves prey on a variety of species, but in Yellowstone they have a significant impact on elk. Prior to wolf reintroduction, elk numbers were extremely high. Moreover, both aspen and willow were disappearing from the northern part of the park, where the elk population was highest.

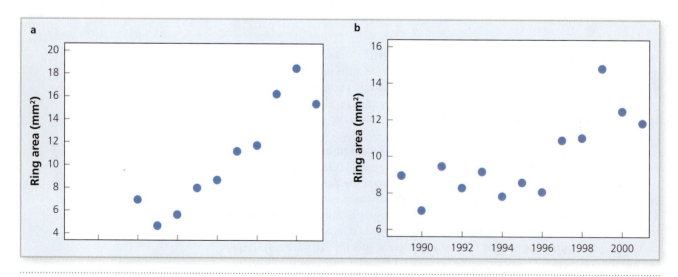

Figure 16.18 A trophic cascade in Yellowstone National Park initiated by the reintroduction of wolves. The decrease in elk numbers and changes in their foraging patterns released aspen and two species of willow, *Salix boothii*, (a) and *S. geyeriana* (b) and increased their abundance on the northern range. The increase in growth rate of willows coincided with the restoration of wolves to Yellowstone (from Beyer et al., 2007). **Analyze: What other factors might explain the increase in aspen?**

After wolves were restored to the ecosystem, researchers noticed the resurgence of young aspen and willow. Beyer et al. (2007) examined the growth patterns of willow (*Salix spp.*) in Yellowstone. After statistically eliminating the effect of precipitation on willow growth rates, they were able to show a marked increase in the annual growth of willows after wolves were restored. Aspen regeneration also increased markedly (Ripple et al., 2001). Thus, it appears that wolves initiated a trophic cascade that not only affects the elk on which they prey but also the primary producers (Figure 16.18). Further study showed that the impact of wolves was not simply on the number of elk; they also changed the behavior of their prey. With wolves present, elk are less likely to forage in thick vegetation, such as willow or aspen thickets, where they cannot see the approach of a wolf. As a result, both aspen and willow were released from the impact of elk browsing when wolves arrived in the ecosystem.

trophic cascade When a top-down process affects multiple lower trophic levels.

THINKING ABOUT ECOLOGY:

Is a trophic cascade more or less likely to occur in complex food webs compared to simple food chains?

KEY CONCEPTS 16.4

- Ecosytems differ in the number of trophic levels they contain.
- We are currently trying to understand the factors that determine the number of trophic levels in an ecosystem.
  - Productivity does not seem to determine how many trophic levels occur.
  - Disturbance may reduce the number of trophic levels by limiting the populations of species high in the food chain.
  - There is also empirical support for a role for ecosystem size: some studies show that larger ecosystems have longer food chains.
- The impact of high-level consumers can cascade down through the food chain to affect lower trophic levels.

QUESTION:

What is the relationship between production and disturbance in determining the number of trophic levels?

Putting It All Together

The blue whale lives in the coldest ocean on the planet as part of an ecosystem powered by tiny photosynthetic algae. This simple food chain—algae supporting krill supporting blue whales—seems remarkable, perhaps unusual. But in fact that food chain is similar in its key elements to all food chains. The productivity of the phytoplankton in the Antarctic Ocean is due to the high nutrient content of those waters. Consequently, the most important grazers in this system also occur in large numbers. They represent a rich energy source for whales. But the whale is an apex predator in this system. No species—except humans—preys on adult blue whales. Sufficient solar energy is captured and passed on to support the whales. But there is insufficient energy to support any predator of the whales. Here in microcosm are the essential elements of energy flow in ecosystems—the structure of this ecosystem is determined by the basic energetic constraints common to all ecosystems.

The basic principles of energy capture and loss and the hierarchical structure of food webs are universal. However, factors such as light, water, and nutrients locally limit primary production. The global variation in these limiting factors accounts for the variation among ecosystems. The result is the range of production shown in Figure 16.11. The biology and ecology of the component species, especially their foraging habits and their size and physiology, introduce yet another level of variation among ecosystem types. Food chains vary in length, in the complexity of the webs to which they belong, and in the importance of omnivory.

Food chains differ, too, in the nature of control—top-down or bottom-up—and the presence of trophic cascades.

Thus, the basic question of this chapter—how does energy flow through ecosystems?—has both a simple and a complex answer. The basic organization of ecosystems and pattern of flow is universal thanks to the laws of thermodynamics and basic principles of biological energy use. The complex answer is that in important ways, energy flow is ecosystem specific. The questions of current interest in this field center on understanding the differences among ecosystems. Which factors—production, transfer efficiency, disturbance, or ecosystem size—are most important in organizing the structure of a particular food web?

Summary

16.1 What Are the Energy Sources in Ecosystems?
- Autotrophs capture energy from inorganic sources: chemoautotrophs use energy-rich electron donors such as H_2S; photoautotrophs capture solar energy via photosynthesis.
- The total rate of energy capture from these inorganic sources is the gross primary production. This is the base source of energy for the rest of the ecosystem.
- The energy of the autotrophs is captured and passed on by the consumers (heterotrophs) in the system.
- Energy flow is hierarchical: energy is passed from one group of organisms to the next through a food chain. Food chains are connected into food webs.

16.2 What Limits Primary Production?
- Aquatic and terrestrial ecosystems vary enormously in primary production.
- Any factor that limits photosynthesis affects primary production.
- Light, water, and nutrients are the most important factors limiting primary production.

16.3 What Controls Energy Flow Through Consumers?
- Energy flow through consumers cannot be 100 percent efficient; energy is lost at each transfer from one trophic level to the next.
- Three factors determine the amount of energy passed from one trophic level to the next: the ingestion, assimilation, and production efficiencies.

16.4 How Does Energy Flow Structure Ecosystems?
- The loss of energy at each trophic transfer results in a common structure in all ecosystems.
- Because of the inevitable loss of energy through the food chain, eventually there will not be enough energy to support another trophic level.
- Ecosystems vary from two to six trophic levels.
- Disturbance and ecosystem size are two important determinants of the number of trophic levels.
- The flow of energy and the populations in each trophic level are controlled by top-down or bottom-up processes.
- Some ecosystems are controlled by trophic cascades, in which the effect of consumers high in the food chain cascade down through lower trophic levels.

Key Terms

accessory pigments p. 360
actual evapotranspiration (AET) p. 358
apex predator p. 352
assimilation efficiency p. 361
autotroph p. 350
biomagnification p. 353
biomass p. 356
bottom-up control p. 368
chemoautotrophs p. 351
cultural eutrophication p. 359

EDDY analysis p. 356
eutrophic p. 358
food web p. 353
gross primary production (GPP) p. 351
heterotroph p. 350
ingestion efficiency p. 361
net primary production (NPP) p. 351
oligotrophic p. 358
omnivory p. 353
primary consumer p. 352

primary production p. 350
production efficiency p. 362
secondary consumer p. 352
secondary production p. 351
stromatolite p. 351
tertiary consumer p. 352
top carnivore p. 352
top-down control p. 368
trophic cascade p. 369
trophic magnification p. 353

Review Questions

1. What was the ecological significance of the evolution of photosynthesis?

2. Why might hydrothermal vents provide clues to the origin of life?

3. Explain how the stable isotopes of nitrogen (^{15}N and ^{14}N) can be used to quantify noninteger trophic position.

4. What are the advantages and disadvantages of the various methods of measuring primary production?

5. Why is the amount of primary production not the key factor determining the number of trophic levels?

6. How would trophic cascades be different in food chains with even versus odd numbers of trophic levels?

7. In what ways is energy flow similar among ecosystems? In what ways does it differ?

8. What is the importance of nonequilibrium ecology in the patterns of energy flow in ecosystems?

9. How do we know that the original selective advantage of C4 photosynthesis differs from its current adaptive value?

10. Why hasn't C4 photosynthesis replaced the C3 pathway in all ecosystems?

Further Reading

Chapin, F.C., et al. 2002. *Principles of terrestrial ecosystem ecology*. New York: Springer.

This book provides an excellent overview of energy flow, especially the poorly understood role of detritivores.

Edwards, E.J., et al. 2010. The origins of C4 grasslands: integrating evolutionary and ecosystem science. *Science* 328:587–591.

This paper is an excellent review of the origin of C4 photosynthesis and the ecology underlying its spread into grasslands worldwide.

Schackell, N., et al. 2009. Decline in top predator body size and changing climate alter trophic structure in an oceanic ecosystem. *Philosophical Transactions of the Royal Society B* 277:1353–1360.

In this study, Schackell et al. examine two important impacts on the trophic structure in a marine ecosystem: climate change and overfishing.

Terborgh, J., and J. Estes. 2010. *Trophic cascades*. Washington, DC: Island Press.

This provides the most up-to-date discussion and analysis of trophic cascades in a wide array of ecosystems.

Chapter 17

Biogeochemical Cycles

On Alaska's North Slope, 160 miles north of the Arctic Circle, methane and carbon dioxide gas bubble up from soil exposed by a small break in the tundra vegetation. The vent is tiny, just a fine stream of molecules, invisible and odorless. There is little drama here—not a fire raging through Yellowstone or a plague of locusts. But elsewhere the effect of this event, and others like it around the world, are dramatic. Wildfires blacken thousands of dry acres in Texas. Hurricanes in unprecedented numbers and intensity strike broad reaches of the Caribbean and the United States. Receding glaciers in Greenland dump hundreds of icebergs into the North Atlantic. In Arctic Alaska a polar bear searches for a den site on the coast because the sea ice is too depleted to provide habitat.

It may be difficult to imagine a connection between Arctic carbon emission, climatic shifts, and spectacular storms thousands of miles away. But those few molecules of methane epitomize the essence of ecology—the cumulative effect of myriad interactions.

The methane escaping from the permafrost is the product of plants that lived there millions of years ago. For all that time, it was sequestered in the Arctic soil, frozen and covered since the last ice age. As the earth gradually warmed, the Arctic permafrost began to thaw. Eventually, in some places, the ground collapsed, forming gullies of exposed soil called **thermokarsts**. The carbon held in the permafrost escapes, mostly in the form of methane. Two factors make this globally significant: methane is a greenhouse gas 25 times more potent than carbon dioxide, and the Arctic tundra holds almost twice as much carbon as the atmosphere. In fact, it holds far more than all the industrial carbon emissions to date. As that carbon is released into the atmosphere, a positive feedback loop is unleashed—the increased carbon increases the temperature, which in turn releases

■ **thermokarsts** Gullies of exposed soil produced by the selective melting of permafrost.

more carbon. This vastly complicates our efforts to mitigate climate change. We need to be concerned not just about industrial emissions and the burning of fossil fuels but also the natural processes of carbon release that result from climate change itself. Of course, the Arctic is just one system so affected. Combine this with the release of carbon from other natural systems and you begin to see the outlines of a major ecological problem.

Carbon emissions from fossil fuels and climate change receive much media attention, especially as climate change becomes linked to other effects, such as the frequency of hurricanes, sea level change, and impacts on agricultural production. The simple part of this phenomenon is the rate of carbon dioxide emission from fossil fuel. And we are developing climate models sufficiently sophisticated to predict the impact of increasing global temperature. But far more complex, and ultimately crucial, are the patterns of storage and movement of carbon within and among ecological systems. Where is most of the world's carbon stored? How is the movement of carbon in the Arctic connected to primary production and decomposition in the tropics? What role does the ocean play in carbon storage or emission? These questions are crucial to our understanding of climate change. They are also central to a full understanding of ecosystem ecology. And they lead us to the central question of this chapter: *How do nutrients move through ecosystems?*

17.1   What Are the Major Patterns of Nutrient Movement?

Both plants and animals require specific nutrients for growth, maintenance, and reproduction. As we saw in Chapter 3, any essential nutrient may limit these processes, and each species has a unique set of minimum requirements. Consequently, each species faces selection pressure to acquire and maintain sufficient nutrients for survival and reproductive success. Some are obtained directly from the physical environment; others are acquired from other organisms, either by consumption or a mutualistic relationship.

Nutrients move through ecosystems in **biogeochemical cycles**, or nutrient cycles, in which elements and compounds are exchanged among the atmosphere, rocks, soil, water, and organisms. The elements that comprise nutrients are neither created nor destroyed; they simply change chemical form as they move through the ecosystem. Thus, the movement of nutrients is cyclic. This is an important contrast with the flow of energy in ecosystems, which is always linear—the sun is a constant source of new energy, but that energy is lost to the ecosystem as it passes up the food chain.

Ecosystem ecologists employ a systems approach to the analysis of nutrient cycles. A simple systems model of a biogeochemical cycle is shown in Figure 17.1. Nutrient cycles contain two key components. **Pools**, or **sinks**, are places that hold nutrients. They can be either biotic or abiotic ecosystem components. Each pool is characterized by the amount of material it holds and its chemical form. **Fluxes** are the pathways of nutrient transfer among pools. They are characterized by the direction and rate of movement. Nutrient cycles may be defined on almost any spatial scale. However, if we are to quantify the pools and fluxes, we must

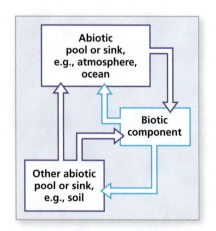

Figure 17.1 Nutrient flow. A generalized model of nutrient flow in an ecosystem. Nutrients are found in pools. Fluxes are the paths of transfer between pools. **Analyze:** How would you define a nutrient cycle that is in equilibrium?

precisely define the ecosystem boundaries. At the global scale, biogeochemical cycles are completely closed; there is no net loss of material from the earth. However, at smaller spatial scales, it is possible to have a net input or output of material from a local nutrient cycle (Figure 17.2).

How Do Carbon, Nitrogen, and Phosphorus Cycle in Ecosystems?

We can describe the biogeochemical cycles of virtually any nutrient important to plants or animals. The carbon, nitrogen, and phosphorus cycles are especially well known because these elements are important to nearly every organism. And together they illustrate the most important aspects of biogeochemical cycles, as well as some of the important differences among them.

The global carbon cycle is driven by the fundamental energy pathways in ecosystems: photosynthesis and respiration (Figure 17.3). Chapter 15 examined the basic flow of energy through ecosystems. This begins with the capture of energy and its storage in the energy-rich bonds of carbohydrates. This chemical energy powers the rest of the trophic levels as the energy of primary producers is consumed and transferred up the food chain. In the process, carbon is released by the energy-yielding biochemistry of respiration. Overall, these central chemical reactions capture or release carbon according to the fundamental equation

$$6CO_2 + 6H_2O \underset{\text{Respiration}}{\overset{\text{Photosynthesis}}{\rightleftarrows}} C_6H_{12}O_6 + 6O_2.$$

biogeochemical cycle (nutrient cycle) The movement of nutrients through ecosystems by exchange among the atmosphere, rocks, soil, water, and organisms.

pool (sink) Components of an ecosystem (biotic or abiotic) that hold nutrients.

flux The pathway of nutrient transfer among pools.

THINKING ABOUT ECOLOGY:

What different ecological questions can be addressed in studies of nutrient cycles at a scale of a few square meters compared to an entire watershed?

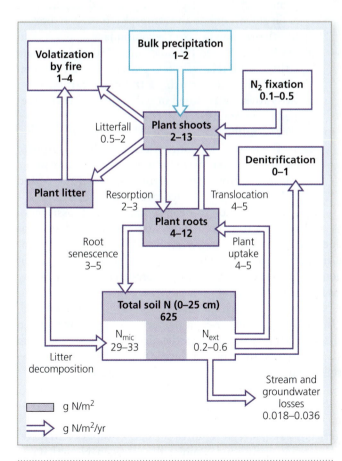

Figure 17.2 Local nutrient cycle. A local nutrient cycle with inputs and exports of nutrients (from Blair et al., 1998). **Analyze:** Can local nutrient cycles be unbalanced?

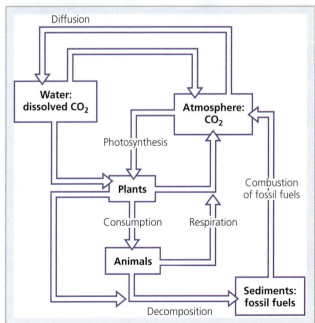

Figure 17.3 Global carbon cycle. The atmosphere is a major pool in this cycle. Photosynthesis and respiration are the key fluxes of carbon to and from organisms. **Analyze: What forms of carbon can be used by plants?**

The atmospheric carbon pool is relatively large—some 720 gigatons of carbon. At present, the CO_2 concentration of the atmosphere is approximately 400 ppm. Consequently, photosynthesizing plants are generally not limited by CO_2 availability.

The aquatic pool of the carbon cycle is more complex (Figure 17.4). Carbon enters aquatic ecosystems not directly by photosynthetic use of atmospheric CO_2 but by the physical diffusion of CO_2 at the surface. Much of this physical flux occurs in the cold waters at high latitude. Ocean currents carry it to deeper waters, which circulate to lower latitudes, where upwelling brings it back to the surface. CO_2 reacts with water to form carbonic acid (H_2CO_3), a weak acid. Carbonic acid dissociates to form other ions according to the pH of the water:

$$CO_2 + H_2O \leftrightarrow H_2CO_3 \leftrightarrow H^+ + HCO_3^- \leftrightarrow 2H^+ + CO_3^{-2}.$$

Carbon, primarily in the form of bicarbonate (HCO_3^-), enters the food chain by the photosynthetic activity of phytoplankton. From these primary producers, it enters a complex web of pools and fluxes.

The carbon cycle is intimately connected to the fundamental energy processes of ecosystems—primary production and respiration. Not surprisingly, the net rate of photosynthesis is not always exactly balanced by the net rate of respiration. Consequently, at various times in geological history, the oxygen and carbon dioxide content of the atmosphere have shifted significantly. Data on global climate and atmospheric carbon dioxide detail these changes over evolutionary time. When plants first colonized the land some 450 million years ago, there were relatively few grazers to exploit them. The result was that much of the fixed carbon from photosynthesis was stored in sediments rather than returned to the atmosphere in the form of CO_2.

For 30 million years, the earth's climate cooled, because the lower CO_2 concentration in the atmosphere reduced the level of natural greenhouse warming. The Carboniferous Period, which spanned the period 350–300 million years ago, was a time of significant addition of carbon to deep sediments and fossil fuels. This was a time of shallow seas and lowland swamps with high plant production. Many of the newly evolved land plants were supported by lignin. The enzymes microbes use to decompose lignin today had not yet evolved. Consequently, a great deal of carbon was stored in sediments. Today we burn that carbon as fossil fuel.

Like the carbon cycle, the nitrogen cycle contains a large atmospheric pool of nitrogen (Figure 17.5). This nitrogen, however, is in the form of molecular nitrogen (N_2), which can be utilized by only a few organisms, the **nitrogen fixers**. This group includes cyanobacteria and bacteria that contain the crucial enzyme, nitrogenase, which can break the triple bond of molecular nitrogen and catalyze the reaction

$$N_2 + 8\,H^+ + 8\,e^- \rightarrow 2\,NH_3 + H_2.$$

This reaction is energetically expensive; it requires the energy input of 16 molecules of ATP to break the strong triple bond in molecular nitrogen. Note that one of the products of the reaction is ammonia, a nitrogen compound that plants can utilize. Nitrogenase is highly conserved among bacteria and cyanobacteria, suggesting a single, ancient origin of the enzyme.

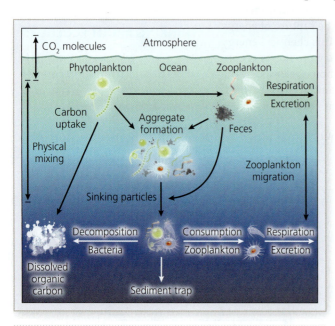

Figure 17.4 Aquatic portion of the carbon cycle. The aquatic portion of the carbon cycle depends on CO_2 that dissolves in water from the atmosphere (from Normile, 2009). **Analyze: What is the potential effect of increasing atmospheric CO_2 on ocean carbonate and bicarbonate?**

nitrogen fixers Organisms that can convert molecular nitrogen into organic form.

algal boom Explosive algal growth due to nutrient addition to an aquatic ecosystem.

residence time The length of time an element stays in a particular pool.

Nitrogen fixation occurs in two important ways in ecosystems. Some is accomplished in aquatic ecosystems by free-living cyanobacteria or nitrogen-fixing bacteria. They utilize the ammonia in their own growth, and it is from them that it enters other species in the system. Nitrogen fixation also occurs as a result of a mutualistic association of nitrogen-fixing bacteria and plants. Alders and legumes are well-known partners of nitrogen-fixing bacteria. Root nodules house the mutualistic bacteria. There they receive carbon sources for their metabolism in exchange for fixed nitrogen. Recall from Chapter 14 that these species are important in some primary succession sequences because they accelerate the addition of nitrogen to the developing system.

We are now beginning to understand that nitrogen fixation by free-living bacteria or the mutualistic relationships in higher plants are not the only important sources of nitrogen to ecosystems. Some lichens, mutualistic associations between fungi and algae, include cyanobacteria in the algal component, which significantly augments nitrogen fixation in some ecosystems. Even termites are responsible for some of the nitrogen that enters biotic pools due to the nitrogen-fixing bacteria they house in their guts.

The nitrogen that enters plants by these processes enters the other biological components of the ecosystem— the consumers who pass it through the food chain in the form of proteins and nucleic acids. From the primary producers and consumers, nitrogen passes to the soil or water via decomposition and excretion. There, a crucial set of chemical conversions takes place. Animals excrete nitrogen in the form of urea, uric acid, and ammonia. The decomposition of proteins releases amino acids. Bacteria convert uric acid, urea, and amino acids to ammonia. This compound is oxidized by other bacteria to form nitrite, which in turn is oxidized to form nitrate, which can once again be taken up by primary producers. Nitrogen can be lost to the biotic component of the ecosystem by the energy-yielding process of denitrification, which converts nitrate to molecular nitrogen.

Nitrogen often limits primary producers, especially in aquatic ecosystems. Coral reefs are among the most productive marine ecosystems, in part because of the unusually high rate of nitrogen fixation there. The addition of nitrogen to a lake can result in explosive growth of algae, known as **algal blooms**, as part of the process of eutrophication (Chapter 15). Intensive agriculture can remove significant amounts of nitrogen from the soil, eventually requiring the input of nitrogenous fertilizer.

Phosphorus is crucial to many aspects of biochemistry. It is an important structural component of nucleic acids as well as cell membranes. It is also central to the function of ATP, the energy currency of the cell. Like nitrogen, phosphorus is a common limiting factor in primary production.

The key pools and fluxes of the phosphorus cycle are shown in Figure 17.6. Note that unlike the carbon and nitrogen cycles, there is no atmospheric pool of phosphorus. Instead, the major pool of this element is in phosphate rock or sediments. Phosphate in these pools may remain unavailable to organisms for thousands or even millions of years. Once phosphorus enters living organisms, it is often recycled there for long periods. The **residence time**, the length of time an element persists in a particular pool, is on the order of thousands of years for phosphorus in the biotic component of an ecosystem. This contrasts with a residence time of approximately 625 years for nitrogen in living organisms.

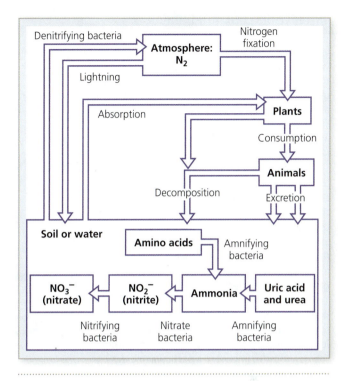

Figure 17.5 Global nitrogen cycle. The global nitrogen cycle is characterized by a large atmospheric pool and specialized organisms that can convert specific nitrogenous compounds. **Analyze:** How is the nitrogen cycle similar to and different from the carbon cycle?

THINKING ABOUT ECOLOGY:

Can you devise an explanation for the fact that nitrogen fixation is limited to prokaryotes?

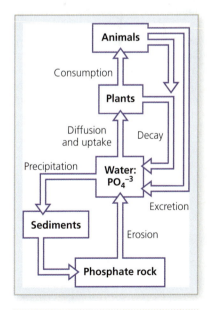

Figure 17.6 Global phosphorous cycle. In the global phosphorus cycle, there is a large pool in phosphorus rock. **Analyze:** How is the phosphorus cycle different from the nitrogen cycle?

Variance and Ecosystem Models

In Chapter 3 we introduced the concept of the variance, S^2, the variability associated with a measurement. The variance depends on the natural variability in the parameter being measured, as well as the sample size. If the sample size is small, a few abnormally large or small values may increase the variance associated with the mean (Figure 1).

There are many situations in which we want to use a mean value in other calculations. For example, if we are trying to model the effect of the release of CO_2 from burning fossil fuels, we must incorporate mean values of CO_2 emissions as well as the means of the various pools and fluxes in the carbon cycle. Obviously, as the variance of these means increases, the models we generate become less accurate.

An important set of models is designed to predict the effect of carbon emissions on global temperatures due to greenhouse effects. Specifically, they address a key question: What will be the quantitative impact of a particular increase or decrease in CO_2 on global temperature? The models are based on the interaction of the carbon cycle, atmospheric CO_2, and climate. We know that global temperature is roughly a logarithmic function of the atmospheric CO_2 concentration. Consequently, we can model the atmospheric CO_2 needed to stabilize greenhouse climate change at a particular level of warming (Calderia et al., 2003). This relationship is described by the equation

$$\frac{P_{stab}}{P_{280}} = 2^{\left(\frac{\Delta T_{stab}}{\Delta T_{2x}}\right)}.$$

In this equation, P_{stab} is the CO_2 concentration (in ppm) at which temperature is stabilized; P_{280} is the preindustrial CO_2 concentration (280 ppm); ΔT_{stab} is the stabilized temperature change; and ΔT_{2x} is the sensitivity of the climate to greenhouse effects. Specifically, ΔT_{2x} is the mean temperature change resulting from a twofold increase in atmospheric CO_2 concentration. Thus, the target CO_2 concentration for stabilizing global temperature increases exponentially with the ratio of the stabilization temperature to the sensitivity.

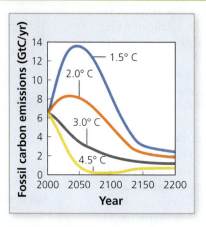

Figure 2 **Allowable CO_2 emissions.** Allowable emissions of CO_2 to produce climate stabilization at 2°C global mean warming (from Calderia, 2003). **Analyze:** Why do the curves decline over time?

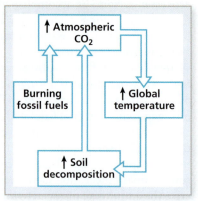

Figure 3 **Positive feedback loop.** A positive feedback loop initiated by burning fossil fuels that leads to increasing global temperature. **Analyze:** Why does an increase in global temperature lead to increased decomposition in soil?

This equation seems relatively straightforward—it allows us to calculate the maximum amount of CO_2 we can allow in the atmosphere without increasing global temperature. However, there is variation inherent in key components of the equation. In particular, we do not know the value of ΔT_{2x} within a factor of three. Estimates range from 1.5°C to 4.5°C. The results of the equation differ markedly depending on which value is correct (Figure 2). If the correct value is 1.5°C, we could allow atmospheric CO_2 to reach 700 ppm. On the other hand, if the value is 4.5°C, we must not allow CO_2 to rise above 380 ppm. This is a crucial difference, given that we have already passed that value.

Of course, this model depends on a detailed understanding of the carbon cycle, especially the measured values of the various pools of carbon, such as the amount in the ocean and that stored in vegetation, as well as the magnitude of the fluxes among compartments. Each of these measurements also contains some error (variance) due to measurement limitations or natural variation.

We continue to refine our understanding of the carbon cycle and the measurements of its pools and fluxes. For example, we now understand that the world's soils contain at least twice the amount of carbon as the atmosphere, much in the form of organic matter. As this material decomposes, additional CO_2 is released. Unfortunately, the rate of decomposition increases with temperature. Thus, a positive feedback loop develops, in which increasing atmospheric CO_2 raises the global temperature, which in turn causes increased decomposition in the soil, releasing even more CO_2 and ultimately increasing the temperature even more (Figure 3).

The variation and uncertainty surrounding each component of the carbon cycle and its operation contributes to the imprecision of our predictive models. The fact of global warming is indisputable. The uncertainty lies in the quantitative predictions we can make about the impact of specific increases or decreases in carbon emissions.

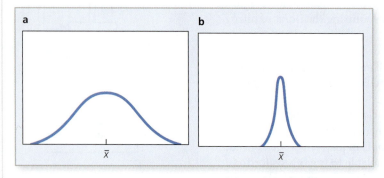

Figure 1 **Variance.** Two measurements with the same mean but large (a) and small (b) variance. **Analyze: For which of these graphs is the mean more representative of the distribution?**

Oxygen and the Origin of Complex Life Forms

We know that the chemical composition of the Earth has changed over geologic and evolutionary time. Moreover, the dynamic nature of the Earth's abiotic systems is a direct result of the activity of living organisms. In fact, the biotic and abiotic components of the Earth's ecosystems have been engaged in an intimate feedback system from the time life first appeared.

The first organisms were simple, composed of one or a few cells. Initially, all were heterotrophs dependent on organic material in the seas or on other organisms. When photosynthesis evolved, it provided a major new source of organic energy and released molecular oxygen into the atmosphere. Ferric iron deposits—that is, oxidized iron rocks—are first found at this time, confirming the presence of oxygen in the atmosphere. It was at this point that the carbon cycle came to resemble more closely the modern pattern, in which carbon is fixed by photosynthesis and released by respiration.

An oxygen-rich atmosphere radically changed the biochemistry and energy processes in living things. The origin of eukaryotes occurred in this new, oxygen-rich world. One of the most significant biochemical differences between the eukaryotes and prokaryotes is that the former use oxygen as a final electron acceptor in respiration, a difference that generates far more energy. In fact, many scientists believe that one of the most profound expansions of animal life was powered by oxygen and the energy it provided (Payne et al., 2009; Dahl et al., 2010).

Approximately 530 million years ago, in an event known as the Cambrian explosion, the diversity of organisms suddenly increased by an order of magnitude. Within a few million years, most of the modern animal phyla had emerged. These new taxa were larger and more complex, their internal structure now composed of tissues and organs. The energetic demands of these new forms for development, activity, and maintenance were served by the energy from photosynthesis and the efficient respiration possible because of the availability of oxygen. In addition, oxygen may have contributed to larger, more complex organisms because it plays an important role in the production of complex polymers, such as collagen, that provide structural support in larger animals.

Figure 1 Devonian predators. Large predatory fish evolved in the Devonian period, when the atmosphere and aquatic systems had higher oxygen concentrations.

A second major increase in the size of organisms occurred approximately 400 million years ago in the Devonian Period. This, too, was associated with a jump in atmospheric oxygen. The Devonian oxygen increase is correlated with the diversification of vascular plants, which likely contributed importantly to this change. Estimates place the oxygen concentration as high as 36 percent—much higher than the present 21 percent. The Devonian is known for the origin of large predatory fish—a body type and ecology that requires a great deal of energy to sustain (Figure 1). Fish are relatively intolerant of hypoxia. Thus, it is not surprising that the origin of large predatory fish coincides with evidence of higher marine oxygen levels (Dahl et al., 2010). Of course, these predators would also require a significant prey base. The numbers and sizes of organisms lower in the food chain also increased at this time.

QUESTION:

How is oxygen production via vascular plants on land connected to the oxygen concentration of marine systems?

Phosphorus becomes available to organisms when phosphate rocks weather and erode. This phosphorus reacts with water to form phosphate (PO_4^{-3}), the form of phosphorus that plants absorb. These reactions, and those from excretion and decomposition, do not require specialized organisms, as in the nitrogen cycle. Phosphate ions can, however, form insoluble compounds with aluminum, iron, and calcium at high or low pH. Consequently, the availability of phosphorus to primary producers, and its limit to production, is determined by soil pH.

Local phosphorus budgets are particularly affected by net input and loss from the system. This is illustrated by the phosphorus budget of the Everglades in South Florida. This ecosystem consists of extensive stands of sawgrass (*Cladium jamaicense*) marsh. The main input of phosphorus to this system is in the sheet flow characteristic of the Everglades. The Everglades were historically an oligotrophic aquatic system in which phosphorus was limiting. Intensive agriculture to the north, primarily in heavily fertilized sugar cane plantations, is shifting this ecosystem toward eutrophication.

KEY CONCEPTS 17.1

- The movement of nutrients through the ecosystem is cyclic.
- Biogeochemical cycles are characterized by pools, or sinks, components of the ecosystem where the nutrients reside, and by fluxes, the pathways of transfer among pools.
- Nutrient cycles can be defined and studied on any spatial scale.

QUESTION:

What are the essential differences among the carbon, nitrogen, and phosphorus cycles?

17.2   What Are the Mechanisms of Nutrient Flux?

Our description of the carbon, nitrogen, and phosphorus cycles included examples of both physical and biological processes that move nutrients through the ecosystem. What general principles of nutrient flux emerge from these specific examples?

What Are the Important Physical Mechanisms of Flux?

The ultimate sources of inorganic nutrients are the abiotic pools in rocks and the atmosphere. Nutrients such as phosphorus, calcium, and potassium have large pools in the Earth's crust. The inorganic forms of these nutrients are found in rocks as **minerals**—elements or chemical compounds that have been formed by geological processes. **Rocks** are aggregates of minerals. Soil is derived from the **parent material**, the underlying substrate, which includes bedrock and deposits of material from elsewhere. **Glacial till** is material deposited by glaciers. In the Great Plains, much of the parent material that ultimately led to the fertile soils of the prairie is glacial till that carried nutrients from bedrock farther north during glacial advances. Glacial material deposited by wind, known as **loess**, also contributed to the mineral richness of the prairie soil.

Soil is a complex substance composed of organic material derived from the breakdown of plants and animals, the mineral nutrients derived from the parent material or elsewhere, water, gases held among the soil particles, and many living organisms. Soil has a vertical structure—it is composed of layers, known as **soil horizons**, with different characteristics and composition (Figure 17.7).

Soil forms over the parent material as a result of both physical and biological processes. The nutrient content of soil is determined partly by the addition of nutrients from the parent material. This occurs by the process of **weathering**. Weathering begins with the physical breakdown of the parent material by freezing and thawing or other mechanical processes. Rocks, with their larger surface area, are vulnerable to chemical weathering in which the constituent nutrients change chemical form.

Another major source of inorganic nutrients in ecosystems is **atmospheric deposition**, the transfer of materials from the atmosphere to terrestrial or aquatic ecosystems. Many elements and chemical compounds adhere to small particles of dust in the atmosphere or dissolve in water vapor. Collectively, these suspended materials are known as **aerosols**. Atmospheric deposition occurs when aerosols reach the surface by gravity or precipitation. This process can move mineral nutrients great distances. There is some evidence that primary production in marine

minerals Elements or chemical compounds formed by geological processes.

rock An aggregate of minerals.

parent material The rock and mineral substrate underlying a region.

glacial till Material deposited by a glacier.

loess Glacial till deposited by wind.

soil horizons The series of layers in soil with different biological and physical properties.

weathering The addition of nutrients from the parent material by physical and chemical processes.

atmospheric deposition The transfer of materials from the atmosphere to terrestrial or aquatic ecosystems.

aerosols Elements and compounds suspended in the atmosphere.

ecosystems is limited by iron. Atmospheric deposition of iron aerosols from land masses is an important source of this limiting nutrient in oceans.

Alluvial deposition is the addition of nutrients to a system by flowing water. This process is important in a number of ecosystem types, particularly when episodic flooding deposits nutrients where they would otherwise be limiting. The fertility of some of the world's most productive agricultural areas, such as those along the Mississippi, Nile, and Mekong River deltas, is a result of this process. Dams that eliminate these deposits may result in a gradual decline in fertility and thus production.

What Are the Important Biological Mechanisms of Flux?

Photosynthesis and nitrogen fixation constitute two critically important biological fluxes moving nutrients from inorganic to organic pools in global biogeochemical cycles. Each requires a specific biochemical process limited to specific groups of organisms—photoautotrophs and nitrogen fixers, respectively. Other fluxes entail the direct absorption by plants or animals of nutrients converted into the appropriate compounds by physical processes.

We are learning that biological mechanisms of flux are complex and interconnected. For example, it has recently been discovered that the leaf-cutter ants and their mutualists play a significant role in both the carbon and nitrogen cycles (Pinto-Tomas et al., 2009). Leaf-cutter ants harvest leaves in tropical forests. They "farm" certain species of fungus on the leaves they collect, providing the fungus with a carbon source and protection (Figure 17.8). The ants harvest the fungus to feed to their larvae. Leaves typically have a low N:C ratio; thus, nitrogen is a limiting nutrient in this system. However, the ants also engage in a mutualistic relationship with nitrogen-fixing bacteria that live in their fungus gardens. Leaf-cutter ant colonies are massive, some of the largest of any social insects. A colony may contain 8 million workers, and their nests may comprise 20 m^3. Each colony contributes approximately 1 kg fixed nitrogen per square meter per year to the rainforest ecosystem. Given the abundance of these colonies, this represents a significant nitrogen addition to these forests.

Local nutrient cycles have both inputs and outputs of organic nutrients. Some of these fluxes occur by physical mechanisms such as aerosol deposition or the downstream flow of nutrients in a river. But biological processes can be important components of local nutrient input and output. For example, salmon provide important inputs of nitrogen and other elements to terrestrial ecosystems. Salmon accumulate 99 percent of their body mass and nutrients in the ocean (Janetski et al., 2009). As they migrate upstream, spawn, and eventually die, they are consumed by a number of terrestrial predators and scavengers, including grizzlies, eagles, ravens, and gulls. In the process, nitrogen and phosphorus accumulated in the open ocean is deposited in nutrient-limited terrestrial ecosystems far away (Figure 17.9). As salmon decompose in the water, they raise the concentration of nitrogen and phosphorus in suspended stream sediments by as much as an order

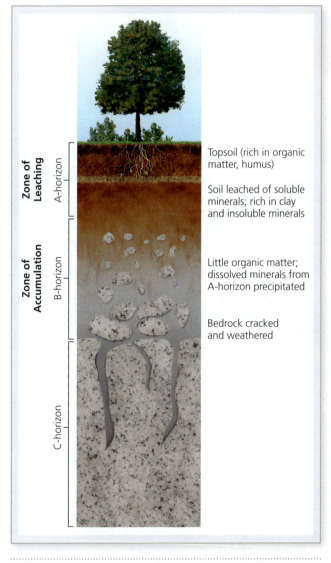

Zone of Leaching
Zone of Accumulation
A-horizon
B-horizon
C-horizon

Topsoil (rich in organic matter, humus)

Soil leached of soluble minerals; rich in clay and insoluble minerals

Little organic matter; dissolved minerals from A-horizon precipitated

Bedrock cracked and weathered

Figure 17.7 Soil horizons. The main soil horizons are differentiated by specific biological and physical characteristics. **Analyze:** How do nutrients in the B and C horizons move into the A horizon?

alluvial deposition The addition of material by flowing water.

Figure 17.8 Leaf-cutter ants. Leaf-cutter ants play a significant role in the nutrient cycles in the rainforest.

THE HUMAN IMPACT

Soil Fertility and Nutrient Cycles

Human agricultural production is intimately tied to the biogeochemical cycles of essential plant nutrients. As we know from Chapter 3, nitrogen, phosphorus, and potassium are the primary limiting macronutrients for plants. Thus, these three nutrients represent the central components of soil fertility and crop production.

Soil constitutes a pool in both the nitrogen and phosphorus cycles. However, soil is far more complex than illustrated in these overviews of the cycles. It is the A horizon that is crucial to agricultural production. This is the zone in which plants are rooted and from which they obtain mineral nutrition. The A horizon is biologically and chemically complex. It contains three essential components: humus, mineral nutrients, and the living organisms important in decomposition and biochemical transformations such as nitrification.

The soils of different biomes differ significantly in texture and nutrient content. These differences underlie the suitability of the various soil types for agriculture. For example, the soils of the North American prairie are among the most fertile in the world. Many river deltas, where floods replenish minerals and organic matter, are also prime agricultural regions.

Unfortunately, many of the world's soil systems are rapidly losing fertility. In effect, human activity, sometimes agriculture itself, has disrupted the basic biogeochemical cycling that maintains soil fertility. To the extent that nitrogen or phosphorus is lost from the A horizon, the fundamental movement of nutrients shifts from cycling to net export. This occurs in two ways. Harvesting crops removes not only plant material but the nutrients it contains. Second, soil erosion removes the A horizon. Agricultural practices disrupt nutrient cycles in other ways as well. For example, the extensive use of pesticides destroys mycorrhizae, important contributors to plant nutritional health.

The Dust Bowl of the mid-1930s represents one of the most spectacular examples of anthropogenic destruction of soil fertility. By the late nineteenth century, much of the grassland of the

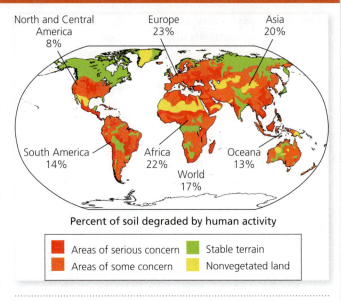

Percent of soil degraded by human activity

| ■ Areas of serious concern | ■ Stable terrain |
| ■ Areas of some concern | ■ Nonvegetated land |

Figure 2 Soil degradation. Regions of concern about soil degradation and loss of fertility (from Enger and Smith, 2000). **Analyze:** What human activities besides agriculture lead to soil degradation?

Great Plains had been plowed and converted to row crops. Deep plowing exposed virtually the entire A horizon to wind and water erosion. And because the land was cropped for only half the year, bare soil was exposed for long periods. The intense drought and crop failures of the 1930s further reduced plant cover. The result was staggering loss of topsoil and its fertility that permanently altered the soil profile and the contents of the A horizon (Figure 1).

The loss of soil and its fertility is now a matter of global concern (Figure 2). Since World War II, some 17 percent of the vegetated land surface of the Earth has experienced some level of soil degradation. The majority of this loss is due to the significant increase in grazing and the erosion it causes. It is estimated that agricultural practices are responsible for about one-quarter of the total amount of soil that has been lost. Of course, the same processes that originally produced these soils are still occurring. However, the rate at which soil forms is slow relative to our ability to destroy it. In the United States, the current rate of soil erosion is approximately 17 metric tons per hectare per year; soil formation occurs at the rate of only 3.7 metric tons per hectare per year. In other parts of the world, soil forms even more slowly. For every inch of topsoil that is lost, crop production declines about 6 percent. This is a serious concern in a world whose population is increasing at approximately 1.1 percent per year.

QUESTION:

How is anthropogenic disruption of nutrient cycles similar to, and different from, the processes that occur in succession?

Figure 1   Dust storms. Enormous dust storms resulted in significant loss of the soil horizon and soil fertility in the 1930s.

of magnitude. This material is transported downstream, where it may enrich the lakes through which the salmon passed during their migration.

Once nutrients are in the biological pools of the ecosystem, fluxes cycle at least some of that material locally. Heterotrophic consumption is central to this movement. Herbivory and carnivorous predation move materials directly from one living organism to another. Nutrients enter the soil by excretion and in the bodies of dead organisms. **Decomposition**, the breakdown of animal excretions or the bodies of dead organisms, provides nutrients in forms that plants can utilize. This particular flux depends on two processes, decomposition itself and **mineralization**, the transformation of organic matter into inorganic nutrients. The initial phases of decomposition depend on the feeding activity of **detritivores**, organisms that consume feces and dead organisms. Vertebrate and invertebrate carrion eaters such as vultures and crows begin the process of breaking down the bodies of dead animals. A host of soil invertebrates and fungi derive energy from decomposing bodies, dead plants, and feces and in the process break the material into smaller pieces, which are consumed by other detritivores. Often there is a specific sequence of organisms that break down a particular type of material, a process known as **degradative succession**.

Mineralization is a chemical process accomplished by bacteria and fungi. Some of these organisms release enzymes into the substrate that break down organic molecules into forms that are then absorbed by the decomposer. The A horizon of soil contains much of this material in the form of **humus**. Humus is a complex mix of organic molecules, produced by the enzymatic breakdown of carbohydrates, lipids, and proteins, that has reached a stable state with no further breakdown.

The process of decomposition and mineralization depends on the nature of the organic material. Lignin and cellulose, important plant structural elements, and chitin, a component of invertebrate exoskeletons, are difficult to break down. We have seen that the evolution of lignin had a significant effect on the carbon cycle. Recent work shows that under certain circumstances lignin actually accelerates decomposition. Lignin is especially susceptible to photodegradation, breakdown caused by absorption of light (Austen and Ballare, 2010). Thus, in ecosystems in which light intensity is high, such as deserts, clearcut or burned forests, and agricultural systems, release of carbon by photodegradation may be high. Moreover, other nutrients protected from decomposers by lignin also become more available.

Biological and physical processes may lead to the net input or loss of nutrients from a local ecosystem. We have seen, for example, that the migration of salmon results in the net transport of nutrients over great distances. Flowing water also leads to this kind of movement and net gain or loss of nutrients. In rivers and streams, the flow of water moves nutrients downstream. The result is that complete nutrient cycles do not occur at any one point in the stream. Rather, a stream contains a **nutrient spiral**, in which nutrients in the form of living organisms and detritus are transported downstream, where they enter new biological systems and then continue downstream in new forms (Figure 17.10). In other words, any given atom eventually completes a nutrient cycle, but it may do so over a great length of stream in several different local ecosystems. The distance an atom requires to complete a cycle is known as the **spiraling length**. Where fluxes are slow and stream flow is fast, spiraling length is long; where fluxes occur rapidly and the stream moves slowly, the spiraling length is shorter.

Figure 17.9 Nutrients in salmon. The nutrients in salmon move from aquatic to terrestrial ecosystems by predators and scavengers.

decomposition The breakdown of animal excretions or the bodies of dead organisms.

mineralization The transformation of organic matter into inorganic forms.

detritivore An organism that consumes feces or dead organisms.

degradative succession A sequence of decomposition processes accomplished by a series of organisms.

humus A mix of organic molecules from the chemical decomposition of carbohydrates, lipids, and proteins.

nutrient spiral The transport of organisms and detritus downstream.

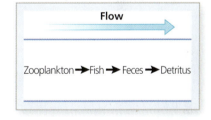

Figure 17.10 Nutrient spiral. A nutrient spiral in a stream results in the movement of nutrients along the length of the stream as material moves from producers to consumers and decomposers. **Analyze: What does this pattern mean for the nature of nutrient cycles in first-order compared to sixth-order streams?**

KEY CONCEPTS 17.2

- Nutrients are added to ecosystems by internal and external physical processes. Weathering of parent material is the major local source of nutrients.

spiraling length The distance an atom requires to complete a nutrient cycle in a nutrient spiral.

- Nutrients from weathered rock may be transported by wind, glaciers, or water. Nutrients are also transported from one ecosystem to another as aerosols.
- Photosynthesis and nitrogen fixation are important biological inputs of carbon and nitrogen, respectively, to ecosystems.
- Decomposition releases organic nutrients into the soil or water, where they can be taken up again.

QUESTION:

What factors would you expect to limit the rate of decomposition?

17.3 How Are Biogeochemical Cycles Controlled?

Biologists place great emphasis on understanding the control of biological processes. Previous chapters have addressed the control of phenomena such as population dynamics and energy flow. Biogeochemical cycles lend themselves to this type of analysis because the movement of nutrients is at least potentially cyclic. This means we can ask how the flux rates and pool sizes are controlled and whether these factors result in cycles that are in balance. For cyclic processes, this is the equivalent of asking if the system has reached a dynamic equilibrium in which fluxes maintain relatively constant pool sizes and there is no net input or output of nutrients.

What Processes Are Limiting in Nutrient Cycles?

In any cyclic process, one or more fluxes can limit the rate of cycling and the content of the pools. Any factor that increases or decreases these critical fluxes can slow or accelerate the cycle. Two generalizations emerge: the rate of input of nutrients from inorganic pools and the rate of decomposition are important limiting processes. The consumption of one organism by another is a key process of nutrient flux. Of course, primary production underlies all trophic structure. Thus, it is logical to infer that any processes that affect trophic structure and energy flow have a controlling effect on nutrient cycles as well. Primary production thus becomes crucially important to the flux of carbon as well as other nutrients.

Although we have presented biogeochemical cycles as discrete entities, they are in fact intimately connected (Elser et al., 2007; Townsend et al., 2011). One important example is the coupling of the nitrogen and phosphorus cycles to the carbon cycle due to those elements' limiting role in primary production. Nitrogen or phosphorus enrichment experiments demonstrate the limiting role these two elements play in terrestrial, freshwater, and marine production (Figure 17.11).

Much current research focuses on interactions in tropical systems because they hold so much carbon and thus play a significant role in the carbon budget of the planet and global warming. Many tropical forests are phosphorus limited because phosphorus is depleted in the ancient, highly weathered soils found there.

In other tropical forests, especially montane forests, nitrogen is the nutrient that limits production and thus the movement of carbon from the atmosphere into biomass. These interactions extend to other cycles as well. Phosphorus availability is determined by both the nitrogen and oxygen cycles. When nitrogen is more abundant than phosphorus, plants allocate more energy to

THINKING ABOUT ECOLOGY:

Does this information suggest that nutrient cycles are controlled by top-down or bottom-up mechanisms? How could you experimentally distinguish between the two possibilities?

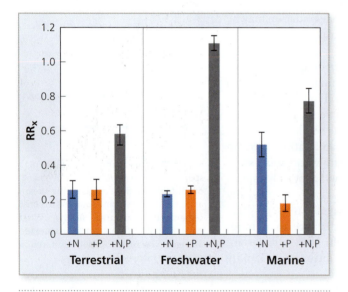

Figure 17.11 Nitrogen and phosphorous enrichment. The response of autotrophs to enrichment by nitrogen, phosphorus, or nitrogen + phosphorus measured by the value RR_x, the ratio of production in enriched treatment divided by the control biomass (from Elser et al., 2007). **Analyze:** Why is RR_x a useful measure of the effect of fertilization?

The Nitrogen Paradox in Tropical Forests

The species diversity and geographic size of tropical forests make their biogeochemical cycles particularly important. An interesting paradox has been described in the nitrogen budget of tropical forests. Nitrogen availability often exceeds demand, yet the nitrogen fixation rates are as high as in any other ecosystem. Why should energetically expensive fixation occur at a high rate when nitrogen is not limiting?

One suggestion to explain the unexpectedly high rates of nitrogen fixation is that the process interacts with and is stimulated by the availability of other macronutrients, especially carbon and phosphorus. Reed et al. (2008) examined these aspects of the control of nitrogen fixation in a Costa Rican forest. Their study concentrated on free-living nitrogen fixation—that accomplished by microorganisms. They addressed two general hypotheses relevant to the nitrogen paradox:

HYPOTHESIS 1: **The rate of nitrogen fixation varies among species and functional components of tropical forest.**

HYPOTHESIS 2: **The rate of nitrogen fixation is determined, in part, by soil phosphorus concentration.**

The researchers sampled eight individuals of six species along a vertical profile from soil to canopy leaves. For each sample, they measured the rate of free-living nitrogen fixation as well as the levels of other nutrients. From these measurements, they could test three predictions that follow from their hypotheses.

PREDICTION 1: **The rates of N-fixation vary along a vertical profile of living leaves, senescent leaves, litter, and soil.**

PREDICTION 2: **Live and senesced canopy-leaf N-fixation varies among tree species.**

PREDICTION 3: **Species-specific differences in N-fixation are associated with phosphorus concentrations in leaves or soil.**

The rate of nitrogen fixation varied across the vertical profile. Generally, rates increased from the canopy to the forest floor,

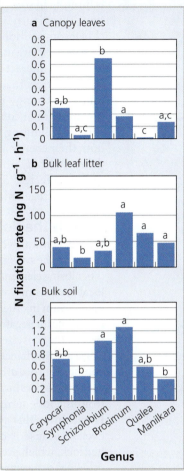

Figure 2 **Nitrogen fixation and total phosphorous content.** The rate of nitrogen fixation as a function of total phosphorus in (a) canopy leaves, (b) leaf litter, and (c) bulk soil (from Reed et al., 2008). **Analyze:** These data suggest that nitrogen fixation is controlled by total phosphorus. How could you determine if, instead, nitrogen fixation controls total phosphorus?

confirming the first hypothesis (Figure 1). Moreover, these rates also varied among tree species. Finally, nitrogen fixation was directly related to total phosphorus content (Figure 2). The differences among species were associated with interspecific differences in phosphorus content. It appears that each species develops a phosphorus "footprint" across its vertical profile: the phosphorus content of the canopy leaves is reflected in the phosphorus content of the litter beneath the tree.

These data suggest that control of biogeochemical processes varies interspecifically and it is tied to the availability of other nutrients. It also suggests that nutrient budgets are not regulated by simple negative feedback mechanisms. If they were, the paradox of nitrogen fixation in excess of demand would not occur. The connection between nitrogen fixation and the phosphorus of the soil is just one of many possible interactions among nutrient cycles that remain to be explored.

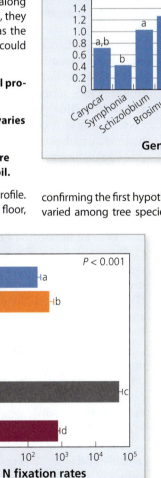

Figure 1 **Nitrogen fixation.** The rate of nitrogen fixation at different points from soil to the canopy (from Reed et al., 2008). **Analyze: Why is so little nitrogen fixed in canopy leaves?**

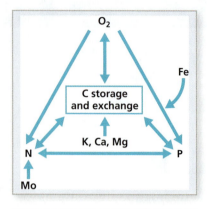

Figure 17.12 Tropical nutrient cycles. The interactions among nutrient cycles in tropical forest (from Townsend et al., 2011). **Analyze:** Why is nitrogen central to this set of interactions?

enzymes that promote the decomposition of organic phosphorus. When oxygen is limiting in soil, phosphorus becomes more available via anaerobic decomposition. This in turn is dependent on interactions with iron (Figure 17.12).

Decomposition determines the rate at which nutrients are released for transfer into new organisms by either heterotrophic or autotrophic processes. Here, too, cycles interact in important ways. In soil, the ratio of carbon to nitrogen (C:N ratio) controls the rate of decomposition. The carbon content of detritus determines the energy available to heterotrophic decomposers—material with high organic carbon content contains much energy. For many decomposers, nitrogen is the key limiting nutrient. For most, the optimum C:N ratio is 10:1, at which both their energy and nitrogen demands are fulfilled. If the C:N ratio of detritus differs radically from this optimum, both microbial growth and the flux of nitrogen are affected. Specifically, if there is insufficient carbon to support the microbes, nitrogen may be released into the soil and perhaps lost from the system.

What Is the Role of Disturbance?

Biogeochemical cycles are in balance when there is no net input or outflow of nutrients from the system. Of course, on the planetary scale, cycles are balanced; the Earth is not losing elements and the amount that arrives via meteors is negligible. However, local systems may or may not be in balance. We have encountered the question of equilibrium in other contexts. For example, in population regulation (Chapter 9), it is possible for negative feedback systems to maintain a relatively constant population size. At the ecosystem level, this kind of tendency toward equilibrium is unlikely because of disturbances. Just as disturbance and stochastic events preclude population equilibrium, disturbance disrupts the balance of biogeochemical cycles.

Some of the best empirical evidence for the importance of disturbance comes from the long-term studies of the Hubbard Brook Experimental Forest in New Hampshire. This system lends itself to experimental studies of biogeochemical cycles because it contains well-defined watersheds (ecosystems) with granite bedrock, which prevents the loss of nutrients to groundwater. Consequently, it is possible to measure all the nutrient inputs from streams and atmospheric deposition and the output via streams. Studies of the nitrogen budget in undisturbed forest show that only about 0.1 percent of the nitrogen content of a watershed is lost by stream flow. Nitrogen cycles tightly within these systems. However, if the forest in a watershed is artificially disturbed by deforestation, stream flow increases substantially. As a result, nutrient export increases by as much as 60-fold (Likens et al., 1970; Figure 17.13). Water that would have been held in an intact system drains into the stream, carrying nutrients with it. The experimental addition of nutrients to watersheds yields similar results. When organic phosphorus (PO_4^{3-}) was added to the streams in a series of watersheds, retention of this nutrient was related to the age of the surrounding forest—that is, the time since the last disturbance (Warren et al., 2007; Figure 17.14).

The effect of disturbance on nutrient budgets occurs largely because of the impact of disturbance on primary producers. Fire or storms that remove the dominant vegetation alter the rate of primary production. This shifts the balance of the carbon cycle toward increased respiration. The shift from living to dead biomass increases the rate of decomposition and thus triggers a massive release of nutrients. Without living plants to absorb those nutrients, they may be lost from the system.

These data and the fact that we now understand that disturbance is common in ecological systems suggest that balanced nutrient cycles are relatively uncommon. Data from a number of ecosystems bear this out—most studies find a net gain or loss of nutrients. Nevertheless, some systems do have balanced nutrient budgets. For example, although streams carry nutrients exported from disturbed terrestrial systems, a number of studies indicate that once nutrients enter the stream, the cycles become balanced over the course of the stream nutrient spiral (Brookshire et al., 2009).

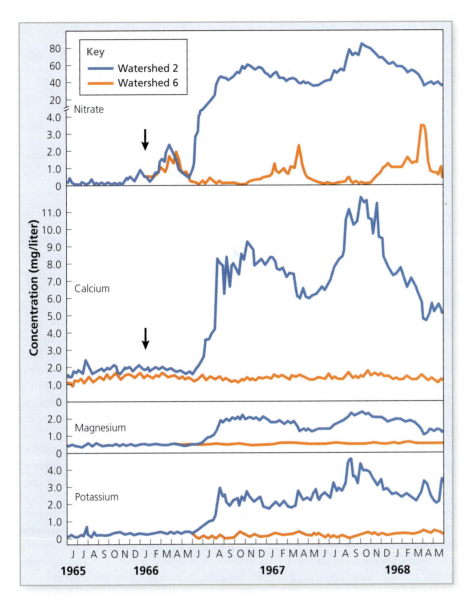

Figure 17.13 **Forest disturbance.** The effect of forest disturbance (logging Watershed 2) on nutrient export from watersheds at the Hubbard Brook Experimental Forest (from Likens et al., 1970). **Analyze:** What biological processes contribute to the loss of nutrients in disturbed systems?

The nutrient budget of an ecosystem also depends on the spatial scale at which the ecosystem is defined. At very small scales, chance effects and local perturbations are likely to disrupt the balance of nutrient cycles, leading to the net loss of nutrients. We expect that equilibrium is rare at the smallest spatial scales. At larger spatial scales, the ecosystem incorporates a mosaic of smaller systems. Although each may operate independently and be heavily influenced by local factors, these many smaller systems may also be interconnected by input and export. Thus, at larger spatial scales, ecosystems approach a function analogous to the metapopulation concept of populations (Chapter 9). None of the local systems comprising the larger ecosystem is in balance, but equilibrium may be approached at the larger spatial scale due to the summed interactions among the smaller component systems.

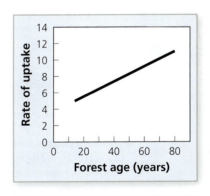

Figure 17.14 **Phosphorous uptake.** The uptake of phosphorus (PO_4^-) increases with the age of the forest (Warren et al., 2007). **Analyze:** Why do you think older forests take up more phosphorus?

KEY CONCEPTS 17.3

- Biogeochemical cycles are controlled by the most limiting fluxes. Often these are the fluxes that support primary production and decomposition.
- Biogeochemical cycles are not independent entities. Each nutrient cycle is connected in some way to others.

■ Disturbance disrupts the balance of biogeochemical cycles. Disturbed systems tend to have net input or export of nutrients.

■ The balance of nutrient budgets depends on the spatial scale at which the system is defined.

QUESTION:

Explain why we are more likely to observe equilibrium in nutrient budgets at larger spatial scales.

Putting It All Together

The release of methane from exposed Arctic permafrost is just one of the millions of fluxes in biogeochemical cycles. But it exemplifies the complex processes by which nutrients move through ecosystems. In this example, carbon moves from one pool, the frozen soil, to another, the atmosphere. The atmospheric carbon may now enter living components of the ecosystem directly by photosynthesis or indirectly when CO_2 dissolves in aquatic systems, where it is taken up by aquatic producers. This carbon, long sequestered in soil, may now cycle rapidly among the pools of the carbon cycle, perhaps thousands of miles away. Some, especially the methane, will remain in the atmosphere, where it will contribute to global warming. This scenario also exemplifies the long-term nature of these processes. The carbon released by this flux is the result of a time in Earth's history when the balance between respiration and photosynthesis, and thus the storage of carbon in soil, was different than it is today.

We can now place this scenario in the broader context of the fundamental question of this chapter: How do nutrients move through ecosystems? They move in cyclic fashion as a result of a combination of physical and biological mechanisms. Nowhere in ecology is the interaction between physical and biological processes so important. Erosion, flooding, sedimentation—an array of physical mechanisms move nutrients from one pool to another. The most fundamental biological processes—production (photosynthesis) and consumption—drive the fluxes and determine the size of the biotic nutrient pools. The size of some pools and flux rates constitute crucial limiting factors in all cycles. Disturbance alters these components and thus shifts the potential balance such that many local cycles have a net input or loss of nutrients.

rate-limiting step A crucial step in a series of fluxes that determines the overall rate of nutrient cycling.

 Summary

17.1 What Are the Major Patterns of Nutrient Movement?

■ The movement of nutrients through ecosystems is cyclic.

■ Nutrient pools occur in biotic and abiotic components of the ecosystem. Nutrient flux occurs by physical-chemical processes or biological mechanisms.

■ The carbon cycle
  • is characterized by a significant atmospheric pool.
  • is driven by photosynthesis and respiration.

■ The nitrogen cycle
  • includes a significant atmospheric pool.
  • requires specialized organisms for many important fluxes.

■ In the phosphorus cycle,
  • there is a large sedimentary pool of inorganic phosphorus.
  • phosphorus becomes available to organisms these rocks erode.

17.2 What Are the Mechanisms of Nutrient Flux?

■ Weathering of rock and parent material releases nutrients that can enter the biological component of the ecosystem.

■ Other nutrients enter biological systems by biological processes such as photosynthesis (carbon) or nitrogen fixation (nitrogen).

■ Decomposition is an important biological process that releases organic nutrients.

17.3 How Are Biogeochemical Cycles Controlled?
- The rate of cycling is determined by the slowest flux, the **rate-limiting step**.
- Decomposition is often a limiting process.
- Disturbance often leads to net gain or loss of nutrients from local ecosystems.

 ## Key Terms

aerosols p. 380	glacial till p. 380	pool p. 375
algal bloom p. 376	humus p. 383	rate-limiting step p. 388
alluvial deposition p. 381	loess p. 380	residence time p. 376
atmospheric deposition p. 380	mineralization p. 383	rock p. 380
biogeochemical cycle p. 375	minerals p. 380	sink p. 375
decomposition p. 383	nitrogen fixers p. 376	soil horizons p. 380
degradative succession p. 383	nutrient cycle p. 375	spiraling length p. 383
detritivore p. 383	nutrient spiral p. 383	thermokarsts p. 374
flux p. 375	parent material p. 380	weathering p. 380

Review Questions

1. What are the components of a systems approach to biogeochemical cycles?

2. What is the fundamental difference between energy flow and nutrient flux?

3. In what key ways are energy flow and nutrient flux similar?

4. What are the essential differences between the movement of carbon, nitrogen, and phosphorus in ecosystems?

5. How are biogeochemical cycles interconnected?

6. What is the effect of spatial scale on the function of nutrient cycles?

7. What factors determine if a nutrient cycle is in balance?

8. What are the key rate-limiting steps in nutrient cycles?

9. What are the major forms of abiotic input of nutrients to ecosystems?

10. What is the role of decomposition in biogeochemical cycles?

Further Reading

Krapivin, V., and C.A. Varostos. 2008. *Biogeochemical cycles in globalization and sustainable development*. New York: Springer.

This book places the theoretical and basic ecology of biogeochemical cycling in the context of human environmental impacts. Soil fertility is directly connected to biogeochemical cycles. Thus, sustainable agricultural development requires maintaining functional nutrient cycles. The book also explores the problems inherent in the transfer of technology adapted to one region to other developing areas.

Letcher, T.M. 2009. *Climate change: observed impacts on earth*. Amsterdam: Elsevier.

This is a nontechnical explication of the data and models of climate change due to industrial carbon emission.

Reed, S., et al. 2011. Functional ecology of free-living nitrogen fixation: a contemporary perspective. *Annual Review of Ecology, Evolution, and Systematics* doi: 10.1146/annurev-ecolsys-102710-145034.

This review summarizes the most recent ecological information on the crucially important process of nitrogen fixation.

Townsend, A.R., et al. 2011. Multi-element regulation of the tropical forest carbon cycle. *Frontiers in Ecology and the Environment* 9:9–17.

The high species diversity of tropical forests and the nutrient cycles that support that diversity have been studied for decades. This paper extends the large body of previous work to the question of the interrelationships among biogeochemical cycles. Although the focus is on tropical forests, interactions among cycles are of general importance.

Part 5

Large-Scale and Applied Ecology

18 Conservation Biology   **19** Landscape Ecology   **20** Human Global Ecology

Deforestation creates a sharp boundary between virgin rain forest and agricultural lands. In these three chapters we examine the scale and intensity of human impacts on the Earth. We apply many of the principles from previous chapters to understand how we can conserve the Earth's species. In Chapter 19, we examine the techniques to measure large-scale ecological patterns. In the final chapter, we analyze the human impact on global ecology.

Chapter 18

Conservation Biology

In the fall of 1813, John James Audubon, famed naturalist and artist, rode from Henderson to Louisville, Kentucky. A few miles beyond the town of Hardensburgh, he encountered passenger pigeons passing overhead "in greater numbers than I thought I had ever seen them before." He dismounted and tallied the flocks as they flew by. After 20 minutes he had marked 163 flocks. They continued to pass for *three full days*. He wrote, "at noon the light . . . was obscured as by an eclipse, the sound a dull roar."

Accounts such as this are common in the notes and journals of nineteenth-century naturalists in the Midwest. The passenger pigeon is believed to have been the single most abundant bird in the world. Estimates place the total population at 3–5 *billion* birds, or as much as 40 percent of the total bird population of the United States. Moreover, they were social species that traveled together in immense flocks like the one Audubon recorded. Their numbers and thus their collective appetite were so great that they were constantly on the move, searching for the next mast trees—the beeches, oaks, and chestnuts whose seeds made up the bulk of their diet. In 1859, J. M. Wheaton wrote of a feeding flock, "those in the rear, finding the ground nearly stripped of nuts, rose above the treetops and alighted in front of the advancing column . . . so rapidly that the whole presented the appearance of a rolling cylinder . . . the noise deafening."

Remarkably, just 101 years after Audubon's account, the last passenger pigeon died in a zoo in Cincinnati. One of the most prolific vertebrate species ever known had gone extinct in a single century. How could this happen so quickly? We understand now that two factors led to this stunning decline. First, as the Midwest forests were cleared for agriculture, vast tracts of virgin forest and the mast they produced were eliminated. Also, mortality from commercial harvest decimated the species. Professional hunters killed them in numbers almost unimaginable and shipped them off to Chicago, Cincinnati, St. Louis, and the East Coast. In 1878, market hunters killed 50,000 nesting birds *per day for five months* at a nesting area near Petosky, Michigan. By the end of the decade, the pigeons were rare. No wild flocks were known by the early twentieth century.

In a sense, this extinction is simple to explain. The gregarious nature of the birds and the unremitting slaughter pushed them past the point from which they could recover. This history demonstrates that any species, no matter how abundant, is vulnerable if exploitation by humans is sufficiently intense. Today, biologists believe we are in the midst of a new mass-extinction event (see Chapter 15, "The Human Impact"), this one caused by humans. We are losing approximately 27,000 species per year as a direct result of human activity. And, of course, not all these species are being hunted to extinction like the passenger pigeon. In fact, the underlying causes of anthropogenic extinction are complex. If we are to reverse this trend, we must address the fundamental question: *What factors cause species extinction?* As you will see, the answer to this question integrates information from many previous chapters.

18.1 What Are the Global Patterns of Species Richness and Extinction?

Our knowledge of the magnitude of the extinction problem is limited by our knowledge of the Earth's biodiversity. This may seem surprising given how intensely we have studied the taxonomy and ecology of plants and animals. Nevertheless, there remain important gaps in our knowledge. For example, we do not yet know, even approximately, how many species inhabit the Earth. Estimates of global species richness range from 30 million to over 100 million species. These are derived by extrapolation from samples in species-rich habitats. Typically, a set of samples of plants or animals is taken from a habitat. The number of species in these samples is then extrapolated to the entire area of the habitat. Global species richness is estimated by adding together the results of many such extrapolations.

The wide range of the estimates of total species diversity makes it difficult to assess the importance of the current extinction rate. Only about 2 million species have been assigned scientific names. Thus, only a tiny fraction of the Earth's species are known to science.

Some taxonomic groups are especially diverse. Human impacts that affect these groups have a disproportionate effect on the rate of species extinction. The insects are known for their high species richness—more than a million species have been named. And within this group, the coleoptera, or beetles, constitute 40 percent of all named insects and some 25 percent of all known animals. We know enough about some other groups to understand that they represent a large proportion of the Earth's biodiversity. The nematodes, small soil-inhabiting roundworms, are numerically abundant across a wide array of habitats. They may be the single most abundant group of animals on Earth, yet their taxonomy is poorly known and there are probably millions of undescribed species. The fungi, too, are numerically abundant, species rich, and very poorly known.

We know from Chapter 15 that there are significant differences among ecoregions in species diversity. Some systems, such as islands and peninsulas, have

THINKING ABOUT ECOLOGY:

If you were to undertake such a sampling analysis, what factors would be important in the design of the sampling? What assumptions underlie the process of extrapolation?

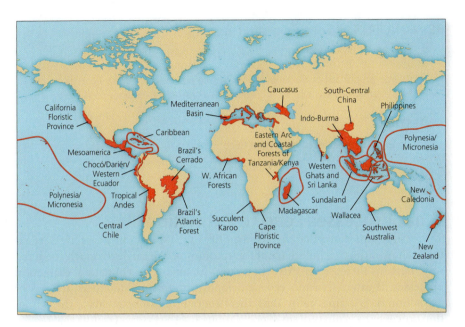

Figure 18.1 Hot spots. Hot spots of species diversity, places where biodiversity is exceptionally high (from Myers et al., 2000). **Analyze:** What is the relationship between endemism and biodiversity hot spots?

fewer species than comparable areas on large mainlands. Other systems, such as coral reefs and tropical forests, hold a disproportionately large fraction of the Earth's species. If we further analyze the spatial patterns of species diversity, we find that within the richest systems, there are "hot spots," regions with even higher numbers of species (Figure 18.1). Some 20 percent of all vascular plant species and 16 percent of terrestrial vertebrates are found in just 25 hot spots (Myers et al., 2000; Brooks et al., 2002). Together these 25 sites represent only 1.4 percent of the land area of the Earth. The present extinction rate reflects, in part, the exaggerated human impact on these biodiversity hot spots.

The human impact on biodiversity hot spots is magnified by the fact that many of the species found in these systems are **endemics**, species found there and nowhere else. These species are of special concern because for them, local extinction is equivalent to global extinction. Some 44 percent of the vascular plants and 35 percent of the terrestrial vertebrates of the hot spots are endemics. This makes the future of these systems crucial to maintaining the world's biodiversity.

■ **endemic** A species found in a particular locale and nowhere else.

Coral reefs hold the highest concentration of marine biodiversity. The key species underlying this diversity are the reef-building corals. These species, which number more than 800 worldwide, provide the nutritional and habitat foundation on which hundreds of other invertebrate and vertebrate species depend. Recent surveys indicate that approximately one-third of the reef-building corals face extinction due to human impacts (Carpenter et al., 2008). Reefs are particularly sensitive because the association of corals and their zooxanthellae is fragile—elevated ocean temperature, pollution, and acidification can destroy this obligate mutualism. Environmental insults cause the zooxanthellae to abandon the coral, which causes coral bleaching. And because so many other species depend on the reef-building corals, the biodiversity cost of their decline is magnified.

If we are to analyze the human impact on biodiversity, we must also consider where human pressure is most intense. It is not surprising that the global biodiversity crisis is tied to the rapid increase in the world's population. As the human population grows, pressure on natural habitats intensifies. There are currently more than 6 billion people, and projections put the number at 8 billion by the middle of this century. Like the Earth's biodiversity, human population growth is not distributed evenly across the globe. Some regions, especially Asia and parts of Africa, are growing at much higher rates than other regions. Human growth there affects some of the most species-rich parts of the globe. In addition, the majority of the human population lives on or near coastlines, where its impact on near-shore marine ecosystems is profound.

KEY CONCEPTS 18.1

- The species richness, or biodiversity, of the Earth is poorly known.
- Species richness varies widely among taxonomic groups.
- A large proportion of global biodiversity is concentrated in a few ecoregions and ecosystem types, especially in tropical terrestrial systems and coral reefs.

QUESTION:

Why are endemic species so important to conservation?

18.2 What Processes Lead to Extinction?

The fossil record shows that the vast majority of the species that ever lived are extinct. Thus, extinction is both natural and common. We can learn much about the ways humans increase the extinction rate with an analysis of natural extinction processes. If external factors reduce a population to small numbers, the risk of extinction rises dramatically. Thus, there are two key components to the extinction process: the ecology of small populations and the external factors that reduce populations to small numbers.

What Are the Effects of Small Population Size?

We know from Chapter 8 that species differ markedly in their population dynamics. Some, for example, typically occur at high density, whereas others are rare. Also important is the fundamental mathematics of population regulation. The intrinsic rate of growth of a population, r, is determined by four factors:

$$r = (b + i) - (d + e)$$

where $b =$ the birth rate, $i =$ the immigration rate, $d =$ the mortality rate, and $e =$ the emigration rate. Immigration and emigration are less important to the extinction process, so we focus our attention on the birth and death rates. Obviously, if $d > b$ for long, the population will decline significantly. Some of the factors that determine b and d act in a density-dependent fashion. If so, they may protect the population from severe declines by increasing the population growth rate when the population is small. We know from Chapter 9 that we can depict this equilibrium process with the equation

$$dN/dt = rN(K-N)/K$$

in which the population growth rate (dN/dt) is determined by the intrinsic rate of growth (r) and the current population size (N). If N is less than the carrying capacity (K), the growth rate is positive; if $N > K$, the growth rate is negative, and if $N = K$, the growth rate is 0. Populations that behave this way are somewhat protected. However, many populations and many species do not have density-dependent regulation. Moreover, external forces can overwhelm those that do.

When a population declines, chance factors become more important. Two significant stochastic processes affect the probability that a population will persist. **Environmental stochasticity** is the unpredictable occurrence of unfavorable abiotic conditions. Physical factors such as fire, flood, and extreme weather may inflict significant mortality or depress reproductive success. As populations grow smaller, they are numerically more vulnerable to such events. Consider a fire that kills 10 percent of the individuals in a population. In a population of 1 million, 100,000 individuals die. Although this is a large number, 900,000 remain. Conversely, with this mortality rate, only 90 individuals would remain from a population of 100. Although just 10 percent of the population succumbs, the effect is significant.

environmental stochasticity The random occurrence of unfavorable environmental conditions.

The spotted owl (*Strix occidentalis*) is well known for its dependence on old-growth conifer forests in the Northwest and for its vulnerability where logging eliminates old-growth habitat (Figure 18.2). Franklin et al. (2002) showed that the multiplicative growth rate, λ, (Chapter 8) varies stochastically in small populations of this species. More importantly, climate strongly affects reproductive traits in this species, which in turn affect λ. These random changes in population growth imposed by climate are most pronounced in marginal habitat. Thus, adverse habitat changes due to logging increase the variation in λ and ultimately the risk of extinction. Environmental stochasticity decreases survival as well, especially in poor-quality habitat.

The second random process, **demographic stochasticity**, is a random change in the demographic variables in a small population. As a population declines, chance events have a greater effect. Many of the demographic parameters we discussed in Chapter 8, such as l_x (the age-specific mortality rate), b_x (the age-specific birth rate), λ, and the sex ratio, are rates or probabilities that we present as mean values. Over time, these parameters vary simply by chance. We expect on average a proportion, l_x, to survive in each age category. But this precise value will not necessarily occur each time interval. When the population is very small, any chance deviation from expected value can have significant consequences.

Consider the impact of small population size on the sex ratio. Most sex-determining mechanisms produce approximately 50 percent males and 50 percent females. If the population is large, deviations from this expectation have minimal effect. However, the smaller the population, the larger the proportional effect of deviation from 50:50. Imagine a population of just eight individuals, four male and four female. If each of the four females produces one offspring, we expect them to produce two males and two females. But at such low numbers it is entirely possible that only males might be produced. In fact, we can calculate the probability of this occurring using the product rule of probability. If each female produces one young, the probability that all four females will produce only males is $\frac{1}{2} \times \frac{1}{2} \times \frac{1}{2} \times \frac{1}{2} = \frac{1}{2}^4 = \frac{1}{16}$, certainly a possibility. If one sex becomes very scarce or absent, reproduction may decline dramatically, putting the population at grave risk of extinction.

Precisely this phenomenon increases the risk of extinction of threatened subpopulations of the southwestern willow flycatcher (*Empidonax traillii extimus*) (Paxton et al., 2002). This federally endangered species occurs in very small populations of about 10 breeding pairs. The sex ratios of these isolated subpopulations do vary, sometimes favoring males, other times favoring females (Table 18.1). Shifts in sex ratio this large in a population this small have exaggerated effects on population growth and hence threaten its long-term survival.

Figure 18.2 Spotted owl. The spotted owl depends on large tracts of old-growth forest.

■ **demographic stochasticity** Random changes in demographic parameters such as l_0 or R_0.

TABLE 18.1 Sex ratios of small populations of the southwestern willow flycatcher. Note the variation in sex ratio among sites. Also note how much greater the variation is when the number of clutches is very small

BREEDING SITE	NUMBER OF FEMALES	NUMBER OF MALES	% FEMALE	NUMBER OF CLUTCHES
Roosevelt	78	53	60	62
San Pedro	18	27	40	22
White Mountains	15	6	71	8
Verde River	0	5	0	3

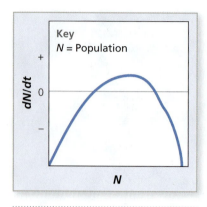

Figure 18.3 Allee effect. Species that experience the Allee effect have a negative growth rate at low population density. **Analyze: What is the minimum population size that can be maintained?**

■ **Allee effect** A phenomenon in which the population growth rate becomes negative below a minimum population size.

■ **fitness Allee effect** A phenomenon in which adult fitness declines below a minimum population size.

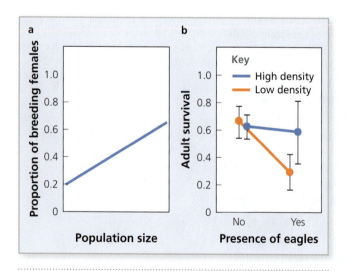

Figure 18.4 Fitness Allee effect. (a) The proportion of breeding females is lowest in small populations. (b) Adults in small populations face greater risk of eagle predation (from Angulo 2007). **Analyze: How is the fitness Allee effect different from the standard Allee effect?**

Other problems may arise due to chance events in small populations. For example, the hairy puccoon (*Lithospermum carolinensis*), a plant native to the tallgrass prairie of the Midwest, has a distylous breeding system. This means there are two mating types: one has long anthers and short pistils; the other has short anthers and long pistils. Each type can mate only with the other mating type. The mating types are determined genetically and are produced in a 50:50 ratio. As is the case in many prairie plants, the puccoon is found in relatively small populations. In some small populations, chance deviation from the expected 50:50 ratio results in a population in which one of the mating types is missing. When this occurs, the remaining types cannot reproduce and the population gradually declines to extinction.

Some species face an additional risk at low population density, known as the **Allee effect**, in which there is a density below which population growth becomes negative (Figure 18.3). In populations in which density-dependent factors operate, the growth rate is high at low density and becomes negative above *K*. But in species with the Allee effect, negative growth also occurs at low density. This is usually due to a failure of reproduction when the population falls below some minimum level. In essence, this represents a reverse density-dependence: rather than reproduction increasing at low density, it declines.

This phenomenon was first described for species that reproduce in large groups. For example, many gulls and seabirds nest in dense colonies. The reproductive physiology of both sexes is stimulated in large aggregations. If the population drops too low, this social stimulation does not occur and reproduction fails. We now understand that other mechanisms can cause the Allee effect. For example, Moller and Legendre (2001) show that it occurs in species in which female choice is a component of the mating system. Apparently, the Allee effect in these species is due to the negative effect of the high probability that in small populations a suitable mate cannot be found. The problem is compounded by the fact that demographic stochasticity also increases in this scenario because so few pairs mate.

Another form of Allee effect, the **fitness Allee effect**, has recently been described (Angulo et al., 2007). The island fox (*Urocyon littoralis*), an endemic species of the Channel Islands off the coast of southern California, exemplifies this phenomenon. In the smallest populations, predation on adults by golden eagles is significantly higher (Figure 18.4). This effect is based, not on reproductive failure, but on decreased adult fitness at low density. The fitness Allee effect is superimposed on a reproductive Allee effect that also occurs in these small populations, further increasing the probability of local extinction.

Chance events in small populations also have genetic consequences. Three genetic phenomena—inbreeding (Chapter 7), genetic drift (Chapter 2), and population bottlenecks—can have deleterious consequences in small populations. We know from Chapter 7 that genetic variation is important to the immediate and long-term health of a population. When genetic variation declines, the population is less able to adapt to environmental change. All three processes reduce genetic variation.

Inbreeding increases in small populations because the probability of mating between relatives increases. Specifically, in small populations the probability that an offspring receives alleles that are identical by descent increases (see Chapter 7, "Do the Math"). Alleles are identical by descent if they are derived from the same ancestral copy of the gene. This occurs when relatives carrying copies of the ancestral gene mate. The inbreeding coefficient (F) measures the probability that two alleles in an individual are identical by descent. Inbreeding is important in conservation biology because in inbred populations, the frequency of homozygotes increases by the factor pqF, whereas the frequency of the heterozygote decreases by the amount $2pqF$. This has two important consequences. First, the increase in homozygosity means that the chance that two deleterious recessive alleles come together is larger. If so, the result can be inbreeding depression, the decline in the fitness of individuals due to homozygosity. Second, the reduction of heterozygotes represents a loss of genetic variation that reduces the capacity to adapt to new environmental conditions.

THE EVOLUTION CONNECTION

Evolution of the Population Growth Rate

This chapter focuses on the ways human activity increases species' risk of extinction. As we change the environment, we also change the selection pressures on animals and plants. Many components of an organism's phenotype play a role in its ability to withstand human impacts on its environment. We know, too, that both the environment and the genotype determine the phenotype. The relative time scales over which these two effects occur have important implications for conservation biology.

Consider one of the most important determinants of population stability—the intrinsic rate of increase, r (Chapter 8). This variable measures the per capita growth rate of the population as determined by the per capita birth and death rates measured when the population is small:

$$r = b_0 - d_0.$$

It is the value of r that underlies the reproductive strategy of the species (Chapter 10). The optimal reproductive strategy balances the allocation of energy to reproduction (b_0) with survival and competitive ability, which directly affect d_0. For some organisms, the optimal strategy is a heavy investment in reproduction at the expense of competitive ability; for others, the reverse is true.

However, the value of r is also affected directly, in ecological time, by the environment. For example, poor food conditions can immediately lower the value of r through compromised reproduction or increased mortality. When human activity reduces the value of r, the population may decline toward extinction. Can selection modify r such that these impacts are mitigated?

There is evidence that suggests that a compensatory evolutionary change in r is unlikely. Experiments with *Drosophila* show that if a population reared for many generations with little crowding and abundant food is switched to a low-food competitive environment, the value of r does not evolve in response to the new conditions (Mueller, 1988; Mueller and Joshi, 2000). In other words, when crowded, the fruit flies are not able to adaptively shift their energy allocation from reproduction to competitive ability. The value of r responded to the immediate ecological conditions, to the detriment of the population. In nature, this would mean that the population might be depressed to the point of local extinction.

It is not surprising that evolutionary change in r is slow relative to the immediate impact of the environment. Our understanding of the process of natural selection allows us to identify the factors that determine the relative speed of evolutionary versus ecological impacts on the value of r. Generation time relative to the time scale of the environmental change is a crucial factor. If the generation time is short compared to the pace of the environmental change, selection can compensate for the changes. Thus, we might expect insects to respond adaptively to climate change, whereas long-lived mammals may not be able to change fast enough.

A second key factor is the amount of genetic variation on which selection can act. There are immediate deleterious effects of inbreeding in small populations. However, the negative impacts extend farther. We know from Chapter 7 that according to Fisher's fundamental theorem, the rate at which the mean fitness of a population increases by natural selection is exactly equal to the additive genetic variation in fitness. Thus, if any factors have reduced the genetic variation in the population, its adaptive response to environmental change will be compromised. This is particularly important because often the human impact on a population and its environment has precisely this effect—the reduction of the genetic variation needed to respond. But the impact carries farther in terms of the species' ability to adapt. The genetic impediment to an adaptive response to human threats can be significant. For example, in the nineteenth century, the North American population of elephant seals (*M. angustirostris*) was reduced by hunting and habitat alterations to less than 30 individuals. Today it has rebounded to more than 20,000. However, it still carries the genetic cost of that tiny population: elephant seals are monomorphic for most genetic loci (Bonnell and Selander, 1974). Thus, there is almost no variation on which selection can act if the environment changes.

QUESTION:

What other factors might increase or decrease the potential for adaptive change in the value of r?

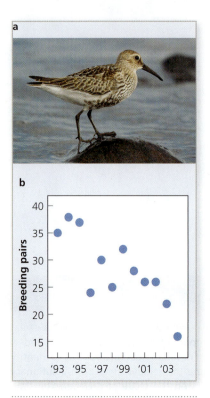

Figure 18.5 European dunlin. European populations of the dunlin are declining.

The dunlin (*Calidris alpina*) is a migratory shorebird. European populations are isolated and many are in decline (Figure 18.5). Blomqvist et al. (2010) document that population decline is associated with increased pairing between related individuals. The more genetically similar members of a pair are, the higher the homozygosity of their offspring and the lower their hatching success (Figure 18.6). Genetic analysis of successfully and unsuccessfully hatched offspring shows that the latter have lower levels of heterozygosity.

Genetic drift, stochastic change in allele frequency, occurs in all populations. However, the effects are most dramatic in small populations. Genetic drift ultimately leads to the chance loss or fixation of alleles, thereby decreasing the genetic variation in the population. The effective population size (N_e; Chapter 2) is a measure of the potential impact of drift. Any factor that reduces the size of the group of individuals that actually mate at random lowers N_e. Thus, a skew in the sex ratio reduces the effective population size and increases the effect of genetic drift. We have already seen that demographic stochasticity can have this effect on sex ratio. N_e also declines if the population experiences a population bottleneck, a decline in numbers and subsequent recovery (Figure 18.7). Even though the population has recovered, the genetic consequences of the low point carry forward in the form of reduced genetic variation.

In much of the mountain West, grizzlies (*Ursus arctos*) occur in small, isolated populations. The total population (N) has declined from an estimated 100,000 individuals in 1800 to less than 1,000 today. Even the large protected area of Yellowstone National Park harbors only 200 individuals. Their remaining habitat is isolated by roads and development, thus reducing the exchange of individuals among subpopulations. These are conditions in which both inbreeding and genetic drift become significant factors. Genetic studies suggest that the effective population size is only 25 percent of the total population size (Allendorf et al., 1991). Simulations indicate the rapid loss of heterozygosity from these populations (Figure 18.8). Molecular analyses of recent samples and museum specimens indicate a decline in N_e and thus genetic variation over time (Miller and Waits, 2003; Figure 18.9).

What Role Does Habitat Play in Extinction?

Each species is adapted to a particular habitat type. Some species are habitat specialists, able to survive and reproduce only in a narrow range of ecological conditions. The pika (*Ochotona princeps*) exemplifies extreme specialization. It

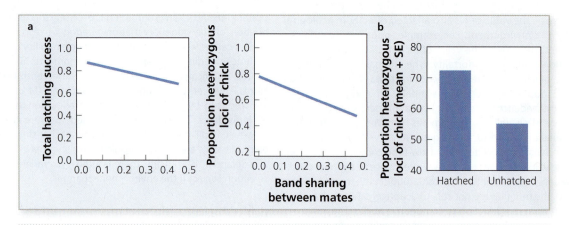

Figure 18.6 Homozygosity versus hatching success. (a) The time course of genetic relatedness and hatching success in the dunlin. Band-sharing is an index of genetic similarity among individuals. (b) The proportion of hatched and unhatched eggs for dunlins with high and low heterozygosity (from Blomqvist et al., 2010). **Analyze: What is the relationship between heterozygosity and inbreeding?**

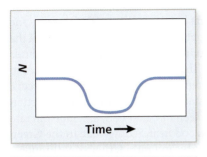

Figure 18.7 Population bottleneck.
A population bottleneck occurs when a population declines to low numbers, then increases again. **Analyze:** Why would the loss of genetic variation during a bottleneck persist even after the population increases again?

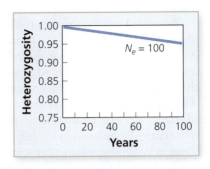

Figure 18.8 Loss of heterozygosity.
Predicted loss of heterozygosity in grizzlies over time due to inbreeding (from Allendorf et al., 1991). **Analyze:** How would you expect the loss of heterozygosity to change if N_e were 200?

inhabits high-altitude talus slopes, large boulder fields, in the Rocky Mountains and Sierra Nevada (Figure 18.10). Dispersing individuals are occasionally found in other habitats, but they can survive and reproduce only on talus piles located near productive meadows where they can forage. Most specialists have a minimum habitat requirement for successful survival and reproduction. For example, a pair of spotted owls requires approximately 2,000 acres of old-growth forest to successfully nest and rear young. If logging destroys part of this old-growth habitat, nesting fails or the pair disperses in search of new habitat.

Not all species are as specialized as pikas and spotted owls. Nevertheless, even species with broad habitat requirements must have tracts of appropriate habitat. For example, many passerine birds inhabit eastern deciduous forest habitat. Within these forests there may be a wide range of microhabitats and tree communities. Birds such as the scarlet tanager, ovenbird, and Swainson's thrush inhabit many of these forest types (Figure 18.11). But they cannot survive in young forest or grassland. As old-growth forests decline, these species, too, are at increasing risk of extinction.

One of the most prominent impacts of human activity is the direct loss of habitat. Agriculture, oil and gas extraction, logging, and urban sprawl are among the most significant causes of habitat loss. As the world's population increases, habitat loss accelerates. Some estimates place the total loss of natural habitats by the year 2050 at 7.5 million km², an area the size of Australia. Of particular concern is the loss of tropical forest, the most species-rich and extensive habitat type on Earth. Approximately half of the world's tropical forests have been destroyed. At the current rate of deforestation, this number will rise to 80 percent by 2030. Satellite images illustrate the pattern and extent of loss: as roads enter virgin forest, a network of other roads and development radiate outward (Figure 18.12).

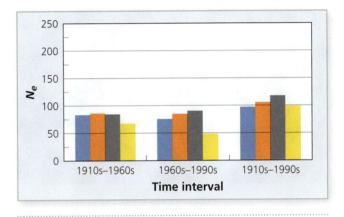

Figure 18.9 N_e over time. Estimates of N_e over time for grizzlies in the greater Yellowstone area (Miller and Waits, 2003). **Analyze:** If N_e is this low (and about 25% of N), what other factors might cause further decline in the grizzly population?

Figure 18.10 Pika. Pikas inhabit isolated talus slopes in the high mountains. Dispersal connects these discrete populations.

Figure 18.11 Eastern deciduous forest birds. Scarlet tanagers (shown here) ovenbirds, and Swainson's thrushes are birds of the eastern deciduous forest. They cannot survive and reproduce in other habitats.

Figure 18.12 Deforestation. As roads enter virgin rainforest, development and deforestation spread.

habitat fragmentation A process in which large tracts of habitat are broken into small, isolated fragments.

Figure 18.13 Arctic National Wildlife Refuge. The Arctic National Wildlife Refuge on the North Slope in Alaska is home to the Porcupine caribou herd.

For migratory species, the problem of habitat loss can be particularly acute. One of the reasons ecologists are so concerned about the impact of oil exploration in the Arctic National Wildlife Refuge is that this region contains crucial calving grounds for the Porcupine caribou herd (Figure 18.13). This species travels long distances to the coastal calving grounds, where there is abundant forage and relief from insects. Although the herd's tundra habitat is largely intact, impacts on this critical portion of that habitat may be significant. As we have seen in Chapter 3, antelope migrate more than 400 miles from the high country in and near Yellowstone to the Green River in Wyoming, where they winter. During this migration, they pass through a narrow canyon where oil and gas exploration threatens their travel route. Many of the passerine birds of the eastern forests winter in Central and South America. They face habitat loss in both their summer breeding range as well as their winter range.

Climate change impacts species in two important ways. If climatic shifts exceed the physiological tolerance of a species, it may succumb from these direct effects. Most mobile species can potentially relocate to new, suitable habitat. However, when climate change alters the broad patterns of habitat distribution, it may be impossible for dispersal to rescue species from extinction. Thus, the pika, the high-altitude-habitat specialist, is declining across much of its range because the warming climate is reducing the available habitat. Its habitat is shifting to higher and higher elevation as the climate warms. But this is ultimately limited by the height of the mountains themselves. A similar phenomenon has been documented for butterfly communities in the mountains of Spain, where species' presence is correlated with specific isotherms (regions of similar temperature), resulting in shifts in the composition of the community as a function of altitude. As climate change pushes those isotherms higher, the butterfly communities shift in response. However, the highest-elevation communities are at risk because they eventually reach the limit of altitudinal migration (Wilson et al., 2007).

This threat affects marine systems by a similar process. Because water temperature is an important component of marine habitats, rising global temperatures push species into new regions, often where other crucial habitat components are absent. Because these effects are so widespread, their impact affects total marine biodiversity (Figure 18.14).

Although the total habitat available to species is crucial, the spatial pattern of that habitat also plays a role in extinction. In addition to simply reducing the total habitat area, human activity often reduces the remaining habitat to small isolated patches, a process known as **habitat fragmentation**. The populations of many species constitute a metapopulation (Chapter 9)—isolated populations connected by dispersal. Habitat fragmentation exaggerates this pattern. The effect is compounded if the habitat destruction includes the dispersal corridors. When this occurs, each habitat patch becomes more isolated. Because immigration to a patch is rare, local extinction is less likely to be rescued by immigration.

This process affects each species individually—each responds differently to habitat loss. But the principles of island biogeography also apply to the set of habitat islands. We know that islands contain fewer species than mainland habitats. Also, the equilibrium number

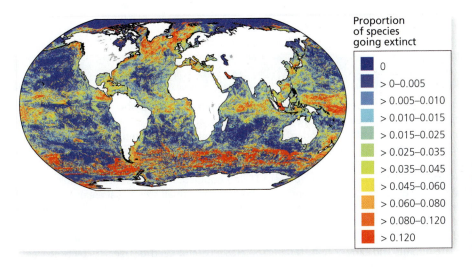

of species on an island is determined by island size through the immigration and extinction rates (see Chapter 15). Fragmentation and the loss of species by the processes of island biogeography lead to a **relaxation fauna or flora** (Figure 18.15). In a relaxation flora or fauna, the new equilibrium number of species is lower than when the system comprised a large tract of continuous habitat. Because the equilibrium species diversity is determined by stochastic events, it is difficult to determine which species will persist and which will go extinct.

relaxation flora/fauna A new, lower equilibrium species richness on habitat islands.

Note in Figure 18.15 that some time is required for the system to reach a new equilibrium species richness. Consequently, there may be a time lag between the habitat alteration and the new equilibrium. The continued loss of species long after fragmentation is known as the **extinction debt**. Some species are particularly susceptible to habitat change and are immediately lost. "Slow species" respond more gradually to these effects, and it may be some years before they disappear. For example, two ground-dwelling woodland birds, the brown treecreeper (*Climacteris picumnus*) and the hooded robin (*Melanodryas cucullata*) of Australia, are now declining toward extinction due to island biogeographic effects despite the fact that their habitat was fragmented more than 100 years ago (Ford et al., 2009).

extinction debt The continued loss of species following fragmentation.

As species are gradually lost from fragmented habitat, new processes accelerate the loss of biodiversity. This is especially true for species that participate in mutualistic associations. Loss of one mutualist may doom the other, especially in obligate relationships. When this happens, the total rate of species extinction may accelerate beyond the rate predicted just from habitat loss or fragmentation. There is evidence that when one mutualist is lost, the other may shift to a more antagonistic relationship with another species. Ultimately, the process compounds the rate of species extinction. This compounding effect is not limited to mutualists. It can also be important for other interactive species, such as parasites. Recall from Chapter 12 that many parasites have complex life cycles in which they depend on a series of intermediate hosts. The loss of one of these crucial intermediates can doom the parasite to extinction as well.

What Anthropogenic Mortality Factors Lead to Extinction?

Habitat destruction and fragmentation ultimately lead to mortality and increased risk of extinction. These effects, however, are indirect—they increase the probability that the problems of small population size come into play. Other kinds of human activity increase mortality directly.

We exploit species for many purposes. Despite the importance of agriculture and livestock, a large proportion of our food comes from the harvest of wild species. Certainly one of the most significant harvests of wild species is the global fishery, which now approaches 100,000,000 tons annually. There are many

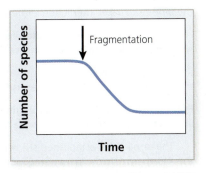

Figure 18.15 Relaxation fauna. A relaxation fauna is a new lower-diversity fauna following habitat fragmentation. **Analyze: What factors determine this new equilibrium?**

Effects of the Decline of Whitebark Pine

Virtually every species on Earth is ecologically connected to other species through interactions such as competition, predation, parasitism, and mutualism (see Chapter 12). Consequently, impacts on one species may ripple through the other species with which it interacts. The conservation implications of mutualisms are particularly important: loss of a mutualist can lead to the decline and even extinction of species that depend on it.

The Clark's nutcracker (*Nucifraga columbiana*) is a facultative mutualist of the whitebark pine (*Pinus albicaulis*) in the Rocky Mountains. The pine is obligately dependent on the nutcracker for dispersal of its large, wingless seeds. In late summer and early fall, nutcrackers harvest seeds from the cones and cache them in the ground for use over the winter. The nutcrackers harvest and store 1.8–5 times the number of seeds needed for the winter. Those lost or unused are the source of new whitebark pine seedlings. Without dispersal by nutcrackers, whitebark pine recruitment falls sharply. Other species also depend heavily on whitebark pine seeds and their high fat content. Red squirrels (*Tamiasciurus hudsonicus*) harvest seeds from cones opened by nutcrackers and cache them on the ground. Grizzlies (*Ursus arctos*) use seeds raided from red squirrel and nutcracker caches in their fall preparation for hibernation. A large proportion of the bear's winter fat store comes from whitebark pine seeds.

The whitebark pine blister rust (*Cronartium ribicola*), an invasive fungal pathogen, attacks and kills whitebark pine. First introduced to North America in 1910, this pathogen has been steadily spreading until it has reached epidemic proportions in some populations. Not only does the rust ultimately kill the tree, it attacks the cone-bearing branches, reducing or eliminating cone production. If nutcrackers abandon a whitebark pine population because cone production is too low, the result is the potential extinction of whitebark pine and the cascading effects of this loss on the rest of the ecosystem.

If we are to manage this problem and mitigate the impact of the rust, it is important to know the dynamics of the nutcracker–whitebark pine interaction. A crucial aspect of this interaction is the relationship between whitebark pine cone production and attraction of nutcrackers.

McKinney et al. (2008) addressed this question in a study of the pine-nutcracker interaction in the Northern Rockies.

HYPOTHESIS: There is a threshold seed production, below which nutcrackers no longer disperse the seeds.

If this hypothesis is correct, when the rust inhibits cone production, nutcrackers should no longer visit the trees.

PREDICTION: Nutcracker seed foraging and seed dispersal should decline as cone production falls.

The researchers established study plots in three ecosystems: the Northern Divide in extreme northwestern Montana, the Bitterroot Mountains, and the Yellowstone ecosystem. In each site, trees were sampled for infection (proportion infected), size (*dbh*—diameter at breast height) and basal area (the cross-sectional area of the tree), mortality (proportion of living trees

per hectare), and cone production (number of cones and seeds per tree). Nutcracker foraging activity was quantified by walking standard transects and counting the number of seeds consumed and the number carried away by nutcrackers.

The three sites differed significantly in the impact of the rust. The Northern Divide population had significantly higher infection and mortality rates and small size (basal area). The Yellowstone site was the healthiest of the three, but the difference from the Bitterroot population was not statistically significant. Rust infection and mortality both decreased cone and seed production. Nutcracker presence and seed dispersal activity fell as a

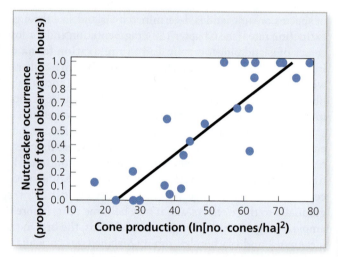

Figure 1 Whitebark pine cone production. Seed dispersal as a function of cone production, which determines nutcracker presence (from McKinney et al., 2008). **Analyze: What are some of the factors that might determine cone production in whitebark pines?**

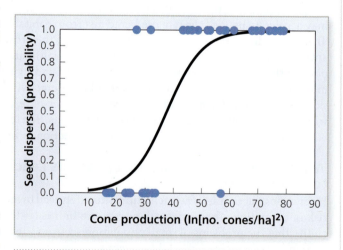

Figure 2 Threshold effect. There is a threshold effect of cone density on seed dispersal (from McKinney et al., 2008). **Analyze: At what level of cone production does this threshold occur?**

Effects of the Decline of Whitebark Pine

function of declining cone production (Figure 1). The y-intercept of this regression suggests that nutcrackers abandon a whitebark pine population if the cone production falls to about 130 cones/ha (ln <cones/ha>^{2}/ha = 23.64). This results in a threshold effect on seed dispersal: there is an abrupt decline in the probability that a seed is dispersed when cone production falls (Figure 2). The Northern Divide populations have approached this threshold. We can expect that over time these populations will become smaller and more locally isolated. The nutcrackers appear not to be at significant risk of extinction because they shift to other species when their preferred whitebark pine seeds are rare. The pine, however, may face local extinction. And because this species is a keystone species in this alpine ecosystem, other species that depend on its seeds may be at risk. Of particular concern are grizzlies, whose populations are already affected by habitat loss and inbreeding.

examples of species that have been harvested to the brink of extinction or beyond. Overfishing has essentially eliminated the once rich North Atlantic cod fishery. Unsustainable harvest has taken its toll on many species. The bison (*Bison bison*), which once numbered in the tens of millions, was reduced to just a handful of individuals by market hunting. The passenger pigeon is perhaps one of the most spectacular examples of overexploitation that directly contributed to its extinction. Fashion can also drive unsustainable harvest, with devastating consequences. In the eighteenth and nineteenth centuries, beaver pelts were a valuable source of felted fur for men's hats. Trappers reduced their number so dramatically across much of western and northern North America that many local populations were driven to extinction. In the early twentieth century, it was the fashion to decorate women's hats with the plumes of egrets. In 1902, 1,400 kilograms of egret plumes were sold in London. By 1911, only 10 egret colonies remained in all of the southeastern United States. Even if the mortality we impose does not push the species to extinction, it may reduce the numbers to the point that other factors, such as inbreeding or demographic stochasticity, become significant.

How do we manage harvests so they are sustainable? The harvest rate that provides the greatest yield without jeopardizing the population is the **maximum sustainable yield**. This value must be calculated from detailed demographic information, including the intrinsic rate of growth and age-specific birth and death rates. In a **fixed-quota harvest**, the harvest is limited to a certain number of individuals by establishing catch limits. When the quota is reached, no more harvest is allowed. This system is often employed in the Pacific salmon fisheries. The alternative is a **fixed-effort harvest**, in which managers limit the harvest by regulating the length of the harvest season, type of gear allowed, and so forth. Obviously, the latter is effective only if there is a strong correlation between effort and actual harvest.

Humans also introduce chemicals into the environment that cause direct mortality. Industrial pollutants such as mercury and lead are lethal even in small doses. In some cases, we aren't aware of the potential effects until long after the pollutants have been released. After World War II, the pesticide DDT came into widespread use in the United States for both mosquito control and to kill agricultural insect pests. It was not until some years later that we understood that this compound has two detrimental characteristics. First, it persists in the environment—it does not break down for years after it has been applied. As a result, the compound accumulates in animals and is passed from prey to predator, gradually becoming more concentrated by the process of biomagnification (see Chapter 16). Second, it disrupts calcium metabolism in birds, causing eggshell thinning and high chick mortality. Many apex predators, such as bald eagles, ospreys, and brown pelicans, were pushed to the brink of extinction by this pollutant.

Pollution mortality also occurs from the accidental release of chemicals. Oil spills are notorious sources of direct mortality. In 1989 the supertanker *Exxon*

maximum sustainable yield The maximum harvest of a population rate that can be maintained over time.

fixed-effort harvest A management strategy in which the harvest is limited by the length of season, means of harvest, or other aspects of the harvest.

fixed-quota harvest A management strategy in which the harvest is limited to a specific number of individuals.

THINKING ABOUT ECOLOGY:

Imagine that you are responsible for setting the harvest quota of salmon at a sustainable level. What information would you need in order to be able to do this? Which information would be most difficult to obtain?

DO THE MATH

Maximum Sustainable Yield

Consider the logistic growth curve in Figure 1 from the standpoint of sustainable harvest. At first glance, you might assume that it would be best to maintain the population at as high a level as possible—that is, close to K. However, if the harvest is to be sustainable, the population must replace the lost individuals. This suggests that not only is the current number important but so is the rate of growth of the population. At any one moment in time, the growth rate is measured by a line tangent to the growth curve. The growth rate of the population changes throughout the course of logistic growth. Note the significant differences in growth rate at different points in the growth curve. Another way of looking at this relationship is to plot the growth rate itself as a function of N (Figure 2).

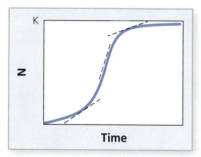

Figure 1 Logistic growth. During logistic growth, the rate of change of the population (dashed lines) changes. **Analyze:** Which line on this graph has the highest growth rate?

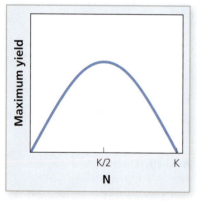

Figure 2 Maximum sustainable yield. The maximum sustainable yield is found in a graph of population growth rate as a function of N. **Analyze:** What would be the short-term and long-term effects of harvesting at rates closer to K?

This figure shows that the recruitment rate to the population is maximal at K/2. This value represents the maximum sustainable yield. If we harvest at a rate that leads to a lower population, growth rate would be high but the number available for harvest would be small. Harvest at higher densities means that more individuals are available but the growth rate is too low to maintain this density.

We can quantify the harvest strategy that will result in maximum sustainable yield with a form of the logistic equation

$$N_{t+1} = N_t + rN_t\left(\frac{K-N_t}{K}\right).$$

The population size at time $t + 1$ equals the current population size (N_t) + new recruits. Recruitment is determined by the growth rate (rN) reduced by an amount proportional to how near K the population stands or (K − N/K).

We can add a term to this equation that quantifies the intensity of harvest. This can be measured in two ways. To incorporate these measures into the logistic equation, we reduce the growth rate by an amount proportional to the harvest. For a fixed-quota harvest, we simply subtract the size of the quota (Q):

$$N_{t+1} = N_t + rN_t\left(\frac{K-N_t}{K}\right) - Q$$

In the equation for a fixed-effort harvest, we also want to decrease the growth rate by the amount of harvest. In this case, the effort (E) is the proportion of the population taken by the allowed methods. The product of that successful effort and the current population is the number of individuals harvested.

$$N_{t+1} = N_t + rN_t\left(\frac{K-N_t}{K}\right) - EN_t$$

The goal in both cases is to set the quota or the effort such that the population is maintained at a value K/2 where recruitment of new individuals is maximized.

Figure 18.16 Deepwater Horizon. The Deepwater Horizon drilling platform exploded and sank, causing a massive oil spill in the Gulf of Mexico.

Valdez ran aground in Prince William Sound, Alaska, releasing some 750,000 barrels of crude oil. This was the largest oil spill in US waters until the Deepwater Horizon drilling platform in the Gulf of Mexico exploded in 2010 (Figure 18.16). The oil itself is toxic, especially to zooplankton and small animals that ingest it directly. Many birds and mammals killed by oil succumb because a coating of oil mats the feathers and fur, causing them to die of hypothermia. Any species at risk of extinction from other causes can be pushed past recovery by such events. Brown pelicans, which were recovering from the effects of DDT, were killed by oil during the Deepwater Horizon spill.

Extinction is a complex process. In fact, it is rare that we can pinpoint a single cause for species extinction. One factor may depress the population to the point that other negative effects compound the mortality rate. As a population declines, these other processes can mutually reinforce one another, driving the population

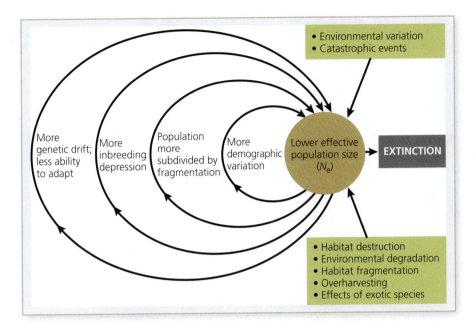

Figure 18.17 Extinction vortex. The extinction vortex is the set of compounding factors that lead a population to extinction. **Analyze: Which components of this process include positive feedback loops that increase the probability of extinction?**

rapidly to extinction. We refer to this downward spiral as an **extinction vortex** (Figure 18.17). The prairie chicken (*Tympanuchus cupido*) population in Illinois exemplifies the process (Westermeier et al., 1998). The habitat of this bird, tallgrass prairie, was nearly eliminated by the conversion of grassland to row crops. The species was also hunted by market hunters, who shipped thousands of birds to the cities. Eventually the population declined from more than 1 million to less than 50, at which point inbreeding and loss of genetic variation reduced reproductive success. The bird had entered a downward spiral toward extinction. At that point, protection from harvest and habitat restoration failed to rescue the population. This is a common characteristic of species trapped in an extinction vortex—alleviating the original cause of decline does not reverse the downward trend.

The extinction vortex and the compounding extinction risk are direct functions of the suite of factors that operate in small populations. Analyses of vertebrates that declined to extinction show that the length of time a population persists is a logarithmic function of its size (Figure 18.18). One generalization

extinction vortex A process in which multiple negative impacts on a population cause the numbers to decline toward extinction.

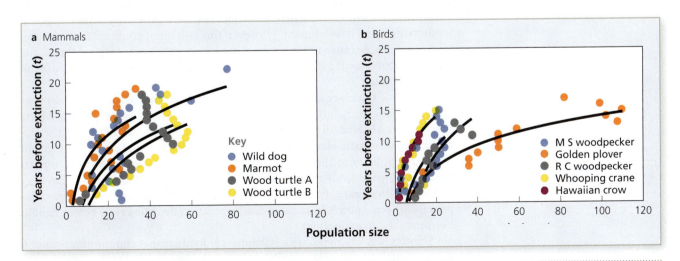

Figure 18.18 Population persistence. Population persistence as a function of population size for (a) mammals and reptiles and (b) birds (from Fagan and Holmes, 2005). **Analyze: What do these graphs suggest about the conservation of Yellowstone grizzlies?**

that has emerged is that once a population declines below 20 individuals, extinction is all but inevitable within 10 years. However, this number is a general estimate. As we have seen, many factors affect a species' vulnerability. The key to species conservation is keeping their populations above the threshold at which they enter the extinction vortex.

KEY CONCEPTS 18.2

- When populations become small, stochastic factors play a larger role in their ecology and genetics. Demographic and environmental stochasticity lead to random and often deleterious changes to factors like population growth rate and sex ratio. In small populations, inbreeding and genetic drift reduce genetic variation, with immediate and long-term consequences.
- Habitat loss is one of the most important causes of extinction. We are destroying natural ecosystems at a tremendous rate, with devastating effects on biodiversity.
- Habitat fragmentation further reduces species diversity, because island-like patches of habitat hold fewer species than larger, contiguous tracts.
- Humans impose direct mortality factors by overharvest and the introduction of pollutants to the environment.
- Extinction is generally the result of a combination of factors. When multiple factors operate, the species can enter an extinction vortex that rapidly and inevitably leads to extinction.

QUESTION:

Why can't density-dependent changes in the birth and death rate compensate for human-caused mortality?

18.3 How Can We Prevent Extinction?

Knowing the primary causes of anthropogenic extinction allows us to develop strategies that reduce the risk for individual species and communities. A successful approach to this problem begins with understanding the differences among species that affect their vulnerability. This knowledge informs the design of measures to protect them.

What Characteristics Put Species at Risk?

Human population pressure and development are the fundamental threats to species. Thus, one important determinant of risk is the proximity to human population centers and development. However, this risk is not confined to industrialized countries. Because the rate of habitat loss is so great in tropical forest, the highest global rates of extinction occur there even though the human population is relatively small.

Several of the mechanisms leading to extinction are based on random processes. Genetic drift, demographic stochasticity, and island biogeographic effects in fragmented habitat all have important random components. Although the *processes* have stochastic elements, the species on which they act differ markedly in their susceptibility to them. Thus, species that typically occur in small populations are more vulnerable than abundant species. We know that some species have historically been rare. For example, the ivory-billed woodpecker (*Campephilus principalis*) was known to occur at low population densities even before habitat loss brought it to extinction.

The core-satellite hypothesis (Chapter 15) predicts that communities contain some species that are widespread and abundant (core species) and others that are rare and found in only a few sites (satellite species). The latter periodically disappear from the community. Satellite species would obviously be more vulnerable to extinction than core species. Also, top carnivores occur in smaller numbers than species in lower trophic levels. Thus, it is not a surprise that species like tigers, eagles, and wolves are endangered whereas many of their prey species are not.

Geographic range is also an important component of extinction risk. The more geographically restricted a species is, the less likely that local extinction can be rescued by immigration from other populations. For small mammals, those less than 3 kg in body size, small geographic range is the best predictor of extinction risk (Davies et al., 2008). Historical factors play an important role in the size of the geographic range. For example, the redwood (*Sequoiadendron giganteum*) was widespread in North America before Pleistocene climate change reduced its range to the moist coastal climate zone of California. A number of species of desert pupfish in the family Cyprinodontidae are restricted to tiny desert pools and wetlands. They were probably isolated there as larger bodies of water dried up at the end of the Pleistocene. Many are now known only from a single locale.

Body size is also an important determinant of extinction risk. Larger species generally require larger home ranges to satisfy their energetic needs. Consequently, habitat reduction disproportionately affects their density. Large species also differ from small species in their reproductive strategies. All other things being equal, they are more likely to adopt a *K*-selected strategy, which emphasizes competitive ability over numbers of offspring (Chapter 10). Thus, elephants and sea turtles, both of which take years to mature and reproduce at low rates, are not able to reproductively respond to population declines. The bristlecone pine (*Pinus longaeva*) of the White Mountains in California occurs in small, isolated, high-altitude populations (Figure 18.19). Some individuals do not begin reproducing until they are hundreds of years old. In this species, climate change and atmospheric pollutants can reduce the population size to the point that reproduction cannot compensate for the loss of individuals.

Another important contributor to a species' vulnerability is its successional status. Many of our impacts are disturbance events that replace climax communities with earlier successional systems. We have seen that spotted owls require old-growth forest of a minimum size for successful nesting. Thus, logging converts climax forest into successional stands uninhabitable by the spotted owl. The tallgrass prairies of the Great Plains have been reduced to less than 1 percent of their original extent by conversion to agriculture and grazing. Many native prairie species occur only in tiny remnant communities. The climax prairie plants cannot compete with nonnative species adapted to exploit disturbance. Invasion of aggressive nonnative species pushes the native prairie species toward extinction.

Typically, early-successional and disturbance-tolerant species are less threatened because they are adapted to disturbance, including human disturbance. However, there are cases in which management strategies designed to improve conditions for climax species threaten rare species that depend on disturbance. Kirtland's warbler (*Dendroica kirtlandii*) is the rarest of the wood warblers (Figure 18.20). This species nests only in stands of jack pine (*Pinus banksiana*) larger than 80 acres. Jack pine is a successional species that develops in single-species stands following fire. The warbler is endangered because we have nearly eliminated the historical pattern of fire in northern conifer forests.

What Conservation Strategies Protect Species?

There are two key components of conservation strategies. First, it is essential that we determine conservation priorities. This includes identifying the species most at risk and those whose presence helps preserve others. Second, we must take

Figure 18.19 Bristlecone pine. The bristlecone pine, which lives for thousands of years, does not begin reproducing until the individual is more than 100 years old.

Figure 18.20 Kirtland's warbler. Kirtland's warbler nests only in stands of jack pine that develop after a fire.

THE HUMAN IMPACT

The Value of Biodiversity

This entire chapter is, of course, devoted to an important human impact—the effect of the human population on extinction. It might seem reasonable in this context to ask: Why should we be concerned about species extinction? Ecologists, ethicists, and economists have all addressed this question. Together they outline the case for conservation of the species with which we cohabit the Earth.

A purely ecological argument centers on the relationship between species diversity and ecological stability. Chapter 15 examines this relationship in detail. Our current understanding is that in many systems, species diversity increases population, trophic, and biogeochemical stability. Like any disturbance, species loss disrupts patterns of energy and nutrient flow and often increases the magnitude of population fluctuation.

Clearly, we benefit economically from many species. Commercial fishing and livestock grazing in native grasslands are examples of direct economic benefits of healthy ecosystems and species populations. In addition, many wild species have enormous economic potential. For example, the majority of modern pharmaceuticals are based on plant compounds. Drug companies continue to search the world's flora for new compounds with biomedical value.

However, there are less direct, less obvious economic benefits to functional ecosystems (Daly, 1977). We refer to these benefits as *ecosystem services*. These benefits extend beyond just the direct value of a fishery or grass to support livestock. They include the monetary value of crucial services that healthy, functional ecosystems provide. For example, the complex trophic structure of wetlands improves water quality by filtering out pollutants, toxins, and even some microorganisms. The direct cost of replacing this service with industrial water treatment would be enormous. Many important agricultural crops are pollinated by insects. Estimates of the dollar value of pollinators range from $5.2 billion to $16.4 billion (Losey and Vaughan, 2006). Native species of insects and vertebrates contribute importantly to agricultural pest control. If their impact were lost, the cost of replacement with chemical control would be approximately $13.6 billion

annually (Losey and Vaughan, 2006). Native ecosystems, especially forests and grasslands, sequester carbon, protecting us from even more greenhouse effects. For example, the more species that comprise a prairie system, the more carbon is sequestered (Tilman, 1999). Not all ecosystem services are so obvious and dramatic. For example, the 100 million head of cattle in the United States produce millions of kilograms of feces per year. Dung beetles are crucial to the processing and decomposition of this material.

Determining the economic benefit of species conservation is not without controversy. First, it is extraordinarily difficult to assign costs to many of these ecosystem services. The calculations require many assumptions and extensive extrapolation. Moreover, in many cases we have no information on the cost of providing these services by our own industrial procedures. Second, some wonder if it might be dangerous to reduce conservation to a purely economic calculation. They reason that if we can assign a dollar value to particular species or systems, we open the door for reasoning that loss of that system is worth it relative to other economic values. If Arctic caribou have a dollar value via ecological service, should we balance that against the economic value of oil exploration that might threaten them?

This leads to another key aspect of the human impact—our ethical responsibility to the natural world and the species around us. We can approach this from two points of view. First is the anthropocentric argument that having healthy, diverse ecosystems provides important benefits to each of us. Many of us simply enjoy interacting with the beauty and diversity of the natural world. Second, some ecological ethicists argue that the natural world and the species that comprise it have an inherent right to exist. It is our responsibility to ensure that our activities do not infringe on this basic right to life.

QUESTION:

How do natural and domesticated species differ in their benefits to humans?

measures that keep populations from declining to the point that they enter the extinction vortex.

The resources we can devote to conservation programs are finite. Thus, it is essential, even if regrettable, that we assign conservation priorities. One way to balance our ethical responsibility to all organisms with the reality that resources must be wisely allocated is to focus on species whose preservation will benefit many others. There are important differences among species that we can use to our advantage.

Keystone species are species that have a disproportionately large impact on the community in which they occur. We have seen (Chapter 13) that keystone predators are important in maintaining community species richness by suppressing the impact of dominant competitors. Thus, conservation efforts that protect keystone species may be magnified by their positive impact on the entire community. Keystone mutualists can also be essential for the preservation of many other species. Pollinators that ensure reproduction and disperse pollen (and thus genes) long distances have exaggerated importance in many systems.

Species whose presence signifies certain ecological conditions are known as **indicator species**. These species can provide an important measure of habitat quality. Wolves require large tracts of pristine habitat with a large prey base. If they are present, we can be confident of a minimum level of habitat quality; if they are absent, we may want to assess the management of the ecosystem. **Umbrella species** are species that require large contiguous areas of high-quality habitat. By focusing conservation efforts on such species, we indirectly benefit many others. Tigers, for example, are umbrella species important to India's conservation strategy. A management program known as Project Tiger aims to preserve large tracts of tiger habitat in different parts of India. To the extent that the tigers are protected and their ecological requirements are satisfied, many other species of plants and animals are protected.

The second component of conservation strategies is the development of strategies that prevent individual species from entering the extinction vortex. This includes analysis of the long-term prospects for the species and the design of reserves to maximize their long-term survival.

Conservation biologists need to assess quantitatively the risk of extinction. A set of techniques known as **population viability analysis (PVA)** provide estimates of the probability that a species will go extinct. Moreover, these models provide a time scale for the estimate. For example, PVA analysis can tell us that a species has X probability of persisting for Y years. This is not just a way to measure how concerned we should be about the population's status; the models allow us to analyze the factors that might increase viability. That is, we can compare the relative contributions of various demographic factors to the risk experienced by the population.

Engen et al. (2001) developed a PVA analysis for the barn swallow (*Hirundo rustica*) in Denmark (Figure 18.21). The model includes empirically measured demographic variables, including the variation in them (demographic stochasticity), to generate probabilities of extinction over various time intervals. Models such as these are valuable to conservation biologists in two ways. They allow comparison of species' risk so that conservation priorities can be established. In addition, they allow managers to develop management plans to preserve species.

Although the population dynamics of each species are unique, and each faces its own set of risks stemming from human activity, recent PVA modeling has shown that for long-term persistence, population sizes must be surprisingly large. Previous work suggested that population sizes in the hundreds were sufficient; this new work shows that populations should be 10 times that large to be safe from extinction (Traill et al., 2010).

In addition to estimating the probability of extinction, it is important to determine a target population density that will prevent extinction. This can be done with **minimum viable population (MVP)** analysis, which determines the smallest population size required for the population to persist for a specified length of time. MVP models typically calculate the expected persistence time for populations of different size (Figure 18.22). MVP analysis is based on detailed demographic data, especially population growth rate. The analysis incorporates ecological and stochastic variation to determine how large the population must be to overcome fluctuations in growth rate. Demographic and ecological differences among species result in a wide range of species-specific MVP values.

THINKING ABOUT ECOLOGY:

Occasionally, the conservation requirements of two endangered species come into conflict—what benefits one may harm the other. Can you devise a methodology or strategy to determine how to resolve such conflicts? What biological factors would you consider? What ethical or other nonbiological factors would you include?

indicator species A species found only under specific ecological conditions.

umbrella species Species that require large tracts of suitable habitat. Protecting them protects many other species.

population viability analysis (PVA) A mathematical technique to analyze the probability that a population can be maintained at a specified size for a specified period.

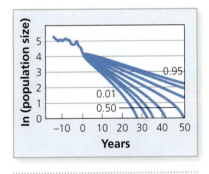

Figure 18.21 Population viability analysis. The results of population viability analysis of the barn swallow. The population trajectory up to the present is shown, as well as possible future trends under different demographic conditions. Each future-trend line shows the probability that the population goes extinct in the time frame shown (from Engen et al., 2001). **Analyze: What factors do you think would lead to the more rapid extinction scenarios?**

minimum viable population (MVP) The minimum population size, identified by a mathematical technique, required for a population to persist for a particular length of time.

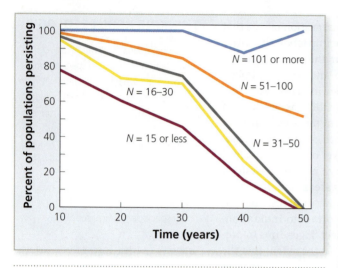

Figure 18.22 MVP predictions. MVP predictions for bighorn sheep populations of different size (from Berger, 1990). **Analyze: Is a population of 110 individuals safe from extinction?**

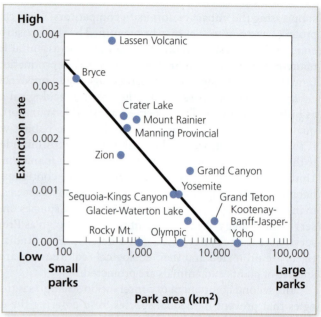

Figure 18.23 Reserves. The extinction rate decreases as a function of reserve size (from Newmark, 1995). **Analyze: What other factors might cause some reserves to fall below the line or above the line?**

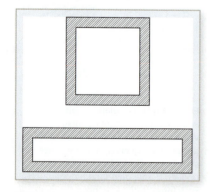

Figure 18.24 Reserve shape. Two reserves with the same total area but different shapes. If the edge effect is the same in both cases, reserve shape determines how much habitat is protected from impacts from outside the reserve. **Analyze: Which reserve has the higher ratio of edge to interior?**

■ **edge effect** An external environmental impact that extends into the habitat.

The other key component of a conservation strategy is the design of preserves. Obviously, protecting large areas of critical habitat is important; size is perhaps the single most crucial characteristic of preserves (Figure 18.23). Conservation biologists have debated the relative merits of single, large reserves versus a series of smaller sites. This has come to be known as the SLOSS debate: single large or several small reserves. Today we understand that each option is appropriate for different habitats and different species.

The SLOSS debate has been beneficial because it has expanded our analysis of reserve design. We now understand that shape and connectivity are also important. Figure 18.24 shows reserves of the same size but whose shape varies. In a long, narrow reserve, for example, much of the habitat is near the edge. Impacts that extend into the reserve from outside are known as **edge effects**. The impact of cowbirds on forest birds exemplifies the importance of minimizing edge effects. Cowbirds are nest parasites that lay eggs in other species' nests. Their offspring reduce the survival of the host offspring by competing with them for food. Many birds of the eastern deciduous forest are in decline due to the loss of mature forests. Cowbirds' preferred habitat is relatively open country; they do not penetrate deeply into forest. Thus, the amount of forest distant from the edge is important: in reserves that have a small perimeter-area ratio, the proportion of interior habitat is larger. Cowbirds will affect fewer nests. When the perimeter-area ratio is small, more nests occur near the edge and cowbirds significantly reduce recruitment of their host species.

When more than one reserve is part of the conservation strategy, dispersal corridors are critical design components. A series of small habitat patches can effectively protect species if there are avenues by which dispersers can move from one patch to another. This essentially replicates a natural metapopulation structure. If one local population declines, immigration can restore it. Moreover, dispersers from other tracts bring in new alleles, mitigating the loss of variation due to inbreeding and genetic drift.

Reserve and corridor design must take into account species-specific behavior and ecology. For example, many species of New World primates travel almost exclusively in the forest canopy; many rarely come to the ground. Roads that penetrate forest reserves prevent the normal movement patterns and can isolate monkeys even when the total forest area is large. Simple solutions such as leaving occasional patches of large canopy trees that bridge the road allow primates to move freely through the entire habitat area.

KEY CONCEPTS 18.3

- Because the threats to populations increase as population size decreases, conservation strategies emphasize mechanisms that maintain large populations.
- Species differ in key traits that affect their susceptibility to human activity. Key attributes include body size, reproductive strategy, natural population level, and tolerance of disturbance.
- An effective conservation strategy requires setting priorities. Species differ in their ecological roles and value to conserving other species.
- Reserves should be designed to maximize population size and minimize the effects of fragmentation. Shape and dispersal corridors are crucial design elements.

QUESTION:

Why is the metapopulation concept so important in conservation biology?

Putting It All Together

The extinction of the passenger pigeon was especially dramatic given the staggering numbers in which the birds once occurred. The example is significant because it demonstrates that no matter how abundant a species might be, humans have the capacity to wipe it out. Hunting wild species for food does not necessarily lead to extinction. But if hunting is pursued with sufficient intensity and across a wide geographic scale, extinction becomes inevitable. The mechanisms that brought the passenger pigeon to extinction were perhaps the simplest described in this chapter. Still, the example is of general importance because it illustrates the importance of scale and intensity. Many human activities—logging, harvest, pollution—have minimal impact if their scale and intensity are small. But as the human population steadily increases and our technical ability increases the scale and intensity of our activity, the impact on the Earth's biodiversity increases proportionately. Moreover, we have reached the point that the extinction rate rivals that of previous mass extinctions in Earth's history. This chapter shows that we have the ecological knowledge to mitigate and in some cases reverse this trend. The science of conservation biology is crucial. So, too, is our collective will to implement change.

Summary

18.1 What Are the Global Patterns of Species Richness and Extinction?

- We do not have a precise measure of Earth's species richness.
- Species are not distributed evenly across the Earth. Some regions, such as rainforests and coral reefs, have high richness; others, such as the open ocean, have low richness.
- Taxonomic groups differ enormously in diversity.
- Some of the most species-rich habitats are also heavily impacted by humans.

18.2 What Processes Lead to Extinction?

- When populations become small, their risk of extinction increases.
- Small populations are susceptible to random changes in population parameters (demographic stochasticity) and random changes in the environment (environmental stochasticity).
- Small populations also suffer from genetic problems such as inbreeding.
- Loss of habitat is a prime contributing factor in species extinction.
- Animals and plants that are harvested at rates higher than can be sustained by reproduction are vulnerable to extinction.

18.3 How Can We Prevent Extinction?

- Species differ in their vulnerability to anthropogenic impacts. Among the factors that contribute to their susceptibility are size, geographic range, and life history.
- Species' responses to disturbance play a significant role in their susceptibility.
- Extinctions can be prevented by allocating resources to strategies that will protect many species. This includes focusing on keystone or umbrella species or species-rich habitats.
- Population viability analysis allows managers to develop quantitative goals for protecting species.

■ Key Terms

Allee effect p. 398
demographic stochasticity p. 397
edge effect p. 412
endemic p. 395
environmental stochasticity p. 396
extinction debt p. 403
extinction vortex p. 407

fitness Allee effect p.398
fixed-effort harvest p. 405
fixed-quota harvest p. 405
habitat fragmentation p. 402
indicator species p. 411
maximum sustainable
    yield p. 405

minimum viable population
    (MVP) p. 411
population viability analysis
    (PVA) p. 411
relaxation flora/fauna p. 403
umbrella species p. 411

Review Questions

1. What is the state of our knowledge of Earth's biodiversity?

2. Why is species diversity important to humans and to ecosystems?

3. What demographic processes become important in small populations?

4. What genetic effects threaten species in small populations?

5. Why do some species have lower growth rates when their population is small?

6. How do species differ in their vulnerability to extinction?

7. Why is it hard to pinpoint a single cause for a species' extinction?

8. What are the key elements in the design of reserves?

9. How is habitat fragmentation related to the problems of small populations? How is it different?

10. What principle underlies the principle of maximum sustained yield?

Further Reading

Brook, B.W., et al. 2008. Synergies among extinction drivers under global change. *Trends in Ecology and Evolution* doi:10.1016/j.tree.2008.03.011.

This review paper synthesizes current research on the interaction of the extinction vortex and climate change.

Burkey, T.M., and D.H. Reed. 2006. The effects of habitat fragmentation on extinction risk: mechanisms and synthesis. *Songklanakarin Journal of Science and Technology* 28:9–37.

This important review paper summarizes current research and understanding of habitat fragmentation.

Quammen, D. 1996. *The song of the dodo: island biogeography in an age of extinction*. New York: Simon and Schuster.

David Quammen is a well-known science writer. This popular book examines the patterns and causes of anthropogenic extinction across many ecoregions.

Talent, J.E. 2012. *Earth and life: global biodiversity, extinction intervals and biogeographic perturbations through time*. New York: Springer.

This large volume synthesizes our knowledge of the history of biodiversity and biogeography across geological time. It provides background for the analysis of current threats to diversity.

Chapter 19

Landscape Ecology

The monarch butterfly (*Danaus plexippus*) is the only North American butterfly to undertake a continental-scale migration. Monarchs that live east of the Rockies begin a southern migration in late summer and early fall that ultimately takes them to the fir forests of Michoacán, Mexico. As remarkable as this long-distance flight is, even more striking is the fact that the disparate eastern populations converge on a tiny region of forest, where they overwinter in just 12 known sites of about 8 acres each. Fir trees in these sites harbor millions of butterflies. Because the entire eastern population relies on these few sites, the monarch's future depends on the health and preservation of the forests. The Mexican government has established reserves to protect both the forests and the butterflies. Monitoring the extent and health of these forests is crucial to this conservation strategy. How can we assess the status of these reserves? In the past, this task would require extensive fieldwork to analyze and map the system. Today, data collected by satellites allows us to measure forest conditions across large geographic areas. Not only do satellites provide photographic information, they can measure the nature and condition of the vegetation. Comparison of satellite data over time shows the gradual deterioration of the forests in the reserves and their increasing isolation (Figure 19.1).

This phenomenon—the migration and tiny winter habitat—illustrates the crucial role of *spatial scale*. Monarchs face a wide array of risks across their extensive summer range. At the other end of the spatial scale are the individual trees on which monarchs gather in winter. These individual trees are embedded in a matrix of disturbance history, forest structure, and human encroachment. Spatial scaling has been an important aspect of our discussion of many ecological phenomena. Here we shift our focus from specific ecological

Figure 19.1 Anganqueo Reserve. Satellite photos of Anganqueo Reserve in 1986 (top) and 2001 (bottom). Red indicates healthy fir forest. The inner white line indicates the monarch reserve.

■ **landscape ecology** The analysis of the patterns of spatial variation in the environment and in ecological processes.

■ **landscape** A region in which ecological factors vary.

processes to the question of spatial variation per se by asking the fundamental question: *How do ecological processes vary geographically?* This question is relevant to basic ecological questions as well as to understanding and mitigating the human impact on species and ecosystems.

19.1 What Is an Ecological Landscape?

The study of the interaction between spatial patterns and ecological processes is known as **landscape ecology**. This field recognizes the importance of spatial heterogeneity in both the environment and the organisms' response to that variation. In fact, we define the **landscape** in terms of variation: a landscape is an area that is heterogeneous in at least one ecological factor of interest.

There are two key components to landscape ecology. First, landscape ecology explicitly addresses the size, shape, and structure of ecologically distinct regions. Second, it focuses on a much larger spatial scale than traditional ecological studies. In much of the population and community ecology we have studied so far, key processes occur locally as individuals respond to one another in a particular place. In landscape ecology we analyze the broad patterns of ecological variation. In previous chapters we have implicitly taken a landscape perspective in topics such as metapopulations, metacommunities, island biogeography, and the difference between local nutrient budgets and global biogeochemical cycles. In this chapter we formalize and, more importantly, quantify that perspective.

Although landscape ecology has recently emerged as a well-defined field of ecological study with its own research journals, ecologists have been addressing spatial heterogeneity for some time. Clements's description of the North American plant communities, or biomes, in the early twentieth century was actually a landscape approach, as was Merriam's identification of the life zones that occur with increasing altitude on San Francisco Peak in Arizona. One of the classic studies of vegetation dynamics was Whittaker's (1956) study of the distribution of vegetation types across elevation, slope, and aspect in the Smoky Mountains (Figure 19.2). Today we recognize that he was pioneering a landscape approach to plant ecology.

What Are the Components of a Landscape?

Look more carefully at the vegetation mosaic in Figure 19.2. How would you describe this pattern? Each vegetation type has certain basic spatial characteristics, such as size, shape, and position in the landscape. A set of specific terms has been developed to describe landscape patterns. Figure 19.3 shows a satellite view of Brazilian rainforest. Note that forest vegetation intermixes with deforested and agricultural regions. Each local ecological situation, natural and disturbed, represents a **cover type**, defined by the main vegetation comprising it. A related term, **patch**, refers to an area that differs in some ecological feature or process from its surroundings. When human disturbance encroaches on the rainforest and breaks it up into smaller island-like patches, we say the rainforest has been **fragmented**.

Fragmentation can have natural causes. However, habitat fragmentation is a common result of human activity. The reduction of large tracts of habitat to small, isolated fragments (islands) is thought to be among the

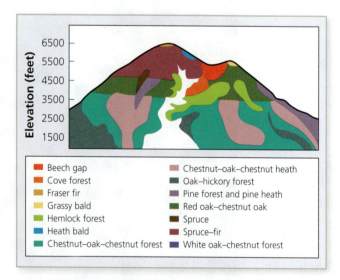

Figure 19.2 Vegetation mosaic in the Smoky Mountains. Abbreviations for vegetation types shown below (from Whittaker, 1956). **Analyze: Why would some of these bands of vegetation be found at a specific altitude, whereas others are found on different slopes?**

Legend:
- Beech gap
- Cove forest
- Fraser fir
- Grassy bald
- Hemlock forest
- Heath bald
- Chestnut–oak–chestnut forest
- Chestnut–oak–chestnut heath
- Oak–hickory forest
- Pine forest and pine heath
- Red oak–chestnut oak
- Spruce
- Spruce–fir
- White oak–chestnut forest

most significant threats to biodiversity (Fischer and Lindenmayer, 2007). Habitat fragmentation represents an important connection between landscape ecology and conservation biology (Chapter 18). We have also seen the connection between habitat islands and the theory of island biogeography (Chapter 13). As a result, conservation biologists must focus their efforts on the population dynamics and threats, and threats to individual species, but also on the statistical effects that emerge at the landscape level.

Each landscape, of course, has a unique pattern of variation. Note, too, that the size of landscape components varies. The spatial extent of cover types, patch size, and the degree of fragmentation all vary among landscapes. Consequently, another important landscape parameter is **spatial scale**, the relative size of landscape components. As we shall see, spatial scale is both a characteristic of the landscape and a component of how we study and quantify the landscape.

What Are the Causes of Landscape Patterns?

In effect, this question poses several related questions. What factors underlie the structure of the landscape? Why do the boundaries between patches occur where they do? Why are some patches large and others small?

We begin to answer these questions with the identification of three additional means of describing a landscape (Levin, 1976). **Local uniqueness** refers to the physical and biological factors that make a particular locale different from others in the landscape. If local abiotic factors such as precipitation or substrate distinguish the site from others in the region, they contribute to local uniqueness. Similarly, local human activity or the presence of a particular predator exemplifies biotic causes of local uniqueness. Spatial patterns that result from variation in the disturbance history or regime are known as **phase differences**. A large proportion of spatial heterogeneity in the landscape is the result of differences in the current stage of succession. The third category, dispersal, is the extent to which the movement of individuals or propagules swamps out local uniqueness and phase differences. This process is analogous to the genetic process of gene flow, in which the movement of alleles increases the genetic similarity of populations that otherwise differ due to local selection or genetic drift.

These three categories are the result of both physical and biological factors. The primary abiotic causes of landscape patterns are climate and **landform**. Landform is a geological term referring to surface shape, elevation, and slope. Together, climate and landform determine the nature of the local soil and substrate, which in turn are central determinants of the vegetation. This is especially true across larger spatial scales such as latitudinal and altitudinal gradients. Also important to the context of landscape ecology is the dynamic nature of climate. As the climate changes, so, too, do landscape patterns. Thus, climate change due to the release of CO_2 from fossil fuels deservedly receives a great deal of attention because of its potential ecological consequences for species and communities.

Landform determines the pattern of flow, specifically of water, and thus the rate and patterns of the movement of nutrients and energy from high to low elevation. Even the movement of seeds or individuals can be affected by landform and gravity. The landform also plays a significant role in the pattern of phase differences across the landscape. Factors such as storm intensity and fire behavior are

Figure 19.3 Brazilian rainforest. This satellite image of Brazilian rainforest shows healthy forest in red and deforested regions in pink.

cover type The vegetation in a defined area.

patch An area that differs from its surroundings in an ecological feature or process.

fragmented Describes a habitat that has been broken up into small patches, generally as a result of human activity.

spatial scale The relative sizes of the components of a landscape.

local uniqueness The abiotic and biotic factors that differentiate an area from others.

phase difference Spatial variation in the landscape due to disturbance patterns.

landform The surface shape, elevation, and slope of a region.

THINKING ABOUT ECOLOGY:

Imagine that you are assigned the task of monitoring elk populations and their communities in the Northern Rocky Mountains. What landscape factors do you think would be important to analyze?

Recovery of the Black-Footed Ferret

Our impact on the landscape has been profound. And if we are to mitigate our impact, we must adopt a landscape perspective. We can illustrate the importance of a landscape conservation perspective with the example of the black-footed ferret (*Mustela nigripes*), a highly endangered species. The black-footed ferret is a specialized predator of prairie dogs. This species was thought to be extinct until a colony was discovered near Meeteetse, Wyoming, in 1981. The 18 surviving individuals

Figure 1 Black-footed ferret. The black-footed ferret is an endangered species in western grasslands.

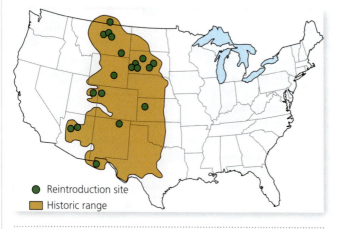

● Reintroduction site
▢ Historic range

Figure 2 Black-footed ferret range. The historic range of the black-footed ferret. Circles indicate reintroduction sites (from Black-Footed Ferret Recovery Implementation Team, 2011). **Analyze: Why is it important to reintroduce ferrets across the entire historical range?**

from this population were used to begin a captive breeding program. In 1991 federal and state agencies began releasing ferrets into the wild. At present, reintroductions have occurred across much of the ferret's historic range.

The initial recovery efforts focused on the most local spatial scale—the sole surviving wild colony. The local habitat was so degraded, and its prairie dog prey base so tenuous, that the only hope was to capture the ferrets and place them in a facility where the needs of each individual could be met. In essence, that patch in the landscape was beyond repair. When the captive population grew to the point that reintroduction to the wild was possible, the focus for reintroduction was still entirely local: a thriving prairie dog town in healthy native grassland. Each recovery site had to be embedded in a broader landscape free from threats such as cattle grazing, or habitat loss due to development. Because both prairie dogs and ferrets are susceptible to sylvatic plague, a form of bubonic plague, each colony had to be relatively isolated from potential disease-bearing rodents.

Once each colony was established and stable, conservationists could begin to increase the spatial scale of their attention in order to address other potential threats. As we know from Chapter 18, small local populations face the threat of inbreeding and loss of genetic variation. This danger was particularly acute in the ferrets, since the entire recovery program was based on just 18 individuals. The scale had to include the connectivity of populations and the potential for gene flow. Because both prairie dogs and ferrets require native grasslands to prosper, the spatial extent of prairie habitat and grassland corridors connecting them are essential.

The recovery program has now progressed to the point that we can begin to manage ferrets at the largest spatial scale—their original geographic range. At this scale, new forms of landscape structure emerge—differences among grassland types, climate, and substrate, as well as shifts in the species of prairie dogs available as prey. Each site must be managed according to its own ecology and potential threats. Nevertheless, the fact that black-footed ferrets now inhabit a broad *landscape* once again—a series of connected populations embedded in a complex grassland matrix sustained by a varied prey base—significantly improves the probability of their long-term survival. The larger and more diverse the landscape, the more this species is protected from the chance events that so dramatically affect local populations.

QUESTION:

What are the ecological and genetic consequences of connections among populations of endangered species?

determined at least in part by landform. For example, in Labrador, lightning ignites fires in spruce-fir forests along ridgelines. The fires burn downslope, finally stopping when they reach the wet valleys below. Birch colonizes these recent burns; thus, the landscape consists of spruce-fir forests on the ridgelines and birch stands on intermediate slopes.

There are also important biotic contributors to landscape patterns. Any of the interactions among organisms that affect presence or absence or population size contribute to the landscape mosaic. To the extent that competition varies in intensity and result, so, too, will there be variation in the structure of biological communities at the landscape level. Factors that affect the outcome of competition, such as the presence or absence of keystone species (see Chapter 12), contribute to this heterogeneity as well. Consumption by predators and herbivores is another important biotic effect on the landscape. For example, moose on Isle Royale in Lake Superior are important browsers. A recent decline in the moose population and thus in browsing intensity caused a significant shift in the spatial pattern of woody plants and nutrients (de Jager and Pastor, 2009). Specifically, as browsing intensity declined, the distribution of browse and nutrients became more randomly distributed across the landscape. In other systems, ecosystem engineers significantly alter the physical habitat and contribute importantly to landscape patterns. For example, beavers dam streams and create ponds with many new aquatic habitats. The depressions excavated by alligators create reservoirs of water that many other species depend on during dry periods.

KEY CONCEPTS 19.1

- Spatial variation, differences from one locality to another, is central to landscape ecology.
- Landscapes are characterized by patches, cover type, fragmentation, and spatial scale.
- Landscapes are determined by physical features such as topography and landform as well as biological processes such as competition and ecosystem engineering.

QUESTION:

What is the difference between a cover type and a patch?

19.2 How Do We Quantify Landscape Patterns?

Ecologists have been addressing landscape-level questions for more than 100 years, as exemplified by the vegetation analyses of Clements and Merriam. Most of these early studies were qualitative rather than quantitative. Although ecologists developed hypotheses to explain the patterns they observed, landscape ecology was primarily a descriptive science. Landscape ecology entered a new era when quantitative methods could be applied to landscape questions.

What Role Does Technology Play?

Two key developments led to rapid progress in landscape ecology: the availability of satellite data and the exponential increase in computing power. Satellites have provided increasingly sophisticated data to landscape ecologists. This information takes two forms. High-resolution satellite photos show sufficient detail that they can be used in quantitative analyses (Figure 19.4).

In addition, satellites can gather data on the reflectance patterns of the components of the landscape. When sunlight strikes the earth, whether it is vegetation, soil, or other surface features, some of that light is absorbed and some is reflected (Figure 19.5).

THE EVOLUTION CONNECTION

Landscape Genetics

The intimate connection between population ecology and population genetics has been a recurring theme in this text. It began with the relationship between the environment and the main mechanisms of evolutionary change: natural selection, genetic drift, and gene flow (Chapter 2). In the last chapter, we explored the impact of human activity on the population genetics of species at risk of extinction. Clearly, the patterns of genetic variation and change reflect spatial patterns in the environment. Thus, it should be no surprise that as the field of landscape ecology has developed, so, too, has the parallel field of landscape genetics.

Landscape ecology has provided us with an increasingly sophisticated understanding of spatial variation in the environment. Barriers to gene flow are central to the spatial patterns of microevolution. Gene flow from one environment to another slows local adaptation. Isolation of small genetic groups amplifies the effect of genetic drift and increases the probability of inbreeding. Landscape genetics focuses on two key landscape features: (1) local environmental differences that determine the selective regime and (2) barriers that disrupt gene flow and isolate populations. The accompanying figure shows the potential effects of landscape features that prevent the movement of individuals and gene flow.

Genetic variation among populations is the result of slight genetic differences among fairly closely-related individuals. Consequently, the molecular information used to analyze these patterns must have high resolution. The two most common techniques are single sequence repeats (SSRs) and amplified fragment length polymorphisms (AFLPs). SSRs are genetic markers that are selectively neutral—they are not affected by selection. They are composed of DNA sequences with a variable number of repeats of two to six nucleotides. AFLPs are often used in plant studies. Many hundreds of DNA fragments scattered across the genome comprise the basic molecular data. Each individual has a unique set of fragments of different length and composition. When separated on an electrophoretic gel, these fragments resemble a bar code—a unique genotype that can be compared to others for the degree of similarity or difference. The parents of any individual can be identified if the "bar code" of the individual can be derived from the combination of two potential parents.

The data derived from these techniques can be used to assess the critical question of landscape genetics: how much gene flow is occurring? Fine-scale genetic information can be used to estimate directly the rate of gene flow. One commonly used technique is the *assignment test*. Individuals are sampled from populations across the landscape mosaic, and the genotype of each is identified. The allele frequencies in each subpopulation are calculated from this information. Statistical procedures then determine which individual in a given population is the best match to the genotypes in another population. That individual is likely a migrant, and the frequency of such migrants defines the rate of gene flow. This information allows us to test hypotheses about the evolutionary history across this landscape as well as the conservation implications of recent landscape changes.

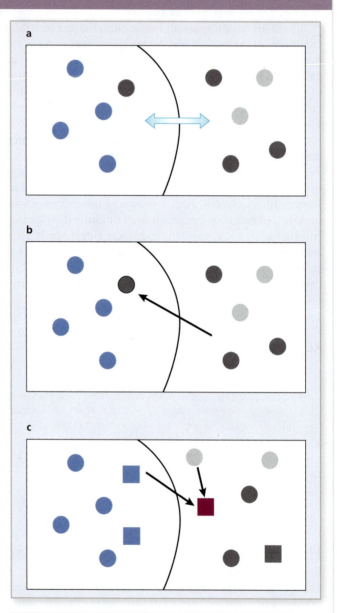

Figure 1 Barriers to gene flow. The nature of barriers to gene flow. The black line refers to a landscape feature, such as a river, that might inhibit gene flow. Filled circles refer to adult individuals; gray shading refers to genetically similar genotypes. In (a) the genetic distance is calculated among all individuals and related to habitat features. In (b) gene flow is assessed by the assignment test (see text). In (c) gene flow is measured by analysis of the parentage of offspring (from Holderegger and Wagner, 2008). **Analyze:** Explain how each of these analyses measures the degree to which gene flow is inhibited.

QUESTION:

How would landscape genetics affect the design of reserves for endangered species?

Figure 19.4 Satellite photo. These satellite photos of Brazilian forests show (a) encroaching development and (b) fires.

Satellites carrying spectrophotometers can measure the reflected light, thus quantifying both the absorption and reflection. Remarkably, the technology has advanced to the point that these spectra can be measured at a 1 m² scale. Because different vegetation types have distinct spectra, these data provide fine-scale measures of the plant cover across wide geographic areas (Figure 19.6). Other important information, such as surface temperature, can also be acquired by satellites.

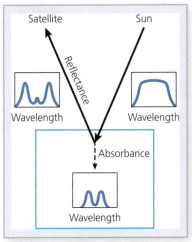

Figure 19.5 Solar radiation. Full-spectrum solar radiation strikes the earth. The vegetation, soil, and substrate absorb certain wavelengths. The rest are reflected. Satellites detect the reflected wavelengths. Each combination of vegetation, soil, and substrate has a unique reflectance pattern. **Analyze: Why would it be important to verify satellite data like this with field observations on the ground?**

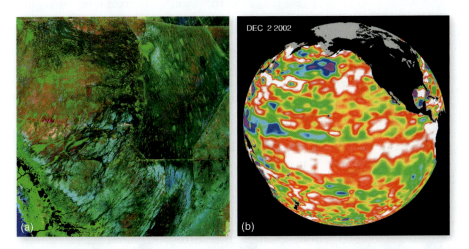

Figure 19.6 Landscape information. Satellite data provide an array of landscape information. (a) Vegetation types in the Everglades of southern Florida. Each color represents a different cover type. (b) Sea surface temperatures. Colors depict different temperatures: white/red = warm; green/yellow = cooler; blue/purple = cold.

Satellite technology produces prodigious amounts of information. Imagine the amount of information generated at a continental scale from the reflectance spectra of 1 m² parcels. Thus, the second crucial technological development was the computing power to handle this information. Of course, the software to handle these data had to develop concurrently. **Geographic Information Systems (GIS)** are among the most important of these software developments. GIS programs organize and analyze the many kinds of spatial data as a series of layers, each composed of a specific kind of data. Imagine that you have spatial data for an ecosystem that includes vegetation type, surface temperatures, ground water resources,

■ **geographic information systems (GIS)**
A software system that analyzes spatial ecological and environmental data by producing a series of overlaid maps, each with the spatial variation in a single variable.

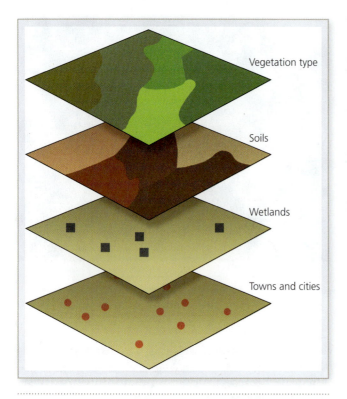

Figure 19.7 GIS map. A GIS map showing layers containing different spatial information. **Analyze: Why is it useful to depict spatial information as a series of layers?**

urban development plans, and soil type. Each data set includes precise spatial locations (maps) for this information. A GIS system develops a map in which each type of information is placed in a different layer (Figure 19.7). Moreover, the software allows us to quantify key aspects of each layer, such as shape or area. It also permits us to analyze the overlap among different layers, which allows us to address landscape-level questions. For example, by analyzing the relationship of the soil and vegetation layers, we can ask if soil is an important determinant of vegetation type. By comparing an urban planning layer with the vegetation layer, we can ask which vegetation types will be most impacted by development.

What Are the Key Measures of Landscape Structure?

The data generated from either photographs or spectral analysis must be assigned to some ecologically relevant category. The reflectance patterns must be shown to represent a specific and useful ecological condition or vegetation type. Often this requires comparing satellite spectral reflectance maps with direct surveys on the ground. In addition, the level of resolution and distinctions across space are determined by the question being asked. Figure 19.8 shows two landscape maps with different resolution that identify different levels of vegetation patterns. The appropriate classification scheme depends on the specific question we hope to answer.

Landscape patterns are the result of variation from one site to another—that is, differences in the occurrence and pattern of patches. We define a patch mathematically as a surface area that differs ecologically from its surroundings. The characteristics of the landscape are measured in cells—surface areas in which the relevant variables such as vegetation type, age, soil characteristics, and so forth—are quantified. Computer algorithms define a patch as a contiguous group of cells with the same mapped characteristics. Thus, for example, we can identify patches of unburned forest located in a recent burn (Figure 19.9).

Figure 19.8 Resolution level. The effect of grain size on the analysis of the Yellowstone vegetation. Grid cells are measured in meters (from Turner, 2005).

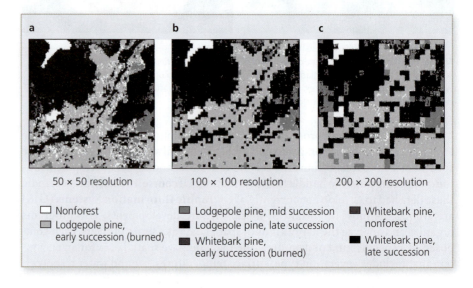

50 × 50 resolution   100 × 100 resolution   200 × 200 resolution

☐ Nonforest

☐ Lodgepole pine, early succession (burned)

▨ Lodgepole pine, mid succession

■ Lodgepole pine, late succession

▨ Whitebark pine, early succession (burned)

▨ Whitebark pine, nonforest

■ Whitebark pine, late succession

Fractal Geometry

The shapes and dimensions of landscape features are central to the study of landscape patterns. We have discussed a number of measures of landscape structure, such as patch size, connectivity, and perimeter/area ratio. Many of these depend in some way on the perimeter of the patch. Fractal geometry connects these various measures of shape.

Imagine that you are measuring the length of the coastline of a lake from an aerial photograph. Now imagine that you measure it with the large ruler shown in Figure 1 (a). This coarse measurement obviously misses some of the small indentations and prominences that would be included if we used a smaller ruler. Thus, an interesting phenomenon emerges: the measurement of a shape depends on the scale at which it is measured. The small-scale measurement reveals variation in the coastline that is missed with the larger ruler.

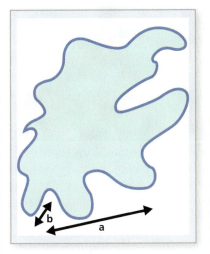

Figure 1 Shoreline of a lake. The measurement of its length will be different if a large ruler (a) is used compared to a smaller ruler (b). **Analyze:** How does the effect of ruler size differ with the shape of the object?

We can quantify this phenomenon with a parameter known as the *fractal dimension (D)*. The fractal dimension is calculated

$$D = \log N / \log r$$

in which *N* is the number of steps (rulers of a certain length) used to measure a pattern length and *r* is the scale ratio—that is, the ratio of the sizes of the measurement rulers. Consider the shapes shown in Figure 2. The distance from A to B measured in units of *x* (that is, a ruler *x* long) in Figure 2 (a) is 4. If we increase the resolution of our measurement by a factor of 3 (*r* = 3), more detail is seen. The fractal dimension D = *log* 4/*log* 3 = 1.2618.

One important characteristic of fractal analysis is that the dimension of a shape is not an integer but a fraction. Traditional Euclidean objects have integer dimensions. A line has dimension 1, an area has dimension 2, and a volume has dimension 3. But an ideal fractal curve has a dimension somewhere between 1 and 2. As the fractal dimension approaches 1, the shape is more like a pure line. As the fractal dimension approaches 2, more variation or uncertainty occurs in its measurement due to deviation from a simple straight line.

There are two important applications of this mathematics in ecology. First, different organisms perceive the landscape differently because they "measure" it at different scales. Consider how two different organisms that utilize the lake shown in Figure 1 relate to its shape. For a large, mobile bird such as a gull, the length of the coastline would be shorter than for a smaller, less

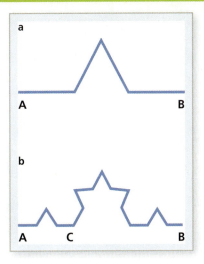

Figure 2 Fractal curves. A pair of fractal curves of increasing complexity (from Turner et al., 2001). **Analyze:** Why is the complexity of a shape important in ecology?

mobile animal like a crayfish. The available habitat for the crayfish might be significantly greater because it "measures" the coastline with a smaller ruler.

Second, fractal dimension is a useful metric of landscape structure. Higher values of *D* (closer to 2) indicate a more complex shape than smaller values (near 1, a simple line). Moreover, the larger the value of *D*, the more rapidly the measured length changes as a function of scale. For example, the fractal dimension of deciduous forests in Mississippi changes as a function of the size of the forest. At approximately 70 hectares, the value of *D* increases dramatically. This means that smaller forests tend to have relatively smooth shapes compared to larger forests. Most of the larger forests are undisturbed and so their boundaries are determined by complex factors such as soil type and slope. In contrast, smaller forests tend to be woodlots created by human activity. These tend to be simpler shapes like squares or rectangles (Krummel et al., 1987).

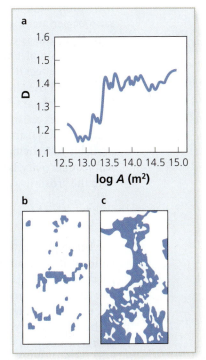

Figure 3 Fractal dimension. The fractal dimension (*D*) of forest patches in Mississippi as a function of patch size (a). Small patches like those in (b) have small values of *D*; large patches like those shown in (c) have larger values of *D* (from Turner et al., 2001; from Krummel et al., 1987). **Analyze:** What does a high or low value of *D* mean?

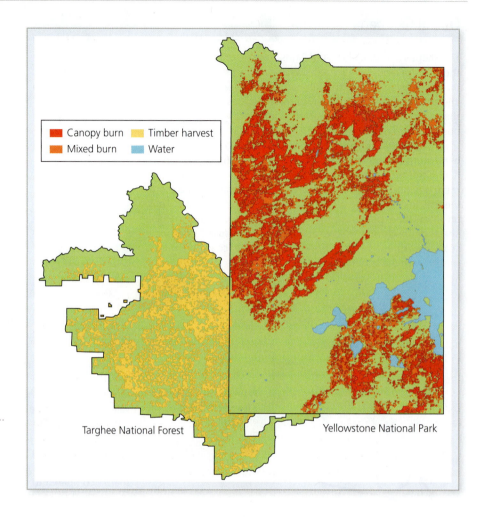

Figure 19.9 Burn intensity. This satellite map of burn intensity in the 1988 Yellowstone fire shows regions of hot fires (red), moderate fire intensity (pink), and unburned areas (green).

Legend:
- Canopy burn
- Mixed burn
- Timber harvest
- Water

Targhee National Forest

Yellowstone National Park

THINKING ABOUT ECOLOGY:

What is the effect of the spatial scale at which measurements are taken and the scale of landscape patterns? For example, what would be the effect on the results if we sampled a few large plots in the vegetation mosaic in Figure 19.8? What would be the result if we sampled many small plots?

proximity A measure of the relative distance between patches.

Once a patch is defined, we can quantify its characteristics in a number of ways. Simple but important metrics are patch area and perimeter. In addition, the ratio of perimeter to area (P/A ratio) conveys information regarding shape. A high P/A ratio is characteristic of a complex shape or one with an elongated boundary. Simpler or more compact patches have lower P/A ratios. In Chapter 18 we saw the effect of the P/A ratio on cowbird parasitism of forest birds. In forests with a high P/A value, there is little interior forest area where nesting birds are safe from cowbird parasitism.

If patches constitute habitat for plants or animals, connections among patches permit dispersal and gene flow. Connectivity is the degree to which one patch is available to individuals in another patch. A related measure, strongly correlated with connectivity, is **proximity**, the relative isolation of patches. We quantify this with the proximity index PX_i for a focal patch (i):

$$PX_i = \sum \frac{S_k}{n_k}$$

For this equation, we define a search distance to other patches on the basis of biologically relevant information such as the species' mobility, ability to detect patches, and so forth. The value S_k is the area of patch k within the search radius. The term n_k measures the distance between the focal patch and the nearest patch k. For example, hummingbirds forage in patches of flowers that contain nectar. The higher the connectivity among patches, the more likely a bird is able to satisfy its energetic needs in the landscape. As connectivity declines, the bird may need to seek a different landscape in which to forage.

Other landscape measures focus on biologically relevant aspects of patches. If we are interested in the spatial dynamics of a particular species or population, its use of and distribution across patches is important. The **proportion of patches occupied** is a simple but vitally important variable. If the landscape contains a number of patch types, we calculate the proportion of each type occupied, p_i, in which i denotes the patch type and varies from 1 to s, the total number of cover types.

It may be important to quantify the heterogeneity of the landscape. **Relative richness (R)** is the total number of cover types as a proportion of the number of possible types:

$$R = S/S_{max} \times 100$$

where S is the number of cover types present and S_{max} is the maximum number possible.

By sampling the ecological heterogeneity of the landscape and the use of that landscape by species of interest, we can begin to understand the fine details of habitat use. For example, Barbaro et al. (2007) studied how birds used a region of varied habitat types in southeastern France consisting of open habitat, pine plantations, deciduous woods, and urban areas. Counts of birds were made at standardized locations throughout the habitat mosaic. At each of these points, the researchers recorded detailed habitat information. Bird species utilized different aspects of the landscape heterogeneity in this region. For example, some, such as Bonelli's warbler (*Phylloscopus bonelli*) and the redstart (*Phoenicurus phoenicurus*), were found most often in deciduous woodland, whereas others, such as the skylark (*Alauda arvensis*) and the red-backed shrike (*Lanius collurio*), used open shrub habitat. Once these patterns of landscape use are quantified in this way, it is possible to address questions such as the connection of populations (and patches) or the possible consequences of increased urban development.

> **proportion of patches occupied** The proportion of available habitat patches occupied by a species.

> **relative richness (R)** The total number of cover types as a proportion of the number of possible types.

....

KEY CONCEPTS 19.2

- Landscape analysis is heavily dependent on two types of technology: satellite images and spectra, and computing power.
- Geographic Information Systems (GIS) integrate the landscape metrics of many types of cover or patches into a single model.
- The central datum of landscape ecology is the patch. Patches are characterized by spatial parameters such as size, shape, and connectivity.

QUESTION:

Which kinds of ecological questions do you imagine are best answered with information on patch size? Which would require information on shape?

....

19.3 What Are the Effects of Landscape Scale and Structure?

A central feature of landscape ecology is its emphasis on the patterns at large spatial scales. We define **scale** as the spatial dimension at which we view an ecological phenomenon or process. Scale has two key components: **grain** and **extent**. *Grain* refers to the finest spatial resolution in a data set. The size of the cells at which we make measurements defines the grain. *Extent* is the total size of the area under study. We must choose a scale appropriate to the question at hand. For example, questions regarding sea surface temperatures in the southern Pacific should have a greater extent than questions about the habitat use of intertidal

> **scale** The spatial dimension at which we view an ecological phenomenon or process.

> **grain** The finest spatial resolution in a data set.

> **extent** The total area of the study.

invertebrates. For both logistical and biological reasons, grain will be larger for the study of sea temperature than for the habitat use of invertebrates.

Landscape ecologists also recognize that a scale important to humans is not necessarily the scale important to the study organism. Consider the example of the population dynamics of butterflies that inhabit patches of flowers in mountain meadows. From a human perspective, we might choose to study a set of patches in a large, open meadow. However, the important mechanisms of butterfly population dynamics may operate entirely within one patch of flowers. At the larger spatial scale, our study might combine a set of distinct populations, perhaps with misleading results.

How Do Ecosystem Processes Change Across the Landscape?

The human impact on global patterns of nutrient cycling and production has stimulated analyses of ecosystem processes at multiple spatial scales. These studies have revealed that it is not just the measurements that are scale dependent; some ecological processes change with scale. We have already seen an example of this in our discussion of nutrient cycling in Chapter 17. Local nutrient budgets may or may not be in balance, depending on local processes that lead to net gain or loss of nutrients. However, at larger spatial scales, and especially at the largest (global) scale, nutrient budgets are in balance.

Primary production is limited mainly by light, water, and nutrients. Satellite measurements of both climate and the vegetation itself allow us to predict the spatial patterns of production. Because spectral reflectance in the near-infrared and red portions of the spectrum is strongly correlated with photosynthesis, we can use these spectra to measure primary production. An important index, the **normalized difference vegetation index (NDVI)**, calculated from the relative red and near-red reflectance, quantifies primary production in both natural vegetation and crops. NDVI is positively correlated with another useful measure of production, the **leaf-area index (LAI)**, which is the ratio of leaf area to the ground surface area. This variable measures not only potential production but also the interception of rainfall by the vegetation. By comparing primary production topography and NDVI or LAI, we can quantify the patterns of primary production across a heterogeneous landscape. We might test hypotheses about production by analyzing its relationship to variables such as soil type and precipitation patterns. This information also allows us to model the potential effect of changes in these variables from either natural or anthropogenic causes.

Remote measurements of biogeochemical processes are not possible at present. Thus, landscape-level studies of nutrient cycling require extensive fieldwork and laboratory analysis. However, the studies that have been done suggest that different nutrient fluxes scale differently across the landscape. For example, Morris and Boerner (1998) examined the patterns of nitrogen mineralization and nitrification at three spatial scales: within a watershed, between adjacent watersheds, and regionally in four sites separated by up to 65 km. The pattern of nitrification did not vary with spatial scale; the pattern at the largest scale was simply an extrapolation of what occurred within a watershed. In contrast, the rate and pattern of mineralization varied significantly even within a watershed.

How Does Landscape Structure Affect Populations and Communities?

Each species' preferred habitat is embedded in a landscape composed of an array of environments. Consequently, the structure of the landscape is central to species' presence/absence and population dynamics. We have thoroughly documented this important relationship in our discussion of population dynamics (Chapter 9) and community organization (Chapter 12). The focus of much of this work is at the

normalized difference vegetation index (NDVI) An index of primary production based on the reflectance in the red and near-red spectrum.

leaf area index (LAI) The ratio of leaf to ground area.

local scale within the landscape, sometimes at the level of the patch. When the spatial scale is expanded, new relationships emerge. For example, the relative abundances of two grassland bird species, the bobolink and savanna sparrow, are determined by the mosaic of different types of habitat patches. Both species require grassland habitat, but their demographic responses depend on how much woodland habitat surrounds the patches of grassland (Renfrew and Ribic, 2008).

We also can identify the effects of competition at the landscape level. Recall that Bowers and Brown (1982) found that competition among granivorous desert rodents is weaker at larger spatial scales (Chapter 12). Recent analyses confirm this result. They are based on the premise that if competition is an important determinant of community structure, certain plants and animals should co-occur less frequently than expected. We quantify this with covariance, a measure of the probability that two species occur together. A negative covariance signifies that they occur together less frequently than expected by chance; co-occurrence more frequent than expected by chance results in a positive covariance. Among 41 plant and animal studies, the majority showed positive rather than negative covariance (Houlihan et al., 2007). However, this analysis also showed a significant effect of

ON THE FRONTLINE

Equilibrium and Fire in Ponderosa Pine Forests

The dry forests of Eastern Washington are dominated by ponderosa pine (*Pinus ponderosa*). The conventional wisdom maintained that these forests persist in a relatively stable equilibrium due to regional climate and a specific disturbance regime. Mature forests, consisting of open parklike stands of ponderosa pine, were assumed to be stable and maintained by high-frequency, low-intensity surface fires. These fires were thought to suppress new saplings but not harm the canopy trees. This was thought to

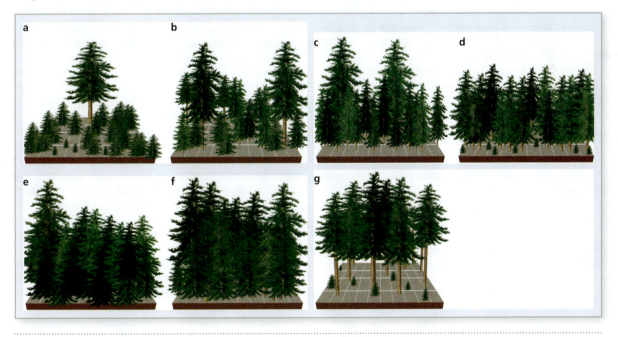

Figure 1 Structural types of ponderosa pine forests. (a) Stand initiation after disturbance. (b) Open-canopy new stem suppression. (c) Closed-canopy new stem suppression. (d) Understory reinitiation. (e) Young multistory forest. (f) Old multistory forest. (g) Old single-story forest. (From Hessburg et al., 2007.) **Analyze: How do these forest types relate to the hypothesis in this experiment?**

continued

ON THE FRONTLINE *continued*

Equilibrium and Fire in Ponderosa Pine Forests

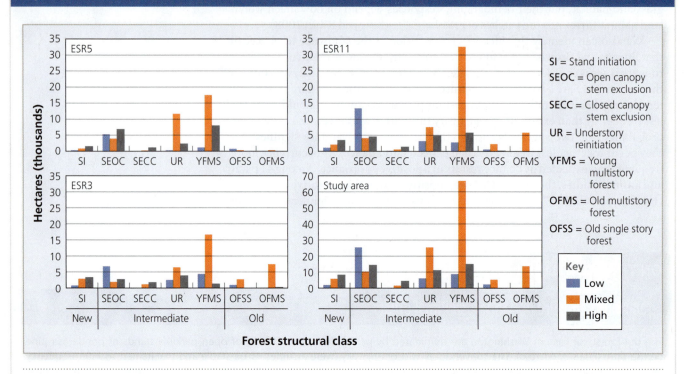

Figure 2 Forest area. The proportions of forest area in various structural classes at four study sites (from Hessburg et al., 2007).

eventually lead to an even-aged overstory structure because the growth rate of older trees slowed and that of younger trees increased, eventually producing a stand of similar-sized trees. This explanation for the structure of these forests had not, however, been rigorously tested. Hessburg et al. (2007) addressed the question: Do frequent, low-intensity fires maintain these forests in equilibrium?

HYPOTHESIS: Frequent, low-intensity fires maintain the structure of mature equilibrium ponderosa pine forests.

PREDICTION 1: The majority of fires in this system are recurring low-intensity fires.

PREDICTION 2: Even-aged stands of mature trees are the typical community structure.

The first data set Hessburg et al. analyzed included quantitative stand characteristics (canopy cover and size-class distribution). The second was a series of early (1930–1940) stereo aerial photographs of the forests. From these photos, the researchers

could estimate key stand variables including overstory canopy cover, species composition, size classes, number of canopy layers, and percentage of dead canopy cover. These data indicated that these forests are composed of seven types of structural classes. The severity of the most recent fire was determined from the overstory canopy proportion and its size classes, the understory size class, and the relative fire tolerance of the cover type.

The results showed that moderate-severity fires were most common across all stands: 47 percent compared to 16 percent low-severity fires and 37 percent high-severity fires. The forest structure was dominated by intermediate-age stands: young multistory, understory reinitiation, and canopy stem exclusion. Even old, single-size-class canopy forests showed evidence of mixed rather than low-severity fire.

Thus, Hessburg et al. rejected the hypothesis that single-age-class canopy ponderosa pine forests are an equilibrium end point maintained by frequent low-intensity fire. Rather, this landscape is a mosaic of forest stands, most of which have a complex structure that results from a history of moderate fire severity.

spatial scale. In the vast majority of studies that examined covariance at more than one spatial scale, negative covariance predominated at the smallest spatial scale. In other words, competition is important locally but less so regionally.

Analyses such as these have stimulated ecologists to consider the idea that the concept of the ecological community is itself scale dependent. At one end of the spectrum is the local community. At this spatial scale, competition and equilibrium are important structuring components. But we can also conceive of communities at much larger spatial scales: sets of plants and animals that coexist and interact regionally (Ricklefs, 2008). The structuring forces of these large-scale communities are quite different from those of local communities. They are more likely to depend on individual species' physiological tolerance, mobility, habitat selection in the broader habitat matrix, and the frequency of disturbance. Recall that one of the most important early debates in community ecology centered on Clements's and Gleason's differing views of the nature of the community and the importance of equilibrium. Disturbance was an essential component of Gleason's view, and thus nonequilibrium ecology traces its roots to the Gleasonian community.

How Do Disturbances Vary Spatially?

Disturbance, phase differences in the language of landscape ecology, is an integral part of most ecosystems. Recall that disturbances vary along three key axes: size, frequency, and severity. Moreover, these characteristics apply across the entire landscape. For example, fire varies in size, frequency, and severity at a landscape scale. Thus, specific sites within the landscape experience different disturbance regimes. The key effect is that at the landscape level, disturbance creates a mosaic of seral stages. The shift in the way we conceive of a community as a function of scale reflects the importance of disturbance as an important source of landscape heterogeneity.

Landscape structure, particularly topography, plays an important role in the mosaic created by disturbance. For example, hurricanes tend to move into New England forests from the southwest. The susceptibility of any given site is determined by its exposure to these tracks. Southeastern slopes, hilltops, and northwest lakeshores typically incur the most damage; valleys protected from the south generally incur the least damage (Foster, 1988). Disturbance itself can affect the frequency of future disturbances. If a forest has a uniform closed canopy, strong winds tend to flow along the canopy surface in horizontal layers known as laminar flow (Figure 19.10). If, however, a previous storm or other disturbance has created a canopy gap, airflow across the gap becomes turbulent. Turbulent flow is much more likely to topple additional trees. As a result, some landscapes contain a linear series of canopy gaps of decreasing age.

The effect of landscape structure on disturbance patterns may persist long after the disturbance. The maps in Figure 19.11 show the pattern of presettlement forest and grassland habitat in Iowa and Illinois. In this region, precipitation is sufficient to support trees. However, tallgrass prairie extended into both states because fires in the extensive grasslands to the west pushed east, killing back forest and stimulating the growth of grasses. Note the different pattern of forest and grassland in the two states. In Iowa, major rivers drain to the southeast, where they join the Mississippi. In Illinois, the major rivers trend to the southwest. River valleys slow or stop the spread of fire. Fires driven by strong southwest winds would encounter river breaks in Iowa. Thus, relatively small reaches of prairie occur in that portion of the landscape. Large open prairie occurred in the middle of Iowa where few rivers were present to halt the fires. In contrast, strong southwest winds could

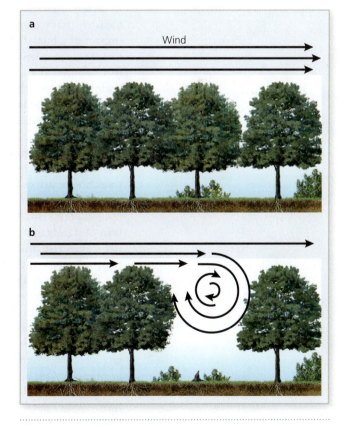

Figure 19.10 Canopy gaps. In forests with a closed canopy (a), laminar flow occurs. Canopy gaps (b) create turbulent flow that can damage other trees. **Analyze: What does this phenomenon mean for the spatial distribution of disturbance in a forest?**

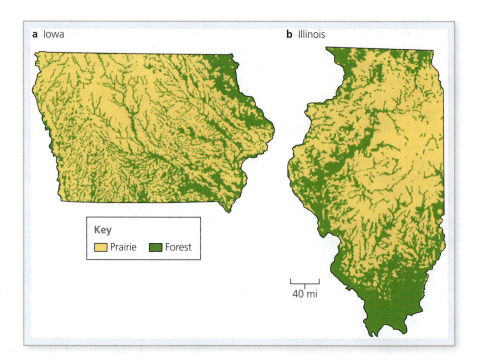

a Iowa   b Illinois

Key
☐ Prairie   ■ Forest

40 mi

Figure 19.11 Prairie vegetation. Map of historical prairie vegetation in Iowa and in Illinois. **Analyze: What role do rivers play in the distribution of prairie?**

push fire parallel to the river valleys in Illinois, resulting in a large central core of grassland.

Size is an important characteristic of disturbance. This variable is a relative measure that scales according to the biological element it impacts. Consider the impact of a hurricane on forest vegetation. If the focus is on the individual, a single tree may live its entire life without experiencing damage from hurricane-force winds. The spatial scale of the individual is small relative to the spatial scale of the disturbance. Once we increase the spatial scale to a stand of trees, the nature of the impact changes. The disturbance can now potentially create spatial heterogeneity if its impact varies across the stand. The result is that in this spatial hierarchy, a disturbance may be disruptive at one level but stabilizing at another. From the perspective of the individual killed by the storm, a hurricane is maximally disruptive. The same may be true if entire stands of trees are destroyed. However, for the entire forest, disturbance results in the summed deaths of a proportion of the individuals across some fraction of the landscape. At this scale, disturbance is stabilizing because it maintains the pattern of stands of different age, reproductive class, and size. Over time, individual stands rise and fall due to disturbance, but the overall forest maintains a relatively constant structure.

The three variables that describe disturbance (size, frequency, and intensity) are themselves dynamic; each is subject to change over time. There is evidence that the size of forest fires in the western United States is increasing. Over the past 10 years, more than 60 fires larger than 100,000 acres have occurred, a sharp increase compared to preceding decades. The frequency of disturbance changes, too. We measure the frequency of disturbance as the probability that any one locale is affected. Thus, the return time for hurricanes in the Caribbean is approximately 60 years. However, that measure is based on the historical pattern. Some climate models suggest that as the earth's temperature increases, the frequency of hurricanes will increase. This indeed seems to be occurring. The intensity of disturbance is dynamic as well. The same climate models that predict more frequent hurricanes also suggest that their intensity will increase. The more frequent forest fires in the West have also been more intense than those in the past.

KEY CONCEPTS 19.3

- Ecological processes occur at different spatial scales. We study scale by our choice of grain and extent.
- Ecosystem processes, community structure, and population dynamics all vary across the landscape. The central processes and mechanisms of each change as a function of spatial scale.
- At larger spatial scales, direct interactions among organisms become less important.
- The role of disturbance varies with spatial scale. Locally, it is an important source of mortality and species presence/absence. At larger scales, it determines the spatial heterogeneity central to landscape ecology.

QUESTION:

What is the effect of spatial scale on the role of chance?

Putting It All Together

The life cycle of the monarch butterfly plays out across much of the eastern United States and Mexico. The butterfly's success depends on a set of factors and conditions that vary widely in this landscape. Reproduction depends on milkweed host plants. As the density, health, and distribution of milkweed patches change, so does the monarch's breeding success. Agricultural practices, pollution, urban sprawl, and an array of other factors determine the status of the host plant population across the landscape. Other sources of mortality include the quality and quantity of adult food sources and hazards to migration. And of course the entire eastern population is vulnerable because of its absolute dependence on a small set of overwintering sites in Mexico.

This species demonstrates the key element of landscape ecology: important ecological processes and life cycle events play out across a spatially heterogeneous region. For each aspect of the life history and migration, a different scale and landscape structure is important.

As illustrative as the monarch is, no one species provides the answer to the central question of the chapter: How do ecological processes vary geographically? Nevertheless, in this chapter we have identified the two components of the landscape that apply to all species and to ecological processes: landscape structure and scale. The two are intimately connected, and thus a full understanding of ecological processes requires analysis of structure across a hierarchy of scales.

Summary

19.1 What Is an Ecological Landscape?
- The concept of the landscape is based on the inherent variation or heterogeneity over space.
- Landscapes are characterized by patches, cover type, fragmentation, and spatial scale.

- Landscape heterogeneity is the result of physical variation, such as topography and climate, and biological variation, such as differences in the species present.

19.2 How Do We Measure and Quantify Landscape Patterns?
- Two technological advancements, the availability of satellite data and increased computing power, were central to the development of landscape analysis.
- Landscape analysis is based on cells, the smallest units in which landscape characteristics are measured.
- Geographic Information Systems (GIS) integrates layers of landscape information in models that can be quantitatively analyzed.
- Size, shape, connectivity, and composition are key measures of landscapes.

19.3 What Are the Effects of Landscape Scale and Structure?
- Spatial scale is composed of grain and extent.
- Community processes, such as primary production or competition, are scale dependent. They vary depending on the spatial scale at which they are analyzed.
- Disturbance results in a heterogeneous landscape, a mosaic of habitat characteristics.
- Chance plays an important role in the spatial scaling of disturbance. At larger spatial scales, direct interactions among organisms become less important; statistical processes become more important.

◼ Key Terms

◼ Review Questions

1. What is the relationship between the terms *landscape*, *patch*, and *landscape structure*?

2. Why is landscape ecology so dependent on technology?

3. What do we mean by the statement "Some ecological processes change with spatial scale"?

4. Why is shape so important in landscape structure?

5. What is the relationship between landscape ecology and conservation biology?

6. How does a landscape approach affect our understanding of fundamental ecological concepts like the community or the population?

7. What is the relationship between patch dynamics and phase differences across the landscape?

8. What is the relationship between landscape structure and gene flow?

9. Explain how reflectance spectra measure important ecological variables.

10. What are the considerations for choosing extent and grain size in landscape studies?

Further Reading

Soulé, M.E., and J. Terborgh. 1999. *Continental conservation*. Washington, DC: Island Press.

> Landscape ecology and conservation biology are intimately connected. In fact, some important areas of landscape ecology were developed specifically for their application to conservation. This volume summarizes a number of research and theoretical treatments of this important interaction.

Turner, M.L., et al. 2001. *Landscape ecology in theory and practice*. New York: Springer.

> This comprehensive book summarizes the key elements of the landscape approach to ecology. It examines the topic from both a theoretical and practical point of view. Turner et al. illustrate the key principles of the field with specific empirical examples.

Turner, M.G. 2005. Landscape ecology: what is the state of the science? *Annual Review of Ecology, Evolution, and Systematics* 36:319–344.

> This review paper examines the growth of landscape ecology. Particular emphasis is placed on the importance of disturbance and succession to spatial heterogeneity.

Whittaker, R.H. 1956. Vegetation of the Great Smoky Mountains. *Ecological Monographs* 26:1–80.

> Whittaker produced a detailed analysis of the spatial patterns of the vegetation in the Smokies and their relationship to microclimate and soil. This classic paper precedes the term "landscape ecology" in the ecological literature but represents an early landscape approach to the spatial analysis of the vegetation.

Chapter 20

Human Global Ecology

In late summer 2005, Hurricane Katrina formed near the Bahamas as a Category 1 hurricane. Its track passed near the Florida Keys, where it entered the Gulf of Mexico. As it passed over the warm waters of the Gulf, its power and size increased. In a few hours it was a Category 5 hurricane with winds over 280 kmph. Katrina made landfall near New Orleans as a huge Category 3 storm. The city was damaged by both wind and the immediate storm surge. But the most significant damage occurred a few hours later when the levees protecting New Orleans catastrophically failed, flooding 80 percent of the city and killing more than 1,800 people. This storm was the sixth most powerful Atlantic storm in history. Its economic cost was higher than any previous storm, and it was one of the five deadliest hurricanes to hit the United States. At its peak power, Katrina was the strongest hurricane ever recorded in the Gulf of Mexico, a distinction lost only a few weeks later when Hurricanes Rita and Wilma exceeded its strength.

Katrina was both a regional and a global event. Its damage was centered on the Gulf Coast. But the links of causation extend far beyond the gulf. Climate change due to elevated atmospheric CO_2 is a well-established fact. Climate models predict that as the Earth warms, the intensity of tropical storms will increase. It is virtually impossible to determine if the strength of a single storm is the direct result of climate change. Nevertheless, Katrina was consistent with the statistical predictions of our current models.

In 2010, the Gulf Coast was visited by another catastrophic event, this one unambiguously caused by human industrial activity. On April 20 of that year, the Deepwater Horizon oil rig exploded, burned, and sank. The surface connection to the active well was severed and oil began pouring into the gulf from the well head more than a mile deep. By the time it was finally capped months later, more than

4.9 million barrels of oil had been released into the gulf. An additional 1.8 million gallons of chemical dispersants had been applied to the massive oil slick.

Subsequent studies of the marine and coastal ecosystems showed the nature and extent of the ecological damage. Some 8,000 birds were killed directly by the oil; estimates of the number killed but not found range five times higher. Oyster beds were inundated and will take at least a decade to recover. Dead dolphins have washed up on the beaches. Marine biologists suspect but have not proved that these deaths were caused by the spill. Keystone species such as starfish and coral suffered immediate and direct mortality. The ripple effects of these losses on the marine communities have not yet fully played out.

In a five-year span, the Gulf Coast experienced two major disasters. Together, their human and economic costs were enormous. Moreover, they illustrate key features of the human impact on the environment. The Deepwater Horizon operated in the gulf, and that's where its ecological impact was felt. The origin of the Katrina disaster is more nebulous—its connection to human activity is statistical rather than direct. More importantly, although the impact of the storm was regional, its ultimate causation may have been global. Human industrial activity is so massive and pervasive that the entire planet, including the climate itself, is affected. Each of these disasters answers the question, how do humans impact ecological systems? However, we cannot comprehensively answer that question in a single chapter. But we can begin to answer it by addressing another: *How do the ecological concepts of the previous 19 chapters contribute to our understanding of the human impact on the environment?*

20.1   What Is the History of the Human Environmental Impact?

The first modern humans probably lived as small bands of hunter-gatherers. The fossil and genetic evidence suggests that humans migrated from Africa to parts of Asia, Europe, and the Middle East over the last 300,000 years. These populations were nomadic; they followed migrating game or the plants they relied on. At that point in our evolutionary and ecological history, the human environmental impact was at its minimum—our numbers were so small and our local impact so ephemeral that our imprint was small. Some 8,000 years ago, agriculture and animal husbandry emerged. This economy was more stable, predictable, and independent of the vagaries of migrating animals and the uncertainty of wild fruits. Two profound ecological effects followed: the human population increased and our impact was concentrated into much smaller spaces. This was a major turning point in our environmental history. As nomadic hunter-gatherers, we were hardly more ecologically important than any other member of the community. Once we settled into permanent agriculture-based systems, we began to affect many species. In clearing land for cultivation, creating irrigation systems, and fighting weeds and pests, we increased our impact markedly and concentrated it in a much smaller area. Increasingly, our goal became control of the natural world, including the elimination of species that interfered with agriculture.

What Is Natural?

There are two opposing but valid answers to this question. In one sense, humans are simply one among the millions of other species. We are a product of evolution in the same way and by the same forces as other species. In this light, one could argue that everything we do is natural, whether it benefits or harms the rest of the natural world.

On the other hand, the human impact on the natural world has been disproportionate. No other species has so modified the ecology and evolution of so many other species. The brief history of early human ecology outlined above suggests that we made the transition from being a part of the natural world to controlling as much of nature as possible. Because of the scale of our impact and the artificial ecology created by our agricultural and urban systems, one could argue that there is a distinction between humans and the rest of the ecological world.

The major types of terrestrial and aquatic communities were described in Chapters 4 and 5. These were presented as the units of vegetation and assemblages of species independent of human habitation and impact. In this sense, they represent the ecology of the "natural" world. However, we can also represent the major ecological units of the Earth in terms of the human footprint and types of human activity. Figure 20.1 shows a map of the **anthromes**, the major

anthrome A major unit of human impact on the Earth; analogous to *biome*.

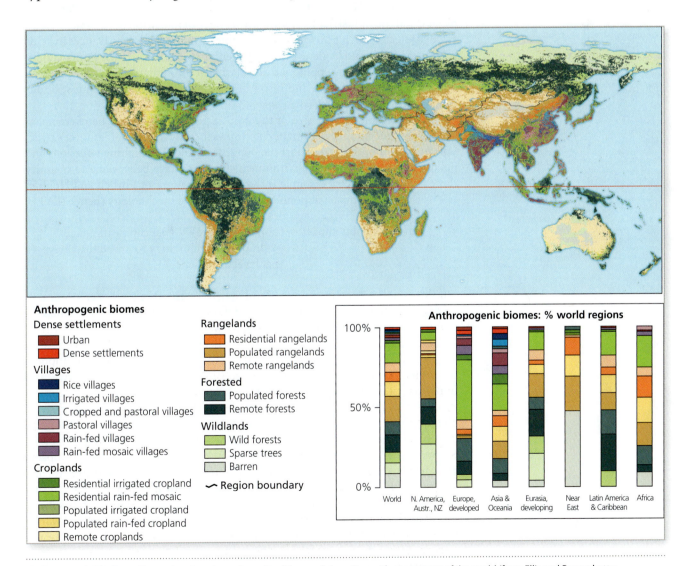

Figure 20.1 Anthromes. The anthromes, the major units of human interaction with ecosystems of the world (from Ellis and Rumankutty, 2008). **Analyze: Which of these anthromes do you think is increasing most rapidly in size?**

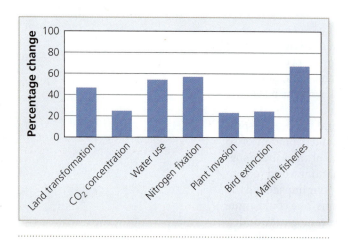

Figure 20.2 Ecological effects. Ecological changes imposed on ecosystems by humans (Vitousek et al., 1997). **Analyze: Which of these changes affect you most directly?**

units of sustained, direct human interaction with ecosystems. This kind of analysis shows that three-fourths of the ice-free land on Earth has been altered by human presence and land use. The remaining wild land is located in the least productive regions, such as the Arctic and deserts.

Our footprint extends to ecological function as well (Figure 20.2). For example, more atmospheric nitrogen is fixed by human agricultural activity than by all other terrestrial species combined. More than half of the accessible surface freshwater is used by humans. We are directly or indirectly responsible for the extinction of nearly 25 percent of all bird species. Consequently, many ecologists now assert that humans are the dominant ecological force on Earth (Vitousek et al., 1997).

What Are the Root Causes of the Human Impact?

The increase in the human impact parallels the exponential increase in the human population itself. Note the tight connection between the human population explosion and the Industrial Revolution. This wave of industrial development had a twofold effect. First, it provided the economic resources and the advances in fields such as medicine and hygiene that facilitated a rapid increase in the human population. Second, our developing industrial capacity itself became a central part of our impact—the pollution and resource exploitation that underlie most of our ecological problems today.

The importance of the size of the human population cannot be overemphasized. The sheer number of humans determines directly the level of resource extraction and consumption to support the population, and the resulting emissions and pollutants that degrade the environment. Attempts have been made to quantify these relationships. One example is the IPAT "equation" of Ehrlich and Holdren (1971) that relates impact to population, affluence, and technology (hence IPAT):

$$\text{Environmental impact} = (\text{population size}) \times (\text{per capita affluence}) \\ \times (\text{impact of technology required to achieve} \\ \text{that level of affluence})$$

This equation is conceptually useful in that it relates the level of environmental degradation to the population size and the level of consumption and emission required to maintain a certain lifestyle. However, it is also misleading in that it suggests that our impact is a simple multiplicative function of population size. This is unlikely to be correct (Hart, 2007). Two factors suggest that a multiplicative equation underestimates the relationship between population size and impact. First, many impacts have a threshold effect—that is, a sudden change occurs when a threshold has been reached. Bleaching of coral reefs is a threshold phenomenon. Reefs are relatively insensitive to thermal stress until the temperature exceeds 27°C. Then, rather suddenly, the reef deteriorates rapidly. Second, synergistic effects among ecological phenomena can result in impacts greater than the sum of the individual problems. As discussed in Chapter 16, one effect of global warming is melting of the permafrost in the Arctic. When this occurs, the large stores of methane sequestered in the soil enter the atmosphere. Methane is a potent greenhouse gas that contributes to additional global warming. Synergies like this create webs of positive feedbacks that exaggerate the impact of single factors.

The human population explosion occurred over a very short period of time. Human activity represents a significant disturbance event that emerged more

THINKING ABOUT ECOLOGY:

How is the human population growth rate affected by cultural, economic, political, and religious factors?

DO THE MATH

The Binomial Probability

The impacts of climate change range from subtle shifts in species' distribution to dramatic consequences such as increases in sea level and destructive hurricanes. In 2005, Hurricane Katrina drove into the Gulf Coast, causing damage from Mississippi to Texas and hundreds of deaths. This storm was unusual in several respects. It intensified rapidly from a Category 1 to a Category 5 storm in just a few hours. It was also a massive storm system—hurricane-force winds extended 120 miles from the eye. Hurricanes draw energy from warm subtropical waters. As ocean surface temperatures increase, climate models predict an increase in the frequency of intense tropical storms. Specifically, the frequency of Category 4–5 storms is expected to double by the end of this century. The recent history of tropical storms is consistent with this prediction.

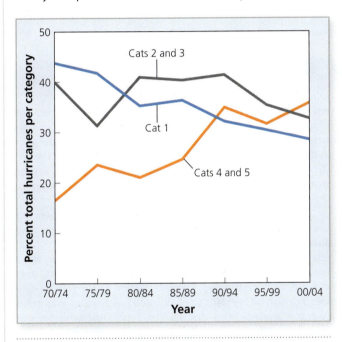

Figure 1 Hurricanes. The recent history of hurricanes (Union of Concerned Scientists). **Analyze: Can we conclude that any single hurricane is the result of higher ocean temperatures?**

The specific case of Hurricane Katrina embedded in these statistical projections raises an important issue: How do we relate a single event to a set of statistical predictions? In other words, if climate change is known to increase the frequency of severe storms, can we say that Hurricane Katrina was the result of climate change? This question draws attention to a significant feature of statistical analysis: statistics convey the behavior and expectations of a *population*. It is one thing to say that severe storms will increase in frequency—this is a prediction about a population of seasons and storms. It is quite another to assert that a specific storm is the product of the forces underlying the statistics. Inherent in any population of storms is variation—complex mechanisms and random variables that determine the outcome of any single event.

Statistical analysis does, however, provide the means for quantitative analysis of the frequency of events. One of the most powerful tools for this is the *binomial distribution*. The binomial distribution applies to events with the following properties:

1. A set of n repeated trials.
2. Each trial can result in just two possible outcomes, referred to as a success or failure. (For example, a hurricane is greater than Category 3 or not.)
3. The probability of success, denoted by P, is the same in each trial.
4. The trials are independent. (One hurricane does not affect other storms.)

We can illustrate the concept with a coin-flipping exercise. Imagine that we flip a coin two times. The binomial variable is the number of heads (successes), and the probability of heads is 0.5. The expected outcome is shown below.

NUMBER OF HEADS	PROBABILITY
0	0.25
1	0.50
2	0.25

The binomial probability is the probability that a binomial process results in exactly x successes. If we know the probability of a Category 4–5 hurricane in any year, what is the probability that there will be exactly x such hurricanes? We can calculate this probability, denoted $b(x; n, P)$—that is, the binomial probability of x events in n trials with a probability of success P

$$b(x; n, P) = {}_nC_x * P^x * (1 - P)^{n-x}$$

where ${}_nC_x$ is the number of combinations of n things taken x at a time.

Now imagine that the probability of a Category 4–5 storm is 0.17 in any year. We can ask a question such as, If the Gulf of Mexico experiences 5 hurricanes in a season, what is the probability of exactly 2 Category 4–5 storms? In this scenario, the number of trials is 5, the number of successes is 2, and the probability of success in any trial is 0.17. The number of combinations of 5 storms with 2 successes is 10 (there are 10 possible combinations of 2 successes in 5 tries). We calculate the probability of 2 Category 4–5 storms:

$$b(2; 5, 0.17) = {}_5C_2 * (0.17)^2 * (0.833)^3$$
$$= 10 * 0.0289 * 0.578 = 0.161.$$

There is a 16.1 percent probability that if 5 storms occur in the Gulf of Mexico, 2 will be Category 4–5 hurricanes. These calculations illustrate the statistical nature of events like major storms. Although we can calculate that the expected frequency and severity of storms will increase with climate change, we cannot assign that cause to any particular storm.

rapidly than ecological and evolutionary processes can react. Just as has been the case throughout evolution, some species are more readily adaptable than others, for the reasons outlined in Chapter 2.

The potential for adaptive response to the human-dominated environment is a major focus of current research in conservation biology. For example, climate change affects the timing and nature of seasonally important events such as snowfall, frosts, and monsoon rains. Moreover, the rate of global temperature change is rapid (Figure 20.3). We find that some species are adapting to these changes; others are not. For example, the European tit (*Parus major*) depends on caterpillars to feed its nestlings. Climate change has led to earlier spring weather across much of the tit's range and a much earlier availability of caterpillars. Many tits cannot breed this early, and the direct result is a decrease in reproductive output. Some, however, are able to adjust the timing of egg laying. These birds are at a clear selective advantage. The inherent variation in the species will perhaps buffer it from rapid change in its environment.

THE EVOLUTION CONNECTION

Genetically Modified Organisms

The practice of artificial selection, in which humans select plants or animals with certain traits, was well known to Darwin. In fact, Darwin used this practice to understand and explain the process of natural selection in which it is the environment, rather than humans, that determines fitness. Humans have artificially selected for desired traits in crop plants, domestic animals, and even their pets for thousands of years. In doing so, we have manipulated the evolutionary process, significantly altering the genotypes and phenotypes of many species.

Recent advances in molecular biology open the door to an even more direct and controversial manipulation of the genotype. It is now possible to transfer specific genes from one organism to another. These *genetically modified organisms (GMOs)* carry genes derived not from selection and crosses among members of a single species but from an entirely different species. For example, the bacterium *Bacillus thuringiensis* produces a toxin that interferes with digestion in many insect species. The toxin is coded for by a single gene. Once the gene was identified, it was possible to remove it from the bacterium and insert it into a crop plant such as corn, thereby producing plants that contain the toxin. This new strain, known as *Bt corn*, is resistant to insect pests.

One of the most widely used herbicides is glyphosate, which is marketed as Roundup. This compound is effective on many weed species, but of course it can harm crop plants as well. Molecular biologists discovered a gene that confers resistance to glyphosate. By inserting this gene into corn or soybeans, they produced strains of these crops resistant to glyphosate. Theoretically, this allows the farmer to kill the weeds but not harm the crops.

The production of GMOs is a significant change in the way we manipulate the evolutionary process. Rather than shape the naturally occurring or induced variation in a single species, we are able to insert single genes from completely unrelated forms into plants or animals. The genes that occur together naturally in a species are the product of a long history of evolution in which

developmental processes and the resulting phenotypes are shaped by natural selection on the entire genome as a coordinated unit. This new technology represents a fundamental shift in the relationship between the gene, the organism, and the environment. In GMOs, single genes instantly appear in genotypes that never housed them before. Completely novel phenotypes, untested by the process of selection, emerge.

Insect- and herbicide-resistant corn has obvious benefits to the farmer and to a growing world population. But there may also be ecological costs. For example, knowing that there is no effect of glyphosate on their corn or beans, farmers apply large amounts of the herbicide to their fields. In the process, they are essentially performing artificial selection on the weeds—any weed genotype resistant to glyphosate will have an enormous selective advantage in that environment. Over time we can expect the effectiveness of glyphosate to decline as resistance inevitably evolves. There is evidence that glyphosate-resistant weeds and Bt-resistant insects are emerging.

Other dangers are theoretically possible but have not been studied enough to evaluate. Many ecologists are concerned about the impact of certain gene products, like the Bt toxin, on nontarget species. In addition, transfer of genes among similar or related plant species does occur, although rarely. It is at least possible that genes intended for crops might transfer to other plants, especially weeds. In addition, the prevalence and economic success of Roundup-ready crops have led to a significant decline in the genetic diversity of corn, cotton, and soybeans. Currently, 25 percent of all the corn and 60 percent of the cotton planted in the United States comprises just a few genotypes.

QUESTION:

What are the potential risks of crops comprising just a handful of genotypes?

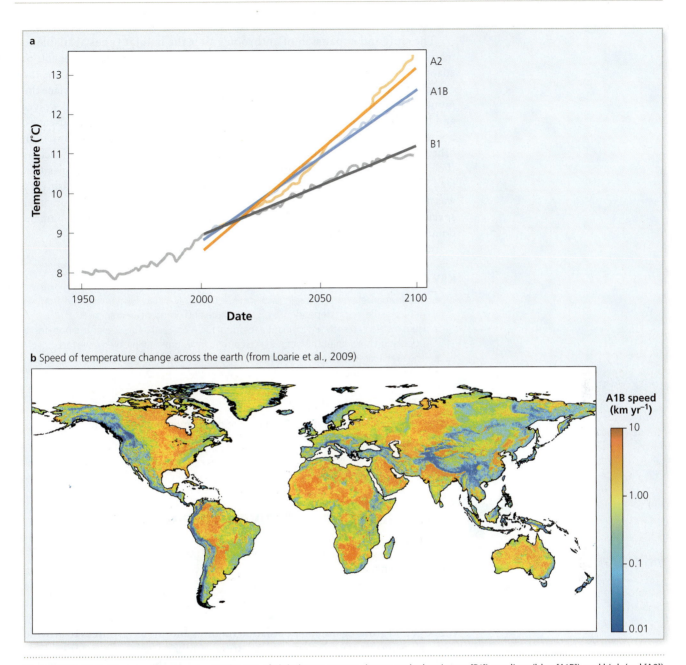

Figure 20.3 Global temperature change. (a) The rate of global temperature change under low (green [B1]), medium (blue [A1B]), and high (red [A2]) CO_2 emission scenarios. (b) The speed of temperature change across the Earth (from Loarie et al., 2009). **Analyze: What hypotheses can you propose for the geographic variation in climate change?**

Other species may not be able to respond in this way. Snowshoe hares (*Lepus americanus*) inhabit high-latitude conifer forests. They change coat color from brown in summer to white in winter. Recall from Chapter 1 that hares are most vulnerable to predation in the spring and fall, when their color change may be slightly ahead of or behind the change in snow cover. Camouflage mismatches— hares that are white when there is no snow or that are brown when there is—are disproportionately eaten by predators. The color shift is timed primarily by photoperiod. However, temperature can accelerate or delay the shift, to some degree. The timing of color change evolved in response to a relatively predictable seasonal pattern of snow cover. As climate change alters that pattern, camouflage mismatches may be more likely. Current research focuses on determining if the plasticity of the color change with temperature is sufficient to protect the species.

The shift and expansion of anthromes alter the habitat types available to many species. As in the case of climate change, some are inherently better able to adapt to new land-use patterns. Raccoons, deer, and coyotes have shown the capacity to thrive in urban and suburban landscapes. Indeed, estimates place the number of coyotes in Chicago at 2,000 individuals. Other species, such as wolves and lynx, do not coexist with human populations in highly altered landscapes. Some species are more tolerant of high levels of pollutants, especially toxins, than others. The variation is due to physiological and ecological difference, such as their position in the food chain. Thus, raptors such as ospreys and eagles that feed high in the food chain were decimated by DDT, whereas their fish prey was not. In a sense, we are imposing a massive artificial-selection regime on other species. We are selecting for those species most tolerant of their new human-dominated environment.

KEY CONCEPTS 20.1

- The human environmental impact increased sharply when humans shifted from hunter-gatherers to dependence on agriculture and domesticated animals.
- Humans are currently a dominant ecological force on Earth. The size of the human population is a significant factor in the current environmental problems we face.
- The human impact is both massive and rapid. This challenges the potential adaptive responses of other species.

QUESTION:

What demographic factors from Chapters 8 and 9 are most important to human population growth?

How Does Human Activity Threaten the Environment?

The human environmental impact stems from two fundamental effects of human population size and activity: (1) the use and depletion of resources and (2) the emission of the waste and by-products of our resource consumption. When the human population was small and isolated, the impact was primarily local. Today, the large and growing human population requires enormous resources to provide food, housing, transportation, and an array of other needs. The magnitude of both resource extraction and emissions is so great that the impacts are now global in scale.

What Are the Important Global Environmental Threats?

There are many human threats to the environment. We will, however, examine a set of examples that illustrate the fundamental ecological principles that apply to the human environment.

The phenomenon of climate change due to CO_2 emissions from burning fossil fuels has received considerable public attention. Moreover, we have discussed the phenomenon and some of its impacts in a variety of contexts throughout this text (see Chapter 3 for a discussion of the basic process). The atmospheric CO_2 concentration is clearly rising (Figure 20.4). Moreover, the increase is tied directly to the Industrial Revolution. We know this because the history of atmospheric carbon dioxide is recorded in glaciers. Small samples of air are trapped during the formation of glacial ice. The prevailing temperature regime is also recorded, because the ratio of the isotopic forms of hydrogen is an index of global temperature at the time the ice forms. Thus, ice cores from deep glaciers allow us to reconstruct

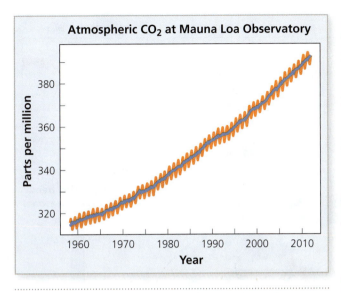

Figure 20.4 **Rise of the atmospheric CO_2 concentration.** The blue line is the yearly mean. The orange line shows the seasonal fluctuation (NOAA, 2014). **Analyze: Why does CO_2 vary seasonally?**

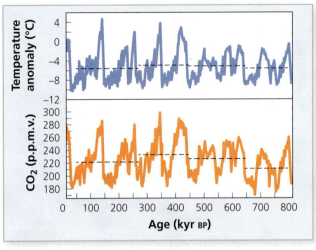

Figure 20.5 **Antarctic ice cores.** The relationship between temperature and CO_2 as revealed by Antarctic ice cores (from Lüthi et al., 2008). **Analyze: Why is it important to establish this correlation?**

the history of temperature and CO_2 in the atmosphere. Studies of Antarctic cores now extend that history back 800,000 years. These data clearly show that the pre-industrial CO_2 concentration of the atmosphere ranged from 172 to 300 ppm (Lüthi et al., 2008), compared to more than 400 ppm today. Moreover, there is a strong correlation between the atmospheric CO_2 concentration and temperature (Figure 20.5). However, CO_2 is not the only important greenhouse gas produced by human industrial activity. Methane, which is produced by livestock digestion and incomplete decomposition in aquatic systems, is a potent greenhouse gas. Climate scientists are confident that these emissions play a dominant role in the climate changes that are currently taking place.

We outlined the fluxes and sinks in the global carbon cycle in Chapter 17. Carbon moves from the atmosphere to the ocean as dissolved CO_2 and it is taken up by plants in the process of photosynthesis. As the carbon dioxide concentration in the atmosphere increases, those two fluxes increase as well, somewhat mitigating the greenhouse effect. However, these processes, especially primary production, are negated by large-scale deforestation. Tropical rainforest sequesters large amounts of carbon. As those forests are destroyed, not only does that flux decrease but decomposition releases additional carbon dioxide.

Current computer models predict two important features of climate change. First, even if we are able to stabilize atmospheric carbon dioxide at current concentrations, climate change will continue for decades (Solomon, 2009). Second, the climatic effect of warming is spatially heterogeneous across the Earth. The temperature will not rise uniformly; rather, different regions will change in unique ways due to prevailing winds, ocean currents, mountain ranges, large bodies of water, potential snow cover, and myriad other factors. Some regions will experience increased drought; others will see more frequent or more severe storms. Still others will even see higher snowfall and precipitation (Figure 20.6). We can also expect the emergence of the no-analog climates and plant communities discussed in Chapter 4. This heterogeneity is one reason climate scientists refer to "climate change" rather than simply "global warming."

We are already seeing the effects of climate change in many ecological communities. Not surprisingly, the impact is most immediate and obvious in communities in which species are near the limit of physiological tolerance, such as in the Arctic or at high altitude. As global warming increases, the range limits of

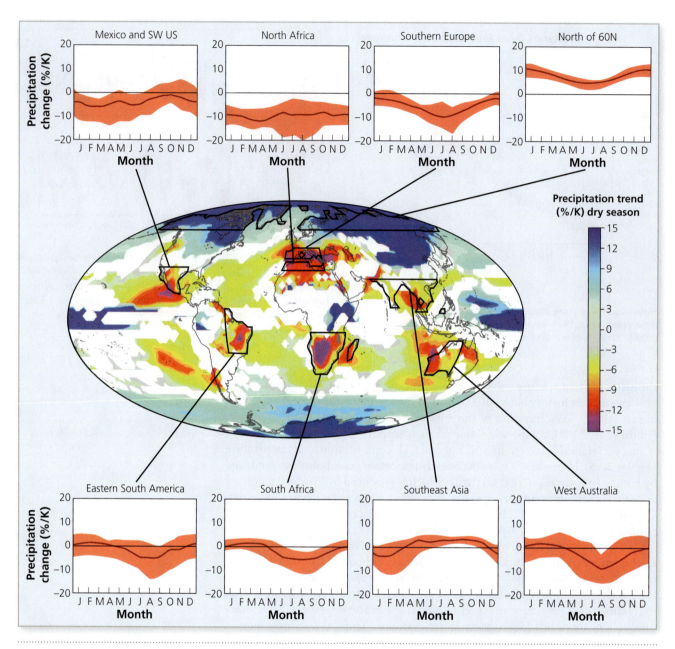

Figure 20.6 Precipitation changes. Changes in precipitation as the Earth warms. The red line in each graph shows the mean predicted increase or decrease in precipitation; the red band shows the confidence intervals around those mean predictions (from Solomon et al., 2009). **Analyze:** Why is it significant that so many temperate regions will experience more drought?

some species are shifting to cooler climates. For example, there is evidence that boreal forest conifers, especially black spruce (*Picea mariana*), are moving northward. Long-term studies of plant altitudinal ranges in the Alps show that many species have shifted to higher elevation (Figure 20.7). From 1905 to 2005, forest plant species shifted their range upwards by 29 meters per decade (Lenoir et al., 2008). Although we expect the most immediate impacts in colder climates, tropical species may ultimately be at even greater risk (Colwell et al., 2008). Many of these species are already near their maximum thermal tolerance but do not have the option of shifting their ranges to higher elevation or latitude. Climate models also predict that the tropics will be not only warmer but drier, which will put many rainforest species adapted to high rainfall and humidity at risk.

Thermal Acclimation in Corals

Many of the world's coral reef communities are in decline. When coral reefs experience stress from factors such as pollution or a significant change in water temperature, the symbiotic relationship between the coral and the dinoflagellate *Symbiodinium* is disrupted. In some cases, the zooxanthellae die. In others, the coral expels the symbionts. Both lead to the death of the coral and the phenomenon of coral bleaching. All that remains are the white skeletons of dead coral. The myriad invertebrates and fish that depend on the coral gradually disappear as well.

Some corals tolerate thermal shifts better than others. Three possible mechanisms explain their higher heat-stress tolerance. First, heat-tolerant strains of coral may arise by natural selection. Second, selection might produce heat-tolerant strains of the symbionts in the genus *Symbiodinium*. In fact, some evidence points to this mechanism. Finally, physiological acclimation of the coral, *Symbiodinium*, or both might increase the thermal tolerance of the community.

Oliver and Palumbi (2012) studied corals able to withstand higher temperatures. They reasoned that the history of exposure to temperature change might play a role.

HYPOTHESIS: Coral communities exposed to fluctuating temperatures should have higher thermal tolerance than communities that experience constant temperature.

The difference might be the result of adaptation or acclimation. They made use of natural variation in the thermal environment on Ofu Island in American Samoa. On the back portion of the reef that occurs there, small lagoon pools experience temperature fluctuation of as much as 6°C over the course of the tidal cycle, whereas the main reef rarely experiences a change of more than one degree. Their experimental design compared thermal-stress tolerance in populations of coral and *Symbiodinium* from stable and fluctuating environments. Their hypothesis made a specific prediction about the populations from these two environments.

PREDICTION: Corals from the thermally fluctuating environment should tolerate higher thermal stress than those from the thermally stable environment.

First, they confirmed the difference in temperature variation among the field sites with long-term detailed temperature measurements. Then they exposed corals and their symbionts collected from the different environments to either constant or fluctuating temperatures. The animals were housed in 50-gallon fiberglass tubs through which water from either the lagoon pools or the main reef circulated. The temperature in each tub was controlled with large aquarium heaters. The control tank replicated the temperature of the native reef; in the experimental treatment, the temperature was elevated 3.5°C for a period of five days, an increase similar to the temperature increase during a major bleaching event documented in Australia. In addition to measuring the physiological differences among coral-*Symbiodinium* samples, they measured the genetic difference among populations of coral and *Symbiodinium* by comparing mtDNA sequences of corals and the sequence of cytochrome B in the dinoflagellate. Finally, they measured the photosynthetic rate of the *Symbiodinium* in the different treatments.

Corals from the larger, less variable pools suffered much higher mortality than those from smaller, variable pools. There was no significant genetic difference among the corals, suggesting that those in the variable pools acclimate to that more variable environment. Two strains of *Symbiodinium* were found in this system. Strain D, which has higher heat tolerance than Strain C2, was found in all the corals of the small, variable pools. The photosynthetic machinery of this strain was not affected by heat. In contrast, Strain C2 suffered a significant decline in photosynthesis under heat stress. Some corals from the large, constant pools contain Strain D. Their response was intermediate—Strain D conferred some tolerance, but they did not survive as well as the strain from the more variable environment. The researchers concluded that in this system, the thermal history of the population determines its susceptibility to high temperature.

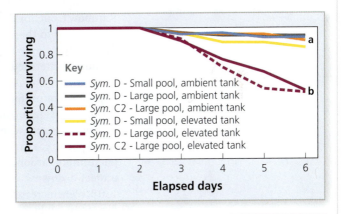

Figure 1 Survival of corals. The survival of corals from different source pools (small = high natural temperature variation; large = small natural temperature variation), with different strains of *Symbiodinium*, and under different experimental treatments (blue = natural temperature of the pool; red = elevated temperature regime). (From Palumbi, 2012.) **Analyze: What do these data mean for the survival of corals as sea temperatures rise?**

Acid precipitation, the decrease in the pH of rain and snow, is another global impact that results directly from industrial emissions, especially from power plants. Some types of coal have significant amounts of sulfur. Although it is technically possible to remove sulfur from the emissions, many power plants still release sulfur into the atmosphere in the form of sulfur dioxide (SO_2). When this

■ **acid precipitation** The decrease in the pH of rain or snow due to human industrial emissions.

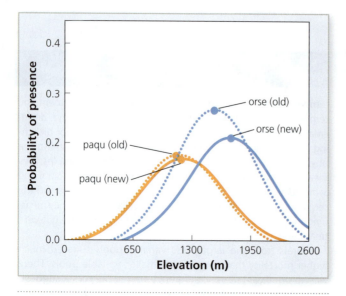

Figure 20.7 Elevational shifts. Elevational shifts in the French Alps of two bird species from 1905 to 2005. Paqu = *Parus quadrifolia*; orse = *Orthilia secunda* (from Lenoir et al., 2008). **Analyze: What do these data mean for sessile species?**

nitrogen deposition The addition of nitrogenous compounds to the ecosystem by agricultural and industrial activity.

compound reaches the upper atmosphere, it reacts with oxygen to form sulfur trioxide:

$$2SO_2 + O_2 \rightarrow 2SO_3.$$

The SO_3 diffuses into water droplets, where it reacts with water to form sulfuric acid:

$$SO_3 + H_2O \rightarrow H_2SO_4.$$

Nitrous oxides also react in the atmosphere to form nitric acid (HNO_3). Normal rainwater is only slightly acid (ph 5.4), due to the formation of the weak acid carbonic acid from the reaction of water and CO_2. In contrast, sulfuric and nitric acids are strong acids. As a result, in parts of the northeastern United States, precipitation now has a pH of 4.1 and some storms have produced rainfall of pH 3.0, nearly 1,000 times more acid than normal rainfall.

The primary ecosystem effect of acid precipitation is a change in aquatic or soil chemistry. Oligotrophic lakes in basins with granitic bedrock have the lowest ability to buffer acid input. Most soils have a relatively high buffering capacity, but it, too, can be exceeded. When the soil reaches this point, hydrogen ions (H^+) released by the acid percolate through the soil. They readily replace other cations that are held on the surface of soil particles. Some of these, such as Ca^{+2}, K^+, and Mg^{+2}, are important plant nutrients. When these cations are released, they leach from the soil. Some toxic cations, such as Al^{+3}, are also released.

In aquatic systems, acids increase the mortality rate of some fish species. Rainbow trout, for example, do not tolerate waters with a pH below 6.0. Lakes are also subject to the effect of acid on nutrient availability and aluminum toxicity. These effects may be compounded by additional inputs of acid from the surrounding watershed.

Marine systems are also experiencing significant acidification. However, this impact is connected to elevated atmospheric CO_2 levels (Doney et al., 2009). Carbon dioxide in the atmosphere dissolves in seawater across the vast surface area of the ocean, where it forms carbonic acid. It takes only a year for the ocean–atmospheric gas exchange to reach equilibrium. As more carbon enters the atmosphere, more carbonic acid forms in the oceans, leading to a decrease in pH. Models project a downward shift in pH of 0.3–0.4 in this century. This represents a 150 percent increase in the H^+ concentration of the ocean. Changes of this magnitude disrupt many biogeochemical cycles. Moreover, acid inhibits the formation of the calcium carbonate skeletons crucial to many marine organisms, especially coral.

Nitrogen deposition is a third important threat to the global environment. In the last century, the total amount of nitrogen in the biosphere has doubled due to industrial and agricultural emissions. The combustion of fossil fuels releases oxidized nitrogen (NO_x) compounds. Reduced reactive species (NH_x) enter aquatic ecosystems by runoff from agricultural fields to which high levels of nitrogenous fertilizers have been applied. Both types of compounds alter the biogeochemistry of nitrogen. In terrestrial systems, there is an initial fertilizing effect. This stimulates primary production and hence the uptake of atmospheric CO_2, potentially mitigating climate change. However, if the capacity of the vegetation to take in and utilize nitrogen is exceeded, microbial activity stimulated by high levels of NH_x compounds produces nitrous oxides. One of these compounds (N_2O) is a greenhouse gas, which negates any earlier benefit of increased production and sequestering of carbon. If nitrate (NO_3^-) reaches a high concentration, it leaches from the soil. In the process, it carries cations important to soil fertility, such as Ca^{+2}, K^+, and Mg^{+2}, with it. In aquatic ecosystems, nitrogen deposition alters lake N:P stoichiometry. If atmospheric nitrogen deposition increases the N:P ratio, phosphorus becomes the limiting nutrient. Phytoplankton diversity drops due to strong selection for the ability to compete for phosphorus and use it efficiently (Elser et al., 2009).

One of the most devastating impacts of nitrogen deposition is the development of "dead zones" in coastal regions exposed to high concentrations of nitrogen compounds. Huge amounts of nitrogen are carried to the ocean by rivers carrying runoff from agricultural systems. The estuary of a 10th-order river like the Mississippi, which drains millions of acres of agricultural land, receives many thousands of tons of nitrogen annually. This massive input causes eutrophication along the coast. As organic material accumulates and falls to the bottom, microbial decomposition increases. This in turn depletes dissolved oxygen in the benthos, with devastating effects on fish and invertebrates. Much

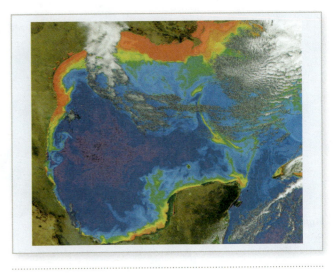

Figure 20.8 Mississippi River dead zone. The dead zone at the mouth of the Mississippi River. Areas in red are anoxic; yellow regions are hypoxic (from Diaz and Rosenberg, 2008).

of the benthos in an area of some 6,000 square miles in the Gulf of Mexico is anoxic and virtually devoid of life (Figure 20.8).

Atmospheric nitrogen is also responsible for depleting the protective layer of ozone high in the Earth's atmosphere. Ozone (O_3) forms in the stratosphere from O_2 in light-mediated reactions. The ozone layer protects the biosphere from high-energy ultraviolet radiation known as UVB. This high-energy radiation destroys the photosynthetic apparatus in plants and causes mutations in DNA. In humans, these mutations are an important cause of skin cancer. A number of industrial compounds that reach the stratosphere destroy ozone. These **ozone-depleting substances (ODSs)** act as catalysts—they participate in the reaction but are not consumed by it:

ozone-depleting substances (ODS) Compounds that catalyze the destruction of atmospheric ozone.

$$ODS + O_3 \rightarrow ODS\text{-}O + O_2$$
$$ODS\text{-}O + O_3 \rightarrow ODS + 2O_2.$$

Each ODS molecule destroys two ozone molecules but is released to participate in the reaction sequence again. The result is a compounding depletion of ozone. Industrial compounds containing chlorine and bromine are important ozone-depleting substances. Recent international agreements have significantly reduced the use and emission of these types of ODS. However, atmospheric nitrogen, especially N_2O, is also a potent ODS, and it continues to enter the atmosphere in enormous quantities (Ravishankara et al., 2009). Significant thinning of the ozone layer is now well documented. The effect is most dramatic over Antarctica in the austral spring, because the ozone-depletion reactions are facilitated by light and low temperature (Figure 20.9).

Habitat loss and conversion of natural communities into anthromes affects a wide range of ecological processes both regionally and globally. The very concept of anthromes demonstrates the scale of human alteration of the landscape. For example, in North America, agricultural land has replaced 99.9 percent of the native grassland or prairie community. The tallgrass prairie (Chapter 4) is now one of the rarest plant communities on the continent. The rate of deforestation in the tropics is staggering. Between 1990 and 2005, Brazil lost more than 400,000 square kilometers of forest. Significant losses are proceeding in virtually every major ecological community.

habitat loss The conversion of natural communities into anthromes.

Desertification, the conversion of plant communities to desert or the expansion of desert into other communities, rivals the loss of tropical rainforest in extent and significance. Drylands—that is, any of the xeric plant communities such as grassland, dry shrubland, and desert—cover some 40 percent of the Earth's land

desertification The conversion of plant communities or agricultural systems to desert.

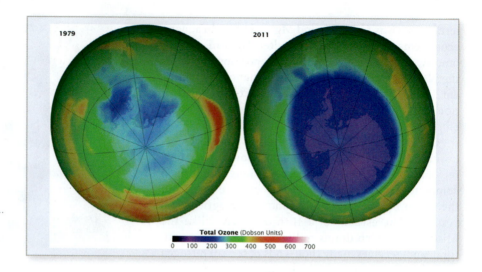

Figure 20.9 Ozone hole. Map of the ozone hole (violet) over Antarctica in the austral spring.

THINKING ABOUT ECOLOGY:

How would you rate the quality of your environment? What are the most important environmental factors that determine your quality of life?

point sources: Discrete, local sources of pollution.

nonpoint sources Releases of pollution across a broad area.

surface and hold nearly the same proportion of the human population. Rapid desertification is proceeding on 20 percent of this land, a scale that affects some 250 million people.

Desertification is a complex process with multiple interacting causes (Okin et al., 2009). Removal or degradation of the natural vegetation occurs by agricultural conversion and as a result of climate change. When xeric plant communities are converted to agriculture, runoff and erosion remove soil and nutrients. Any salts in irrigation water are concentrated in the soil by evaporation. Together these processes ultimately destroy the agricultural system and any remaining natural vegetation. This leads to still more erosion and the propagation of desert across larger areas (Figure 20.10). Desertification is also promoted by global processes, especially climate change. As the Earth warms, the mid-latitudes experience more frequent and prolonged drought. This destroys xeric agricultural systems and shifts dryland plant communities to desert.

What Determines the Spatial Scale of Human Impacts?

The previous section outlined some of the most significant global environmental issues. Of course, not every human impact becomes a global problem. Why do some affect the entire planet?

The nature of emissions plays a major role in the scaling of pollution. Emissions that emanate from a single, discrete source are known as **point sources**. The smokestack of a power plant or a drainpipe from a factory are potential point sources of pollution. Because point sources are discrete and identifiable, they are generally easier to monitor and regulate. Of course, it is entirely possible for point-source pollution to have global effects. The summed emissions may produce a global effect. Thus, carbon dioxide emissions are ultimately from point sources, but the number of these sources and the volume of their emissions are enormous. **Nonpoint sources** are releases that occur across a large area. For example, excess fertilizer that enters rivers and streams does so from hundreds or even thousands of acres of agricultural fields. The dead zone in the Gulf of Mexico is the result of this kind of nonpoint source pollution.

Figure 20.10 Desert propagation. Propagation of desert downwind (yellow lines) from center-pivot irrigation systems (from Okin et al., 2009).

In some cases, regulations have shifted local impacts to a much larger scale. The sulfur and nitrogen emissions from power plants can cause serious local air pollution. In order to ameliorate the effect on the nearby population, much taller smokestacks were installed in many plants. This relieved the local pollution by ejecting the sulfur dioxide and nitrous oxides much higher into the air, where it was carried away. However, the result was a shift in scale: local pollution became a regional problem as the precursors of acid precipitation were carried hundreds of miles from the point sources.

The pattern of movement and migration of animals connects disparate ecosystems, and this, too, affects scaling. For example, five species of Pacific salmon spend years in the open ocean of the North Pacific before returning to their natal streams to reproduce. The dead and dying salmon are eaten by bears, eagles, ravens, and many other species. In the process, nutrients move from the marine system to terrestrial ecosystems many miles away. A large proportion of these salmon spawn in the rivers of the vast wilderness ecosystem of the Bristol Bay area of Alaska (Figure 20.11). The proposed Pebble gold mine in the region would use cyanide to extract gold from soil and rock. The effluent of this process would be held in settling ponds, theoretically keeping the toxins from entering the ecosystem. However, in this seismically active region it is impossible to guarantee that releases will not occur. The immediate area of the mine is ecologically connected to a much wider region, including the North Pacific Ocean, by rivers and mobile species such as salmon, waterfowl, and shorebirds. The result is that local pollutants potentially affect distant ecosystems.

Another important component of scaling is the difference between local exploitation of a single species and statistical processes that affect species diversity. The Atlantic cod fishery off the coast of Newfoundland exemplifies the overexploitation of a regional resource. When Europeans reached North America, they found a staggeringly large population of cod in the waters of eastern Canada. This fishery has been all but wiped out by technical advances in fishing and years of overexploitation. The collapse of the cod population was regional; there are healthy populations of cod elsewhere in the Atlantic. Contrast this with the impact of rainforest destruction in the Amazon basin on species extinction. In Chapter 18 we analyzed the statistical relationship between habitat fragmentation and species diversity. When habitats are reduced to small island-like patches, the number of species drops to a new and lower equilibrium number known as a relaxation fauna. We can predict the proportion of species lost but not which species will disappear.

Of course, direct exploitation of a single species can result in global extinction. Humans have directly caused the demise of a number of species in this way. In North America, the passenger pigeon (*Ectopistes migratorius*), the Labrador duck (*Camptorhynchus labradorius*), and the great auk (*Pinguinus impennis*) were all hunted to extinction (Figure 20.12). Each of these species differed in geographic range and ecology, but none was able to withstand the spatial scale and intensity of harvest.

Figure 20.11 Bristol Bay rivers. The rivers that flow into Bristol Bay, Alaska, are vitally important breeding habitat for salmon. There is concern that a proposed gold mine could poison these rivers.

Figure 20.12 Auk and Labrador duck. The auk and Labrador duck were hunted to extinction early in the history of Europeans in North America.

KEY CONCEPTS 20.2

- The human population and our industrial activity are so large that many environmental issues affect the entire planet. Some globally important environmental issues are climate change, acid precipitation, ozone depletion, disruption of the nitrogen cycle, and desertification.
- Many human environmental impacts arise first as local problems. They become global due to the sheer magnitude of emissions, complex ecological interactions, and the connections among distant ecosystems.

QUESTION:

What environmental impacts are significant in your community?

20.3 What Ecological Principles Can We Apply to Environmental Problems?

Each environmental issue has a unique cause, effect, and solution. These solutions require detailed local knowledge and creative and case-specific approaches. Nevertheless, we have reached the point in our study of ecology that we can identify some general principles that might guide our approach to environmental problems.

Ecological systems do not have intrinsic homeostatic mechanisms. This concept is crucially important to understanding human global ecology. Ecological systems occur high in the hierarchy of biological systems. Figure 1.5 of Chapter 1 depicts this traditional organization of biology from the smallest units, such as cells and organelles, through the individual and eventually to the population, community, and ecosystem. An important change occurs at the level of the individual. Selection favors physiological mechanisms in the individual that ensure homeostasis. If temperature, osmotic conditions, or other chemical-physical factors exceed certain limits, there are mechanisms in the physiology and biochemistry of the individual that bring them back to tolerable levels. The fitness of the individual depends on these homeostatic processes. However, a crucially important evolutionary principle arises in this context. Natural selection does not operate at levels above the individual (with the exception of some cases of kin selection). Thus, selection does not produce homeostatic mechanisms in populations or communities. There may be processes that reduce a population that exceeds its carrying capacity or that shape the development of a plant community after disturbance. But they are the collective result of natural selection operating independently on all the interacting species. The lack of homoestasis in ecological systems is one reason nonequilibrium processes are important in ecology.

We are beginning to explore the role of nonequilibrium ecology in environmental science (Moore et al., 2008). Disturbance is an important component of nonequilibrium processes such as population dynamics and plant community structure and organization. We now understand that "the balance of nature" is ephemeral; although periods of equilibrium may occur, most systems are in stages of recovery from disturbance events much of the time. Consequently, if we are to manage and restore ecological systems, we must shift our focus from specific end points of these processes to the ecologically important *processes* in each system.

One danger that emerges from nonequilibrium ecology is the idea that if disturbance is "natural," human impacts can be ignored or accepted as simply a natural process. Of course, this ignores the immense scale and intensity of human activity relative to the disturbances that occur in natural ecological systems. Some ecologists have suggested that we distinguish between disturbance and **degradation**, the latter denoting human disruption of ecological processes.

Another important principle is the effect of species diversity on the function of ecological systems. The conservationist and wildlife biologist Aldo Leopold wrote, "The first rule of intelligent tinkering is to keep all the parts." His metaphor emphasizes the importance of the complete set of species, associations, habitat types, and interactions that make up healthy ecosystems. Consequently, concern about widespread extinction extends beyond the aesthetic or ethical consequences of anthropogenic extinction of our fellow species. Ecologists are still developing a full understanding of the relationship between species diversity, ecological complexity, and ecosystem function. But examples of their environmental importance are emerging (Ives et al., 2007). For example, recent analyses show that as ocean species diversity declines, the economically important resources collapse and water quality decreases (Worm et al., 2006). The rapid loss of species from marine systems is thus a serious concern, given the many ecosystem services provided by the oceans. To the extent that diversity, complexity, and ecosystem function are interconnected, we reduce species diversity at our peril.

degradation Anthropogenic disruption of ecological processes.

THE HUMAN IMPACT

The Relationship of Humans to the Natural World

It might seem odd to include a separate feature on the human impact in a chapter devoted entirely to that subject. Over the course of this book, "The Human Impact" features have addressed the direct practical application of the ecological principles and concepts in each chapter. Here, we have the opportunity to consider the human impact in a broader context—the nature of our relationship to the natural world and to our fellow species on Earth.

For much of human history, we have seen ourselves as separate from and independent of the natural world. A number of factors contribute to this view. First, we see ourselves as special, in some ways different from, and perhaps superior to, other living things. Our intelligence does indeed distinguish us from other species and drives our impact on the Earth. Second, we tend to see the rest of the natural world in terms of resources important to us—as a source of food, energy, housing, medicines, and even recreation. This perspective is reinforced by ecological analyses that identify the ecosystem services that natural areas provide. Intact functional habitats filter and cleanse the water we drink, absorb and sequester pollutants, store carbon that would otherwise contribute to climate change, and carry out scores of other processes from which we benefit.

Many philosophers and historians agree that our negative impact on the biosphere is a direct result of a worldview in which we see ourselves as separate from nature. This separation makes it more likely that we will view nature as a commodity, a set of resources we can own, use, manipulate, and even destroy, to serve our own interests. Is there an alternative to this anthropocentric view of our relationship to nature?

Aldo Leopold's seminal book *A Sand County Almanac* offers an alternative view. Leopold proposed a concept known as the *land ethic*. In his view, humans are an integral part of natural communities. He argues that once we place ourselves in this context, once we embed ourselves in the natural world, it follows that what we do to nature we do to ourselves. Ethical restraint and concern for ecological health develop from a combination of self-interest and the recognition that rights extend beyond ourselves to the other members of the natural community to which we belong. Leopold's is not the only answer to the question posed above. The literature on the subject is large. Many philosophical and theological treatments of the question lead to an ethical position similar to Leopold's. From these analyses we must find the basis for an approach to our relationship to the natural world and to our environment. The quality of our lives and even our survival depend on it. We will live in harmony with the natural world only when we adopt a scientifically sound and ethically defensible relationship with the Earth.

Figure 1 Aldo Leopold. Aldo Leopold, author of *A Sand County Almanac*, developed the concept of the land ethic.

QUESTION:

How would you describe your own relationship to the natural world?

Another fundamental ecological principle relevant to environmental concerns is the finite nature of most important resources. Many of the ecological mechanisms we have examined are governed by some kind of resource limitation. For example, population growth changes when the number of individuals exceeds the carrying capacity. Competition is intrinsically connected to limiting resources. The equilibrium theory of island biogeography is based on resource limits on islands that shift the extinction rate as a function of the number of species present. Limits also apply to the human interaction with the environment. There is a limit to the amount of arable land, freshwater for irrigation or direct human use, and the numbers of animals we can harvest. We use the atmosphere and oceans to dilute our emissions, but there is a limit to the amounts they can absorb. And ultimately there is a limit to the number of people the Earth can accommodate. That our world is finite seems obvious, but too often we ignore this most fundamental principle (Figure 20.13).

Finally, ecology is a science of interaction. The centrality of interaction formed our definition of ecology in Chapter 1 and underlies virtually every ecological

THINKING ABOUT ECOLOGY:

Resources are finite because the Earth is for the most part a closed system. However, the planet does receive a steady input of solar energy. Which of our most pressing environmental problems are, and which are not, affected by this external input?

Figure 20.13 Earth. Earth as seen from the moon. In this context we can appreciate the finite nature of the Earth's resources.

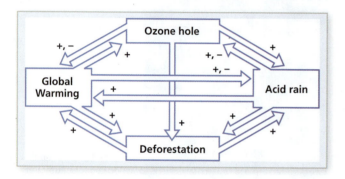

Figure 20.14 Synergies. Synergies (+ signs) among global environmental impacts (from Hart, 2007). **Analyze: How are deforestation, global warming, and acid rain connected?**

concept we have analyzed. We must therefore expect that only in the rarest cases will the human impact be local and isolated. The very nature of ecological systems means that the changes we impose will extend outward to many other systems and processes. We discussed the causes and effects of global environmental issues such as climate change and acid rain. However, many of these processes interact with one another, often in synergistic ways (Figure 20.14).

Regional impacts expand and interact with other ecosystems as well. The potential impact of pollution from the Pebble gold mine in Alaska is just one example of this concept. Consider the web of interactions among the songbirds inhabiting the deciduous forests of eastern North America and the neotropical rainforest of South America. This community of species such as ovenbirds, scarlet tanagers, and hermit thrushes breed in northern deciduous forests. As we cut and fragment those forests, the total breeding habitat is diminished and the remaining forests become isolated islands of habitat. These species reproduce with much lower success at the forest edge than in the interior. Predators such as raccoons and skunks, and nest parasites such as cowbirds, have greater access to them at the margins of their habitat. Thus, as the northern forests are fragmented, the birds' reproductive success declines. These songbirds migrate to the rainforests of the Amazon Basin. Not only are they vulnerable to environmental degradation along their migration route, their winter survival declines as a function of the deforestation of their winter habitat. Each of the processes that determine their survival and reproductive success is connected to other species and to human activity across a huge geographic scale.

KEY CONCEPTS 20.3

- The emergence of nonequilibrium ecology affects how we think about human environmental impacts.
- The resources on which we depend are finite. There is a limit to the number of individuals that can be sustainably harvested. There is a limit to the buffering capacity and diluting power of the environment.
- Species diversity and ecological complexity affect the normal function of ecosystems.
- Interaction is a central feature of ecological systems. Consequently, few human impacts are isolated. Their effects extend to other systems.

QUESTION:

What is the connection between nonequilibrium ecology, species diversity, and stability?

Putting It All Together

The two environmental disasters that struck the Gulf of Mexico exemplify key features of the human impact on the environment. Hurricane Katrina was consistent with the effect of carbon emission and the climate change we know results from the greenhouse effect. However, we are limited to a statistical connection between that storm and human industry; we cannot infer a direct causal link for this one storm. In contrast, the explosion of the Deepwater Horizon, and the subsequent pollution of the gulf, exemplifies a clearly assignable impact from a specific event. However, the long-term effects of this event are as complex as the factors that led to Katrina. Some, like the immediate bird mortality, are clear. Others, like the population trends of fish and shrimp, will play out over a long period. As ecologists, we understand that a complex web of interactions has been disrupted. Although some effects will be immediate and obvious, others will be more subtle and may not be manifest for years.

It is the complexity of ecological systems that fascinates ecologists and draws us to the study of the natural world. But this same complexity challenges our ability to preserve and restore natural systems. The environmental problems imposed by our burgeoning population may seem intimidating. Nevertheless, the tools we require are available to us. The ecological concepts developed in the preceding chapters can lead us to solutions to the challenges we face. Moreover, ecology is an evolving science. As new principles emerge and new understanding develops, we will be able to develop novel approaches to our environmental problems. These are crucial tasks because as our population grows, healthy, functional, sustainable ecological systems become ever scarcer. We lose them at our peril, for they support our economies, keep us healthy, and contribute in ways direct and indirect, tangible and intangible, to the quality of our lives.

 Summary

20.1 What Is the History of the Human Environmental Impact?
- The human impact increased when we shifted from a hunter-gatherer mode to agriculture.
- Humans impact nearly every aspect of global ecology.
- The human impact is directly related to our population size.

20.2 How Does Human Activity Threaten the Environment?
- Significant human impacts include
  - Increased CO_2 and climate change
  - Acid precipitation
  - Nitrogen deposition
  - Habitat loss
  - Desertification

- Many of these impacts are so rapid that species cannot respond.
- Pollutants enter the environment from point and nonpoint sources.

20.3 What Ecological Principles Can We Apply to Environmental Problems?
- Ecological systems do not have intrinsic homeostatic mechanisms that mitigate damage.
- Chance events, including natural disturbance, interact with human impacts.
- The resources of the Earth are finite.
- Ecology is a science of connections and interactions. Human impacts disrupt these interactions with wide-ranging effects.

Key Terms

acid precipitation p. 447
anthrome p. 439
degradation p. 452
desertification p. 449

habitat loss p. 449
nitrogen deposition p. 448
nonpoint sources p. 450

ozone-depleting substances (ODSs) p. 449
point sources p. 450

Review Questions

1. What are the key events in the history of the human environmental impact? When did these events occur?

2. What is the importance of threshold effects and synergies?

3. What kinds of factors shift local environmental problems to a global scale?

4. Why is climate change so ecologically significant?

5. What role does natural selection play in the ecological response to the human impact?

6. What fundamental ecological principles are important to understanding and mitigating our impact?

7. Why do some ecologists distinguish between disturbance and degradation?

8. How is the human impact "natural"? How is it "unnatural"?

9. Why is the human population size and growth rate so important?

10. Some environmental issues are much more difficult to rectify than others. What factors determine the difficulty in identifying, analyzing, and solving environmental problems?

Further Reading

Goudie, A. 2006. *The human impact on the natural environment*. 6th ed. Hoboken, NJ: Wiley-Blackwell.

Goudie summarizes the current state of the global environment and the most important human threats. This book is part of a series that has become an important compendium of the most recent environmental research.

Leopold, A.S. 1949. *A Sand County almanac*. New York: Oxford University Press.

This classic book provided the philosophical, biological, and ethical basis for the development of modern environmentalism and conservation. The impact of this book was profound— nearly every treatment of the ethics and philosophy of conservation is based on, or responds to, Leopold's ideas.

Letcher, T.M. 2009. *Climate change: observed impacts on planet earth*. Amsterdam: Elsevier.

This book compiles the current state of knowledge about climate change, including its causes, history, evidence for it, and its ecological impact. Each section is written by experts actively engaged in climate change research.

Wilson, E.O. 2003. *The future of life*. New York: Vintage Books.

Wilson, an important twentieth-century ecologist, explores the main environmental issues we confront in the new century. The topics he explores range from the population explosion to the loss of biodiversity. The book is a valuable compendium of human environmental impacts.

Figure Analysis: Answer Key

The following answers are provided for the **Analyze** questions posed in the captions of the corresponding figures in each chapter.

Chapter 1

1.3 Many other interactions are possible, such as parasites of any of these species, other physical factors such as floods, and interactions among the hare's food plants.

1.6 There are many such properties that emerge in the transition from one level of the hierarchy to the next. For example, some groups of cells and tissues produce nerve impulses—a mechanism of communication within the body.

1.7 No, we provisionally accept a hypothesis we cannot reject.

1.9 The lines for young and mature forest would not differ.

Chapter 2

2.4 Competition begins where the line for population crosses the line for food production. (This assumes that the food production is scaled in terms of per capita food availability.)

2.5 The genotype is the genetic makeup of an individual; the allele frequency is the proportion of an allele in the entire population.

2.6 Yes. The frequency of a dominant allele that is selected against will decline rapidly.

2.7 No. The value of p may increase or decrease.

2.8 The allele may be both entering and leaving the gene pool. If they enter and leave at the same rate, there is no net movement of alleles. *Net* movement refers to the situation when the immigration of certain genotypes exceeds its emigration or its emigration exceeds its immigration.

2.9 The frequency of the dominant allele would decline rapidly to zero.

2.10 The depth of the beak plays a significant role in the pressure exerted by the bill and thus in its ability to crack open a seed.

2.11 Many phenotypic factors can affect human height, including nutrition, disease, and toxins.

2.13 Yes. Natural selection operates on heritable variation.

THE HUMAN IMPACT The Evolution of Resistance in Pathogens and Pests

Fig. 1 It may take some time for mutations for resistance to arise. But once they do, they rapidly increase in frequency.

ON THE FRONTLINE Genetic Drift in a Desert Plant

Fig. 1 The allele frequency would make a rapid shift right at the boundary between the slopes.

2.14 There is an equal chance that an allele will be lost or fixed. Thus, an equal number of populations will experience loss and fixation.

2.15 Yes. Although the effects of drift are reduced in large populations, it occurs in large populations. Given sufficient time, loss or fixation can occur in a large population.

2.19 To move to that point, the fitness would have to decrease. Natural selection cannot move allele frequencies to regions of lower fitness.

Chapter 3

3.2 Salmon: the salmon's body temperature would change with the ocean temperature. Killer whale: the whale's temperature would be constant across a wide range of ocean temperatures.

3.3 Most factors have inherent variation around the mean. Random factors tend to occur equally above and below the mean.

DO THE MATH Working with Bell Curves

Fig. 1 The variance is a parameter that is similar to the average difference between each value and the mean. The more different the values are from the mean, the higher the variance.

Fig. 2 The narrow bell curve has a small variance; the broader bell curve has a high variance.

3.5 In both cases the organism must be able to respond to or survive different environments. Unpredictable changes in the environment are particularly difficult for the organism to adapt to. Often there are no cues the organism can use to predict a change in the environment.

3.8 Across a range of body size, survival is higher in hibernators than nonhibernators.

3.10 Any factor that caused an increased energetic cost to flight, such as headwinds or predator attacks, might increase heart rate. Tailwinds or periods when the bird can glide might decrease heart rate.

3.12 One advantage is that it requires less energy investment by the animal. One cost is that the animal must be able to locate an appropriate environment for passive warming. It might also be vulnerable to predation during this activity.

3.14   Often the variation in temperature is greater at higher elevation. This would select for the ability to acclimate to temperature change.

3.15   Behaviors that limit heat loss—such as curling up to reduce the exposed surface area, fluffing feathers, or seeking places that are less thermally stressful—would help maintain the animal in its thermal neutral zone.

3.16   In a concurrent system, the temperature differential between the two vessels decreases all along their paths. Because heat flow is determined by the temperature differential, this reduces the flow of heat from the vessels back to the body and reduces the retention of heat.

3.18   In laminar flow, new cooler air is constantly passing over the leaf. This maintains the maximum temperature differential between the leaf and the air and thus maximizes heat loss.

3.21   A freshwater fish placed in saltwater would tend to lose water to its environment.

3.23   If the stomata are closed to avoid water loss, the concentration of CO_2 in the leaf decreases. The higher affinity of the C_4 enzyme for CO_2 means that the plant can continue to fix carbon when the concentration drops.

ON THE FRONTLINE   The Effect of Salinity on Photosynthesis

Fig. 2   Yes. The difference in net photosynthesis between high and low salinity is greatest at high light intensity.

Chapter 4

4.1   At each elevation, the north slope is cooler than the south slope. Thus, the vegetation zones occur at lower elevations on the north side of the peak.

4.2   When sunlight strikes the ground at a 90° angle, all the energy is concentrated in a small area. The smaller the incident angle, the larger the area over which the solar energy is absorbed.

4.3   In the Northern Hemisphere, the days are longer when the Earth tilts toward the sun. The sun crosses the sky more directly overhead in summer, which increases the daily temperature. In winter its path is lower, further south, and sunlight strikes the earth at a shallower angle and for a shorter time each day.

DO THE MATH   The Concept of Statistical Significance

Fig. 1   Because the values of each group overlap, the means of the two populations may not

actually differ; they might be the result of chance sampling errors.

Fig. 2   Large values of t occur when the differences between the means are relatively large and the variance in each population is relatively small. This is less likely under the null hypothesis. Thus, large values of t lead us to reject the null hypothesis.

4.4   This air has lost much of its moisture in precipitation near the equator. Also, as the air descends, it warms, and warm air is able to hold more moisture, thus decreasing precipitation.

4.6   Like mid-latitude desertification, the air descending the lee side of a mountain range has lost much of its moisture, and as it warms, it can hold more moisture, thus decreasing precipitation.

THE EVOLUTION CONNECTION   Historical Effects on Species Distributions

Fig. 1   There might be more opportunity for gene flow among populations on a large continent than on small continents separated by oceans. Each small continent might have different environmental conditions that select for different kinds of adaptations.

Fig. 2   Each of three possible scenarios (a, b, and c) could arise from this single phylogeny. The key is when in the phylogenetic history the range arises. For example, in (c) each closely related species pair has a different range. These differences arose early in the history of each species pair and persist after they diverge. In contrast, in (a) the ranges diverge at the time that each species pair splits.

4.8   It is the specific events, freezes lasting more than 36 hours, that cause mortality. This parameter is not captured in the mean temperature. It is possible to have a high mean temperature but still have 36-hour freezes.

THE HUMAN IMPACT   Global Warming and Ecological Communities

Fig. 1   Much of the naturally occurring CO_2 in the atmosphere is derived from the respiration of plants and animals, which releases CO_2. Manmade sources include the combustion of fossil fuels.

Fig. 2   Natural variation in temperature is less at low latitude. Thus, low-latitude plants are less adapted to temperature variation and probably would be affected more by anthropogenic climate change.

ON THE FRONTLINE Fire and Grazing in the Prairie Community

Fig. 2 The overlap of communities G (four-year burn) and Q (grazed) indicate similar community composition.

4.10 If the species cannot physiologically tolerate the conditions, dispersal and biotic interactions are irrelevant.

4.11 One possibility is that the Himalaya, a massive mountain range in northern India, blocks dispersal and creates different environments.

4.12 Many of the biomes are arranged in latitudinal or longitudinal bands. We know that temperature decreases with latitude. In North America, rain shadow effects determine precipitation in longitudinal bands from west to east. These patterns suggest the importance of climate.

4.13 Each species is probably responding to its own specific abiotic and biotic limits and requirements. Thus, their range limits differ even though all are found in the prairie biome.

4.14 The plant's growth and reproduction might occur in the months January–March and November–December, when there is adequate moisture. It might be advantageous for the plant to be dormant during the dry period of May–September.

4.15 Certain combinations of temperature and moisture are benign; others are stressful for plants. As temperature changes, the limitation of moisture changes, and vice versa. Thus, there will be unique polygons of temperature-moisture at which different groups of plants can survive.

Chapter 5

5.1 Virtually all the components may be affected by climate change. Wind patterns and temperature affect both evaporation and precipitation, and these processes drive the movement of water.

5.2 Each of the highest streams is order 1. Each of the streams formed by their confluence is order 2.

5.3 Oxygen is added to the stream by photosynthesis and as dissolved O_2 from the air. Respiration of organisms, including decomposers, reduces the oxygen concentration. The temperature of the water determines how much oxygen it can hold.

ON THE FRONTLINE The Relationship Between the Environment and Dispersal in Aquatic Plant Communities

Fig. 1 It looks as though the line will level off beyond 160 km. At some point, distant communities may be maximally distinct and the similarity reaches a minimum value that no longer changes.

5.4 Water is maximally dense at 4°C. As a result, as the upper water cools, it sinks and mixes the lake. Also, because ice at 0°C is less dense than water just above freezing, the lake freezes from the top.

5.5 The quantity and quality (wavelengths) of light and water temperature change with depth. Wave action and tides affect near-shore communities. Ocean currents affect each of these zones by moving nutrients.

5.6 Photosynthesis occurs only in the upper, photic zone. It is the carbon fixed in the upper part of the water column that sinks and decomposes.

DO THE MATH Linear Regression

Fig. 1 Other factors besides latitude determine temperature. Also, some points may be affected by chance sampling error.

Fig. 2 No. The correlation means that the two variables are related, but correlation does not necessarily mean that the two variables are causally related.

5.7 As the wind moves across the surface, it pushes the surface water in the direction of the wind. This water is replaced by upwelling of deeper waters.

5.8 The direct effect of wind is most prominent at the surface. Friction reduces the deflection at greater depths.

5.10 The relative sizes of the zones are determined by the shape of the pond. Steep sides reduce the size of the littoral zone and increase the size of the pelagic and profundal zones. If the slope of the shore is gradual, the littoral zone will be larger.

5.11 Ocean temperature changes with latitude, and temperature drives many other physical factors, including ocean currents.

5.12 Marine biomes are characterized primarily by physical factors rather than by the dominant vegetation type.

THE EVOLUTION CONNECTION Not All Characteristics Are Adaptive

Fig. 1 The forelimbs are used for locomotion—forward motion as well as changing direction and depth.

Fig. 2 There are several possible advantages. An aquatic larval stage can take advantage of the productivity available in aquatic systems. Also, in high-altitude streams the winter temperatures may be less extreme than those on land.

Chapter 6

THE EVOLUTION CONNECTION Measuring the Contributions of Environment and Genetics to Behavior

Fig. 1 Environmental factors, including random processes, also affect the phenotype.

6.5 Dimorphic features other than size difference can lead to sexual selection. Also, female choice may be more intense in some species than others.

6.7 The symmetry of the tail is relatively conspicuous. If it correlates with male fitness, it is a good trait for females to use to assess potential mates.

6.8 Short tails are less attractive to females. Long tails may be an impediment to movement, including predator avoidance.

6.10 Energy investment through parental care may accelerate development. This would be an advantage in short-lived small pools.

ON THE FRONTLINE The Nature of Habitat Selection

Fig. 2 Different factors or processes that affected the territories in previous years but not during the experiments might introduce error.

6.11 Any changes in the distribution or amount of important resources, such as food, nest sites, or mates, might cause the home ranges to change.

6.13 The distribution of resources and their daily availability could change the portion of the home range used each day. For example, if resources appear and disappear from parts of the home range, it could affect the pattern of use.

6.14 Not necessarily. Larger male territories might include more females even if they are not part of harems.

6.15 Yes. Selection requires variation on which to act.

6.16 The total amount of dispersal would increase with density. The proportion of dispersers would remain relatively constant.

6.17 The similar slopes imply that regardless of the relative size of the litters produced by different groups, the quantitative impact of inbreeding is the same.

DO THE MATH Using Hardy-Weinberg Analysis to Study Inbreeding

Fig. 3 Each parent has two alleles. The probability that an offspring will receive either is ½.

6.18 Yes. As the number of mating males increases, the number of different alleles that an offspring can receive from the male also increases. In this case, the value 1 in the male contribution ("male has just one set to give") becomes less than 1, thus reducing the coefficient of relationship among sisters.

Chapter 7

7.3 A species that requires several different microhabitats would be more likely to be coarse-grained because its travels would take it to a number of habitat types. A species that can survive in a single habitat might be more likely to be fine-grained.

7.4 "Variation" simply refers to the fact that the environment is not constant. It might change on a daily or seasonal basis. "Unpredictability" refers to the fact that the environment changes but in a random or irregular pattern.

7.6 The trend might be reduced in migratory species if they encounter other habitats where shorter wings are advantageous.

7.8 In some races the color patterns might provide camouflage. Others are brightly colored and could be aposematic.

7.9 The difference between shade and sun leaves has an important environmental component; it is not strictly genetically controlled.

ON THE FRONTLINE Melanism and Phenotypic Plasticity in Beetles

Fig. 2 Dark, melanistic coloration absorbs heat.

7.10 If selection favors the rare morph, it will increase in frequency. At some point it will become the more frequent morph. Selection will then favor the other morph that is now rare.

7.12 Smaller plants may flower earlier during the short growing season. Winds are less strong near the ground, which may facilitate the movements of their pollinators. In winter, short plants may be covered by snow and thus protected from wind and cold.

7.13 Not necessarily. It may be that some other factor that is correlated with temperature is important.

DO THE MATH Measuring Genetic Variation

Fig. 1 The variation within populations is greater than the variation between populations.

Chapter 8

8.2 If an individual mouse encounters several of these habitat types, it represents a coarse-grained habitat.

8.4 Many factors can determine the dispersion pattern. The distribution of important resources such as food and nest sites, concentrations of predators, and favorable abiotic factors contribute to the dispersion pattern. Positive or negative interactions among conspecifics can also determine the proximity of individuals to one another and thus the dispersion pattern.

8.5 The dispersion of chollas could be determined by many factors, including local soil moisture conditions, the presence of competitors, or the dispersal patterns of seeds.

8.6 The Poisson distribution applies to rare, random events.

8.7 A larger or smaller grid size can change the apparent dispersion pattern. For example, if any of the squares in the top panel with many individuals were overlaid with many much smaller squares, the dispersion might appear random. If in the bottom panel a much smaller set of squares were used, many would have no individuals and the dispersion might appear random.

8.8 Each of the species could be randomly dispersed—each appears at least somewhat random. You cannot tell for certain without mathematical analysis.

8.9 Heavy seeds that are attractive to seed predators and that simply fall near the tree would be most affected by this process. Any mechanism (wind or animal dispersal) that carries them farther from the parent tree would reduce the opportunity for predator concentrations near the parent.

8.10 Each peak in the numbers of individuals represents the establishment of a new forest following fire. Thus, for example, in (a) a fire occurred around 1500. In (c) two fires occurred, one in about 1500 and another around 1800.

8.11 Many mortality factors increase with age. For example, the cumulative effect of injuries or increased parasite load can play a role directly or by increasing the risk of predation. If an individual loses social status in old age, it might not have access to important food resources.

8.12 By using % of life span on the x-axis, we can compare the survivorship curves of species with different total life spans on the same graph.

ON THE FRONTLINE Herbivores and Trillium Sex Ratio

Fig. 1 It would be difficult to even obtain graphs like these because deer avoid jack-in-the-pulpit.

Fig. 2 The effect is indirect: it occurs through the effect of deer browsing on trillium and its impact on soil quality.

8.13 Habitat quality is an important factor in mortality and thus should affect the survivorship curve. Males and females differ in the amount and timing of mortality risks. Males incur mortality from mating activities such as fighting other males or guarding harems. Females experience mortality on the nest or from the effort of caring for young.

8.15 Many factors could ultimately limit the expansion of the dove, including abiotic conditions, potential competitors, and the geographic range of important food resources.

8.16 It would be a straight line.

8.17 There is mortality in each season. But in the nonreproductive season, the mortality is not offset by births. As a result, the population declines seasonally.

THE HUMAN IMPACT Human Reproductive Rates

Fig. 1 Males and females face different age-specific mortality risks. Females may experience higher mortality during the reproductive years. Male mortality is affected by culture-specific factors such as the nature of the work environment and even war.

Fig. 2 The death rate is determined by factors such as nutrition and disease. The birth rate declines because the desired number of children declines. The decline in mortality generally precedes the decline in birth rate, leading to a large difference between the two.

8.18 Reproductive parameters such as the number of offspring produced, the age of sexual maturity, and the number of reproductive events per lifetime contribute to r.

8.19 The total population at $t = N_0 + N_1 + N_2 + N_3$.

8.20 The association between these variables is measured with the correlation coefficient.

Chapter 9

9.1 If an animal's health is compromised by poor nutrition, it might be less able to fight parasitic infection.

9.2 No. It is possible for some regulatory factors to operate top-down and others to operate bottom-up.

9.4 No. Additional information would be necessary to determine if the crash was from mortality, reproductive failure, or both.

9.5   It is possible for mortality to decrease with density up to a certain point if there are advantages to being in a large population, such as decreased predation or increased success in locating food. However, most of these kinds of advantages pertain more to group size than total population size.

9.6   As the parasites are gradually eliminated from the grouse population, their effect is reduced and the population fluctuates less.

9.7 K for most species probably changes over time. We assume a constant K over the period of our analysis.

9.8   If the response time of the negative feedback system is rapid, it can lead to a relatively constant population. In other words, if mortality and reproductive changes occur soon after the population size changes, the result will be a stable population. Population cycles are more likely to occur if there is a time lag in the negative feedback mechanisms.

9.9   Past events may be more important to current mortality or reproduction than recent events. For example, it is possible that the number of eggs produced by a female is determined by her fat stores at the time she begins reproductive development rather than immediately before mating.

9.10 Density-independent factors tend to be external factors that are not affected by the population. Examples include abiotic factors such as weather, fire, and soil chemistry.

9.11 There are potentially some direct effects, but the more significant impact appears to occur through the food supply.

9.12 The variation in rainfall is a density-independent factor. Its effect can be great enough that it overrides all other factors, including density-dependent factors.

9.13 The scaling of the y-axis determines how large the population fluctuations *appear*. If the scale is expanded, the apparent size of the fluctuations is smaller than if the scale is reduced.

9.14 It is not possible to tell from these graphs. It could be that as the hares increase in number, the lynx population increases in response. Or it could be that when the lynx population increases, it drives down the hare population. These processes are not mutually exclusive.

9.16 It appears that both populations and especially the moose are in a phase of increased instability. Now that we know many of the factors that affect both populations, it is possible to predict their future trends. But this requires detailed knowledge of the condition of the vegetation, parasites, pathogens, and so forth.

ON THE FRONTLINE Long-Term Studies of Song Sparrow Population Dynamics

Fig. 1 There could be direct mortality effects of rainfall, especially if it causes hypothermia in the young.

9.17 Local extinction has a relatively small effect on the total metapopulation size. However, if the extinction occurs in prime habitat, dispersal from other areas to this region will occur. Thus, the numbers and movements of the entire metapopulation can be affected.

9.18 The chains of small islands provide "stepping stones" for dispersal from one region to another. The effect is to increase the probability of successful dispersal.

9.19 In general, any of the factors that operate in density-dependent fashion tend to decrease the variation in the population size. Density-independent factors, especially if their effects are random and significant, tend to increase the variation in the population size.

Chapter 10

10.1   The optimal allocation strategy balances the costs and benefits of each "decision" in this diagram such that the total genetic contribution to the next generation is maximal.

10.2   Not necessarily. The curve could be skewed to one tail or the other.

10.3   The effect of natural selection is greatest when V_x is maximum.

10.4   Direct development requires the genetic control of only a single path of development. No energy is required to reconfigure the organism from one body form to another. Larval stages can, however, take advantage of energy available in certain environments and can reduce competition between juveniles and adults.

10.9   The organism faces none of the mortality costs of reproduction, and increased size often decreases mortality.

10.10 There might not be sufficient food for larger clutches. They also might be more conspicuous and thus vulnerable to predators.

10.11 The sham operations are controls that demonstrate that the effect of the treatment was not due to the effects of the surgery without the change in egg number.

10.12 There is no additional cost in terms of weaning weight to litters greater than four. If the young can survive at this weaning weight, the larger litter size is not a disadvantage.

10.13 The trade-off between clutch size and off-spring size follows a different trend in these two groups. In the teiids (the whiptail lizards), there is less difference in clutch size between small and large offspring; in the horned lizards, there is a greater difference.

10.14 The Northern Hemisphere land mass and habitats extend to greater latitude than the southern continents. Any latitudinal effect would be exaggerated at the higher latitudes available in the Northern Hemisphere.

ON THE FRONTLINE The Importance of Predation to Clutch Size

Fig. 1 The location of the nest in the habitat might determine how conspicuous it is or how accessible it is to predators. Its construction might also determine how conspicuous it is in the background of the habitat.

Fig. 2 Because body size, nest size, and clutch size are all related, it was necessary to remove the effect of body size in order to establish the relationship between just clutch size and nest size.

10.15 The three axes are probably not entirely independent. For example, competitive ability and stress due to limiting resources are interrelated. Similarly, competition is an important component in the recovery of a plant community from disturbance.

THE HUMAN IMPACT The Effect of Life History on the Risk of Extinction

Fig. 1 Elasticity analysis measures how responsive one variable is to change in another. A variable that changes more rapidly with another is elastic.

10.16 Not necessarily. Components of each strategy could be phenotypically plastic and change in different environments.

10.17 The point where the lines cross represents the degree of unpredictability at which the optimal strategy switches from iteroparity to big bang reproduction.

10.18 It is presumed that variation in reproductive success is the selective force and the proportion of seeds that germinate is the adaptive response.

10.19 The population on South Georgia experiences many complete reproductive failures. If the birds invested significantly in every reproductive event, they might lose the entire

investment in some years. This selects for a bet-hedging strategy.

Chapter 11

11.2 It is indirect competition: the fish are not competing over the time at which to feed; they are competing over good feeding sites. The salmon shift their activity pattern to get access to feeding sites.

11.3 Over time the density would not decrease and plant size would not increase.

11.4 No. There could be other interactions by which *Semibalanus* limits *Chthamalus*. The photographic evidence of *Semibalanus* directly interfering with the growth of *Chthamalus* is important to demonstrate the direct competitive interaction.

11.5 The dotted line represents the reproductive performance of each species based solely on the number of individuals of that species present. It thus represents the null hypothesis—seed production in the absence of competition.

11.6 Intraspecific competition is demonstrated by the fact that the growth of each species declines as a function of the density of conspecifics. By itself, this graph does not demonstrate whether intraspecific competition is greater than interspecific competition. Additional information is needed to distinguish their relative importance.

11.8 The data suggest that tundra plants partition nitrogen resources according to the relative abundance of different forms of nitrogen.

11.9 The volume represents the total size of the niche. The distance between the niche volumes represents differences in resource use by the two species.

11.10 Competition is most intense in the region of niche overlap.

ON THE FRONTLINE Distinguishing Resource and Nonresource Competition

Fig. 1 Asymmetrical root production refers to differences in the performance of the roots that experience different conditions.

11.11 Plants that experienced disturbance grew taller and their seeds dispersed farther.

THE HUMAN IMPACT Competition and Invasive Species

Fig. 1 The invasive population tolerated herbivory better than native populations. The invasive population was also better able to exploit high resources.

11.12 Not necessarily. One species might maintain the same total niche size in competition, whereas the realized niche of the other is smaller than its fundamental niche.

11.13 It would suggest that the two cannot coexist but some factor, including perhaps chance, determines which species wins.

11.14 The speed of the niche shift will depend on any factors that accelerate or retard the rate of adaptive change: the intensity of selection, the genetic variation on which selection acts, and whether drift or gene flow opposes the action of selection.

11.15 The control, the niches in the absence of competition, is the resource use of each species in allopatry.

11.16 Any factor that selects for a shift in the nature of the song could account for these results. For example, the acoustic environment in the sympatric site might select for either high- or low-frequency songs.

DO THE MATH   How to Quantify Niche Overlap

Fig. 1 This allows us to calculate the discrete proportion of each resource type used by the species.

11.17 *P. aurelia* always wins over *P. caudatum*. In the graph, Species 1 always wins over Species 2. Therefore, Species 1 is *P. aurelia*.

11.18 The relative sizes of K and α determine the x- and y-intercepts of the isoclines. This determines their relative positions and slopes and thus the result of competition.

11.19 The isoclines are not linear functions of the relative numbers of the two species; at some densities the effect of the other species is greater, and at other densities it is less. This suggests a density-dependent phenomenon—that α is not constant.

Chapter 12

12.2 If a third species is similar to one of the other two, it might accelerate coevolution. Imagine a predator-prey system. A second predator that is adapting to the prey would constitute another selective force on the prey, increasing its adaptations to its predators. However, in this scenario, a second prey species, very different from the first, might slow the adaptive changes in the predator because it is faced with different and perhaps conflicting selection pressure.

DO THE MATH   How Long Should a Predator Remain in a Patch?

Fig. 1 The tangent line represents the rate of energy gain per unit time at that point on the curve.

Optimal foraging theory determines the strategy that maximizes this rate.

Fig. 2 The slopes of these lines are lower than the slopes for other strategies.

Fig. 3 The maximal energy gain in a low-energy patch (the staying time at the tangent line) occurs at a shorter staying time than it does in a high-energy patch.

12.7 The lines are parallel because as lizard size increases, *both* the maximum and minimum prey size increase.

12.8 Yes, it is theoretically possible, although it seems to occur rarely.

12.9 If handling time is small, the predator can consume many prey items very quickly. As prey density increases, the rate of consumption can increase rapidly.

12.10 The handling time for moose is large. At some prey density, wolf consumption of moose is not limited by access to moose but by satiation and the long handling time.

ON THE FRONTLINE   Eyespots as Deterrents to Predation

Fig. 2 The experimental eyespots were chosen to determine whether spots in patterns that do not specifically resemble eyes can deter predation.

Fig. 3 Large spots and multiple spots were the most effective deterrents of predation.

12.13 Neither is an exact mimic of the coral snake. However, each mimics some aspect of the coral snake. The ancestors of the two species of king snake perhaps had different color patterns; each was modified by selection to match a slightly different element of the coral snake pattern.

12.14 Perhaps, but it also might simply be due to the decrease in resources available in trees that have been defoliated.

12.15 One important benefit is the ability to use intermediate hosts to produce large numbers of infective stages. This increases the probability of an infective stage finding a host. One cost might be the requirement for all intermediate hosts to be present and sufficiently abundant for the parasite to complete the life cycle.

THE EVOLUTION CONNECTION   The Phylogenetic Relationship of Parasites and Hosts

Fig. 1 The phylogeny constructed by parasites differs from the phylogeny based on the genetic relationships of the primates. In other words, some parasites are found on unrelated primates. This suggests they have switched hosts.

Fig. 2 It suggests that the host switching in this group is not evolutionarily difficult—it has apparently happened multiple times among groups of birds and mammals.

12.19 It might be difficult to experimentally remove the features of the acacia that benefit ants without perhaps causing the ants to abandon the plant. One could obtain indirect evidence of the benefits to the ants by comparing the survival of related ants that do not inhabit acacias.

12.23 When many species interact in complex webs such as this, it suggests that conservation efforts will require that all components of the community must be protected. If essential species are lost, other species might disappear as well. Eventually, the community might collapse.

Chapter 13

DO THE MATH The Null Hypothesis in Community Structure

Fig. 1 In the neutral model, any species that can tolerate the physical conditions can be a member of the community. In that case, it is dispersal that brings species into the community and environmental filtering sorts out which can stay.

13.1 The sizes of coexisting species would be random; some species of similar size would coexist.

13.2 Competitive interactions in a species-rich community would probably form a network. With so many potential competitors, it would be unlikely that they could be arrayed in a long hierarchy.

13.3 If some species can also use other resources, they might be able to overlap in their feeding microhabitats. Or, in some microhabitats, the resources might be sufficiently abundant that species with significant overlap can still coexist.

13.4 Increased overlap will be limited by resource levels: if resources are in very short supply, little overlap can be tolerated. Decreased niche breadth is limited by the species' ability to obtain sufficient resources. If the niche is too narrow, there may not be enough resources available to sustain the species. Ultimately, there is a limit to the range of resources available in a niche. As a result, there is a limit to the increase in the range

of resources that can be used to avoid competition.

13.5 No, each resource may be limiting if it is the only one available. But the species' use of the combination of two resources may reduce the degree to which it is resource limited.

13.6 Generally, the tropics are a more benign abiotic environment. As a result, one might expect that high-latitude species that are more directly affected by the physical environment would be more subject to environmental filtering.

13.8 Environmental filtering would select for species with similar traits that enable them to tolerate the conditions. Phylogenetically related species might be more likely to share these traits.

13.9 You would predict that the species that have the most distinct foraging niches are probably the most phylogenetically similar.

13.10 Phylogenetic distance is often negatively correlated with niche similarity: the more phylogenetically distinct a species is, the less similar it is likely to be to other species in the community. Thus, phylogenetic distance and co-occurrence are positively correlated.

THE EVOLUTION CONNECTION Community Heritability and Structure

Fig. 1 The plant and its herbivores are engaged in a series of adaptations and counteradaptations. If the plant develops a novel and effective means of protection, the plant may essentially be freed from herbivory unless and until the grazers can adapt to the new defense.

13.12 The substrate algae compete with the invertebrates for space. When the kelp shades the substrate algae and decreases its cover, the invertebrates are released from competition.

13.14 Yes, the two prey species may also compete with one another.

ON THE FRONTLINE Niche Shifts After Competitive Release

Fig. 1 Prediction 1: TNW increases due to increases in individual niche breadths. In Figure B, the niche of each species increases, and the summed niches lead to increased TNW. In Figure C, the niches are broader with greater overlap, but TNW does not increase. Prediction 3: Increased TNW but not individual niche widths. In Figure D, the decreased niche overlap among species causes TNW to increase.

Fig. 2 Release from competition with trout led to increased specialization, and TNW increased. This is consistent with Prediction 3. Release from sculpin competition did not significantly change TNW. This is consistent with Prediction 2.

13.15 Core species should have broad environmental tolerance. They should be competitively dominant. Satellite species might be expected to be sensitive to the abiotic and biotic environments—that is, they can persist only under specific conditions that may be short-lived.

13.16 Not all of the communities in a metacommunity are necessarily structured in the same way. In some, the niche relationships of species may permit stable coexistence. In others, due to resource limitation or overlap in the niches of the species present, local extinction due to competitive exclusion may periodically occur. This would be a nonequilibrium process. In addition, disturbance may alter some of the communities at random, shifting the species present and their niche relationships.

Chapter 14

14.4 The longer the time since a fire, the more fuel accumulates. Also, any mortality factors that leave dead, standing trees provide a conduit for flames to reach the canopy to form a crown fire.

14.6 Strong wave action can affect corals in shallower water. Also, sediments carried by wind may have an impact. The massive influx of rainwater may temporarily affect salinity and ocean temperature.

14.8 Many of these storms follow tracks that run from the southwest to the northeast.

14.9 They would fall far out on the intensity axis.

14.10 We must assume that the processes of succession have been the same over the entire time span of the chronosequence. That is, we assume that the same sequence of communities occurred in the past as occurs today.

14.15 Large and intense disturbance tends to be relatively rare. Thus, the current view of disturbance emphasizes the importance of frequent, low-intensity, small-scale events.

14.16 Chance is probably more important early in succession because the specific plants that colonize affect the future course of succession.

14.17 Small-scale disturbance and rapid recovery relative to the disturbance frequency are more likely to lead to an equilibrium community.

14.18 The boundary of the prairie peninsula was probably unstable because it depended on the pattern of prairie fires. Fires pushed the boundary east; in the absence of fire, trees moved further west. The ecotone probably shifted frequently.

■ **ON THE FRONTLINE** Equilibrium and Climax in Spruce-Fir Forests

Fig. 1 Any attributes that make fir a superior competitor under climax conditions, such as shade tolerance or the ability to acquire space and nutrients, might lead to its numerical dominance. Other factors, such as herbivore resistance or allelopathy, could play a role.

■ **DO THE MATH** Markov Chains

Fig. 1 In a Markov chain, transition probabilities depend only on the current state. In some cases, it is difficult or impossible for a state to change to certain other states. In succession, it is unlikely that a state will revert to a previous state unless a disturbance occurs.

14.19 As dead plant material decomposes, nutrients are released. Immediately after disturbance, there may be few plants to absorb them. The nutrients can be carried from the system by water that leaches them from the soil.

14.20 The developing plant community casts shade that reduces light input below the vegetation. As succession proceeds and the number and diversity of species increase, nutrients are more rapidly absorbed by the vegetation.

14.21 At the intersection of the lines, the early and late species have the same positive energy balance. Neither has a competitive advantage at this point.

14.22 The leaves on a tree do not occur in clear, discrete layers. The layering of leaves and the interception of sunlight by them is a statistical phenomenon.

14.23 Vegetation tends to reduce the daily temperature variation—shade cools the ground during the day, and the vegetation reduces heat loss from the soil at night. In disturbed sites, especially where most of the vegetation has been removed, the daily temperature fluctuations are not modified in these ways.

14.25 This model uses the same concept of the difference between early and late species' ability to obtain positive resource balances to drive successional change. It extends the previous model by considering two key resources: light and nutrients.

14.26 Often, early colonists are found in very high numbers. This may attract species-specific herbivores.

Chapter 15

15.1 The optimal sampling effort is the point where the curve levels off. Before that point, the sampling is insufficient to discover all the species present. After that point, additional sampling will discover no new species.

15.2 Abundances range from 2^1 (=2) to 2^8 (=256).

15.3 The line with slope 0 indicates that regardless of abundance rank, the proportional abundance does not change. This means that even abundant species (by rank) do not represent a large proportion of the total number of species. This can be true only if species are relatively equally abundant. On the other hand, if the proportional abundance changes significantly at higher abundance classes, it means that the most abundant ranks represent a high proportion of all species abundance—a few species are numerically dominant.

15.4 The data suggest that a few species dominate the community.

15.5 Each taxonomic group will respond differently to the abiotic and biotic conditions of the region.

15.6 Because endemic species are found in only one place, local extinction is the same as global extinction.

15.8 This kind of line is derived from sampling the number of species on islands of different size. The y-intercept denotes the number of species found on the smallest islands. The sample data might indicate that an island of size 0 should have some number of species. Obviously, this cannot be so—the line must actually pass through the origin.

ON THE FRONTLINE **What Determines the Length of a Food Chain?**

Fig. 1 Some may perform key ecological functions. Ecosystem engineers and keystone predators would be examples.

15.9 This would require experimental manipulation of plant diversity in one experiment and fungal diversity in another.

THE HUMAN IMPACT **The Sixth Mass Extinction**

Fig. 1 If many species before the Cambrian explosion were soft-bodied, they would not have left many fossils. The origin of many new species with hard exoskeletons, which are more likely to be preserved as fossils, would contribute to the sudden change in biodiversity.

Fig. 2 About 5–7 million families per million years.

Fig. 3 You would need to be able to represent the extinction rate in the same units as the historical mass extinctions.

15.10 Yes. It might be that until a threshold number of species is present, competition is not intense enough to cause extinction.

15.11 The distance of an island should not change the resource levels of the island. Thus, there should be no effect of distance on the extinction rate.

15.12 There should be no turnover—once a species arrives on the island, it remains there, under a nonequilibrium model.

15.13 No. Large, near islands should recolonize more rapidly than small, distant islands.

THE EVOLUTION CONNECTION **How Species Arise**

Fig. 1 Hybrids are likely to contain genes from each of the two environments. This might make them less fit in either habitat.

15.14 There should be more potential bird niches in more-diverse vegetation structure.

15.16 The Court Jester model requires random disturbances. These will be manifest on larger spatial and temporal scales.

15.17 When disturbance is rare, competitively dominant species may reach high density and cause the local extinction of other species. We have seen this in other contexts—invertebrates in the rocky intertidal and succession in kelp forests.

15.18 Fire is beneficial to many grasses, whose meristem is belowground, where it is protected from fire. Woody plants, which do not have this adaptation, are often suppressed or killed by fire. Thus, fire promotes grasslands and inhibits forests.

Chapter 16

16.3 In the Antarctic winter, when light levels are low, photosynthesis is reduced. Krill numbers decline and there is less energy available to whales. Many whales undertake seasonal migrations because of the changing availability of energy.

16.6 Energy enters some ecosystems as organic matter produced in other ecosystems. For example, river systems receive energy inputs of organic matter from land. Ultimately, however, all energy in ecosystems is derived from autotrophs.

16.7 One hypothesis is that the number of whales is so small that there is not sufficient energy in that trophic level to support another, higher trophic level.

16.8 There is more energy available to humans that consume anchovies than tuna.

16.9 Because other species consume krill, not all of the energy in krill is available to blue whales. This might reduce the numbers of blue whales.

16.10 In marine systems, the highest production occurs along the coasts; the open ocean is relatively less productive. In terrestrial systems, production is lower in desert regions and at high elevation and latitude.

16.11 Evaporation is generally higher at higher temperatures. High precipitation results in high evaporation from soil and more water available to plants. This, in turn, leads to higher transpiration.

THE EVOLUTION CONNECTION The Evolution of C4 Photosynthesis

Fig. 1 The higher affinity of the enzyme for CO_2 in C4 photosynthesis means that C4 plants would be able to fix carbon at lower atmospheric levels of CO_2.

16.13 Any energy that does not go directly to growth or reproduction—that is, energy that goes into maintenance activities such as homeothermy—is lost.

16.14 The effect is due to the formulas for surface area ($4 \Pi r^2$) and volume ($4/3 \Pi r^3$). The smaller the radius is, the greater the difference between the surface area and volume.

DO THE MATH Allometry

Fig. 1 In positive allometry, $a > 1$. This means that the value of y increases with increasing values of x. In negative allometry, $a < 1$; that is, the exponent is a fraction smaller than 1.0. In that case, the value of y decreases as a function of x.

Fig. 2 The value of $a < 1$, so this is a negative allometry.

16.15 This has puzzled ecologists for some time. Most ecologists infer that energy is limiting in many systems; there are few systems with large numbers of trophic levels. The narrow range suggests that ecosystems are constrained by energy input and efficiencies or some other factor.

ON THE FRONTLINE The Length of Food Chains

Fig. 1 Productivity hypothesis: No effect of size; therefore, slope = 0. Production limits trophic levels; therefore, there is a positive slope of food chain length and production. Ecosystem size hypothesis: Ecosystem size determines food chain length; therefore, there should be a positive slope of food chain length and size. No effect of production; thus, slope of food chain length and production = 0. Productive-space hypothesis: Production and ecosystem size interact positively; thus, food chain length should increase with both production and size.

Fig. 2 These data support the ecosystem size hypothesis.

Fig. 3 A change in trophic position means that in the largest ecosystems, some species feed higher in the food chain. This suggests that food chain length increases in large systems because of these shifts rather than the presence of a new higher consumer.

16.16 According to Pimm's hypothesis, species at higher trophic levels do not recover from disturbance quickly. As a result, disturbed systems have shorter food chains. Thus, the variation in consumer abundance is the key determinant of food chain length—more variable populations should lead to shorter food chain lengths.

16.17 Yes. Each decrease in the numbers in a trophic level leads to an increase in the numbers in the trophic level below it. That increase, in turn, leads to a decrease in the next-lower trophic level. The result could be a sequence of alternating increases and decreases in even- and odd-numbered trophic levels.

16.18 Aspen might be limited by any factors that affect mortality and reproduction, such as climatic changes, pathogens, fire, or competition from other plants.

Chapter 17

17.1 An equilibrium cycle could be defined as a system in which the amounts in each pool remain relatively constant over time. Fluxes might change to maintain that balance.

17.2 Yes. If there is net import or export of nutrients, a local nutrient cycle can be unbalanced.

17.3 Plants primarily use CO_2 and bicarbonate, HCO_3^-.

17.4 Dissolved CO_2 from the atmosphere is the main source of carbon in the oceans. As atmospheric CO_2 increases, ocean carbonate and bicarbonate should increase.

DO THE MATH   Variance and Ecosystem Models

Fig. 1 The mean in (b) is more representative because the variation around the mean is smaller.

Fig. 2 As more total carbon is added to the atmosphere over time, the amount of new carbon that can be released decreases.

Fig. 3 Many decomposers are poikilotherms. At higher temperatures, they will be more active.

17.5 Both have a large gaseous pool. However, the nitrogen cycle involves forms of nitrogen that can be used and converted only by specific groups of organisms.

17.6 The phosphorus cycle does not include a gaseous component. Much phosphorus is stored in sediments.

17.7 They can move upward as geological processes push lower soil horizons closer to the surface and new soil forms.

THE HUMAN IMPACT   Soil Fertility and Nutrient Cycles

Fig. 2 Irrigation can lead to increased soil salinity. Intense logging activities can compact soil and cause nutrient loss. Intense grazing can also compact soil and deplete nutrients.

17.10 The sixth-order stream will accumulate greater amounts of nutrients from upstream.

17.11 RR_x measures the production relative to unfertilized plants—the proportion of the production that is due to fertilization.

ON THE FRONTLINE   The Nitrogen Paradox in Tropical Forests

Fig. 1 Nitrogen fixation is not done by plant leaves. It requires cyanobacteria, which are much more numerous in soil.

Fig. 2 It would be necessary to experimentally vary nitrogen fixation and measure the effect on total phosphorus.

17.12 Nitrogen availability is a main factor in the allocation of energy to the acquisition of other nutrients.

17.13 Logging removes trees that capture soil nutrients. When trees are removed, nutrients may leach from the soil. Decomposition of remaining dead biomass can lead to nutrient export. Also, disturbance of the soil decomposers may allow nutrients to leach from the system.

17.14 As the forest develops, biomass increases, and the increased plant biomass requires phosphorus.

Chapter 18

18.1 Many of the hot spots are regions with a high degree of endemism.

18.3 The minimum population that can be maintained is the value of N where the left side of the curve reaches 0. Below that value, population growth is negative.

18.4 The fitness Allee effect results, not from reproductive failure, but from increased adult mortality at low density.

18.6 Inbreeding tends to reduce the frequency of heterozygotes in the population.

18.7 During the period of low numbers, alleles and genetic variation are lost. They will recover only by new mutations or gene flow from another population.

18.8 The decline in heterozygosity would be slower.

18.9 The effect of genetic drift also increases as N_e decreases. This could lead to additional losses of genetic diversity.

18.14 Shallow waters are more susceptible to temperature increases. Ocean temperature is a complex phenomenon that is heavily dependent on the patterns of ocean currents. In some climate models, cool currents that flow along some coasts will no longer bring cool water to those regions.

18.15 The new equilibrium is determined by the equilibrium between extinction and colonization. These are in turn determined by the size of the habitat patch and its distance from other patches.

ON THE FRONTLINE   Effects of the Decline of Whitebark Pine

Fig. 1 The rust infection is the factor analyzed in this study. However, other environmental variables, including temperature, precipitation, fire, and other pathogens and parasites, can affect cone production.

Fig. 2 The threshold is between the values 30 and 50 $(\ln[\text{cones/ha}]^2)$.

DO THE MATH   Maximum Sustainable Yield

Fig. 1 The line in the middle, steep part of the graph has the highest slope and therefore the highest rate of change.

Fig. 2 If we harvest close to K, there will be a short-term abundance. However, because the growth rate is lower near K, this harvest rate cannot be maintained.

18.17 Population isolation and inbreeding depression drive the population down. This, in turn, increases the potential for isolation and for inbreeding. Thus, isolation and inbreeding are part of a positive feedback system that increases the probability of extinction.

18.18 The estimated Yellowstone grizzly population is less than 1,000. These graphs do not extend to populations that large. However, if they are extrapolated even to 1,000, they suggest that grizzlies are at considerable risk of extinction within a century. If the population is much lower than 1,000, the risk is obviously considerably greater.

18.21 Habitat loss is probably the most significant factor in the current decline. However, once the population declines to a certain point, any of the factors of the extinction vortex can come into play.

18.22 The graph suggests that the probability of persisting is near 100 percent. However, this analysis includes stochastic processes; thus, we cannot say for certain that a population this large is safe.

18.23 Factors that cause reserves to fall below the line are those factors that are different from the size of the reserve per se. These might include its isolation from other reserves, the absence of dispersal corridors, and threats and impacts that cross the boundary to affect the reserve.

18.24 The long, narrow reserve has a higher ratio of edge to interior.

Chapter 19

19.2 The vegetation responds to patterns of moisture and temperature. These abiotic factors will vary with altitude and west- versus east-facing slopes. The nature of the substrate varies on the mountain as well, and this may determine soil moisture.

THE HUMAN IMPACT   Recovery of the Black-Footed Ferret

Fig. 2 If the ferret is restored in many widely separated places, it may be more likely to survive local environmental problems. The species' geographic genetic variation and local adaptation are also potentially restored.

THE EVOLUTION CONNECTION   Landscape Genetics

Fig. 1 (a) This method makes the broad population genetic comparison of allele frequencies in the two habitats.
(b) The assignment test measures the probability that an individual is an immigrant into that habitat. By applying the test to many

individuals, it is possible to calculate the rate of gene flow.
(c) The probable parents of each individual are determined from its genotype. Because individuals and their parents can be identified and assigned to one of the habitats, it is possible to measure the number of individuals that emigrate or immigrate and thus the rate of gene flow.

19.5 It is not generally possible to predict theoretically what the reflectance pattern will be from a specific type of vegetation. Thus, it is important to relate reflectance patterns to information from the ground.

19.7 Each layer can be examined and analyzed separately. More important, multiple layers can be analyzed together to examine how different spatial parameters overlap or relate to one another.

DO THE MATH   Fractal Geometry

Fig. 1 The more irregular the shape, the greater the effect of ruler size.

Fig. 2 The complexity of a shape determines important aspects of the spatial distribution of habitat. For example, shape determines the perimeter/area ratio. This determines the relative amounts of edge and interior habitat.

Fig. 3 High values of D indicate a complex shape.

ON THE FRONTLINE   Equilibrium and Fire in Ponderosa Pine Forests

Fig. 1 If these systems are maintained by low-intensity fires, Prediction 2 states that even-aged stands of mature trees are the typical community structure. Each of these structures represents a stage in the development of an even-aged stand of mature trees.

19.10 If one canopy tree goes down, it can increase the probability that adjacent trees fall as well. Over time, one might see lines of canopy gaps through the forest formed by turbulent winds.

19.11 Occasionally, a large fire could jump a river, but in general, rivers were barriers to the spread of prairie fires.

Chapter 20

20.1 Urban areas and rangeland are among the most rapidly growing anthromes.

20.2 Your answer will depend on where you live and the nature of your interaction with the natural world. Have you considered what aspects of the natural world you value?

DO THE MATH The Binomial Probability

Fig. 1 No. The increase in hurricane frequency and intensity is a statistical phenomenon.

20.3 There are many patterns in this map, and many hypotheses could be devised to explain one or more of them. In most cases, you would need to research the factors that determine climate for a particular area before you could devise a specific hypothesis. For example, it appears that major mountain ranges (e.g., Cascades, Sierra Nevada, Andes, and Himalaya) have lower rates of temperature change. You might hypothesize that altitude effects on temperature are so significant that they exceed the effect of global temperature change.

20.4 In the winter months, photosynthetic production and thus the uptake of CO_2 by plants decreases. This affects the seasonal pattern of CO_2 concentration in the atmosphere: atmospheric CO_2 decreases during periods of active photosynthesis; it increases when plants are inactive.

20.5 This correlation establishes that the relationship between atmospheric CO_2 and temperature has existed for thousands of years. The current relationship is not just an anomaly or a random correlation.

ON THE FRONTLINE Thermal Acclimation in Corals

Fig. 1 Any corals that have previously experienced significant temperature fluctuations are more likely to survive changes in ocean temperature.

20.6 Many of the world's most productive agricultural areas are found in temperate regions.

20.7 Sessile species will have a hard time surviving significant increases in temperature. If they cannot adapt, it may be difficult for them to shift their ranges fast enough to survive.

20.14 These processes are intimately connected in many ways. For example, deforestation increases CO_2 in the atmosphere because (a) fewer plants are taking up CO_2 and (b) decaying plant biomass releases CO_2. This contributes to global warming. Acid precipitation can kill forest trees, the effect of which is similar to deforestation. Global warming changes the geographic patterns of precipitation. One consequence is that acid precipitation increases in regions where it was previously less significant.

Glossary

Note: The number at the end of each definition indicates the chapter in which the term first appears.

Absorptivity The tendency of an object to absorb radiation. (3)

Accessory pigments Pigments other than chlorophyll that capture light of different wavelengths. (16)

Acclimation An individual's physiological adjustment to challenging abiotic conditions. (3)

Acid precipitation The decrease in the pH of rain or snow due to human industrial emissions. (20)

Active predator A predator that actively moves through the habitat searching for prey. (12)

Actual evapotranspiration (AET) The total amount of water that evaporates and transpires from a region per unit time. (16)

Adaptation A trait that increases an individual's fitness in a specific environment. (1)

Adaptive landscape A graphical representation of the fitnesses associated with different genotypes in a population. (2)

Adaptive radiation The evolutionary diversification of a group of species, often in newly colonized regions, into a number of new species in new niches. (3)

Adiabatic cooling The decrease in air temperature that occurs at higher elevation or altitude. (4)

Adventitious roots Roots produced at or above the soil surface. (3)

Aerosols Elements and compounds suspended in the atmosphere. (17)

Aestivation A period of torpor or hibernation to avoid heat and water stress. (3)

Age structure The distribution of individuals according to their ages. (8)

Algal bloom Explosive algal growth due to nutrient addition to an aquatic ecosystem. (17)

Allee effect A phenomenon in which the population growth rate becomes negative below a minimum population size. (18)

Allele frequencies The proportion each allele represents in the gene pool. (2)

Allelopathic Describes a plant that produces and releases chemicals that inhibit the growth of nearby individuals of the same or another species. (11)

Allocation, principle of The concept that an adaptation to one selective factor may preclude or reduce adaptations to others. (3)

Allochthonous sources Organic energy that enters the water from terrestrial systems. (5)

Allopatric populations Populations occurring in different places. (11)

Allopatric speciation model A mechanism of speciation based on separation of populations by a physical barrier. (15)

Allozyme An alternate form of a gene that differs in amino acid composition and thus in the size and electrical charge of the molecule. (2)

Alluvial deposition The addition of material by flowing water. (17)

Alpha diversity Local species richness. (15)

Altruistic behavior Any behavior that increases the fitness of others at the expense of that altruist's fitness. (6)

Ambush predator A predator that remains motionless, waiting for a prey item to pass close enough for capture. (12)

Anthrome A major unit of human impact on the Earth; analogous to *biome*. (20)

Apomixis A plant mating system in which individuals reproduce asexually. (6)

Aposematic coloration Distinctive, conspicuous color patterns that signal that the individual is toxic or unpalatable. (12)

Apparency theory A theory for the evolution of specialized herbivores. Conspicuous plants are protected by compounds that reduce herbivore growth; inconspicuous plants are protected by acutely toxic compounds. The latter selects for specialized herbivores. (12)

Apparent competition When two similar species are both negatively affected by a shared predator or herbivore. The negative impact on both species mimics the effect of competition. (13)

Aquifer An underground reservoir of water. (5)

Asexual reproduction A mode of reproduction in which offspring arise from a single parent and inherit its genes directly. (10)

Aspect The direction a mountain slope faces. (4)

Assignment test A technique that uses genotype information from the individuals to determine the amount of gene flow across a landscape. (19)

Assimilation efficiency The amount of energy assimilated by the consumer relative to the amount ingested. (16)

Atmospheric deposition The transfer of materials from the atmosphere to terrestrial or aquatic ecosystems. (17)

Atoll A circular coral reef surrounded by open ocean. (5)

Autochthonous sources Sources of organic energy derived from photosynthesis within the aquatic system. (5)

Autogamy A plant mating system in which individuals self-fertilize. (6)

Autotroph An organism that obtains energy from inorganic sources. (16)

Barrier reef A coral reef that develops parallel to the shore and forms a lagoon between the reef and shore. (5)

Bateman's principle The idea that males experience greater variance in reproductive success than females. (6)

Batesian mimic A benign species that resembles a noxious species. (12)

Behavioral thermoregulation Behaviors that allow the organism to seek and use external factors such as sunlight or warm rocks to change its internal temperature. (3)

Bell curve (normal curve) A graph of the frequency distribution of events across a spectrum of conditions, in which there is high frequency at intermediate levels and low frequency at extreme levels. (3)

Benthic zone The deepest waters and substrate of a lake. (5)

Beta diversity Regional species richness of nearby communities. (15)

Bet-hedging theory The idea that organisms that experience unpredictable environments should reduce variation in fitness by spreading out the risk of reproductive failure. (10)

Bioaccumulation The absorption of chemical compounds from the physical environment. (16)

Biodiversity The total biological diversity of a community. (15)

Biogeochemical cycle (nutrient cycle) The movement of nutrients through ecosystems by exchange among the atmosphere, rocks, soil, water, and organisms. (17)

Biogeographical realms Regions that share species and geographic history. (4)

Biological pump The movement of carbon from the surface of the ocean to deeper waters due to living organisms and physical processes. (5)

Biomass The total mass of organic material of biological origin. (16)

Biome The basic plant community types, characterized by a specific vegetation, that occur in a particular region. (4)

Biosphere (ecosphere) All the ecosystems of the Earth. (1)

Bog A small wetland that has no outlet, typically found in northern (glaciated) regions. (5)

Bottom-up control A process in which organisms lower in the food chain control the density and diversity of higher trophic levels. (9)

C_3 pathway A photosynthetic pathway in which carbon is fixed as a 3-carbon molecule by RuBP carboxylase. (3)

C_4 pathway An alternate photosynthetic pathway in which carbon is fixed as a 4-carbon molecule by an enzyme with high affinity for CO_2. (3)

Calorie The amount of heat required to raise one gram of water $1°C$. (3)

Carrying capacity The number of individuals that can be supported by the resources available to a population. (9)

Character displacement A shift in the niches of competing species that reduces competition among them. Natural selection favors those individuals in each species whose resource use overlaps less with that of the other species. (11)

Chemoautotrophs Organisms that derive energy from the oxidation of electron donors. (16)

Chronosequence A series of communities arrayed linearly from young to old and increasing distance from recent or current disturbance. (14)

Clade A species and its descendants. (4)

Clementsian paradigm The concept that mature, climate-determined plant communities resist change. Most are at equilibrium most of the time; disturbance is rare. (14)

Climate diagram A graphical representation of climate showing the seasonal temperature and precipitation changes of a region. (4)

Climax community The final stage of succession that can replace itself if no new disturbance occurs. (14)

Coarse-grained variation Spatial variation in the environment that is large relative to the mobility of the organism. (7)

Coefficient of inbreeding (F) The probability that two alleles are identical by descent. (6)

Coefficient of relationship (r) The proportion of genes shared by two individuals. (6)

Coevolution The reciprocal selective effect of two or more species that interact intimately. (12)

Coevolutionary arms race A set of adaptations and counteradaptations in exploitative interaction. (12)

Cohort life table A life table based on following a single cohort from birth to the death of the last individual. (8)

Cold-acclimation proteins (CAPs) Proteins that protect the organism from sudden decreases in temperature. (3)

Cole's paradox The surprisingly small difference between the total lifetime reproductive outputs of annuals and perennials. (10)

Common garden experiment An experimental design that distinguishes between genetic variation and phenotypic plasticity. Phenotypically different plants from different environments are grown in a common garden. Differences that persist are genetic; similar phenotypes in the common garden indicate phenotypic plasticity. (7)

Communication Any action by one individual that alters the probability of a behavior in another. (6)

Community heritability The proportion of the variation in a community characteristic explained by genetic variation in a particular species. (13)

Compensation point The depth at which photosynthesis exactly balances respiration. (5)

Compensatory ability The ability of a plant to alter tissue production to counteract herbivory. (12)

Competition coefficient A measure of the competitive effect of one species on another, determined by the overlap in their resource use. (11)

Competitive exclusion The extinction of one species due to competition from another. (11)

Competitive hierarchy Linear competitive dominance relationships among species; species A dominates species B, which dominates species C, etc. (13)

Competitive network A web of competitive interactions in which species subordinate to some competitors dominate others. (13)

Competitive plant In Grime's classification of plant strategies, a plant that is limited more by competition than other factors. (10)

Conduction The transfer of heat between two bodies in physical contact. (3)

Connectivity The link between two or more populations by dispersal of individuals. (8)

Continental climate The climate in regions far from large bodies of water, usually characterized by large temperature fluctuations. (4)

Convection The transfer of heat from an object to a moving fluid or gas. (3)

Convergent evolution Similar adaptations that arise independently in unrelated groups. (16)

Cooperative hunting A form of group hunting in which the group develops tactics that may include anticipation of prey behavior and division of labor. (12)

Coral bleaching The death of coral caused by expulsion of the zooxanthellae, which leaves the white coral skeleton. (3)

Core-satellite hypothesis Describes a community composed of (1) species that are either rare or common and (2) core species with high abundance and frequency, or satellite species found at low abundance and in few sites. (13)

Correlation coefficient A statistical measure of the strength of the association between two variables. (5)

Cost of meiosis The reduction in genetic contribution to the offspring in sexual species. (10)

Countercurrent heat exchanger An arrangement of vessels such that the flow is in opposite directions, maximizing the temperature differential between the fluids and thus maximizing heat exchange between them. (3)

Court Jester model The idea that species diversity is a product of stochastic disturbance events. (15)

Cover type The vegetation in a defined area. (19)

Critical temperature The maximum and minimum temperature an organism can experience without expending energy to maintain a constant internal temperature. (3)

Crypticity A mechanism employed by prey species to avoid detection by predators. Cryptic species appear to the predator as a random sample of the background. (12)

Cultural eutrophication Eutrophication caused by anthropogenic inputs of nutrients. (16)

Deceptive pollination A set of floral characteristics that falsely advertise a reward to pollinator visitors. (12)

Decomposition The breakdown of animal excretions or the bodies of dead organisms. (17)

Degradation Anthropogenic disruption of ecological processes. (20)

Degradative succession A sequence of decomposition processes accomplished by a series of organisms. (17)

Deme The evolutionary population unit; a group of randomly mating individuals. (8)

Demographic stochasticity Random changes in demographic parameters such as l_x or R_0. (18)

Demographic transition The changes in the relative size of the birth and death rates in human population from preindustrial to industrial societies. (8)

Demography The quantitative description of a population and its characteristics. (8)

Density-dependent factor Population processes that change with the population density. (9)

Density-independent factor Population processes that are not affected by density. (9)

Desertification The conversion of plant communities or agricultural systems to desert.

Detritivore An organism that consumes feces or dead organisms. (17)

Diapause An embryonic stage in which the embryo does not implant in the uterus and suspends development. (10)

Diffuse competition The summed effects of all competitors. (11)

Dioecious Describes a plant in which male and female function occur in separate individuals. (8)

Direct development A process in which the adult develops from the fertilized egg without a larval stage. (10)

Directional selection A form of selection in which one tail of the phenotypic bell curve is favored. (2)

Disclimax A community that persists and appears to be a climax because it replaces itself due to recurring disturbance. (14)

Dispersal The one-way movement of an individual from the natal area. (6)

Dispersion The pattern of spatial distribution of species in the habitat. (8)

Disruptive selection A form of selection in which the two tails of the phenotypic bell curve are favored. (2)

Ecological (Hutchinsonian) niche The set of biological and physical resources that determine the growth, survival, and reproduction of a species. (11)

Ecological community A group of coexisting species. (1)

Ecological disturbance A physical or biological factor that alters the structure and species composition of the community. (14)

Ecological stability The tendency of community composition and function to remain constant. (15)

Ecology The study of the interactions between an organism and its biological and physical environment. (1)

Ecosystem All the biotic and abiotic components of a community. (1)

Ecosystem services Natural processes that contribute tangible benefits to humans such as flood control, filtering pollutants, and natural pest control. (18)

Ecotype A genetically distinct population that is adapted to local environmental conditions. (2)

Ectotherm An organism whose body temperature is determined by the external environment. (3)

Eddy analysis A technique to measure primary production by measuring the rate of uptake of CO_2 from the atmosphere. (16)

Edge effect An external environmental impact that extends into the habitat. (18)

Effective population size (Ne) The subset of the total population that mates at random. (2)

Ekman spiral A pattern of water movement in which wind and the Coriolis force cause water to deflect differentially with depth. (5)

Emergents Aquatic plants that are rooted but protrude above the water. (5)

Emissivity The tendency of an object to emit radiation. (3)

Endemic A species found in a particular locale and nowhere else. (18)

Endotherm An organism whose internal temperature is maintained by metabolic activity. (3)

Energy maximizer A species for which selection favors those individuals that obtain the maximum total amount of energy. (12)

Environmental filtering The presence or absence of species according to their physiological tolerance of abiotic conditions. (13)

Environmental science The study of the human impact on the environment. (1)

Environmental stochasticity The random occurrence of unfavorable environmental conditions. (18)

Epilimnion The warmer, well-mixed upper waters of a lake that lie above the thermocline. (5)

Equilibrium theory of island biogeography An explanation of species richness on islands based on a balance between colonization and extinction. (15)

Estuary A partially enclosed body of water connected directly to the ocean that occurs near the mouth of a river. (5)

Eusocial Describes a complex social system in which there is division of labor or castes, a high level of cooperation, and sometimes altruism. (6)

Eutrophic Describes aquatic systems in which nutrients have accumulated and stimulated high rates of plant photosynthetic production. (5)

Eutrophication The accumulation of nutrients in a body of water and its stimulation of photosynthetic production. (5)

Evolutionarily stable strategy (ESS) A mutation or adaptation that can increase in frequency in the population when it is rare. (7)

Evolutionary fitness The survival and reproduction of a particular individual as determined by its characteristics. (1)

Evolutionary theory of virulence The theory that selection favors decreased virulence in the parasite because this increases the probability that the parasite can exploit the host and reproduce. (12)

Evolutionary trade-offs The idea that many traits that confer a fitness advantage with respect to one aspect of the environment can also have a fitness cost relative to another. (2)

Exploitation competition A mechanism of competition in which one species reduces the amount or availability of the limiting resource. (11)

Exploitative interaction One species benefits by exploiting another as a food source. (12)

Extent The total area of the study. (19)

Extinction debt The continued loss of species following fragmentation. (18)

Extinction vortex A process in which multiple negative impacts on a population cause the numbers to decline toward extinction. (18)

Facilitation A mechanism in which the presence of one species increases the probability that another species is present. (13)

Facilitation model A model of successional change in which early colonists improve conditions for other species that subsequently colonize. (14)

Facultative hibernator An organism that does not have to become torpid but does so during harsh conditions and can arouse quickly if conditions change. (3)

Facultative mutualism A mutualistic interaction that is not absolutely required for either species. (12)

Fall turnover The mixing of the hypolimnion and epilimnion in the fall. (5)

Falsifiable Describes a hypothesis that can be proven incorrect by data or observation. (1)

Fecundity The number of offspring produced. It can be measured for a single reproductive event or over the life span. (10)

Female choice A mechanism of sexual selection in which females choose mates on the basis of their physical or behavioral characteristics. (6)

Female defense polygyny A form of polygyny that occurs when groups of females are guarded by males. (6)

Field capacity The amount of water held in the soil. (3)

Fine-grained variation Spatial variation in the environment that is small relative to the mobility of the organism. (7)

Fisheries biology The study of management practices that ensure the sustainable harvest and ecological health of fish populations. (1)

Fisher's Fundamental Theorem The principle that the rate of change in fitness by natural selection is equal to the additive genetic variance in the population. (7)

Fitness The ability of an individual to survive and reproduce relative to other individuals in the population. (2)

Fitness Allee effect A phenomenon in which adult fitness declines below a minimum population size. (18)

Fixed-effort harvest A management strategy in which the harvest is limited by the length of season, means of harvest, or other aspects of the harvest. (18)

Fixed-quota harvest A management strategy in which the harvest is limited to a specific number of individuals. (18)

Floaters Unrooted aquatic plants that float on the lake surface. (5)

Flora The plant species of a region. (4)

Floral advertisement A set of floral characteristics that advertise that the flower provides a reward to pollinator visitors. (12)

Flux The pathway of nutrient transfer among pools. (17)

Food chain The hierarchical pattern of energy transfer from primary producers to consumers. (9)

Food web The connection of many food chains. (16)

Forestry The study of management practices that ensure the sustainable harvest and ecological health of forests. (1)

Fractal dimension (D) A measure of the complexity of a shape determined by the difference in measurement depending on the scale at which the shape is measured. (19)

Fragmented Describes a habitat that has been broken up into small patches, generally as a result of human activity. (19)

Frequency-dependent selection A form of natural selection in which the fitness of a gene is determined by its frequency in the population. (7)

Fringing reef A coral reef that projects directly from shore and develops on rock. (5)

Fugitive species Poor competitors that exploit disturbances, where they face fewer competitors. (11)

Functional niche A descriptive definition of the niche as the ecological role of the species in the community. (11)

Functional response A predator response to increased prey numbers in which each predator consumes more prey. (12)

Fundamental niche The niche in the absence of competition. (11)

Gamma diversity The landscape-level species richness. (15)

Gene A sequence of DNA that encodes the amino acid sequence of a specific protein. (2)

Gene flow The net movement of alleles to or from a population. (2)

Gene pool The sum total of all alleles in a population. (2)

Genet A genetically distinct individual in a plant population. Genets may comprise many individuals, especially when reproduction is by cloning. (8)

Genetic drift Random changes in allele frequencies. (2)

Genetic similarity The measure of the proportion of alleles shared by two populations. (7)

Genetically modified organism (GMO) An organism in which a gene from another species has been inserted into the genome. (20)

Genotype The genetic makeup of the individual that, in concert with the environment, determines the phenotype. (2)

Geographic information systems (GIS) A software system that analyzes spatial ecological and environmental data by producing a series of overlaid maps, each with the spatial variation in a single variable. (19)

Geographic range The region in which a particular species is found. (4)

Ghost of competition past An explanation for communities with niches that appear to be randomly associated because competition eliminated some species in the past. (13)

Glacial till Material deposited by a glacier. (17)

Grain The finest spatial resolution in a data set. (19)

Gross primary production (GPP) The total rate of energy acquisition by autotrophs. (16)

Group hunting Predation by a group of conspecifics. (12)

Group selection A process analogous to natural selection in which the population (rather than the individual) is the object on which selection acts. (9)

Guild A group of ecologically similar, coexisting species. (13)

Gular fluttering Rapid throat movement in birds to lose heat by evaporative cooling. (3)

Habitat The abiotic and biotic characteristics of the place where an organism lives. (6)

Habitat complexity Variation in the abiotic, topographic, or functional components of the habitat. (15)

Habitat fragmentation A process in which large tracts of habitat are broken into small, isolated fragments. (18)

Habitat loss The conversion of natural communities into anthromes. (20)

Handling time The time required for a predator to capture, subdue, and consume a prey item. (12)

Hardy-Weinberg equilibrium A mathematical representation of the genotype frequencies of a population in which the allele and genotype frequencies are not changing. (2)

Heat Thermal energy that can be transferred from one body to another. (3)

Heat of vaporization The decrease in an object's heat content by the evaporation of water. (3)

Heat shock proteins (HSPs) Proteins that protect the organism from sudden increases in temperature. (3)

Herbivore An animal that obtains energy by consuming plants. (12)

Heritability A measure of the proportion of the phenotypic variation for a trait that is determined by additive effects of its genes. (2)

Heritability in the broad sense The proportion of the total phenotypic variation that is due to genetic differences among individuals. (6)

Heritability in the narrow sense (h^2) The proportion of phenotypic variation that is due to additive genetic variation. (6)

Heterotroph An organism that obtains energy from organic compounds. (16)

Hibernation An extended form of torpor. (3)

Home range The portion of the habitat used by an individual on a daily or seasonal basis. (6)

Homeostasis Regulatory mechanisms that maintain an organism's physiological parameters within specific limits. (1)

Homeotherm An organism whose body temperature physiologically is regulated within narrow limits. (3)

Humus A mix of organic molecules from the chemical decomposition of carbohydrates, lipids, and proteins. (17)

Hydrologic cycle The movement of water among the living and nonliving components of the Earth. (5)

Hydroperiod The timing and length of periods of the presence and absence of water in a wetland. (5)

Hypertrophic Highly eutrophic aquatic systems with massive photosynthetic production. (5)

Hypolimnion The denser, colder water in a lake that lies below the thermocline. (5)

Hypoxia Low oxygen concentration. (5)

Identical by descent Describes two copies of a gene that are derived from the same ancestral copy of the gene. (6)

Inbreeding depression A decrease in fitness due to mating among related individuals. (6)

Inclusive fitness A concept of fitness based on the relative ability of an individual to transmit its genes or copies of them to the next generation. (6)

Index of relative abundance A quantitative measure of the relative size of a population using indirect evidence of the presence or absence of individuals. (8)

Indicator species A species found only under specific ecological conditions. (18)

Infiltration The movement of water into the pores between soil particles. (5)

Ingestion efficiency The amount of energy ingested relative to the amount available. (16)

Inhibition model A model of successional change in which early colonists inhibit colonization by new species. (14)

Instantaneous growth rate (also, intrinsic rate of increase) The potential growth rate of a population based on the difference between the per capita birth and death rates, measured at small population size. (8)

Interference competition A mechanism of competition in which one species actively inhibits another from obtaining the resource. (11)

Intermediate disturbance hypothesis The concept that species diversity is low where disturbance is either frequent or infrequent and maximal at intermediate levels. (15)

Intermediate host A host of a parasite in which the parasite produces forms that can locate and infect a different host species. (12)

Interspecific competition The interaction among two or more species over a limiting resource that results in a decrease in the population size of at least one of the species. (11)

Intraspecific competition Competition among members of the same species. (11)

Intraspecific variation Genetic or phenotypic variation within a species. (7)

Invasive species An introduced species that can multiply rapidly and expand its range. (13)

Isocline (in the Lotka-Volterra models) The line on a graph of the population size of species 1 and 2 that represents zero population growth of one of the species. (11)

Iteroparous Describes a reproductive strategy in which females reproduce more than once over the life span. (10)

Kettle hole A depression in the substrate created by glaciers. (5)

Keystone mutualist A species whose participation in webs of mutualistic interactions is vital to the function of the community. (12)

Keystone predator A predator whose presence is central to the structure of the community. (13)

Keystone structure Biological or physical habitat structure that provides important resources to a species. (15)

Krumholz growth form A growth form of trees at high elevation characterized by a mat of branches near the ground and a single, often bare, shoot at snow level. (4)

K-selection A reproductive strategy in which females maximize fitness by producing relatively few young but invest a great deal of energy in each.

Land ethic Aldo Leopold's concept that humans are a part of the natural world and connected ecologically and ethically to other members of the community. (20)

Landform The surface shape, elevation, and slope of a region. (19)

Landscape A region in which ecological factors vary. (19)

Landscape ecology The analysis of the patterns of spatial variation in the environment and in ecological processes. (19)

Langmuir cell A pattern of circular water motion near the surface of bodies of water. (5)

Leaf area index (LAI) The ratio of leaf to ground area. (19)

Least squares regression A technique of linear regression in which the best line is defined as that line for which the square of the sum of the distance between each point and the line is the smallest. (5)

Lek-mating species A species that uses a form of polygyny and female choice in which males gather to display to females on traditional places. (6)

Lentic systems Aquatic ecosystems composed of ponds and lakes. (5)

Life-dinner principle The concept that selection pressure is greater on the prey than on the predator. (12)

Life expectancy (e_x) The mean expectation of further life for an individual of age x. (8)

Life history The combination of development and growth, the life span, and the timing and quantity of reproduction. (10)

Life history strategy The suite of life history traits adaptive in a particular ecological context. (10)

Life span The maximum age to which an individual can live. (8)

Life table A table showing the numbers of individuals of different ages and their age-specific mortality and reproductive rates. (8)

Life zone A band of vegetation associated with a specific altitude on a mountain. (4)

Light gap (canopy gap) An opening in the tree canopy in a forest. (14)

Limiting soma theory The hypothesis that senescence is caused because resources devoted to reproduction are not available to maintain somatic tissue. (10)

Limnetic zone The region of a lake above the compensation point. (5)

Lincoln index A method for determining population size by marking and recapturing portions of a population. (8)

Linear regression A statistical procedure that determines the parameters of the best line representing the relationship between two variables. (5)

Littoral zone The near-shore, shallow region of a lake. (5)

Local uniqueness The abiotic and biotic factors that differentiate an area from others. (19)

Loess Glacial till deposited by wind. (17)

Lotic systems Aquatic ecosystems composed of flowing water. (5)

Lottery model A model in which species diversity is the result of disturbance followed by colonization. (15)

Male-male competition A form of sexual selection in which males compete with one another directly or indirectly for access to females. (6)

Maritime climate The climate near a large body of water, usually characterized by a narrow range of temperature. (4)

Markov chain A mathematical phenomenon in which a set of state conditions occur at discrete times. The probability of transition from one state to another is independent of the current state. (14)

Marsh A wetland in which the dominant vegetation is herbaceous. (5)

Mating system The length of relationships between males and females, the relative contributions of males and females to parental care, and the number of mates an individual copulates with. (6)

Matric pressure The attraction and adhesion of water on surfaces. (3)

Maximum sustainable yield The maximum harvest of a population rate that can be maintained over time. (18)

Mean The average value in a data set. (3)

Measure of central tendency A statistical parameter that characterizes a data set by the average or most common value. (3)

Median The value in a data set for which the number of larger and smaller values is equal. (3)

Mesic species Plant species that require higher levels of moisture. (4)

Metabolic water Water produced by oxidative metabolism. (3)

Metacommunity A set of communities with similar composition connected by dispersal. (13)

Metamorphic development A developmental pattern that includes a larval stage. (10)

Metapopulation A group of populations connected to varying degrees by dispersal. (9)

Microhabitat The subset of the habitat that differs in important abiotic and biotic characteristics. (6)

Mid-latitude desertification A phenomenon in which warm, dry air is found at approximately 30° N and 30° S, resulting in a desert climate. (4)

Migration A seasonal movement from one region to another and back. (3)

Mineralization The transformation of organic matter into inorganic forms. (17)

Minerals Elements or chemical compounds formed by geological processes. (17)

Minimum viable population (MVP) The minimum population size, identified by a mathematical technique, required for a population to persist for a particular length of time. (18)

Mobile link species A species whose mobility across the landscape to a series of dependent mutualists is important to the normal function of the community. (12)

Modular organism An organism that develops by repetitive patterns of growth of body parts. (8)

Monoecious Describes a plant in which male and female function occur in the same individual. (8)

Monogamy A mating system in which each male mates with a single female. (6)

Mullerian mimic A species that resembles other noxious species. (12)

Multiplicative growth rate The factor by which the current population size increases in each time period. (8)

Mutation A random change in the DNA sequence of a gene. (2)

Mutation pressure Changes in allele frequency due to the origin of new alleles in the population. (2)

Mycorrhizae The mutualistic interaction between plants and fungi. (12)

Natural history Qualitative or observational study of organisms in their natural environment. (1)

Natural selection The increase in the frequency of individuals with inherited traits that increase their fitness relative to other individuals. (2)

Negative allometry (hypoallometry) An allometric relationship ($y = bx^a$) in which the exponent a $<$ 1. (16)

Neoteny The attainment of functional sexuality by larval forms that do not undergo metamorphosis. (10)

Net primary production (NPP) GPP minus the losses to respiration. (16)

Net reproductive rate (R_0) The average number of individuals produced by a female in her life span. (8)

Neutral model The concept that communities are random assemblages of species physiologically tolerant of the abiotic conditions. (13)

Niche breadth The variation among individuals in a population in their resource use. (11)

Niche overlap (α) The extent of common use of a resource by two or more species. (11)

Niche separation The difference between two niches measured by the difference in the mean use of a resource. (11)

Niche space A multidimensional representation of a species' use of a set of resources. (13)

Niche-based model A model of community structure in which biological processes, especially competition, act on species' niches to permit coexistence. (13)

Nitrogen deposition The addition of nitrogenous compounds to the ecosystem by agricultural and industrial activity. (20)

Nitrogen fixers Organisms that can convert molecular nitrogen into organic form. (17)

No-analog climate A climate that differs from any of the Earth's currently known climates. (4)

No-analog community A group of species unlike any other current ecological community. (4)

Non-Darwinian evolution Genetic drift. (2)

Nonpoint sources Releases of pollution across a broad area. (20)

Nonshivering thermogenesis The use of brown fat to increase heat production. (3)

Normalized difference vegetation index (NDVI) An index of primary production based on the reflectance in the red and near-red spectrum. (19)

Null hypothesis (H_0) The hypothesis that observed differences are due to chance sampling error. (4)

Numerical response An increase in predator number by reproduction or immigration in response to an increase in prey numbers. (12)

Nutrient spiral The transport of organisms and detritus downstream. (17)

Nutritional mutualism Describes an interaction in which one or both species provide nutrition for the other. (12)

Obligate hibernator An organism that must seasonally hibernate. (3)

Obligate mutualism Describes interaction that is necessary for the survival and reproduction of at least one of the species involved. (12)

Oceanic domains Marine regions characterized by similar climate and currents. (5)

Old field succession The sere following abandonment of an agricultural field. (14)

Oligotrophic Describes aquatic ecosystems with low nutrient content and low rates of photosynthetic production. (5)

Omnivory Feeding at more than one trophic level. (16)

Optimal foraging theory A body of theory that tries to identify the best foraging strategy by which an organism can maximize total energy or energy per unit time. (12)

Organismal level The level of the biological hierarchy in which the focus is on the individual. (1)

Osmoconformers Organisms that allow their internal water balance and solute concentration to vary with the external conditions. (3)

Osmoregulators Organisms that maintain their internal water balance and solute concentration within narrow limits. (3)

Outbreeding depression A decrease in fitness that arises when genes from two locally adapted populations are mixed, because the new combinations of alleles break up adaptive combinations. (7)

Outcrossing A plant mating system in which mating occurs between different individuals. (6)

Ozone-depleting substances (ODS) Compounds that catalyze the destruction of atmospheric ozone. (20)

Parasite A species that lives in or on another species and obtains energy from the other species' living tissue. (12)

Parasitoid An insect whose larvae develop in the body of another arthropod and consume its tissue. (12)

Parent material The rock and mineral substrate underlying a region. (4)

Parental investment The total parental expenditure of energy on offspring through the numbers and size of offspring, and their care, feeding, and defense. (10)

Patch An area that differs from its surroundings in an ecological feature or process. (19)

Patch dynamics A concept of community structure in which a region is a mosaic of sites, each with its own disturbance history and successional status. (14)

Pelagic zone The open, deep water of a lake distant from the shore. (5)

Periphyton (aufwuchs) Algae attached to the substrate in a lotic system. (5)

Permanent wilting point When the water potential of the soil is so low that water cannot be extracted by the roots. (3)

Persistent organic pollutants (POPs) Organic compounds that are slow to decompose. (16)

Phase difference Spatial variation in the landscape due to disturbance patterns. (19)

Phenotype The characteristics including morphology, physiology, and behavior of an individual. (2)

Phenotypic plasticity The ability of an organism to produce different phenotypes in different environments. (2)

Philopatric Describes individuals that do not disperse. (6)

Photorespiration A pathway that occurs in C_3 plants when the CO_2 levels fall so low that RuBP carboxylase picks up O_2 instead of CO_2. (3)

Phylogenetic clustering The relationship among coexisting species under environmental filtering. (13)

Phylogenetic conservatism A reduced amount of changes in a clade due to constraints on its evolution. (4)

Phylogenetic overdispersion The relationship among coexisting species when competition is the dominant structuring mechanism. (13)

Physical (abiotic) factors The physical conditions that affect an organism's growth and survival. (3)

Physical resources The energy and inorganic materials an organism requires. (3)

Phytochemical coevolution theory A coevolutionary interaction between plants and their herbivores in which plant secondary compounds select for specialized herbivores that can counteract these deterrents. (12)

Pioneer species The initial colonists following a disturbance. (14)

Plant secondary compounds Compounds produced by plants that deter herbivores. (12)

Pleiotropic effects The actions of a single gene that affect more than one phenotypic trait. (10)

Poikilotherm An organism whose internal temperature varies, often in response to the external temperature. (3)

Point sources Discrete, local sources of pollution. (20)

Poisson distribution A statistical distribution in which rare events occur at random. It can be used to analyze dispersion. (8)

Pollination syndrome A set of flower characteristics associated with a particular type of pollinator. (12)

Polyandry A mating system in which each female mates with more than one male. (6)

Polygyny A mating system in which males mate with more than one female. (6)

Pool (sink) Components of an ecosystem (biotic or abiotic) that hold nutrients. (17)

Population A group of individuals of a single species inhabiting a particular area. (8)

Population bottleneck A phenomenon in which a population declines to small numbers, then expands again. (7)

Population density The number of individuals per unit area or volume. (8)

Population ecology The study of the interactions between a group of individuals of a given species and the environment. (1)

Population genetics A field of genetics that analyzes the dynamics of genes in an entire population. (2)

Population regulation Mechanisms that determine population size by density-dependent processes. (9)

Population stability The state of a population that fluctuates within narrow limits. (9)

Population viability analysis (PVA) A mathematical technique to analyze the probability that a population can be maintained at a specified size for a specified period. (18)

Positive allometry (hyperallometry) An allometric relationship ($y = bx^a$) in which the exponent a $<$ 1. (16)

Post-mating isolating mechanisms Reproductive isolating mechanisms that prevent the development or sexual function of hybrids. (15)

Preadaptation A process in which the initial adaptive value of a feature is different from its current function. (16)

Predator A species that obtains energy by capturing, killing, and consuming other organisms. (12)

Predator swamping A prey strategy in which the per capita predation rate is reduced at high prey density. (12)

Prediction The result or observation we expect if a hypothesis is true. (1)

Preemptive competition A mechanism of competition in which a plant establishes access to resources by establishing itself and occupying space. (11)

Pre-mating isolating mechanisms Reproductive isolating mechanisms that prevent mating. (15)

Pre-saturation dispersers Individuals that disperse before the habitat is filled or resources are limiting. (6)

Pressure potential The water energy due to pressure exerted as water is forced from one place to another. (3)

Primary consumer An organism that consumes primary producers. (16)

Primary production Energy acquisition by autotrophs. (16)

Primary succession The pattern of succession that occurs following a major disturbance that reverts the environment to nearly abiotic conditions. (14)

Production efficiency The amount of energy devoted to new tissue (growth and reproduction) relative to the amount assimilated. (16)

Profundal zone The region of a lake below the compensation point. (5)

Proportion of patches occupied The proportion of available habitat patches occupied by a species. (19)

Protandrous Describes a pattern of development in which male function develops first. (10)

Protandry A developmental phenomenon in plants in which the male parts of the flower mature before the female parts. (6)

Protection mutualism The protection of one species by another. (12)

Protogynous Describes a pattern of development in which female function develops first. (6)

Proximate factor A direct or immediate cause of a biological process or phenomenon. (1)

Proximity A measure of the relative distance between patches. (19)

Pseudoflowers Structures produced in a plant infected by a rust parasite that superficially resemble the plant's flowers but contain the spores of the rust. (12)

Radiation The transfer of heat between two objects not in physical contact. (3)

Rain shadow The tendency of the lee side of mountain ranges to be drier than the windward side. (4)

Ramet A physiologically distinct individual in a plant population. (8)

Range management The study of management practices that ensure the health and viability of grassland habitats used by domestic or wild animals. (1)

Rank-abundance curve A graph of the log proportional abundance of each species as a function of the rank of its abundance. (15)

Rate-limiting step A crucial step in a series of fluxes that determines the overall rate of nutrient cycling. (17)

Reactive oxygen species (ROS) Oxygen ions and peroxides produced in normal metabolic activity that are toxic to the cell. (10)

Realized niche The subset of the fundamental niche that results from competition. (11)

Red Queen hypothesis The idea that the environment changes faster than adaptations can arise by natural selection. (2)

Relative richness (R) The total number of cover types as a proportion of the number of possible types. (19)

Relaxation flora/fauna A new, lower equilibrium species richness on habitat islands. (18)

Reproductive isolating mechanisms Characteristics of the organism that prevent mating with individuals of a different species. (15)

Reproductive value (V_x) The expected reproductive contribution to the next generation of an individual of age x. (10)

Residence time The length of time an element stays in a particular pool. (17)

Residual reproductive value The remaining reproductive value of an individual after the current reproductive effort. (10)

Resilience The ability of a community to return to equilibrium following disturbance. (15)

Resistance The tendency of a community to remain constant when disturbed. (15)

Resting stage A developmental stage in which the organism is dormant, inactive, and often resistant to harsh environmental conditions. (10)

River A stream of at least sixth order. (5)

River continuum Ecological and hydrological changes such as energy sources, gradient, and flow that occur along the course of a stream. (5)

Rock An aggregate of minerals. (17)

Rocky intertidal A community that develops on rocky shorelines between the high and low tide lines. (5)

r-selection A reproductive strategy in which females maximize their fitness by producing large numbers of offspring but invest little energy in each.

Ruderal In Grime's classification of plant strategies, a plant that experiences low stress but high levels of disturbance of the habitat. (10)

Sample size The number of observations in a data set. (3)

Saturation dispersal Dispersal of individuals that occurs when the habitat is filled or resources are limiting. (6)

Scale The spatial dimension at which we view an ecological phenomenon or process. (19)

Search time The time required for a predator to locate and identify an item as food. (12)

Secondarily aquatic A species whose recent ancestors were terrestrial that has returned to an aquatic lifestyle. (5)

Secondary consumer An organism that consumes primary consumers. (16)

Secondary production The acquisition of energy by heterotrophs. (16)

Secondary succession The pattern of succession that occurs when following a disturbance that leaves the soil and some parts of the community in place. (14)

Selection coefficient (s) The proportion of a genotype that is not represented in the next generation due to death or reproductive failure. (2)

Selection pressure The environmental factors (biotic and abiotic) that determine fitness. (1)

Self-incompatibility A phenomenon in plants in which individuals cannot self-fertilize. (6)

Self-regulation Internal density-dependent processes that operate independently of outside biotic or abiotic forces. (9)

Self-thinning A phenomenon in plants in which individuals at high density have smaller population size but larger individuals. (11)

Semelparous (big bang reproduction) A reproductive strategy in which a female reproduces once in the life span. (10)

Senescence Degenerative changes that increase the probability as age increases. (10)

Sequential hermaphroditism A complex life cycle in which the sex of the individual changes over the course of the life span. (10)

Seral stage A specific community that occurs during the process of succession. (14)

Sere The sequence of stages or community types that occur during succession. (14)

Serial monogamy Monogamy that lasts for just one breeding season. (6)

Sex ratio The proportions of males and females in a population. (8)

Sexual dimorphism Differences in size, coloration, or morphology between males and females. (6)

Sexual selection A form of natural selection in which traits that enhance the ability of one sex (usually males) to compete for or attract mates are favored. (6)

Shannon/Weiner index of species diversity A measure of diversity that incorporates both the number of species and their evenness. (15)

Sheet flow The movement of water in broad sheets across the land. (5)

Social system The organization of a group of individuals in terms of group size and composition, cooperation among individuals, and the mating system. (6)

Soil horizons The series of layers in soil with different biological and physical properties. (17)

Solitary predator A predator that hunts alone. (12)

Spatial scale The relative sizes of the components of a landscape. (19)

Speciation The process by which new species arise. (15)

Species A group of interbreeding organisms with an isolated gene pool. (7)

Species diversity The combination of the number of species (species richness) and their relative abundance (species evenness). (15)

Species endemism The number of species found in one place and nowhere else. (15)

Species evenness The relative abundance of each species in a community; high evenness means most species occur in similar numbers; low evenness means that a few species are numerically dominant. (15)

Species richness The total number of species in a community. (15)

Species-area effect The increase in species richness as a function of area. (15)

Specific heat capacity The amount of heat a substance requires to raise one gram 1°C. (3)

Spiraling length The distance an atom requires to complete a nutrient cycle in a nutrient spiral. (17)

Spring turnover Mixing of lake waters in the spring due to wind and warming lake temperature. (5)

Stabilizing selection A form of selection in which the central portion of the phenotypic bell curve is favored. (2)

Standard deviation The square root of the variance. (3)

Standard error The standard deviation divided by the square root of the mean. (3)

Static life table A life table based on a sample of the population and the distribution of individuals of different age. (8)

Statistical significance The conclusion that two data sets differ for reasons other than chance sampling error. (4)

Stimulus-response Specific behaviors elicited by a specific stimulus. (6)

Stream order The hierarchical position of a stream in a lotic system based on the number of tributaries that form a stream or river. (5)

Stress-tolerant plant In Grime's classification of plant strategies, a plant that inhabits physically demanding environments. (10)

Stromatolites Fossil biofilms of cyanobacteria. (16)

Student's t test A statistical test used to measure the significance of the difference between two means. (4)

Submergents Aquatic plants that are rooted in the bottom but do not extend above the surface. (5)

Subspecies A local, phenotypically distinct form. *Race* and *subspecies* are synonymous. (7)

Succession The sequence of changes in a community following disturbance. (14)

Survivorship curve A plot of the $\log l_x$ as a function of age. (8)

Swamp A wetland in which the dominant vegetation is woody. (5)

Sympatric Describes populations that occur in the same place. (11)

Teleology The assignment of purpose or intent in evolution and adaptation. (10)

Terminal lake A lake basin with no outlet. (5)

Territoriality Exclusive use of a portion of the home range. (6)

Tertiary consumer An organism that consumes secondary consumers. (16)

Test statistic A variable that measures the differences between data sets to test the null hypothesis. (4)

Theory of r- and K-selection The division of life history strategies into species that experience competition, whose strategy emphasizes offspring quality (K-selection), and those that experience disturbance but abundant resources, whose strategy emphasizes offspring quantity (r-selection). (10)

Thermal neutral zone The range of temperature between the upper and lower critical temperatures. (3)

Thermocline A layer of a lake where the temperature and density of the water change rapidly. (5)

Thermogenic Describes plants that can raise their internal temperature above ambient. (3)

Thermokarst Gullies of exposed soil produced by the selective melting of permafrost. (17)

Thermophile An organism that tolerates high temperature. (3)

Tide pool A depression in the rocky intertidal that holds water at low tide. (5)

Time maximizer A species for which selection favors those individuals that obtain the most energy per unit time. (12)

Tolerance, law of The concept that there are upper and lower bounds to the physical factors within which an organism can survive. (3)

Tolerance model A model of successional change in which early species change the abiotic conditions and are replaced by new species able to tolerate those conditions. (14)

Top carnivore, or apex predator The final trophic level in a food chain. (16)

Top-down control A process in which organisms higher in the food chain control the density and diversity of lower trophic levels. (9)

Torpor A state of decreased physiological function during periods of harsh conditions.

Total fertility rate (TFR) The number of children a human female is expected to produce in her lifetime. It is the same as the general term, R_0. (8)

Total niche width (TNW) The summed resource use of all the members of a population. (13)

Transect A sampling method of measuring the presence or absence of individuals along systematic paths through the habitat. (8)

Transition (Leslie) matrix A method of predicting population growth from the probabilities of transition from each age class to the next and the age-specific reproductive rate. (8)

Transpiration The movement of water through a plant, including evaporation from the stem and leaves. (5)

Transportation mutualism A mutualistic interaction in which one species provides transport of gametes or individuals for another. (12)

Tree associations Combinations of tree species with similar climatic and soil requirements that are often found together. (4)

Tree line The altitude or latitude at which trees can no longer survive. (4)

Trophic cascade When a top-down process affects multiple lower trophic levels. (16)

Trophic level A feeding level; one level in a food chain. (9)

Trophic magnification, or biomagnification The increased concentration of chemical elements or compounds at higher levels in a food chain. (16)

Turgor pressure Pressure potential caused by the influx of water to a cell with a rigid cell wall. (3)

Type I error A statistical error that results from rejecting the null hypothesis when it is in fact true. (4)

Ultimate factor The deeper cause of a phenomenon that explains why it occurs. (1)

Umbrella species Species that require large tracts of suitable habitat. Protecting them protects many other species. (18)

Unitary organisms Organisms that exist as separate and distinct individuals. (8)

Variance A statistical measure of the variation surrounding the mean value in a data set. (3)

Vegetation The form of the plant life in a region. (4)

Water potential An energy gradient between two systems caused by their relative water and solute concentration. Water flows from high energy (high water potential) to low energy. (3)

Weathering The addition of nutrients from the parent material by physical and chemical processes. (17)

Wetland A semiaquatic habitat that may permanently or periodically contain water. (5)

Wildlife management The study of the methods and principles by which we can maintain viable populations of wildlife species. (1)

Xeric species Plant species that are adapted to a dry climate. (4)

Xerophyte A plant that tolerates hot, dry environments. (3)

References

Ackerly, D.D., et al. 2006. Niche evolution and adaptive radiation: testing the order of trait divergence. *Ecology* 87:S50–S61.

Ainley, D.G., et al. 2006. Competition among penguins and cetaceans reveals trophic cascades in the Western Ross Sea, Antarctica. *Ecology* 87:2080–2093.

Alessa, L., and F.S. Chapin III. 2008. Anthropogenic biomes: a key contribution to earth-system science. *Trends in Ecology and Evolution* 23:529–531.

Alford, R.A., and R.N. Harris. 1988. Effects of larval growth history on anuran metamorphosis. *American Naturalist* 131:91–106.

Allendorf, F.W., et al. 1991. Estimation of effective population size of grizzly bears by computer simulation. *Proceedings of the Fourth International Congress of Systematic and Evolutionary Biology* 2:650–654.

Alvarez-Filip, L., et al. 2009. Flattening of Caribbean coral reefs: region-wide declines in architectural complexity. *Proceedings of the Royal Society of London, Series B: Biological Sciences* 276:3019–3025.

Anderson, J.T., et al. 2009. Limited flooding tolerance of juveniles restricts the distribution of adults in an understory shrub (*Itea virginica*; Iteaceae). *American Journal of Botany* 96:1603–1611.

Angilletta, M.J. Jr., et al. 2009. Spatial dynamics of nesting behavior: lizards shift microhabitats to construct nests with beneficial thermal properties. *Ecology* 90:2933–2939.

Angulo, E., et al. 2007. Double Allee effects and extinction in the island fox. *Conservation Biology* 21:1082–1091.

Anthony, K.R.N., et al. 2008. Ocean acidification causes bleaching and productivity loss in coral reef builders. *Proceedings of the National Academy of Sciences of the United States of America* 105:1742–1744

Aplet, G.H., et al. 1988. Patterns of community dynamics in Colorado Engelmann spruce-subalpine fir forests. *Ecology* 69:312–319.

Arbogast, B.S., et al. 2006. The origin and diversification of Galápagos mockingbirds. *Evolution* 60:370–382.

Ardia, D.R. 2005. Individual quality mediates trade-offs between reproductive effort and immune function in tree swallows. *Journal of Animal Ecology* 74:517–524.

Arkema, K.K., et al. 2009. Direct and indirect effects of giant kelp determine benthic community structure and dynamics. *Ecology* 39:3126–3137.

Armbrecht, I., et al. 2004. Enigmatic biodiversity correlations: ant diversity responds to diverse resources. *Science* 304:284–286.

Arnqvist, G., and T. Nilsson. 2000. The evolution of polyandry: multiple mating and female fitness in insects. *Animal Behaviour* 60:145–164.

Ashmole, N.P. 1963. The regulation of numbers of tropical oceanic birds. *Ibis* 116:217–219.

Augerot, X. 2005. *Atlas of Pacific Salmon: The First Map-Based Status Assessment of Salmon in the North Pacific.* Berkeley: University of California Press.

Austin, A.T., and C.L. Ballare. 2010. Dual role of lignin in plant litter decomposition in terrestrial ecosystems. *Proceedings of the National Academy of Sciences of the United States of America* 107: 4618–4622.

Avise, J.C., et al. 2008. In the light of evolution II: biodiversity and extinction. *Proceedings of the National Academy of Sciences of the United States of America* 105:11453–11457.

Avise, J.C., and J.E. Mank. 2009. Evolutionary perspectives on hermaphroditism in fishes. *Sexual Development* 3:152–163.

Ayala, F.J. 1969. Experimental invalidation of the principle of competitive exclusion. *Nature* 224:1076.

Báez, S., et al. 2006. Bottom-up regulation of plant community structure in an aridland ecosystem. *Ecology* 87:2746–2754.

Bailey, J.K., et al. 2006. Importance of species interactions to community heritability: a genetic basis to trophic-level interactions. *Ecology Letters* 9:78–85.

Balmer, O., et al. 2009. Intraspecific competition between co-infecting parasite strains enhances host survival in African trypanosomes. *Ecology* 90:3367–3378.

Balvanera, P., et al. 2006. Quantifying the evidence for biodiversity effects on ecosystem functioning and services. *Ecology Letters* 9:1146–1156.

Baptista, L.F., and L. Petrinovich. 1984. Song development in the white-crowned sparrow: social factors and sex differences. *Animal Behaviour* 32:172–181.

Barbaro, L., et al. 2007. The spatial distribution of birds and carabid beetles in pine plantation forests: the role of landscape composition and structure. *Journal of Biogeography* 34:652–664.

Barknosky, A.D., et al. 2011. Has the earth's sixth mass extinction already arrived? *Nature* 471:51–57.

Bartel, R.A., et al. 2010. Ecosystem engineers maintain a rare species of butterfly and increase plant diversity. *Oikos Journal* 119:883–890.

Bass, M.S., et al. 2010. Global conservation significance of Ecuador's Yasuni National Park. *PLOS One* 5:e8767.

Bateman, A.J. 1948. Intra-sexual selection in *Drosophila*. *Heredity* 2:349–368.

Bazzaz., F.A. 1979. The physiological ecology of plant succession. *Annual Review of Ecology, Evolution, and Systematics* 10:351–371.

Bazzaz, F.A. 1996. *Plants in Changing Environments*. Cambridge: Cambridge University Press.

Beer, C., et al. 2010. Terrestrial gross carbon dioxide uptake: global distribution and covariation with climate. *Science* 329:834.

Benton, M.J. 2009. The Red Queen and the Court Jester: species diversity and the role of biotic and abiotic factors through time. *Science* 323:728–732.

Bennett, R.F., and R.E. Lenski. 2007. An experimental test of evolutionary trade-offs during temperature regulation. *Proceedings of the National Academy of Sciences of the United States of America* 104:8649–8654.

Berger, J. 1990. Persistence of different-sized populations: an empirical assessment of rapid extinctions in bighorn sheep. *Conservation Biology* 4:91–98

Berger, S., et al. 2007. Behavioral and physiological adjustments to new predators in an endemic island species, the Galápagos marine iguana. *Hormones and Behavior* 52:653–663.

Bernhardt, P., and R. Edens-Meier. 2010. What we think we know vs. what we need to know about orchid pollination and conservation: *Cypripedium* L. as a model lineage. *Botanical Review* 76:204–219.

Berthold P., et al. 1990. Steuerung und potentielle Evolutionsgeschwindigkeit des obligaten Teilzieherverhaltens: Ergebnisse eines Zweiweg-Selektionsexperiments mit der Mönchsgrasmücke (*Sylvia atricapilla*). *Journal of Ornithology* 131: 33–45.

Beyer, H.L., et al. 2007. Willow on Yellowstone's northern range: evidence for a trophic cascade? *Ecological Applications* 17:1563–1571.

Bianucci, L., and T.E. Martin. 2010. Can selection on nest size from nest predation explain the latitudinal gradient in clutch size? *Journal of Animal Ecology* 79:1086–1092.

Bisson, I., et al. 2009. Evidence for repeated independent evolution of migration in the largest family of bats. *PLOS One* 4(10):e7504.

Blanchet, S., et al. 2008. The effects of abiotic factors and intraspecific versus interspecific competition on the diel activity patterns of Atlantic salmon (*Salmo salar*) fry. *Canadian Journal of Fisheries and Aquatic Sciences* 65:1545–1553.

Blomqvist, D., et al. 2010. Trapped in the extinction vortex? Strong genetic effects in a declining vertebrate population. *BMC Evolutionary Biology* 10:33–43.

Bocak, L., et al. 2008. Multiple ancient origins of neoteny in Lycidae (Coleoptera): consequences for ecology and macroevolution. *Proceedings of the Royal Society of London, Series B: Biological Sciences* 275:2015–2023.

Bohn, T., et al. 2008. Competitive exclusion after invasion? *Biological Invasions* 10:359–368.

Bolnick, D.I., et al. 2010. Ecological release from interspecific competition leads to decoupled changes in population and individual niche width. *Proceedings of the Royal Society of London, Series B: Biological Sciences* 277:1789–1797.

Bonnell, M.L., and R.K. Selander. 1974. Elephant seals: genetic variation and near extinction. *Science* 184:908–909.

Boose, E.R., et al. 1994. Hurricane impacts to tropical and temperate forest landscapes. *Ecological Monographs* 64:369–400.

Börger, L., et al. 2008. Are there general mechanisms of animal home range behaviour? A review and prospectus for future research. *Ecology Letters* 11:637–650.

Bowers, M.A., and J.H. Brown. 1982. Body size and coexistence in desert rodents: chance or community structure? *Ecology* 63:391–400.

Braatne, J.H., and L.C. Bliss. 1999. Comparative physiological ecology of lupines colonizing early successional habitats on Mount St. Helens. *Ecology* 80:891–907.

Brandle, M., et al. 2002. Dietary niche breadth for Central European birds: correlations with species-specific traits. *Evolutionary Ecology Research* 4:643–657.

Bronstein, J.L., et al. 2006. The evolution of plant-insect mutualisms. *New Phytologist* 172:412–428.

Brookshire, E.N.J., et al. 2009. Maintenance of terrestrial nutrient loss signatures during in-stream transport. *Ecology* 90:293–299.

Brooks, T.M., et al. 2002. Habitat loss and extinction in the hotspots of biodiversity. *Conservation Biology* 16:909–923.

Brown, J.H., and D.W. Davidson. 1977. Competition between seed-eating rodents and ants in desert ecosystems. *Science* 196:880–882.

Brown, J.H., and M.V. Lomolino. 1998. *Biogeography*, 2nd edition. Sunderland, MA: Sinauer Associates.

Brown, J.L., et al. 2008. Divergence in parental care, habitat selection and larval life history between two species of Peruvian poison frogs: an experimental analysis. *Journal of Evolutionary Biology* 21:1534–1543.

Bruck, J.N., and J.M. Mateo. 2010. How habitat features shape ground squirrel (*Urocitellus beldingi*) navigation. *Journal of Comparative Psychology* 124:176–186.

Buckling, A., and M.A. Brockhurst. 2008. Kin selection and the evolution of virulence. *Heredity* 100:484–488.

Burba, G., and S.L. Forman. 2008. Eddy covariance method. In *Encyclopedia of Earth*, edited by C.J. Cleveland, 356–387. Washington, DC: Environmental Information Coalition, National Council for Science and the Environment.

Butchart, S.H., et al. 2010. Global biodiversity: indicators of recent declines. *Science* 328:1164–1168.

Byrne, P.G., and J.S. Keogh. 2009. Extreme sequential polyandry insures against nest failure in a frog. *Proceedings of the Royal Society of London, Series B: Biological Sciences* 276:115–120.

Cain, A.J., and P.M. Sheppard. 1954. Natural selection in *Cepea. Genetics* 39:89–116.

Calderia, K., et al. 2003. Climate sensitivity uncertainty and the need for energy without CO_2 emission. *Science* 299:2052–2054.

Calsbeek, R. 2008. Experimental evidence that competitive and habitat use shape the individual fitness surface. *Journal of Evolutionary Biology* 22:97–108.

Cameron, G.N. 1973. Effect of litter size on postnatal growth and survival in the desert woodrat. *Journal of Mammalogy* 54:489–493.

Capers, R.S., et al. 2009. The relative importance of local conditions and regional processes in structuring aquatic plant communities. *Freshwater Biology* doi:10.1111/j.1365-2427.2009.02328.x.

Carey, H.V., et al. 2009. Mammalian hibernation: cellular and molecular responses to depressed metabolism and low temperature. *Physiological Reviews* 83:1153–1181.

Caro, A., et al. 2010. Ecological convergence in a rocky intertidal shore metacommunity despite high spatial variability in recruitment regimes. *Proceedings of the National Academy of Sciences of the United States of America* 107:18528–18532.

Carpenter, K.E., et al. 2008. One-third of reef-building corals face elevated extinction risk from climate change and local impacts. *Science* 321:560–563.

Casey, A.E. 2011. Genetic parentage and local population structure in the socially monogamous upland sandpiper. *Condor* 113:119–128.

Caspari, R., and L. Sang-Hee. 2004. Older age becomes common late in human evolution. *Proceedings of the National Academy of Sciences of the United States of America* 101:10895–10900.

Cavender-Bares, J., et al. 2004. Phylogenetic overdispersion in Floridian oak communities. *American Naturalist* 163:823–843.

Caveieres, G., and P. Sabat. 2008. Geographic variation in response to thermal acclimation in rufous-colored sparrows: are physiological flexibility and environmental heterogeneity correlated? *Functional Ecology* 22:509–515.

Charmantier, A., et al. 2008. Adaptive phenotypic plasticity in response to climate change in a wild bird population. *Science* 320:800–803.

Charnov, E.L. 1976. Optimal foraging: the marginal value theorem. *Theoretical Population Biology* 9:129–136.

Charnov, E.L., and W.M. Schaffer. 1973. Life-history consequences of natural selection: Cole's result revisited. *American Naturalist* 107:791–793.

Chase, M.K., et al. 2005. Effects of weather and population density on reproductive success and population dynamics in a song sparrow (*Melospiza melodia*) population: a long-term study. *Auk* 122:571–592.

Chen, B-M, et al. 2009. Effects of the invasive plant *Mikania micrantha* H.B.K. on soil nitrogen availability through allelopathy in South China. *Biological Invasions* 11:1291–1299.

Chen, L., et al. 1997. Evolution of antifreeze glycoprotein gene from a trypsinogen gene in Antarctic notothenioid fish. *Proceedings of the National Academy of Sciences of the United States of America* 94:3811–3816.

Cheung, W.W.L., et al. 2009. Projecting global marine biodiversity impacts under climate change scenarios. *Fish and Fisheries* 10:235–251.

Childs, D.Z., et al. 2010. Evolutionary bet-hedging in the real world: empirical evidence and challenges revealed by plants. *Proceedings of the Royal Society of London, Series B: Biological Sciences* 277: 3055–3064.

Churchill, T.A., and K.B. Storey. 1995. Metabolic effects of dehydration on an aquatic frog, *Rana pipiens. Journal of Experimental Biology* 198:147–154.

Clark, D.A., et al. 2001. Net primary production in tropical forests: an evaluation and synthesis of existing field data. *Ecological Applications* 11:371–384.

Clausen, J.D., et al. 1948. Experimental studies on the nature of species. III. Environmental responses to climatic races of *Achillea. Carnegie Institute of Washington Publication no. 581*. Washington, DC.

Clements, F.E. 1916. Plant succession: analysis of the development of the vegetation. *Carnegie Institute of Washington Publications* 242:1–512.

Clements, F.E. 1936. Nature and structure of the climax. *Journal of Ecology* 24:252–284.

Cohen, D. 1966. Optimising reproduction in a randomly varying environment. *Journal of Theoretical Biology* 12:110–129.

Cole, L.C. 1954. The population consequences of life history phenomena. *Quarterly Review of Biology* 29:103–137.

Collins, S.L., and S.M. Glenn. 1991. Importance of spatial and temporal dynamics in species regional abundance and distribution. *Ecology* 72:654–664.

Colmer, T.D., and L. Voesenek. 2009. Flooding tolerance: suites of plant traits in variable environments. *Functional Plant Biology* 36:665–681.

Colwell, R.K., et al. 2008. Global warming, elevational range shifts, and lowland biotic attrition in the wet tropics. *Science* 322:258–261.

Condon, M.A., et al. 2008. Hidden neotropical diversity: greater than the sum of its parts. *Science* 320:928–932.

Connell, J.H. 1961. The influence of instraspciic competition and other factors on the distribution of the barnacle *Chthalamus stellatus*. *Ecology* 42:710–723.

Connell, J.H. 1971. On the role of natural enemies in preventing competitive exclusion in some marine animals and in rainforest trees. In *Dynamics of Populations*, edited by P.J. den Boer and G.R. Gradwell, 298–312. Wageningen, The Netherlands: Center for Agricultural Publishing and Documentation. .

Connell, J.H. 1978. Diversity in tropical rain forests and coral reefs. *Science* 199:1302–1310.

Connell, J.H., and R.O. Slatyer. 1977. Mechanisms of succession in natural communities and their role in community stability and organization. *American Naturalist* 111:1119–1144.

Cooper, N., et al. A common tendency for phylogenetic overdispersion in mammalian assemblages. *Proceedings of the Royal Society of London, Series B: Biological Sciences* 275:2031–2037.

Corliss, J.B., et al. 1979. Submarine thermal springs on the Galápagos rift. *Science* 203:1073–1083.

Cornell, H.V., and B.A. Hawkins. 2003. Herbivore responses to plant secondary compounds: a test of phytochemical coevolution theory. *American Naturalist* 161:507–522.

Cornwallis, C.K., et al. 2009. Routes to indirect fitness in cooperatively breeding vertebrates: kin discrimination and limited dispersal. *Journal of Evolutionary Biology* 22:2445–2447.

Costa, G.C., et al. 2008. Optimal foraging constrains macroecological patterns: body size and dietary niche breadth in lizards. *Global Ecology and Biogeography* 17:670–677.

Coughenour, M.B., and F.J. Singer. 1996. Elk population processes in Yellowstone National Park under the policy of natural regulation. *Ecological Applications* 6:573–593.

Coulson, T., et al. 2006. Estimating individual contributions to population growth: evolutionary fitness in ecological time. *Proceedings of the Royal Society of London, Series B: Biological Sciences* 273:547–555.

Cowen, R.K., and S. Sponaugle. 2009. Larval dispersal and marine population connectivity. *Annual Review of Marine Science* 1:443–466.

Cowles, H.C. 1901. The physiographic ecology of Chicago and vicinity: a study of the origin, development and classification of plant societies. *Botanical Gazette* 31:170–177.

Courchamp F., and D.W. Macdonald. 2001. Crucial importance of pack size in the African wild dog *Lycaon pictus*. *Animal Conservation* 4:169–174.

Cox, R.M., and R. Calsbeek. 2010. Severe costs of reproduction persist in *Anolis* lizards despite the evolution of a single-egg clutch. *Evolution* doi:10.1111/j.1558-5646.2009.00906.x.

Creel, S., and N.M. Creel. 2002. *The African Wild Dog: Behavior, Ecology, and Conservation*. Princeton, NJ: Princeton University Press.

Creel, S., et al. 2005. Elk alter habitat selection as an antipredator response to wolves. *Ecology* 86:3387–3397.

Crispo, E. 2008. Modifying effects of phenotypic plasticity on interactions among natural selection, adaptation and gene flow. *Journal of Evolutionary Biology* 21:1460–1469.

Crozier, L.G., et al. 2008. Potential responses to climate change in organisms with complex life histories: evolution and plasticity in Pacific salmon. *Evolutionary Applications* 252–270.

Crutsinger, G.M., et al. 2010. Genetic variation within a dominant shrub species determines plant species colonization in a coastal dune ecosystem. *Ecology* 91:1237–1243.

Curran, S.P., and G. Ruvkun. 2007. Lifespan regulation by evolutionarily conserved genes essential for viability. *PLOS Genetics* 3(4):e56. doi:10.1371/journal.pgen.0030056.

Dahl, T.W., et al. 2010. Devonian rise in atmospheric oxygen correlated to the radiations of terrestrial plants and large predatory fish. *Proceedings of the National Academy of Sciences of the United States of America* 42:17911–17915.

Daily, G. 1997. *Nature's Services: Societal Dependence on Natural Ecosystems*. Washington, DC: Island Press.

Dale, B.W., et al. 1994. Functional response of wolves preying on barren-ground caribou in a multiple-prey ecosystem. *Journal of Animal Ecology* 63:644–652.

Danchin, E., and R.H. Wagner. 2009. Inclusive heritability: combining genetic and non-genetic information to study animal behavior and culture. *Oikos Journal* 119:210–218.

Dangremond, E.M., et al. 2010. Apparent competition with an invasive plant hastens the extinction of an endangered lupine. *Ecology* 91:2261–2271.

Darwin, C. 1839. *Journal of Researches into the Geology and Natural History of the Various Countries Visited by HMS Beagle*. London: Colburn.

Darwin, C. 1861. *On the Origin of Species by Means of Natural Selection*, 3rd edition. London: Murray.

Darwin, C. 1871. *The Descent of Man and Selection in Relation to Sex*. London: Murray.

Dausman, K.H., et al. 2004. Physiology: hibernation in a tropical primate. *Nature* 429:835–826.

Davies, T.J., et al. 2008. Phylogenetic trees and the future of mammalian biodiversity. *Proceedings of the National Academy of Sciences of the United States of America* 105:11556–11563.

Davis, M.A. 2005. Biotic globalization: does competition from introduced species threaten biodiversity? *Bioscience* 53:481–489.

Davis, M.B. 1983. Quaternary history of deciduous forests of Eastern North America and Europe. *Annals of the Missouri Botanical Garden* 70:550–563.

Delgiudice, G.D., et al. 1997. Trends of winter nutrition, ticks, and numbers of moose on Isle Royale. J. *Journal of Wildlife Management* 61:895–903.

de Jager, D.R., and J. Pastor. 2009. Declines in moose population density at Isle Royale National Park, MI USA and accompanied changes in landscape patterns. *Landscape Ecology* 24:1389–1403.

de Jong, G. 2005. Evolution of phenotypic plasticity: patterns of plasticity and the emergence of ecotypes. *New Phytologist* 166:101–117.

DeLuca, T.H., et al. 2008. Ecosystem feedbacks and nitrogen fixation in boreal forests. *Science* 320: 1181–1183.

Desjardins, J.K., and R.D. Fenald. 2009. Fish sex: why so diverse? *Current Opinion in Neurobiology* 19:1–6.

Desrochers, A. 2010. Morphological response of songbirds to 100 years of landscape change in North America. *Ecology* 91:1577–1582.

Diamond, J.D. 1969. Comparison of faunal equilibrium turnover rates on the Channel Islands of California. *Proceedings of the National Academy of Sciences of the United States of America* 69:3199–3203.

Des Roches, S., et al. 2013. Ecological and evolutionary effects of stickleback on community structure. *PLOS One* 8(4):e59644. doi:10.1371/journal.pone.0059644.

Diaz, R.J., and R. Rosenberg 2008. Spreading dead zones and consequences for marine ecosystems. *Science* 321:926–929.

Doney, S.C., et al. 2009. Ocean acidification: the other CO_2 problem. *Annual Review of Marine Science* 1:169–192.

Donner, D.M., et al. 2010. Patch dynamics and the timing of colonization-abandonment events by male Kirtland's warblers in an early successional habitat. *Biological Conservation* 143:1159–1167.

Dowdall, J.T., et al. 2012. Local adaptation and the evolution of phenotypics plasticity in Trinidadian guppies (*Poecilia reticulate*). *Evolution* doi: 10.1111/j.1558–5646.2012.01694.x.

Dowling, D.K., and L.W. Simmons. 2009. Reactive oxygen species as universal constraints in life-history evolution. *Proceedings of the Royal Society of London, Series B:Biological Sciences* 10:124–136.

Dudley, S.A., and A.L. File 2007. Kin recognition in an annual plant. *Biology Letters* 3:435–438.

Dugger, K.M., et al. 2010. Survival differences and the effect of environmental instability on breeding dispersal in an Adélie penguin meta-population. *Proceedings of the National Academy of Sciences of the United States of America* 107:12375–12380.

Dybznski, R., et al. 2008. Soil fertility increases with plant species diversity in a long-term biodiversity experiment. *Oecologia* 158:85–93.

Edwards, E.J., et al. 2010. The origins of C4 grasslands: integrating evolutionary and ecosystem. *Science* 328:587–591.

Ehrlich, P.R., and J.P. Holdren. 1971. Impact of population growth. *Science* 171:1212–1217.

Ehrlich, P.L., and P.H. Raven. 1964. Butterflies and plants: a study in coevolution. *Evolution* 18:586–608.

Elias, M., et al. 2008. Mutualistic interactions drive ecological niche convergence in a diverse butterfly community. *PLOS Biology* 6(12):e300 doi:10.1371/journal.pbio.0060300.

Ellis, E.C., and N. Ramankutty. 2008. Putting people on the map: anthropogenic biomes of the world. *Frontiers in Ecology and the Environment* doi:10.1890/070062.

Ellwood, M.D.F., et al. 2009. Stochastic and deterministic processes jointly structure tropical arthropod communities. *Ecology Letters* 12:277–284.

Elser, J.J., et al. 2007. Global analysis of nitrogen and phosphorus limitation of primary producers in freshwater, marine and terrestrial ecosystems. *Ecology Letters* doi: 10.1111/j.1461-0248.2007.01113.x.

Elser, J.J., et al. 2009. Shifts in lake N:P stoichiometry and nutrient limitation driven by atmospheric nitrogen deposition. *Science* 326:835.

Elton, C.S. 1927. *Animal Ecology.* New York: Macmillan.

Emerson, B.C., and P. Oromi. 2005. Diversification of the forest beetle genus *Turphius* on the Canary Islands, and the evolutionary origins of island endemics. *Evolution* 59:586–596.

Emerson, B.C., and R.G. Gillespie. 2008. Phylogenetic analysis of community assembly and structure over space and time. *Trends in Ecology and Evolution* doi:10.1016/j.tree.2008.07.005.

Emlen, S.T., and L.W. Oring. 1977. Ecology, sexual selection, and the evolution of mating systems. *Science* 197:215–223.

Emson, R.H., and R.J. Faller-Fritsch. 1976. An experimental investigation into the effect of crevice availability on abundance and size structure in a population of *Littorina rudis* (Maton): Gastropoda: Prosobranchia. *Journal of Experimental Marine Biology and Ecology* 23:285–297.

Endler, J.A. 1983. Natural and sexual selection on color patterns in *Poecilia reticulata*. *Environmental Biology of Fishes* 9:173–190.

Engen, S., et al. 2001. Stochastic population dynamics and time to extinction of a declining population of barn swallows. *Journal of Animal Ecology* 70:789–797.

Enger, E.D., and B.F. Smith. 2000. *Environmental Science*. New York: McGraw-Hill.

Erwin, D.H. 2008. Macroevolution of ecosystem engineering, niche construction and diversity. *Trends in Ecology and Evolution* 23:304–310.

Evans, C.W., et al. 2010. How do Antarctic notothenioid fishes cope with internal ice? A novel function for antifreeze glycoproteins. *Antarctic Science* 23: 57–64.

Ewald, P.W. 2004. Evolution of virulence. *Infectious Disease Clinics of North America* 18:1–15.

Fagan, W.F., and E.E. Holmes. 2006. Quantifying the extinction vortex. *Ecology Letters* 9:51–60.

Fakheran, S., et al. 2010. Adaptation and extinction in experimentally fragmented landscapes. *Proceedings of the National Academy of Sciences of the United States of America* 107:19120–19125.

Falkowski, P., et al. 2000. The global carbon cycle: a test of our knowledge of earth as a system. *Science* 290:291.

Falkowski, P., et al. 2008. The microbial engines that drive earth's biogeochemical cycles. *Science* 320:1034.

Fastie, C.L. 1995. Causes and ecosystem consequences of multiple pathways of primary succession at Glacier Bay, Alaska. *Ecology* 76:1899–1916.

Faulkes, C.G., and N.C. Bennett. 2001. Family values: group dynamics and social control of reproduction in African mole rats. *Trends in Ecology and Evolution* 16:184–190.

Feeny, P. P. 1976. Plant apparency and chemical defense. In *Biochemical Interactions Between Plants and Insects*, edited by J.W. Wallace and R.L. Mansell. Plenum, New York.

Feller, G., and C. Gerday. 2003. Psychrophilic enzymes: hot topics in cold adaptation. *Nature Reviews Microbiology* 1:200–208.

Fields, P.A., et al. 2008. Function of muscle-type lactate dehydrogenase and citrate synthase of the Galápagos marine iguana *Amblyrhynchus cristatus*, in relation to temperature. *Comparative Biochemistry and Physiology, Part B: Biochemistry and Molecular Biology* 150:62–73.

Filotas, E., et al. 2010. The effect of positive interactions on community structure in a multi-species metacommunity model along an environmental gradient. *Ecological Modelling* 221:885–894.

Finch, C.E. 1990. *Longevity, Senescence and the Genome*. Chicago: University of Chicago Press.

Fischer, J., and D.B. Lindenmayer. 2007. Landscape modification and habitat fragmentation: a synthesis. *Global Ecology and Biogeography* 16:265–280.

Fisher, R.A. 1930. *The Genetical Theory of Natural Selection*. Oxford, UK: Clarendon Press.

Fletcher, R.J. 2008. Social information and community dynamics: nontarget effects from stimulating social cues for management. *Ecological Applications* 18:1764–1773.

Fisher, R.A. 1930. *The Genetical Theory of Natural Selection*. New York: Oxford University Press.

Ford, H.A., et al. 2009. Extinction debt or habitat change? Ongoing losses of woodland birds in northeastern New South Wales, Australia. *Biological Conservation* 142:3182–3190.

Forrester, G.E., et al. 2006. Assessing the magnitude of intra- and interspecific competition in two coral reef fishes. *Oecologia* 148:632–640.

Foster, D.R. 1988. Species and stand response to catastrophic wind in central New England USA. *Journal of Ecology* 76:135–151.

Foster, D.R., and G.A. King. 1986. Vegetation pattern and diversity in SE Labrador, Canada: *Betula papyrifera* (birch) forest development in relation to fire history and phsiography. *Journal of Ecology* 74:465–483.

Frank, S.A., and P. Schmid-Hempel. 2008. Mechanisms of pathogenesis and the evolution of parasite virulence. *Journal of Evolutionary Biology* 21:396–404.

Franklin, A.B., et al. 2000. Climate, habitat quality, and fitness in northern spotted owl populations in northwestern California. *Ecological Monographs* 70:539–590.

Fraser, C., et al. 2009. The bacterial species challenge: making sense of genetic and ecological diversity. *Science* 323:741–746.

Freestone, A.L., et al. 2010. Impacts of predation on marine epifaunal communities across latitude. Paper presented at the 95th ESA Annual Meeting, Pittsburgh.

Gaba, S., and D. Ebert. 2009. Time-shift experiments as a tool to study antagonistic coevolution. *Trends in Ecology and Evolution* doi:10.1016/j .tree.2008.11.005.

Gault, A., et al. 2008. Consumers' taste for rarity drives sturgeons to extinction. *Conservation Letters* 1:199–207.

Garcia-Verdugo, C., et al. 2009. Phenotypic plasticity and integration across the canopy of *Olea europaea* subsp. *guanchica* (Oleaceae) in populations with different wind exposures. *American Journal of Botany* 96:1454–1461.

Gause, G.F. 1934. *The Struggle for Existence*. New York: Williams and Wilkins.

Gedan, K.B., et al. 2009. Small-mammal herbivore control of secondary succession in New England tidal marshes. *Ecology* 90:430–440.

Ghalambor, C.K., et al. 2007. Adaptive versus non-adaptive phenotypic plasticity and the potential for contemporary adaptation in new environments. *Functional Ecology* 21:394–407.

Gibbs, A.G., et al. 1997. Physiological mechanisms of evolved desiccation resistance in *Drosophila melanogaster*. *Journal of Experimental Biology* 200:1821–1832.

Gibbs, H.L., and P.R. Grant. 1987. Ecological consequences of an exceptionally strong El Niño on Darwin's finches. *Ecology* 68:1735–1746.

Gigord, L.D.B., et al. 2001. Negative frequency-dependent selection maintains a dramatic flower color polymorphism in the rewardless orchid *Dactylorhiza sambucina* (L.) Soo. *Proceedings of the National Academy of Sciences of the United States of America* 98:6253–6355.

Gill, F.B., and L.L. Wolf. 1975. Economics of feeding territoriality in the golden-winged sunbird. *Ecology* 56:333–345.

Gleason, H.A. 1926. The individualistic concept of the plant association. *Torrey Botanical Club Bulletin* 53:7–26.

Gomez, J.P., et al 2010. A phylogenetic approach to disentangling the role of competition and habitat filtering in community assembly of neotropical forest birds. *Journal of Animal Ecology* 79:1181–1192.

Gosline, A.K., and F.H. Rodd. 2008. Predator-induced plasticity in guppy (*Poecilia reticulata*) life history traits. *Aquatic Ecology* 42:693–699.

Gotelli, N.J. 2008. *A Primer of Ecology.* Sunderland, MA: Sinauer Associates.

Gould, S.J. 1980. *The Panda's Thumb.* New York: Norton.

Gotzenberger, L., et al. 2011. Ecological assembly rules in plant communities—approaches, patterns and prospects. *Biological Reviews* doi:10.1111/j.1469-185X.2011.00187.x.

Grant, B.R., and P.R. Grant. 1993. Evolution of Darwin's finches caused by a rare climatic event. *Proceedings of the Royal Society of London, Series B: Biological Sciences* 251:111–117.

Grant, P.R., and B.R. Grant. 2006. Evolution of character displacement in Darwin's finches. *Science* 313:224–226.

Grant, P.R., and B.R. Grant. 2008. *How and Why Species Multiply: The Radiation of Darwin's Finches.* Princeton, NJ: Princeton University Press.

Grant, P.R., and B.R. Grant. 2009. Evolution of Darwin's finches. Paper presented at the 2009 Kyoto Prize Workshop in Basic Sciences.

Gravel, D., et al. 2010. Patch dynamics, persistence, and species coexistence in metaecosystems. *American Naturalist* 176:289–302.

Greene, H.W., and R.W. McDiarmid. 1981. Coral snake mimicry: does it occur? *Science* 213:1207–1211.

Greenwood, P.J. 1983. Mating systems and the evolutionary consequences of dispersal. In *The Ecology of Animal Movement*, edited by I.R. Swingland and P.J. Greenwood, 1140–1162. New York: Oxford University Press.

Grime, P.J. 1977. Biodiversity and ecosystem function: the debate deepens. *Science* 277:1260–1261.

Grinnell, J. 1917. The niche relationships of the California thrasher. *Auk* 21:364–382.

Hall, M., and J.C.S. Bueno. 2007. Within-plant signaling by volatiles leads to induction and priming of an indirect plant defense in nature. *Proceedings of the National Academy of Sciences of the United States of America* 104:5467–5472.

Halperin, B.S., et al. 2006. Strong top-down control in southern California kelp forest ecosystems. *Science* 312:1230–1232.

Hamilton, W.D. 1964. The genetical evolution of social behavior. *Journal of Theoretical Biology* 7:1–52.

Hanski, I. 1982. Dynamics of regional distribution: core and satellite species hypothesis. *Oikos Journal* 38:210–221.

Harris, R.B., and F.W. Allendorf. 1989. Genetically effective population size of large mammals: assessment of estimators. *Conservation Biology* 3:181–191.

Harrison, S., et al. 1988. Distribution of the bay checkerspot butterfly, *Euphydryas editha bayensis*: evidence for a metapopulation. *American Naturalist* 132:360–382.

Hart, A., and M. Begon. 1982. The status of general life-history strategy theories, illustrated in winkles. *Oecologia* 52:37–42.

Harte, J. 2007. Human population as a dynamic factor in environmental degradation. *Population and Environment* 28:223–236.

Hasler, J.F., and M. Banks. 1975. Reproductive performance and growth in captive collared lemmings (*Discrostonyx groenlandicus*). *Canadian Journal of Zoology* 53:777–787.

Hastings, J.R., and R.M. Turner. 1965. *The Changing Mile.* Tucson: University of Arizona Press.

Hatchwell, B.J., and J. Komdeur. 2000. Ecological constraints, life history traits and the evolution of cooperative breeding. *Animal Behaviour* 59:1079–1086.

Hazandy, A., et al. 2009. Net primary productivity of forest trees: a review of current issues. *Pertanika Journal of Tropical Agricultural Science* 32:111–123.

Heckle, C., et al. 2010. Nonconsumptive effects of a generalist ungulate herbivore drive decline of unpalatable forest herbs. *Ecology* 91:319–326.

Hedin, L.O., et al. 2009. The nitrogen paradox in tropical forest ecosystems. *Annual Review of Ecology, Evolution, and Systematics* 40:613–635.

Heppell, S.S. 1998. Application of life-history theory and population model analysis to turtle conservation. *Copeia* 1998(2):367–375.

Hessburg, P.F., et al. 2007. Re-examining fire severity relations in pre-management era mixed conifer forests: inferences from landscape patterns of forest structure. *Landscape Ecology* 22:5–24.

Hill, R.W., and G.A. Wyse. 1989. *Animal Physiology.* New York: Harper and Row.

Hoberg, E.P., and D.R. Brooks. 2008. A macro-evolutionary mosaic: episodic host-switching, geographical colonization and diversification in complex host-parasite systems. *Journal of Biogeography* 35:1533–1550.

Hohochka, P.W., and G.N. Somero. 2002. *Biochemical adaptation, mechanism and process in physiological evolution.* New York: Oxford University Press.

Holderegger, R., and H.H. Wagner. Landscape genetics. *BioScience* 58:199–207.

Hollander, J., and R.K. Butlin. 2010. The adaptive value of phenotypic plasticity in two ecotypes of a marine gastropod. *BMC Evolutionary Biology* 10:333–339.

Holling, C.S. 1959. The components of predation as revealed by a study of small mammal predation of the European pine sawfly. *Canadian Entomologist* 91:293–320.

Hooten, M.B., and C.K. Wikle. 2008. A hierarchical Bayesian non-linear spatio-temporal model for the spread of invasive species with application for the Eurasian collared-dove. *Environmental and Ecological Statistics* 15:59–70.

Houlahan, J.E., et al. 2007. Compensatory dynamics are rare in natural ecological communities. *Proceedings of the National Academy of Sciences of the United States of America* 104:3273–3277.

Howe, E., and W.L. Baker. Landscape heterogeneity and disturbance interactions in a subalpine watershed in Northern Colorado, USA. *Annals of the Association of American Geographers* 93:797–813.

Hubert, C., et al. 2009. A constant flux of diverse thermophilic bacteria into the cold Arctic seabed. *Science* 325:1541–1544.

Hudson, P.J., et al. 1998. Prevention of population cycles by parasite removal. *Science* 282:2256–2258.

Hughes, C., and R. Eastwood. 2006. Island radiation on a continental scale: exceptional rates of plant diversification after uplift of the Andes. *Proceedings of the National Academy of Sciences of the United States of America* 103:10334–10339.

Hughes, W.O., et al. 2008. Ancestral monogamy shows kin selection is key to the evolution of eusociality. *Science* 320:1213–1216.

Huey, R.G. 1991. Physiological consequences of habitat selection. *American Naturalist* 137:S91–S115.

Hutchinson, G.E. 1957. Concluding remarks. *Cold Spring Harbor Symposium on Quantitative Biology* 22:415–427.

Hutchinson, G.E. 1965. *The Ecological Theater and the Evolutionary Play.* New Haven, CT: Yale University Press.

Hutchinson, G.E. 1978. *An Introduction to Population Ecology.* New Haven, CT: Yale University Press.

Ito-Inaba, Y., et al., 2009. What is critical for plant thermogenesis? Differences in mitochondrial activity and protein expression between thermogenic and non-thermogeneic skunk cabbages. *Planta* 231:121–130.

Ives, A.R., and S.R. Carpenter. 2007. Stability and diversty of ecosystems. *Science* 317:58–62.

Ives, A.R., et al. 2007. Stability and diversity of ecosystems. *Science* 317:58–62.

Jablonski, D., et al. 2006. Out of the tropics: evolutionary dynamics of the latitudinal diversity gradient. *Science* 314:102–106.

Janetski, D.J., et al. 2009. Pacific salmon effects on stream ecosystems: a quantitative synthesis. *Oecologia* 159:583–595.

Janzen, D.H. 1966. Coevolution of mutualism between ants and acacias in Central America. *Evolution* 20:249–275.

Janzen, D.H. 1970. Herbivores and the number of tree species in tropical forests. *American Naturalist* 102:592–595.

Jetz, W., et al. 2008. The worldwide variation in avian clutch size across species and space. *PLOS Biology* 6(12):e303. doi:10.1371/journal.pbio .0060303.

Jiang, L., and S.N. Patel. 2008. Community assembly in the presence of disturbance: a microcosm experiment. *Ecology* 89:1931–1940.

Jimenez, J.A., et al. 1994. An experimental study of inbreeding depression in a natural habitat. *Science* 266:271–73.

Joern, A. 2005. Disturbance by fire frequency and bison grazing modulate grasshopper assemblages in tallgrass prairie. *Ecology* 86:861–873.

Johnson, D.W., et al. 2005. The effects of wildfire, salvage logging and post-fire N-fixation on the nutrient budgets of a Sierran forest. *Forest Ecology and Management* 220:155–165.

Johnson, E.A., and K. Miyanishi, eds. 2007. *Plant Disturbance Ecology.* Amsterdam, The Netherlands: Elsevier.

Johnson, E.A., and K. Miyanishi. 2008. Testing the assumptions of chronosequences in succession. *Ecology Letters* 11: 419–431.

Johnson, K.P., et al. 2009. Competition promotes the evolution of host generalists in obligate parasites. *Proceedings of the Royal Society of London, Series B: Biological Sciences* 276:3921–3926.

Johnson, M.T.J., and A.A. Agrawal. 2005. Plant genotype and environment interact to shape diverse arthropod community on evening primrose (*Oenothera biennis*). *Ecology* 86:874–885.

Johnson, M.T.J., and J.R. Stinchcombe. 2007. An emerging synthesis between community ecology and evolutionary biology. *Trends in Ecology and Evolution* doi:10.1016/j.tree.2007.01.014.

Johnson, Z.I., et al. 2006. Niche partitioning among *Prochlorococcus* ecotypes along ocean-scale environmental gradients. *Science* 311:1737–1740.

Jones, C.D., et al. 2003. Uncertainty in climate-carbon-cycle projections associated with the sensitivity of soil respiration to temperature. *Tellus* 55B:642–648.

Karatayev, A.Y., et al. 2008. Invaders and not a random selection of species. *Biological Invasions* doi 10.1007/s10530-009-9498-0.

Karl, D.M., et al. 1980. Deep-sea primary production at the Galapagos hydrothermal vents. *Science* 207:1345–1347.

Kaustuv, R., et al. 2009. A macroevolutionary perspective on species range limits. *Proceedings of the Royal Society of London, Series B: Biological Sciences* 276:1485–1493.

Kay, K.M., and R.D. Sargent. 2009. The role of animal pollination in plant speciation: integrating ecology, geography, and genetics. *Annual Review of Ecology, Evolution, and Systematics* 40:637–656.

Kozlowski, T.T. 1984. Plant responses to flooding of soil. *BioScience* 34:162–168.

Keith, L.B. 1963. *Wildlife's Ten-Year Cycle.* Madison: University of Wisconsin Press.

Keller, L.F., et al. 2008. Testing evolutionary models of senescence in a natural population: age and inbreeding effects on fitness components in song sparrows. *Proceedings of the Royal Society of London, Series B: Biological Sciences* 275:597–604.

Kelly, C.D. 2008. The interrelationship between resource-holding potential, resource-value and reproductive success in territorial males: how much variation can we explain? *Behavioral Ecology and Sociobiology* 62:855–871.

Kelt, D.A., and D.H. Van Vuren. 2001. The ecology and macroecology of mammalian home range area. *American Naturalist* 157:637–645.

Kenyon, C., et al. 1993. A *C. elegans* mutant that lives twice as long as the wild type. *Nature* 366:461–464.

Kerkhoff, A.J., and B.J. Enquist. 2006. Ecosystem allometry: the scaling of nutrient stocks and primary productivity across plant communities. *Ecology Letters* 9:419–427.

Kettlewell, B. 1973. *The Evolution of Melanism.* New York: Oxford University Press.

Kiers, E.T., et al. 2010. Mutualism in a changing world: an evolutionary perspective. *Ecology Letters* 13:1459–1474.

Kikvidze, Z., et al. 2005. Linking patterns and processes in alpine plant communities: a global study. *Ecology* 86:1395–1400.

Kipfmueller, K.F., and W.L. Baker. 2009. A fire history of a subalpine forest in south-eastern Wyoming, USA. *Journal of Biogeography* 27:71–85.

Kirschel, A.N.G., et al. 2009. Character displacement of song and morphology in African tinkerbirds. *Proceedings of the National Academy of Sciences of the United States of America* 106:8256–8261.

Kitching, R.L. 2006. Crafting the pieces of the diversity jigsaw puzzle. *Science* 313:1055–1057.

Kitchner, A.C. 1999. Tiger distribution, phenotypic variation and conservation issues. In *Riding the Tiger: Tiger Conservation in Human-Dominated Landscapes,* edited by J. Seidensticker et al., 19–40 Cambridge, UK: Cambridge University Press.

Koricheva, J., et al. 2009. Effects of mycorrhizal fungi on insect herbivores: a meta-analysis. *Ecology* 90:2088–2097.

Kozlowski, T.T. 1984. Plant responses to flooding of soil. *BioScience* 34:162–168.

Kraft, N.J.B., et al. 2007. Trait evolution, community assembly, and the phylogenetic structure of ecological communities. *American Naturalist* 170:271–283.

Kraft, N.J.B., et al. 2008. Functional traits and niche-based tree community assembly in an Amazonian forest. *Science* 322:580–582.

Kratayev, A.Y., et al. 2009. Invaders are not a random selection of species. *Biological Invasions* doi: 10.1007/s10530-009-9498-0.

Krebs, C.J. 2002. Two complementary paradigms for analyzing population dynamics. *Philosophical Transactions from the Royal Society of London, Series B: Biological Sciences* 357:1211–1219.

Kruger, O., et al. 2009. Does coevolution promote species richness in parasitic cuckoos? *Proceedings of the Royal Society of London, Series B: Biological Sciences* 276:3871–3879.

Krummel, J.R., et al. 1987. Landscape patterns in a disturbed environment. *Oikos Journal* 48:321–324.

Kruuk, L.E.B., et al. 2002. Antler size in red deer: heritability and selection but no evolution. *Evolution* 56:1683–1695.

Lack, D. 1954. The evolution of reproductive rates. In *Evolution as a Process,* edited by J.S. Huxley et al., 156–172. Lawrence, KS: Allen.

Lacy, R.C., et al. 1996. Hierarchical analysis of inbreeding depression in *Peromyscus polionotus. Evolution* 50:2187–2200.

Lake, P.S. 2000. Disturbance, patchiness, and diversity in streams. *Journal of the North American Benthological Society* 19:573–592.

Lamb, E.G., and J.F. Cahill Jr. 2008. When competition does not matter: grassland diversity and community composition. *American Naturalist* 171:777–787.

Lan, G., et al. 2010. Spatial dispersion patterns of trees in a tropical rainforest in Xishuangbana, southwest China. *Ecological Research* 25:46–55.

Lardner, B. 2000. Morphological and life history responses to predators in larvae of seven anurans. *Oikos Journal* 88:169–180.

Lawton, J.H. 1992. There are not 10 million kinds of population dynamics. *Oikos Journal* 63:337–338.

Lawton, R.O., and F.E. Putz. 1988. Natural disturbance and gap-phase regeneration in a wind-exposed tropical cloud forest. *Ecology* 69:764–777.

Layman, C.A., et al. 2007. Can stable isotope ratios provide for community-wide measures of trophic structure? *Ecology* 88:42–48.

Lenoir, J., et al. 2008. A significant upward shift in plant species optimum elevation during the 20th century. *Science* 320:1768–1771.

Le Roux, A., et al. 2009. Vigilance behaviour and fitness consequences: comparing solitary foraging and an obligate group-foraging mammal. *Behavioral Ecology and Sociobiology* 63:1097–1107.

Leverich, W. J., and D.A. Levin. 1979. Age-specific survivorship and reproduction in *Phlox drummondi*. *American Naturalist* 113:881–893.

Levine, J.M. 2000. Species diversity and biological invasions: relating local process to community pattern. *Science* 288:852–854.

Levins, D.A. 1968. *Evolution in Changing Environments*. Princeton, NJ: Princeton University Press.

Levins, S.A. 1976. Spatial patterning and the structure of ecological communities. *Lectures on Mathematics in the Life Sciences* 8:1–35.

Levins, S.A., et al. 1993. *Patch Dynamics: Lecture Notes in Biomathematics*. New York: Springer-Verlag.

Lewis, J.M., et al. 2008. Mapping uncharted water: exploratory analysis, visualization, and clustering of oceanographic data. *Proceedings of Seventh International Conference on Machine Learning and Applications* 388–395

Lidicker, W.Z. 1986. An overview of dispersal in non-volant small mammals. In *Migration Mechanisms and Adaptive Significance*, edited by M.A. Rankin, 569–585. *Contributions in Marine Science, University of Texas* 27.

Liebold, M.A., and M.A. McPeek. 2006. Coexistence of the niche and neutral perspectives in community ecology. *Ecology* 87:1399–1410.

Likens, G.H., et al. 1970. Effects of cutting and herbicide treatment on nutrient budgets in the Hubbard Brook Watershed Ecosystem. *Ecological Monographs* 40:23–47.

Likens, G.E. 1985. An experimental approach for the study of ecosystems. *Journal of Ecology* 73:380–396.

Lima, M. 2007. Locust plagues, climate variation, and the rhythms of nature. *Proceedings of the National Academy of Sciences of the United States of America* 104:15972–15973.

Litchman, E., and C.A. Klausmeir. 2008. Trait-based community ecology of phytoplankton. *Annual Review of Ecology, Evolution, and Systematics* 39: 615–639.

Lively, C.M. 1996. Host-parasite coevolution and sex. *BioScience* 46:107–114.

Lonsdale, W.M., and A.R. Watkinson 1983. Light and self-thinning. *New Phytologist* 90:399–418.

Lopez-Hoffman, L., et al. 2007. Salinity and light interactively affect neotropical mangrove seedlings at the leaf and whole plant levels. *Oecologia* 150: 545–556.

Lortie, C.J., et al. 2004. Rethinking plant community theory. *Oikos Journal* 107:433–438.

Losey, J.E., and M. Vaughan. 2006. Economic value of ecological services provided by insects. *BioScience* 56:311–323.

Lotka, A.J. 1925. *The Elements of Physical Biology*. Baltimore, MD: Williams and Wilkins.

Lovette, I., et al. 2006. Simultaneous effects of phylogenetic niche conservatism and competition on avian community structure. *Ecology* 87:S14–S28.

Lüthi, D., et al. 2008. High-resolution carbon dioxide concentration record 650,000–800,000 years before present. *Nature* 453:379.

MacArthur, R.H. 1958. Population ecology of some warblers of northeastern coniferous forests. *Ecology* 39:533–536.

MacArthur, R.H., and E.O. Wilson. 1967. An equilibrium theory of insular zoogeography. *Evolution* 17:373–387.

MacArthur, R.H., and E.O Wilson. 1967. *The Theory of Island Biogeography*. Princeton, NJ: Princeton University Press.

Maestre, F.T., et al. 2009. Refining the stress-gradient hypothesis for competition and facilitation in plant communities. *Journal of Ecology* 97:199–205.

Mahowald, N. 2011. Aerosol indirect effect on biogeochemical cycles and climate. *Science* 334:794.

Malthus, T.R. 1798. *An Essay on the Principle of Population*.

Marler, P. 1970. Bird song and speech development: could there be parallels? *American Scientist* 58:669–673.

Marshall, D.R., and S.K. Jain 1969. Interferences in pure and mixed populations of *Avena fatua* and *Avena barbata*. *Journal of Ecology* 57:251–270.

Marshall, J.P., et al. 2009. Intrinsic and extrinsic sources of variation in the dynamics of large herbivore populations. *Canadian Journal of Zoology* 87:103–111.

Martin, J.H., et al. 1994. Testing the iron hypothesis in ecosystems of the equatorial Pacific Ocean. *Nature* 371:123–129.

Mayhew, P.J., et al. 2008. A long-term association between global temperature and biodiversity, origination and extinction in the fossil record. *Proceedings of the Royal Society of London, Series B: Biological Sciences* 275:47–53.

McCairns, R.J., and L. Bernatchez. 2009. Adaptive divergence between freshwater and marine sticklebacks: insights into the role of phenotypic plasticity from an integrated analysis of candidate gene expression. *Evolution* 64:1029–1047.

McCloughlin, P.D., et al. 2007. Lifetime reproductive success and composition of the home range in a large herbivore. *Ecology* 88:3192–3201.

McDowell, N., et al. 2009. Mechanisms of plant survival and mortality during drought: why do some plants survive while others succumb to drought? *New Phytologist* 178:719–739.

McFarland, W.N., et al. 1985. *Vertebrate Life*. New York: Macmillan.

McKane, R.B., et al. 2002. Resource-based niches provide a basis for plant species diversity and dominance in Arctic tundra. *Nature* 415:68–71.

McKinney, S., et al. 2009. Invasive pathogen threatens bird-pine mutualism: implications for sustaining a high-elevation ecosystem. *Ecological Applications* 19:597–607.

McMenamin, S.K., and E.A. Hadly. 2010. Developmental dynamics of *Ambystoma tigrinum* in a changing landscape. *Ecology* 10:10.

McWilliams S.R., et al. 2004. Flying, fasting, and feeding in birds during migration: a nutritional and physiological perspective. *Journal of Avian Biology* 35:377–393.

Mech, L.D. 2007. Possible use of foresight, understanding and planning by wolves hunting muskoxen. *Arctic* 60:145–149.

Medel, R., et al. 2003. Pollinator-mediated selection on the nectar guide phenotype in the Andean monkey flower, *Mimulus luteus*. *Ecology* 84:1721–1732.

Meiners, S.J. 2007. Apparent competition: an impact of exotic shrub invasion on tree regeneration. *Biological Invasions* 9:849–855.

Meléndez-Ackerman, E., et al. 1997. Hummingbirds behavior and mechanisms of selection on flower color in *Ipomopsis*. *Ecology* 78:2532–2541.

Menge, D.N.L., and L.O Hedin. 2009. Nitrogen fixation in different biogeochemical niches along a 120,000-year chronosequence in New Zealand. *Ecology* 90:2190–2201.

Messier, C., et al. 2009. Resource and non-resource root competition effects of grasses on early versus late-successional trees. *Journal of Ecology* 97:548–554.

Miller, C.R., and L. P. Waits. 2003. The history of effective population size and genetic diversity in the Yellowstone grizzly (*Ursos arctos*): implications for conservation. *Proceedings of the National Academy of Sciences of the United States of America* 100:4334–4339.

Mitchell, R.J., et al. 2009. Ecology and evolution of plant-pollinator interactions. *Annals of Botany* 103:1355–1363.

Mitchie, L.J., et al. 2010. Melanic through nature or nurture: genetic polymorphism and phenotypic plasticity in *Harmonia axyridis*. *Journal of Evolutionary Biology* 23:1699–2707.

Mittlebach, G.G., et al. 2007 Evolution and the latitudinal diversity gradient: speciation, extinction, biogeography. *Ecology Letters* 10:315–331.

Moller, A.P., and S. Legendre 2001. Allee effect, sexual selection and demographic stochasticity. *Oikos Journal* 92:27–34.

Monaghan, P., et al. 2008. The evolutionary ecology of senescence. *Functional Ecology* 22:371–378.

Montesinos, A., et al. Demographic and genetic patterns of variation among populations of *Arabadopsis thaliana* from contrasting native environments. *PLoS One* 4:e7213.

Moore, J.W., et al. 2007. Biotic control of stream fluxes: spawning salmon drive nutrient and matter export. *Ecology* 88:1278–1291.

Moore, S.A., et al. 2008 Diversity in current ecological thinking: implications for environmental management. *Environmental Management* doi: 10.1007//s0267-008-9187-2.

Morea, C.R., et al. 2000. *Home range and movement of alligators in the Everglades*. Poster presented at the Greater Everglades Ecosystem Restoration Conference, December 2000.

Morin, X., et al. 2007. Process-based modeling of species' distributions: what limits temperate tree species' range boundaries? *Ecology* 88:2280–2291.

Morris, S.J., and R.E.J. Boerner. 1998. Landscape patterns of nitrogen mineralization and nitrification in southern Ohio hardwood forests. *Landscape Ecology* 13:215–224.

Mueller, L.D. 1988. Evolution of competitive ability in *Drosophila* by density-dependent natural selection. *Proceedings of the National Academy of Sciences of the United States of America* 85:4383–4386.

Mueller, L.D., and A. Joshi. 2000. *Stability in Model Populations*. Princeton, NJ: Princeton University Press.

Mullan, C., et al. 2008. Secondary succesdional dynamics in estuarine marshes across landscape-scale salinity gradients. *Ecology* 89:2889–2899.

Mullen, L., et al. 2009. Natural selection along an environmental gradient: a classic cline in mouse pigmentation. *Proceedings of the Royal Society of London, Series B: Biological Sciences* 276 3809–3818.

Munch, S.B., and S. Salinas. 2009. Latitudinal variation in lifespan within species is explained by the metabolic theory of ecology. *Proceedings of the National Academy of Sciences of the United States of America* 106:13860–13864.

Murie, A. 1944. The Wolves of Mt McKinley. Fauna of the National Parks of the US. Fauna Series No 5 US Dept Interior, NPS.

Myers, N., et al. 2000. Biodiversity hotspots for conservation priorities. *Nature* 403:853–858.

Naidoo, R., et al. 2008. Global mapping of ecosystem services and conservation priorities. *Proceedings of the National Academy of Sciences of the United States of America* 105:9495–9500.

NOAA Earth System Research Laboratory. Global Monitoring Division.

Nei, M. 1972. Genetic distance between populations. *American Naturalist* 949:283–292.

Nelson, B.W., et al. 1994. Forest disturbance by large blowdowns in the Brazilian Amazon. *Ecology* 75:853–858.

Nevoux, M., et al. 2010. Bet-hedging response to environmental variability, an intraspecific comparison. *Ecology* 91:24416–2427.

Newmark, W.D. 1995. Extinction of mammal populations in western North American national parks. *Conservation Biology* 9:512–527.

Noe, G. B., and D.L. Childers. 2007. Phosphorus budgets in Everglades wetland ecosystems: the effect of hydrology and nutrient enrichment. *Wetlands Ecology and Management* 15:189–205.

Nooker, J.K., and B.K. Sandercock. 2008. Phenotypic correlates and survival consequences of male mating success in lek-mating greater prairie chickens (*Tympanuchus cupido*). *Behavioral Ecology and Sociobiology* 62:1377–1388.

Normile, D. 2009. Round and round: a guide to the carbon cycle. *Science* 325:1642–1645.

Novotny, V., et al. 2006. Why are there so many species of herbivorous insects in tropical rainforests? *Science* doi:10.1126/science.1129237.

Nunssey, D.H., et al. 2007. The evolutionary ecology of individual phenotypic plasticity in wild populations. *European Society for Evolutionary Biology* 20:831–844.

Olofsson, H., et al. 2009. Bet-hedging as an evolutionary game: the trade-off between egg size and number. *Proceedings of the Royal Society of London, Series B: Biological Sciences* 276:2963–2969.

Olson, J.S. 1958. Rates of succession and soil changes on southern Lake Michigan sand dunes. *Botanical Gazette* 119:125–170

Okin, G.S., et al. 2009. Do changes in connectivity explain desertification? *BioScience* 59:237–244.

Oliver, T.A., and S.R. Palumbi 2011. Do fluctuating temperature environments elevate coral thermal tolerance? *Coral Reefs* doi:10.1007//s00338-011-0721-y.

Orme, C.D., et al. 2005. Global hotspots of species richness are not congruent with endemism or threat. *Nature* 436:1016–1019.

Paetkau, D., et al. 2008. Variation in genetic diversity across the range of North American brown bears. *Conservation Biology* doi: 10.1111/j.1523-1739 .1998.96457.x.

Paine, R.T. 1966. Food web complexity and species diversity. *American Naturalist* 100:65–75.

Palkovacs, E.P., and A.P. Hendry. 2010. Eco-evolutionary dynamics: intertwining ecological and evolutionary processes in contemporary time. *Biology Reports* doi:10.3410/B2-1.

Palmer, T.M., et al. 2008. Breakdown of an ant-plant mutualism follows the loss of large herbivores from an African savanna. *Science* 319:192–194.

Pärn, H., et al. 2009. Sex-specific fitness correlates of dispersal in a house sparrow metapopulation. *Journal of Animal Ecology* 78:1216–1225.

Parry, M.L., et al. 2004. Effects of climate change on global food production under SRES emissions and socio-economic scenarios. *Global Environmental Change* 14:53–67.

Patette, S., and A.C. Laliberté. 2008. Primary succession of subarctic vegetation and soil on the fast-rising coast of eastern Hudson Bay, Canada. *Journal of Biogeography* 35:1989–1999.

Pauw, A. 2007. Collapse of a pollination web in small conservation areas. *Ecology* 88:1759–1769.

Paxton, E.H., et al. 2002. Nestling sex ratio in the southwestern willow flycatcher. *Condor* 104:877–881.

Payette, S., and A.C. Laliberté. 2008. Primary succession of subarctic vegetation and soil on the fast-rising coast of eastern Hudson Bay, Canada. *Journal of Biogeography* 35:1989–1999.

Payne, J.L. 2009. Two-phase increase in the maximum size of life over 3.5 billion years reflects biological innovation and environmental opportunity. *Proceedings of the National Academy of Sciences of the United States of America* 106:24–27.

Pearman, P.B., et al. 2007. Niche dynamics in space and time. *Trends in Ecology and Evolution* 23:149–158.

Pergams, O.R., and R.C. Lacy. 2007. Rapid morphological and genetic change in Chicago-area *Peromyscus*. *Molecular Ecology* doi:10.1111/j/ 1365-294X.2007.03517.

Perry, G.W. 2002. Landscapes, space and equilibrium: shifting viewpoints. *Progress in Physical Geography* 26:229–259.

Peterson, D.W., and P.B. Reich. 2008. Fire frequency and tree canopy structure influence plant species diversity in a forest-grassland ecotone. *Plant Ecology* 194:5–16.

Peterson, R.O. 1999. Wolf-moose interaction on Isle Royale: the end of natural regulation? *Ecological Applications* 9:1–16.

Pimentel, D., et al. 2005. Update on the environmental and economic costs associated with alien-invasive species in the United States. *Ecological Economics* 52:273–288.

Pimm, S.L. 1982. *Food Webs*. London: Chapman and Hall.

Pimm, S.L. 1988. Energy flow and trophic structure. In *Concepts of Ecosystem Ecology*, edited by L.R. Pomeroy and J.J. Albers. New York: Springer-Verlag.

Pinto, A.A., et al. 2009. Symbiotic nitrogen fixation in the fungus gardens of leaf-cutter ants. *Science* 326:1120.

Poage, N.J., et al. 2009. Influences of climate, fire, and topography on contemporary age structure patterns of Douglas fir at 205 old forest sites in western Oregon. *Canadian Journal of Forest Research* 39:1518–1530.

Post, D.M., et al. 2000. Ecosystem size determines food-chain length in lakes. *Nature* 405:1047–1049.

Post, D.M. 2002. Using stable isotopes to estimate trophic position: eutrophication and recovery in experimental lakes: implications for lake management. *Science* 184:897–899.

Poulin, R. 2007. Are there general laws in parasite ecology? *Parasitology* 134:763–776.

Poutinen, M.L. 2004. Tropical cyclones in the Great Barrier Reef, Australia, 1910–1999: a first step toward characterizing the disturbance regime. *Australian Geographical Studies* 42:378–392.

Price, T.D., and M. Kirkpatrick. 2009. Evolutionarily stable range limits set by interspecific competition. *Proceedings of the Royal Society of London, Series B: Biological Sciences* 276:1429–1434.

Prugh, L.R., et al. 2008. Effect of habitat area and isolation on fragmented animal populations. *Proceedings of the National Academy of Sciences of the United States of America* 105:2070–2077.

Ptacnik, R., et al. 2008. Diversity predicts stability and resource use efficiency in natural phytoplankton communities. *Proceedings of the National Academy of Sciences of the United States of America* 105:513–5138.

Pulido, F., et al. 1996. Frequency of migrants and migratory activity are genetically correlated in a bird population: evolutionary implications. *Proceedings of the National Academy of Sciences of the United States of America* 93:14642–14647.

Pyron, R.A., and F.T. Burbink. 2009. Can the tropical conservatism hypothesis explain temperate species richness patterns? An inverse latitudinal biodiversity gradient in the New World snake tribe Lampropelini. *Global Ecology and Biogeography* 18:406–415.

Quinn, T.P., et al. 2009. Transportation of Pacific salmon carcasses from streams to riparian forests by bears. *Canadian Journal of Zoology* 87:195–203.

Rafaelli, D.G., and R.N. Hughes. 1978. The effects of crevice size and availability on populations of *Littorina rudis* and *Littorina neritoides*. *Journal of Animal Ecology* 47:71–83.

Raff, R.A. 2008. Origins of the other metazoan body plans: the evolution of larval forms. *Philosophical Transactions of the Royal Society of London, Series B: Biological Sciences* 363:1473–1479.

Raven, J.A., et al. 2005. *Ocean Acidification Due to Increasing Atmospheric Carbon Dioxide*. London: The Royal Society.

Ravishankara, A.R., et al. 2009. Nitrous oxide: the dominant ozone-depleting substance emitted in the 21st century. *Science* 326:123.

Raybaud, V., et al. 2009. Similar patterns of community organization characterize distinct groups of different trophic levels in the plankton of the NW Mediterranean Sea. *Biogeosciences* 6:431–438.

Raymond, J., and D. Segre. 2006. The effect of oxygen on biochemical networks and the evolution of complex life. *Science* 311:1764–1767.

Recher, H.F. 1969. Bird species diversity and habitat diversity in Australia and North America. *American Naturalist* 103:75–80.

Reed, A.W., and N.A. Slade. 2008. Density-dependent recruitment in grassland small mammals. *Journal of Animal Ecology* 77:57–65.

Reed, D.H., et al. 2003. Estimates of minimum viable population sizes for vertebrates and factors influencing those estimates. *Biological Conservation* 113:23–34.

Reed, S.C., et al. 2008. Tree species control rates of free-living nitrogen fixation in a tropical rain forest. *Ecology* 89:2924–2934.

Renfrew, R.B., and C.A. Ribic. 2008. Multi-scale models of grassland passerine abundance in a fragmented system in Wisconsin. *Landscape Ecology* 23:181–193.

Reynolds, J.F., et al. 2007. Global desertification: building a science for dryland development. *Science* 316:847.

Reznick, D., et al. 2002. R- and K-selection revisited: the role of population regulation in life history evolution. *Ecology* 83:1509–1520.

Ricklefs, R.E. 2008. The evolution of senescence from a comparative perspective. *Functional Ecology* 22:379–392.

Ricklefs, R.E. 2008. Disintegration of the ecological community. *American Naturalist* 172:741–750.

Ripple, W.J., et al. 2001. Trophic cascades among wolves, elk and aspen on Yellowstone National Park's northern range. *Biological Conservation* 102:227–234.

Robinson, W.D., et al. 2010. Integrating concepts and technologies to advance the study of bird migration. *Frontiers in Ecology and the Environment* 8:354–361.

Rodel, H.G., et al. Optimal litter size for individual growth of European rabbits depends on their thermal environment. *Oecologia* 155:677–689.

Rodhouse, T.J., et al. 2010. Distribution of American pikas in a low-elevation lava landscape: conservation implications from the range periphery. *Journal of Mammalogy* 91:1287–1299.

Rogers, J.A., et al. 2010. Geographic variation in a facultative mutualism: consequences for local arthropod composition and diversity. *Oecologia* doi:10.1007/s00442-010-1584-6.

Romme, W.H., and D.G. Despain. 1989. Historical perspectives on the Yellowstone fires of 1988. *BioScience* 39:695–699.

Rose, M.R., et al. 1998. *Adaptation*. Amsterdam, The Netherlands: Academic Press.

Rosene, W. 1969. *The Bobwhite Quail: Its Life and Management*. New Brunswick, NJ: Rutgers University Press.

Roukolainen, L., et al. 2009. When can we distinguish between neutral and non-neutral processes in community dynamics under ecological drift? *Ecology Letters* doi: 10.1111/j.1461-0248.2009.01346.x.

Roy, B.A. 1994. The effects of pathogen-induced pseudoflowers and buttercups on each other's insect visitation. *Ecology* 75:352–358.

Royer, D.L., et al. 2009. Ecology of leaf-teeth: a multi-site analysis from an Australian subtropical rainforest. *American Journal of Botany* 96:738–750.

Rosenzweig, M.L. 1968. Net primary productivity of terrestrial environments: predictions from climatological data. *American Naturalist* 102:67–84.

Rosenzweig, M.L. 1992. Species diversity gradients: we know more and less than we thought. *Journal of Mammalogy* 73:715–730.

Running, S.W., et al. 1989. Mapping regional forest evapotranspiration and phtosunthesis by coupling satellite data with ecosystem simulation. *Ecology* 70:1090–1101.

Saccheri, I., et al. 1996. Severe inbreeding depression and rapid fitness rebound in the butterfly *Bicyclus anynana* (Satyridae). *Evolution* 50:2000–2013.

Sale, P.F. 1979. Recruitment, loss and coexistence in a guild of territorial reef fishes. *Oecologia* 42:159–177.

Salo, J. et al. 1986. River dynamics and the diversity of Amazon lowland forest. *Nature* 322:254–258.

Sambatti, J.B., and K.J. Rice. 2006. Local adaptation, patterns of selection, and gene flow in the Californian serpentine sunflower (*Helianthus exilis*). *Evolution* 60:696–710.

Sanders, N.J., et al. 2007. Assembly rules of ground-foraging ant assemblages are contingent on disturbance, habitat, and spatial scale. *Journal of Biogeography* 34:1632–1641.

Schaefer, H.M., and G.D. Ruxton. 2009. Deception in plants: mimicry or perceptual exploitation? *Trends in Ecology and Evolution* doi:10.1016/j.tree.2009.06.006.

Schechter, S.P., and T.D. Bruns. 2008. Serpentine and non-serpentine ecotypes of *Collinsia sparsiflora* associated with distinct arbuscular mycorrhizal fungal assemblages. *Molecular Ecology* 17:3198–3210.

Schemske, D.W., and P. Bierzychudek. 2007. Spatial differentiation for flower color in the desert annual *Linanthus parryae*: was Wright right? *Evolution* 61:2528–2543.

Schindler, D.W. 1974. Eutrophication and recovery in experimental lakes: implications for lake management. *Science* 184:897–899.

Schmitz, O.J., et al. 2003. Ecosystem responses to global climate change: moving beyond color mapping. *BioScience* 53:1199–1205.

Schochat, E., et al. 2004. Linking optimal foraging behavior to bird community structure in an urban-desert landscape: field experiments with artificial food patches. *American Naturalist* 164:232–243.

Schluter, D., et al. 1985. Ecological character displacement in Darwin's finches. *Science* 227:1056–1059.

Seehausen, O. 2006. African cichlid fish: a model system in adaptive radiation research. *Proceedings of the Royal Society of London, Series B: Biological Sciences* 273:1987–1998.

Selander, R.K. 1965. On mating systems and sexual dimorphism. *American Naturalist* 105:400–437.

Shefferson, R.P., et al. 2003. Life history trade-offs in a rare orchid: the costs of flowering, dormancy, and sprouting. *Ecology* 84:1199–1206.

Shelford, V.W. 1913. Animal communities in temperate America. *Bulletin of the Geographical Society of Chicago* 5:1–368.

Sherman, P., et al. 1991. *Biology of the Naked Mole Rat*. Princeton, NJ: Princeton University Press.

Simberloff, D.S., and E.O. Wilson. 1969. Experimental zoogeography of islands the colonization of empty islands. *Ecology* 50:278–296.

Skarin, A., et al. 2004. Insect avoidance may override human disturbances in reindeer habitat selection. *Rangifer* 24:95–103.

Soldhi, N.S., et al. 2010. The state and conservation of Southeast Asian biodiversity. *Biodiversity and Conservation* 19:317–328.

Solomon, S., et al. 2009. Irreversible climate change due to carbon dioxide emissions. *Proceedings of the National Academy of Sciences of the United States of America* 106:1704–1709.

Soons, M.B., and J.M. Bullock. 2008. Non-random seed abscission, long-distance wind dispersal and plant migration rates. *Journal of Ecology* 96:581–590.

Sousa, W.P. 1979. Experimental investigations of disturbance and ecological succession in a rocky intertidal algal community. *Ecological Monographs* 49:227–254.

Spacojevic, M.J., et al. 2010. Fire and grazing in a mesic tallgrass prairie: impacts on plant species and functional traits. *Ecology* 91:1651–1659.

Sprules, W.G. 1974. The adaptive significance of paedogeneisis in North American species of *Ambystoma*. *Canadian Journal of Zoology* 52:393–400.

Stein, A., and N. Georgiadis. 2008. Spatial statistics to quantify patterns of herd dispersion in a savanna herbivore community. In *Resource Ecology: Spatial and Temporal Dynamics of Foraging*, edited by H.H.T. Prins and F.v. Langevelde. New York: Springer-Verlag.

Stevens, G.C., et al. 2000. Muscle temperature in free-swimming giant Atlantic bluefin tuna (*Thunnus thunnus* L.). *Journal of Thermal Biology* 25:419–423.

Stevens, M., et al. 2008. Conspicuousness, not eye mimicry, makes "eyespots" effective antipredator signals. *Behavioral Ecology* doi:10.1093/beheco/arm162.

Stige, L.C., et al. 2007. Thousand-year-long Chinese time series reveals climatic forcing of decadal locust dynamics. *Proceedings of the National Academy of Sciences of the United States of America* 104:16188–16193.

Stowe, K. 1998. Experimental evolution of resistance in *Brassica rapa*: correlated response of tolerance in lines selected for glucosinate content. *Evolution* 52:703–712.

Stuchbury, B.J.M., and E.S. Morton. 2008. Recent advances in the behavioral ecology of tropical birds. *Wilson Journal of Ornithology* 120:26–37.

Subramaniam, B., and M.D. Rausher. 2000. Balancing selection on a floral polymorphism. *Evolution* 54:691–695.

Svensson, J.R., et al. 2007. Maximum species richness at intermediate frequencies of disturbance: consistency among levels of productivity. *Ecology* 88:830–838.

Szabolcs, L., et al. 2009. Clutch size determination in shorebirds: revisiting incubation limitation in the pied avocet (*Recurviros avosetta*). *Journal of Animal Ecology* 78:396–405.

Takamoto, G., et al. 2008. Ecosystem size, but not disturbance, determines food chain length on islands of the Bahamas. *Ecology* 89:3001–3007.

Templeton, C.N., et al. 2007. Allometry of alarm calls: black-capped chickadees encode information about predator size. *Science* 308:1934–1937.

Tews, J., et al. 2004. Animal species diversity driven by habitat heterogeneity/diversity: the importance of keystone structures. *Journal of Biogeography* 31:79–92.

Thrall, P.H., et al. 2006. Coevolution of symbiotic mutualists and parasites in a community context. *Trends in Ecology and Evolution* 22:120–126.

Thompson, J.N. 2006. Mutualistic webs of species. *Science* 312:372–373.

Thompson, R.M., et al. 2007. Trophic levels and trophic tangles: the prevalence of omnivory in real food webs. *Ecology* 88:612–617.

Tilman, D. 1985. The resource ratio hypothesis of plant succession. *American Naturalist* 125:827–852.

Tilman, D. 1988. *Plant Strategies and the Dynamics and Structure of Plant Communities*. Princeton, NJ: Princeton University Press.

Tilman, D. 1999. The ecological consequences of change in biodiversity: a search for general principles. *Ecology* 80:1455–1474.

Tilman, D., et al. 2006. Biodiversity and ecosystem stability in a decade-long grassland experiment. *Nature* 441:629–632.

Townsend, A.R., et al. 2008. The biochemical heterogeneity of tropical forests. *Trends in Ecology and Evolution* doi:10.1016/j.tree.2008.04.009.

Townsend, A.R., et al. 2011. Multi-element regulation of the tropical forest carbon cycle. *Frontiers in Ecology and the Environment* 9:9–17.

Townsend, C.R., et al. 1998. Disturbance, resources supply, and food-web architecture in streams. *Ecology Letters* 1:200–209.

Tracy, C.R., et al. 2006. The importance of physiological ecology in conservation biology. *Integrative and Comparative Biology* 46:1191–1205.

Traill, L.W., et al. 2010. Pragmatic population viability targets in a rapidly changing world. *Biological Conservation* 143:28–34.

Trillmich, F., and J.B.W. Wolf. 2008. Parent-offspring and sibling conflict in the Galápagos fur seals and sea lions. *Behavioral Ecology and Sociobiology* 62:363–375.

Trunbore, S.E., et al. 1996. Rapid exchange between soil carbon and atmospheric carbon dioxide driven by temperature change. *Science* 272:393–396.

Tucker, B., et al. 1991. The influence of snow depth and hardness on winter habitat selection by caribou on the southwest coast of Newfloundland. *Rangifer Special Issue* 7:160–163.

Turbill, C., et al. 2011. Hibernation is associated with increased survival and the evolution of slow life histories among mammals. *Proceedings of the Royal Society of London, Series B: Biological Sciences* 278:3355–3363.

Turner, M.G., et al. 1997. Fires, hurricanes, and volcanoes: comparing large disturbances. *BioScience* 47:758–768.

Turner, M.L., et al. 2001. *Landscape Ecology in Theory and Practice*. New York: Springer-Verlag.

Tyerman, J.G., et al. 2008. Experimental demonstration of ecological character displacement. *Evolutionary Biology* 8:34–45.

Valle, C.A., and M.C. Coulter. 1987. Present status of the flightless cormorant, Galapagos penguin and greater flamingo populations in the Galapagos Islands, Ecuador, after the 1982–83 El Niño. *Condor* 89:276–281.

Valverde, T., and J. Silvertown. 1997. Canopy closure rate and forest structure. *Ecology* 78:1555–1562.

Vanclay, J.K., and P.J. Sands. 2009. Calibrating the self-thinning frontier. *Forest Ecology and Management* doi:10.1016/j.foreco.2009.09.045.

Van der Heijden, M.G.A., et al. 1998. Mycorrhizal fungal diversity determines plant biodiversity,

References **499**

ecosystem variability, and productivity. *Nature* 396:69–72.

Vander Zanden, M.J., and W.W. Fetzer. 2007. Global patterns of aquatic food chain length. *Oikos Journal* 116:1378–1388.

Vanpé, C., et al. 2009. Access to mates in a territorial ungulate is determined by the size of a male's territory, but not by its habitat quality. *Journal of Animal Ecology* 78:42–51.

Vazquez, D.P., et al. 2007. Species abundance and asymmetric interaction strength in ecological networks. *Oikos Journal* 116:1120–1127.

Vellend, M., et al. 2006. Extinction debt of forest plants persists for more than a century following habitat fragmentation. *Ecology* 87:542–548.

Violee, C., et al. 2010 Experimental demonstration of the importance of competition under disturbance. *Proceedings of the National Academy of Sciences of the United States of America* 107:12925–12929.

Vitousek, P.M., et al. 1997. Human domination of earth's ecosystems. *Science* 277:494–498.

Volterra, V. 1926. Variations and fluctuations of the numbers of individuals of animal species living together. Reprinted in 1931. In R.N. Chapman, *Animal Ecology*. New York: McGraw-Hill.

Walters, A.W., and D.M. Post. 2008. An experimental disturbance alters fish size structure but not food chain length in streams. *Ecology* 89:3261–3267.

Wang, Z., et al. 2009. Temperature dependence, spatial scale, and tree species diversity in eastern Asia and North America. *Proceedings of the National Academy of Sciences of the United States of America* 106: 13388–13392.

Waples, R.S., et al. 2009. Evolutionary history, habitat disturbance regimes, and anthropogenic changes: what do these mean for resilience of Pacific salmon populations? *Ecology and Society* 14(1):3.

Warne, R.W., and E.L. Charnov. 2008. Reproductive allometry and the size-number trade-off for lizards. *American Naturalist* 172(3) doi:10 1086/589880.

Warnecke, L., et al. 2008. Torpor and basking in a small arid sound marsupial. *Naturwissenschaften* 95:73–78.

Warner, R.R. 1982. Mating systems, sex change and sexual demography in the rainbow wrasse, *Thalassoma lucasanum. Copeia* 1982(3):653–661.

Warren, D.L., et al. 2008. Environmental niche equivalency versus conservatism: quantitative approaches to niche evolution. *Evolution* 62: 2868–2883.

Warren, D.R., et al. 2007. Forest age, wood and nutrient dynamics in headwater streams of the Hubbard Brook Experimental Forest, NH. *Earth Surface Processes and Landform* 32:1154–1163.

Waser, N.M., and M.V. Price. 1994. Optimal outcrossing in *Ipomopsis aggregata*: seed set and offspring fitness. *Evolution* 43:1097–1109.

Weiblen, G.D., et al. 2006. Phylogenetic dispersion of host use in a tropical insect herbivore community. *Ecology* 87:S62–S75.

West, S.A., and A. Gardner. 2010. Altruism, spite, and greenbeards. *Science* 327:1341–1344.

Weatherhead, P.J., et al. 2011. Latitudinal variation in thermal ecology of North American ratsnakes and its implications for the effect of climate warming on snakes. *Journal of Thermal Biology* doi:10.1016/j .jtherbio.2011.03.008.

Webb, W.S., et al. 1978. Primary production and water use in native forest, grassland and desert ecosystems. *Ecology* 59:1239–1347.

Weber, P. 1994. Resistance to pesticides is growing. In *Vital Signs*, edited by L. Brown et al. Washington, DC: World Watch Institute.

Weir, J.T., and D. Schluter. 2007. The latitudinal gradient in recent speciation and extinction rates of birds and mammals. *Science* 315:1574–1576.

Westemeier, R.L., et al. 1998. Tracking the long-term decline and recover of an isolated population. *Science* 282:1695–1698.

White, M.A., and G.E. Host. 2008. Forest disturbance frequency and patch structure from pre-European settlement to present in the mixed forest province of Minnesota, USA. *Canadian Journal of Forest Research* 38:2212–2226.

Whiting, M.F., et al. 2008. A molecular phylogeny of fleas (Insecta:Siphonaptera): origins and host associations. *Cladistics* 24:1–31.

Whittaker, R.H. 1956. Vegetation of the Great Smoky Mountains. *Ecological Monographs* 26:1–80.

Wikelski, M., and C. Thom. 2000. Marine iguanas shrink to survive El Niño. *Nature* 403:37–38.

Wildt, D.E., et al. 1987. Reproductive and genetic consequences of founding isolated lion populations. *Nature* 329:328–331.

Wilf, P., et al. 2006. Decoupled plant and insect diversity after the end-Cretaceous extinction. *Science* 313:1112–1115.

Williams, G.C. 1966. Natural selection, the costs of reproduction and a refinement of Lack's principle. *American Naturalist* 100:687–690.

Williams, J.W., and S.T. Jackson 2007. Novel climates, no-analog communities and ecological surprises. *Frontiers in Ecology and the Environment* 5:475–482.

Williams, S.E., et al. 2008. Toward an integrated framework for assessing the vulnerability of species to climate change. *PLoS Biology* 6 e25.

Wilson, R.J., et al. 2007. An elevational shift in butterfly species richness and composition accompanying recent climate change. *Global Change Biology* 13:1873–1887.

Woods, K. 2000. Long-term change and spatial pattern in a late-successional hemlock-northern hardwood Forest. *Journal of Ecology* 88:267–282.

Worm, B., et al. 2006. Impacts on biodiversity loss on ocean ecosystem services. *Science* 314:787–790.

Wright, S. 1931. The analysis of variance and the correlations between relatives with respect to deviations from an optimum. *Journal of Genetics* 30:243–256.

Wynne-Edwards, V.C. 1962. *Animal Dispersion in Relation to Social Behavior.* Edinburgh, Scotland: Oliver and Boyd.

Van Kleunen, M., and M. Fischer. 2005. Constraints on the evolution of adaptive phenotypic plasticity in plants. *New Phytologist* 166:49–60.

Yamagishi, S., et al. 2001. Extreme endemic radiation of the Malagasy vangas (Aves: Passeriformes). *Journal of Molecular Evolution* 53:39–46.

Yu-Peng, G., et al. 2010. Phenotypic plasticity rather than locally adapted ecotypes allows the invasive alligator weed to colonize a wide range of habitats. *Biological Invasions* 12:179–189.

Zackrisson, O., et al. 2004. Nitrogen fixation increases with successional age in boreal forests. *Ecology* 85:3327–3334.

Zhu, X., et al. 2008. What is the maximum efficiency with which photosynthesis can convert solar energy into biomass? *Current Opinion in Biotechnology* 19:153–159.

Zillio, T., and F. He. 2009. Inferring species abundance distribution across spatial scales. *Oikos Journal* 119:71–80.

Zou, J., et al. 2007. Increased competitive ability and herbivory tolerance in the invasive plant *Sapium sebiferum. Biological Invasions* doi:10.1007/s10530-007-9130-0.

Credits

Transcriptome Reprogramming Underlies Floral Mimicry Induced by the Rust Fungus *Puccinia monoica* in *Boechera stricta*. PLoS ONE 8(9): e75293. doi:10.1371/journal .pone.0075293; **12.17** David T. Krohne; **12.18** Angel DeBilio/Shutterstock.com; **12.20** David T. Krohne; **12.21** Sam K. Tran/Science Source; **The Human Impact, p. 271, Figure 1** David T. Krohne

Chapter 13   Chapter Opening Photo Levin, Ted/ Animals Animals; **The Human Impact, p. 284, Figure 1** idome/Shutterstock.com; **13.11** Ethan Daniels/ Shutterstock.com; **13.13** Thomas and Pat Leeson/ Science Source

Chapter 14   Chapter Opening Photo M.M./ Shutterstock.com; **14.1** David T. Krohne; **14.2** David T. Krohne; **14.3** David T. Krohne; **14.5** David T. Krohne; **14.7** David T. Krohne; **14.11a** David T. Krohne; **14.11b** David T. Krohne; **14.12** Francois Gohier/Science Source; **14.13** Melinda Fawver/Shutterstock.com; Madlen/ Shutterstock.com; Pi-Lens/Shutterstock.com; eedology/ Shutterstock.com; iofoto/Shutterstock.com; **14.14** Dr. Gilbert S. Grant/Science Source; **The Human Impact, p. 315, Figure 1** David T. Krohne; **14.24** Peter Scoones/ Science Source

Part 4   ODM/Shutterstock.com

Chapter 15   Chapter Opening Photo iStockphoto/ samcam; **15.7** David Littschwager/National Geographic Creative; **The Human Impact, p. 337, Figure 1** Science Source; **15.15** Graeme Shannon/Shutterstock.com

Chapter 16   Chapter Opening Photo Christopher Swann/Science Source; **16.1** James Steinberg/Science Source; **16.2** British Antarctic Survey/Science Source;

16.4 Sebastian Kaulitzki/Shutterstock.com; **16.5** Rob Bayer/Shutterstock.com; **The Human Impact, p. 355, Figure 1** David T. Krohne; **16.12** Jerome Wexler/Science Source

Chapter 17   Chapter Opening Photo Art Wolfe/ Science Source; **The Evolution Connection, p. 379, Figure 1** Catmando/Shuttershock.com; **17.8** Pablo Hidalgo – Fotos593/Shuttershock.com; **The Human Impact, p. 382, Figure 1** Ahmad A Atwah/Shutterstock .com; **17.9** iStockphoto/Frank Leung

Part 5   luoman/Shutterstock.com

Chapter 18   Chapter Opening Photo Stephen J. Krasemann/Science Source; **18.2** C.M.Corcoran/ Shutterstock.com; **18.5** Fexel/Shutterstock.com; **18.10** Tom Reichner/Shutterstock.com; **18.11** Jay Ondreicka/ Shutterstock.com; **18.12** PRILL/Shutterstock.com; **18.13** Lowell Georgia/Science Source; **18.16** Carrie Vonderhaar/ Ocean Futures Society/Getty Images; **18.19** Paul W. Landau/Shutterstock.com; **18.20** G. Ronald Austing/ Science Source

Chapter 19   Chapter Opening Photo Frans Lanting/ MINT Images/Science Source; **19.1** Visible Earth/NASA; **19.3** Visible Earth/NASA; **The Human Impact, p. 420, Figure 1** Thomas & Pat Leeson/Science Source; **19.4a** NASA; **19.4b** NASA; **19.6a** NASA; **19.6b** NASA

Chapter 20   Chapter Opening Photo NASA/ Science Source; **20.8** Science Source; **20.9** NASA; **20.11** Fletcher & Baylis/Science Source; **20.12** Marzolino/ Shutterstock.com; Richard Ellis/Science Source; **The Human Impact, p. 453, Figure 1** Aldo Leopold Foundation; **20.13** NASA

Index

503